Fundamentals of Foundation Engineering

This book aims to introduce the principle and design of various foundations, covering shallow foundations, mat foundations, earth retaining structures, excavations, pile foundations, and slope stability. Since the analysis and design of a foundation are based on the soil properties under short-term (undrained) or long-term (drained) conditions, the assessment of soil properties from the geotechnical site investigation and the concept of drained or undrained soil properties are discussed in the first two chapters. Foundation elements transfer various load combinations from the superstructure to the underlying soils or rocks. The load transfer mechanisms, vertical stress or earth pressure distributions, and failure modes of each foundation type are clearly explained in this book. After understanding the soil responses subjected to the loadings from the foundation, the design methods, required factors of safety, and improvement measures for each foundation type are elaborated.

This book presents both theoretical explication and practical applications for readers to easily comprehend the theoretical background, design methods, and practical applications and considerations. Each chapter provides relevant exercise examples and a problem set for self-practice. The analysis methods introduced in the book can be applied in actual analysis and design as they contain the most up-to-date knowledge of foundation design. This book is suitable for teachers and students to use in foundation engineering courses and engineers who are engaged in foundation design to create a technically sound, construction-feasible, and economical design of the foundation system.

Fundamentals of Foundation Engineering

Chang-Yu Ou, Kuo-Hsin Yang, Fuchen Teng, Jiunn-Shyang Chiou, Chih-Wei Lu, An-Jui Li, Jianye Ching, Jui-Tang Liao

CRC Press is an imprint of the
Taylor & Francis Group, an **informa** business
A BALKEMA BOOK

Designed cover image: The Taipei 101, once the tallest building in the world, is located in Taipei City. Below the surface are thick soft clays with SPT-N values between 2 and 10, and below the clays are andesite formations. The foundation must carry the load of the 101-story tower above ground and the 6-story podium. In this case, pile foundations are used to transfer the load to the rock formation. The foundation and excavation is designed by Sino Geotechnology, Inc., Taiwan.

First published 2024
by CRC Press/Balkema
4 Park Square, Milton Park, Abingdon, Oxon, OX14 4RN

e-mail: enquiries@taylorandfrancis.com
www.routledge.com—www.taylorandfrancis.com

CRC Press/Balkema is an imprint of the Taylor & Francis Group, an informa business

ISBN: 978-1-032-39498-5 (hbk)
ISBN: 978-1-032-39497-8 (pbk)
ISBN: 978-1-003-35001-9 (ebk)

DOI: 10.1201/9781003350019

Typeset in Times New Roman
by Apex CoVantage, LLC

Contents

Preface

The most important consideration in the design of a structure such as a building, bridge, retaining wall, or geotechnical structure is a stable foundation. An unstable foundation puts the safety of the entire structure at risk. The analysis and design of a foundation are closely related to the basic properties of soils, especially their drained and undrained behaviors. The method for analysis and design introduced in this book focuses particularly on the drained or undrained behavior of the target soil. Chapter 2 elucidates the concept and analysis methods for soils under drained and undrained conditions. The concept of drained or undrained properties associated with necessary soil parameters and analysis methods is also applied to the analysis and design methods presented in each chapter.

Since a large number of studies related to the analysis and design methods for foundation have been developed, this book is unable to cover them all. Therefore, for teaching purposes, only basic principles of analysis and design are provided in this book, followed by an introduction to their application. If students learn the principles of analysis and design from this book, then in future practical design work, they can perform a sound analysis or design considering relevant design codes.

Generally, the method for foundation design includes the allowable stress method, strength design method, and method considering various limit states. Chapter 2 explains the basic concepts of these design methods. Although the allowable stress method is a mainstream method in foundation design, the strength design method has been adopted to design upper structures. Obviously, it is not reasonable to adopt different design methods for the upper structure and foundation, especially considering the greater uncertainties involved in the soil parameters than in the parameters of structures. Consequently, an increasing number of country building codes adopt more rigorous strength design methods and even more advanced limiting state design methods. The limiting strength design method considers the design at various limiting states, for example, considering ultimate strength, deformation, or earthquakes.

To make it easy for students to understand the basic principles of basic analysis and design, each chapter introduces the basic principles of basic design from the perspective of the allowable stress method. If readers can understand the concepts of the strength design method and limit state design method, they will be able to understand the implications better behind the design codes when they must use them in actual design work.

The topics for foundation design cover a wide range, and relevant studies are becoming increasingly advanced. It is not easy for a single author to be an expert on every topic. Therefore, eight distinguished scholars from different fields contribute to the contents of this book. The book chapters and the contributing authors are arranged as follows. Chapter 1 is about geotechnical site investigation, written by Dr. Chih-Wei Lu and Dr. Jui-Tang Liao. Chapter 2 is about principles of foundation design, written by Dr. Chang-Yu Ou, Dr. Jianye Ching, and Dr. Jiunn-Shyang Chiou. Chapter 3 is about shallow foundations, written by Dr. Jiunn-Shyang Chiou. Chapter 4 is about lateral earth pressure, written by Dr. Chang-Yu Ou. Chapter 5 is about earth retaining structures, written by Dr. Kuo-Hsin Yang. Chapter 6 is about excavation, written by Dr. Chang-Yu Ou. Chapter 7 is about pile foundations, written by Dr. Fuchen Teng. And Chapter 8 is about slope stability, written by Dr. An-Jui Li. Each author is an expert in the field corresponding to the book chapter he wrote. All authors have tried their best to make the content of this book accessible, accurate, and current. All chapters in this book have been reviewed and approved by all authors.

About the authors

Chang-Yu Ou is a chair professor of the Department of Civil and Construction Engineering at National Taiwan University of Science and Technology (NTUST), Taipei, Taiwan. He received his doctoral degree from Stanford University, USA, in 1987. His areas of interest are deep excavations, soil behavior, soft ground tunneling, and ground improvement. He has published more than 200 journal and conference papers. He also has published two monograph books regarding deep excavations in English by Taylor & Francis in 2006 and 2022, respectively, and three deep excavation books in Chinese.

Kuo-Hsin Yang is a professor in the Geotechnical Engineering Program of the Department of Civil Engineering at the National Taiwan University (NTU). He completed his PhD at the University of Texas at Austin. His research interests involve slope stability and earth retaining structures, application of geosynthetics, geotechnical engineering modeling, and geodisaster engineering. He has many years' experience in research and practice on the analysis, design, and case study of slopes and earth retaining structures using both numerical modeling and physical tests.

Fuchen Teng is an associate professor of the Department of Civil and Construction Engineering at National Taiwan University of Science and Technology (NTUST), Taipei, Taiwan. He received his doctoral degree from NTUST in 2011. Before joining the Department of Civil and Construction Engineering, he worked as a post-doctoral fellow and engineer at Northwestern University and Sinotech Engineering Consultants during 2013–2016. He is a member of the ATC6 of the International Society of Soil Mechanics and Geotechnical Engineering. His areas of interest are ground improvement, soil stress-strain-strength behaviors, soil–structure interaction problems, and geological disposal of rad-wastes.

Jiunn-Shyang Chiou is an associate professor of the Department of Civil Engineering at NTU. He received his doctoral degree from NTU in 2001. He was a research fellow of the National Center for Research on Earthquake Engineering, Taiwan. He was a geotechnical engineer of Sinotech Engineering Consultants Ltd. His areas of interest are geotechnical earthquake engineering, foundation engineering, seismic design of structure foundations, and soil–structure interaction analysis. He has published more than 60 journal and conference papers.

Chih-Wei Lu is a professor at NTUST. He completed his Ph. at Kyoto University, Japan. Dr. Lu worked as a geotechnical engineer in Moh and Associates Inc., experiencing an engineering practice in foundation design, geotechnical investigation, and soil liquefaction potential assessment. His research interests involve soil dynamics and soil liquefaction, effective stress analysis on interaction of superstructure–foundation–soil system, and innovative design for geotechnical structures.

An-Jui Li received his PhD degree from the University of Western Australia in 2009. Before joining NTUST in 2017 as an associate professor, he worked as a lecturer at Central Queensland University and Deakin University in Australia for 7+ years. An-Jui teaches geotechnical engineering units. His primary area of research includes rock mechanics, excavation, and slope stability. Currently, he has published 50+ journal and conference articles.

Jianye Ching is a distinguished professor in the Department of Civil Engineering at NTU. He obtained his PhD in 2002 in the University of California at Berkeley. His main research interests are geotechnical reliability analysis and reliability-based design, basic uncertainties in soil properties, random fields and spatial variability, reliability-based geotechnical design codes, and probabilistic site characterization. He is the author or co-author of more than 100 publications in international journals.

Jui-Tang Liao received his PhD from NTUST. He established Land Engineering Consultant Co. Ltd. in 1988. He possesses a 35-year experience in engineering and also was an adjunct associate professor of the Department of Civil and Construction Engineering at NTUST. His specialties are geological investigation, geotechnical monitoring, and landslide mitigation.

Chapter 1

Geotechnical site investigation

1.1 Introduction

Geotechnical engineering problems vary due to different project types and geological conditions. Thus, purposes and characteristics of construction should be recognized sufficiently to develop a complete site investigation plan for performing the optimum engineering solution and thereby assist in solving potential geotechnical problems before, during, and after the construction work. A function of foundation that stabilizes structures is to transfer superstructure loads to the soil or rock or to stabilize temporary excavations. It is known that not only is risk of foundation construction higher for subsurface structures than surface structures during the construction period but also that foundation design expenditures could be considerably reduced because of an integral site investigation performed for the construction work cautiously.

The scope of foundation engineering covers generally building, underground excavation, road or railway, dam, slope engineering, etc. The most significant components of a building foundation are permanent substructures and temporary stabilization structures for excavations. There are diverse temporary stabilization methods for excavations, such as internal bracings, retaining walls, and ground anchors. After the temporary excavation face is stabilized, a permanent substructure can be constructed, such as a mat foundation, pile foundation, or caisson foundation. Based on the stable subsurface structure, the subsequent superstructure is constructed afterward.

However, since most foundation projects are underground projects, they are not visibly exposed. Given underground geological conditions or groundwater conditions, along with their variable distribution characteristics, to a geotechnical engineer, planning and designing are essential. The geological survey should first be executed to investigate the geological and groundwater conditions of different sites in each project and should be used to thoroughly consider the potential problems and possible damage modes that may be encountered in subsequent construction and even after completion to proactively respond to them.

Providing the best solution for engineering design is the most important task for geotechnical engineers. Factors such as the economy, safety, and feasibility of construction are taken into account for obtaining the best solution. When determining the solution, a cost-optimal plan is required for engineering project, but safety should always be prioritized. Safety of different conditions, including safety during construction and long-term safety under heavy rain or earthquake conditions after construction, is required to be analyzed. There are a few examples of appropriate site investigation not being performed for the geotechnical design, thus causing damages, shown in this chapter to remind the readers of the importance of site investigation.

Figure 1.1 shows a problem when conducting basement excavation work in soft clay, which contains a very soft clay layer classified as CL (low-plasticity clay) or CH (high-plasticity clay)

DOI: 10.1201/9781003350019-1

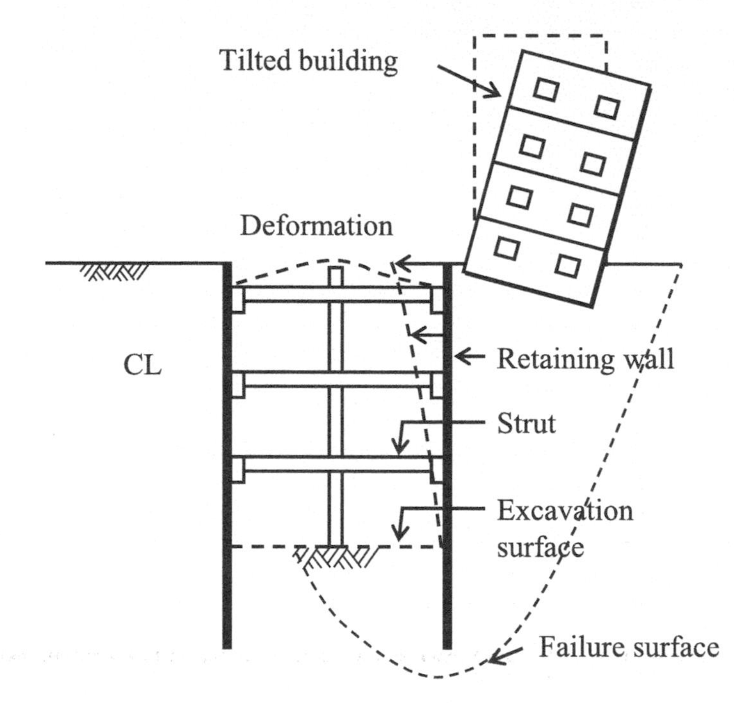

Figure 1.1 Tilt in the adjacent building due to an inadequate support system during excavation.

by the unified soil classification system, and the standard penetration test SPT-N value is less than 4. A braced retaining system was adopted. However, an excavation failure occurred, which caused a serious tilt in adjacent buildings during excavation. The cause of the failure was related to the insufficient understanding of the soil properties below the surface, resulting in inadequate design of braced retaining system. To avoid such disasters, better site investigation must be carried out first to determine the major design parameters, such as groundwater condition, compressibility, and strength of soils. Then, a reasonable stability analysis should be performed, and a safe braced retaining system should be designed.

Figure 1.2 shows several three-story houses built on loose sandy ground. The soil profile is classified as SM (silty sand) and SP (poorly graded sand), and the SPT-N value is smaller than 10. The houses were founded safely until several years after construction. The foundation of the shallow-founded houses experienced severe soil liquefaction in an earthquake, and the occurrence of liquefaction-induced lateral spread due to loss of bearing capacity in the adjacent retaining wall caused permanent inclination of the houses. Soil liquefaction is easily triggered

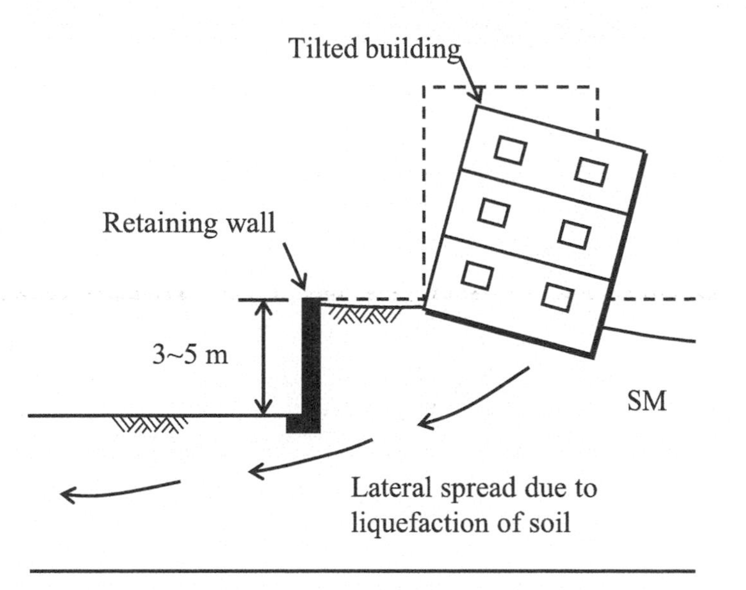

Figure 1.2 Tilt in building due to liquefaction of loose silty sand (N <10).

in loose sandy soils during seismic motion, in which the effective stress of the soils reduces considerably due to excess pore water pressure. It told us a lesson that a more accurate site investigation should be carried out to investigate the SPT-N of the soils and groundwater depth, and laboratory tests of general physical property of soils should be conducted to consider construction on soil liquefiable sites. To ensure reasonable safety against soil liquefaction, construction work in seismic hazard zones has proved to be critical.

Figure 1.3 is a case of a hillside community in northern Taiwan. The anchored retaining wall experienced failure during Typhoon Winnie in Taiwan in August 1997. The site is located on a dip slope with interbedded sandstone and shale. It is observed that when the downslope force increased because the rising groundwater became larger than the resistance force, slope sliding occurred, resulting in instantaneous failure of the anchored retaining wall. Furthermore, the

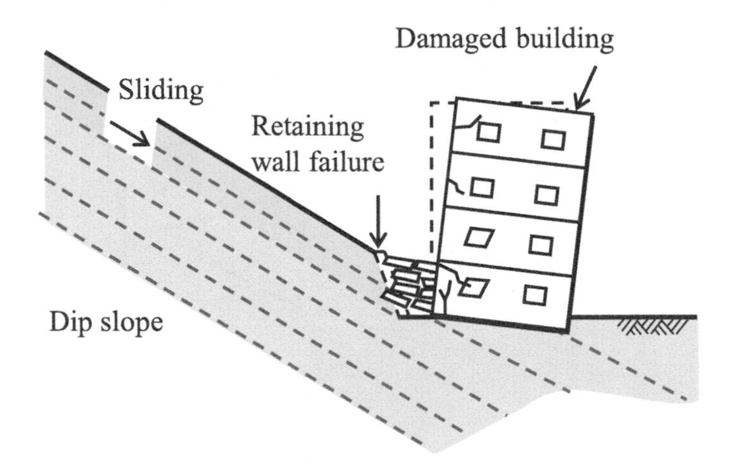

Figure 1.3 Dip slope slides at Lincoln community (hillside residential community).

sliding rock mass of the dip slope quickly rushed into the houses and led to the damage and tilt of the house, resulting in 28 persons' deaths. The failure of the anchored retaining wall is attributed to the insufficient understanding of geological and groundwater conditions.

The discontinuity type and groundwater conditions of the slope can significantly affect slope stability. Therefore, this failed case showed that when designing a retaining wall of man-made slopes, detailed information about the rock type, discontinuity type, strength, and groundwater level in the site investigation is needed. Then, a reliable stability analysis of the slope and retaining wall can be conducted to ensure slope safety under unfavorable conditions, such as heavy rain or earthquakes, to prevent serious landslides.

1.2 Plan for subsurface exploration

The design of the building foundation and basement excavation requires detailed background information for each site, including geological conditions, groundwater conditions, strength of geotechnical material, and so on. Given the unique geological conditions of each site, as well as the fact that the characteristics and scale of different engineering projects are not the same, such as the building height, excavation depth, or condition of the adjacent house, each case poses a different engineering problem, which needs to be resolved individually in a subsurface exploration plan, readers can refer to relevant reference (Geotechnical Engineering Office, 2017).

1.2.1 Investigation methods

Table 1.1 shows a comprehensive list of investigation methods employed for field investigation, including data collection, surface geology surveys, subsurface exploration (soil boring and geophysical prospecting), groundwater investigation, in situ and laboratory tests, and monitoring. Generally, the data collection, surface geological survey, and soil boring in subsurface exploration are often executed in a civil engineering project.

Data collection includes collecting aerial photo, topography, and geological information. Subsurface geological surveys include investigating the landforms, geology, and topography of surfaces in the field. Both are used to preliminarily develop a geological model of the study area for filtering potential problems and planning more detailed investigations.

Table 1.1 also lists commonly applied investigation methods in engineering practice. The timing for the method application is entirely dependent on the scale, complexity, and cost of the construction project. A detailed field investigation is carried out, from which reliable geological conditions at the construction site can be captured. Readers can fere to the relevant references (Welman and Head, 1983).

1.2.2 Site investigation layout

To evaluate the investigation area, the first step is to prepare a site investigation plan. For example, when executing a construction project, one of the major commonly employed methods is soil boring. Several factors should be considered:

1. Investigation area
   The investigation area generally covers the entire building site. However, if the area is surrounded by neighboring buildings, a wider area of the buildings should be investigated for potential damage from construction. In this case, the investigation area will be widened

Table 1.1 Engineering types and investigation methods

Investigation method		*Surface survey*		*Geophysical survey*		*Borehole drilling and sampling test*						*Groundwater investigation*				*Field test*	*Laboratory test*
Engineering types		*Data collection*	*Reconnaissance*	*Seismic survey*	*Electrical and electromagnetic survey*	*Drilling and sampling*	*Standard penetration test*	*Cone penetration test*	*Vane shear test*	*Borehole televiewer*	*Lateral load test*	*Observation well and piezometer*	*Field permeability test*	*Pumping test*	*Water logging test*	*Plate loading test*	
Building	Shallow foundation	○	–	–	–	○	○	○	–	–	–	○	–	–	–	○	○
	Pile foundation	○	–	–	–	○	○	△	–	△	△	○	–	–	–	–	○
	Foundation excavation	○	–	–	–	○	○	△	○	–	–	○	○	○	–	–	○
Underground structure	Buried structure	○	○	–	–	○	○	△	△	–	△	△	–	△	–	–	○
	Large-scale underground structure	○	○	○	○	○	–	–	–	–	○	△	○	–	△	–	○
Road or railway	Cut	○	○	○	○	○	○	–	–	–	–	△	–	–	–	○	○
	Fill	○	○	△	–	○	○	○	–	–	–	△	–	–	–	○	○
	Tunnel	○	○	○	○	○	○	–	–	△	–	○	○	○	△	–	○
	Bridge	○	○	–	–	○	○	–	–	–	○	△	△	–	–	○	○
Dam	Dam foundation	○	○	○	○	○	△	–	–	–	○	△	–	–	○	–	○
	Dam	○	○	–	–	○	–	–	–	–	–	–	–	–	–	–	○
Slope		○	○	○	○	○	○	–	–	△	△	–	○	○	△	○	–

Legend: ○, often used; △, sometimes used; –, rarely

to include the entire influential construction area. Additionally, the foundation type of the neighboring buildings, the geological condition, and the use and material of the pipeline should be included in the investigation.

2. The number of soil borings
   The number of soil borings should be determined to obtain a complete picture of the subsurface soil condition in the project. Generally, at least two soil borings are required for an investigation site and three for a slope area. According to the author's experience, one additional soil boring per 600 m^2 is deployed, and the total number of soil borings is adjustable when the site is over 6,000 m^2. Notably, appropriate additional soil borings are required for the site of the variable ground profile.
3. Others
   Vital buildings, such as buildings in public places, and the safety of these buildings are particularly important and therefore require a more detailed site investigation. If the geological condition of the site is simple and filled with uniform gravel and rock layers, then the investigation may only need to fit to a minimum requirement. On the other hand, if the geological condition is complex, the investigation should be executed in more detail.

1.2.3 *Depth of subsurface exploration*

In an investigation plan, determination of subsurface depth is required. The principle of the determination is that the investigated depth should be able to provide sufficient information for examining the safety of the construction work and of the complete structure. In engineering practice, the minimum boring depth aims at reaching where the net increase in vertical effective stress $\Delta\sigma_0'$ under a foundation with depth, as shown in Figure 1.4, appearing limited influence, can be determined by the following rules:

1. Determine the net increase in effective stress $\Delta\sigma_0'$ under a foundation with depth.
2. Estimate the variation of the vertical effective stress σ_v' with depth.
3. Determine the depth ($D = D_1$) at which the effective stress increase $\Delta\sigma_0'$ is equal to (1/10) q as q equals estimated net stress on the foundation.
4. Determine the depth ($D = D_2$) at which $(\Delta\sigma_0' / \sigma_v') = 0.05$.
5. Determine the depth ($D = D_3$) which is the distance from the lower face of the foundation to bedrock, as shown in Figure 1.4.
6. Choose the smaller of the three depths (D_1, D_2, and D_3) for determining minimum depth of boring.
7. Explore minimum 3 m below the bearing layer for pile foundation, as shown in Figure 1.4.

1.3 Borehole exploration

Borehole drilling is one of the most commonly used methods to carry out geological investigations and has three purposes: profile determination, soil sampling, and monitoring system installation. Usually, a few boring holes after being soil-sampled can assist to determine the soil profile in the investigated area by an experienced geotechnical engineer. The soil boring work, including boring machines, rods, and samplers, aims to explore different geological materials, such as soils and rocks. The soil boring methods often employed on plains sites are the wash boring method combined with the split tube or thin-walled tube method. On slope sites, the rotary method is used to sample the soils. Moreover, in engineering practice, the wash boring

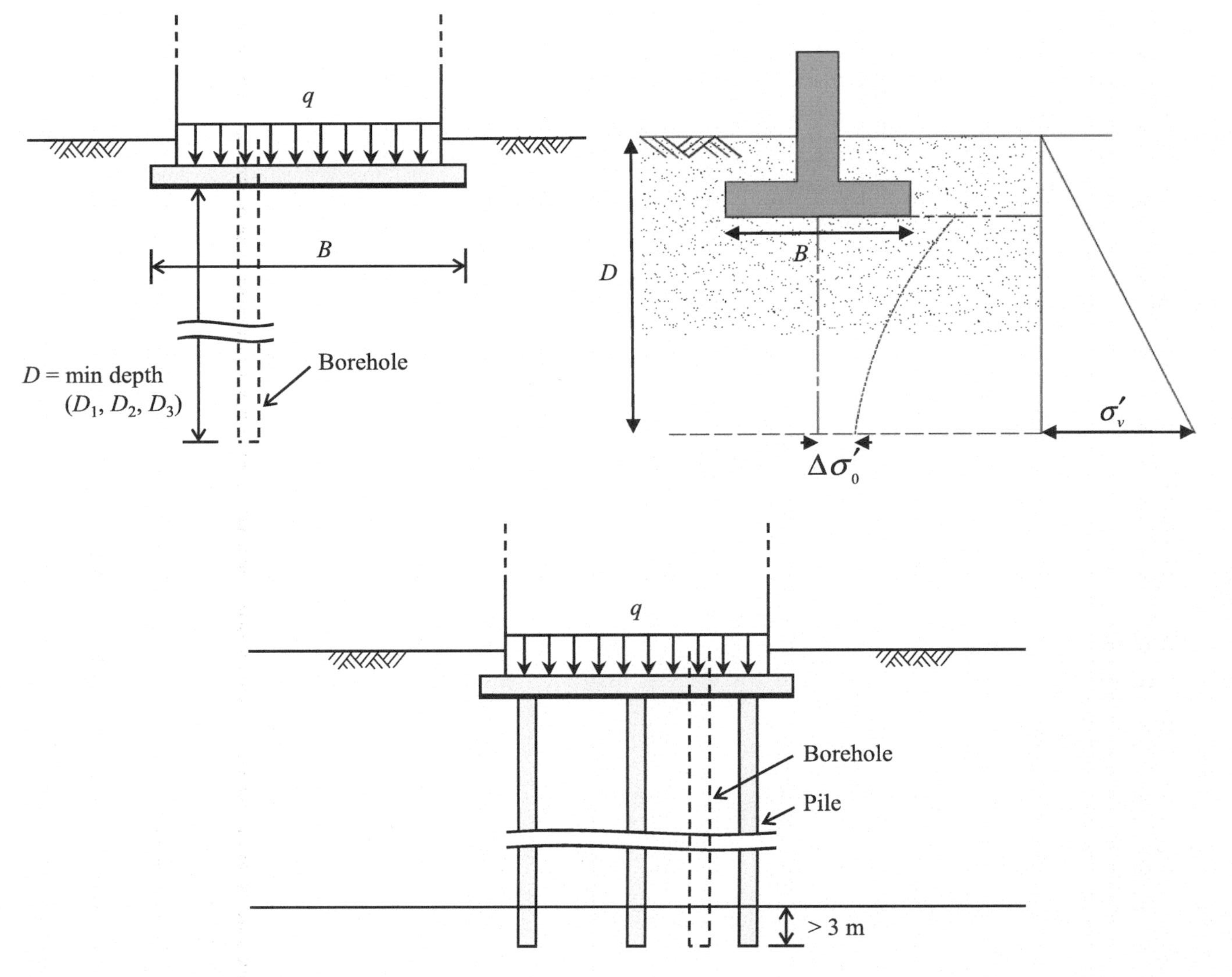

Figure 1.4 Preliminary determination of borehole depth.

method is often combined with the field test method of the standard penetration method (SPT) using a split-spoon sampler, introduced in the following sections. For clarifying the function of the penetration methods and the samplers to the readers, this chapter firstly introduces the bore hole exploration in this section and samplers in a later section, readers can also refer to the references (Hvorslev, 1949).

1.3.1 Wash boring method

A fishtail drill bit (see Figure 1.5) in the wash boring method is commonly employed for penetrating soil washing and boring until it reaches the scheduled sampling borehole depth. A split-spoon sampler or thin-walled tube is then used to sample the soil for different purposes that will be introduced in the following sections. A schematic picture of wash boring operation in Figure 1.6 shows that wash boring uses a hollow drilled rod with a sharp chisel, of which water is pressurized in the drill rod, raised, dropped, and rotated and controlled from the ground. The soil water slurry is forced up to the ground surface and is circulated into the drill rod. A horizontal water jet is usually used to avoid direct downward washing that disturbs the sample when being applied with sampling methods. The sampling interval is usually 1 m or 1.5 m but should be less than 2 m. Denser intervals are necessarily for soil profiles with significant changes.

1.3.2 Rotary drilling method

Rotary drilling is the use of a sharp rotating drill bit to penetrate through the soil or rock layers in the ground by cutting or crushing with an exerted continuous downward pressure that changes with the soil formation, the hardness of the rocks, and the depth. The face discharge of the drill bit, as shown in Figure 1.7, needs to be maintained sharp in rotating and penetrating process. It becomes one of the major costs of this work; therefore, the selection of the drill bit type with the corresponding formation requires wisdom and experience of the contractors and workers. Also, for keeping the hole clear of cutting and reclamation, and the drill bit lubricated and less heated, a circulation of fluid is usually applied in this drilling work. A schematic picture of a rotary drilling method can be referred in Figure 1.8

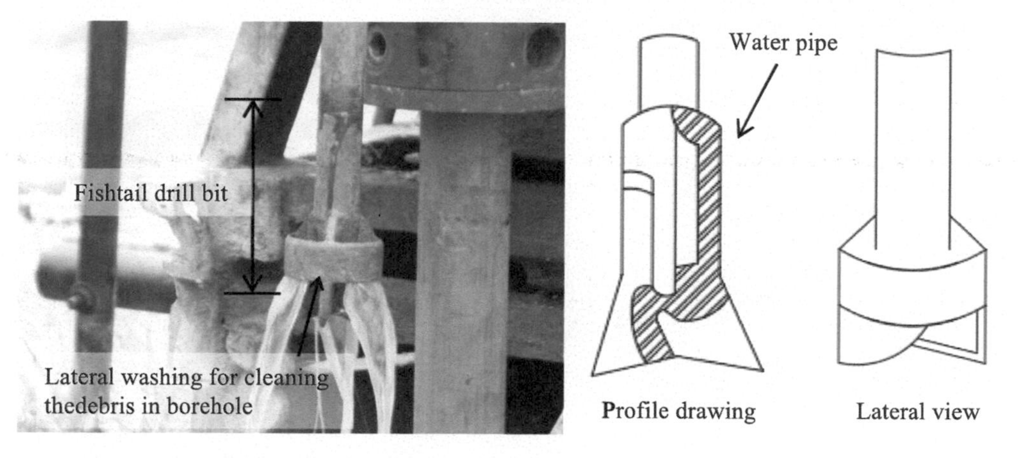

Figure 1.5 Fishtail drill bit.

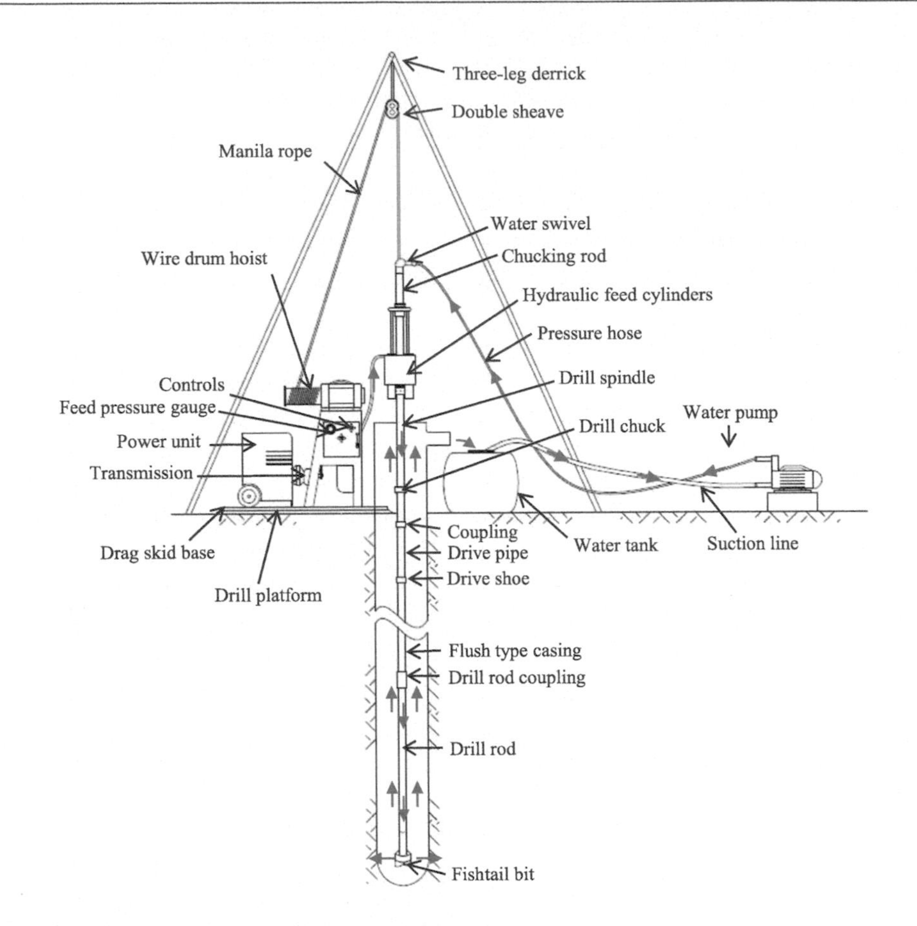

Figure 1.6 Schematic of wash boring operation.

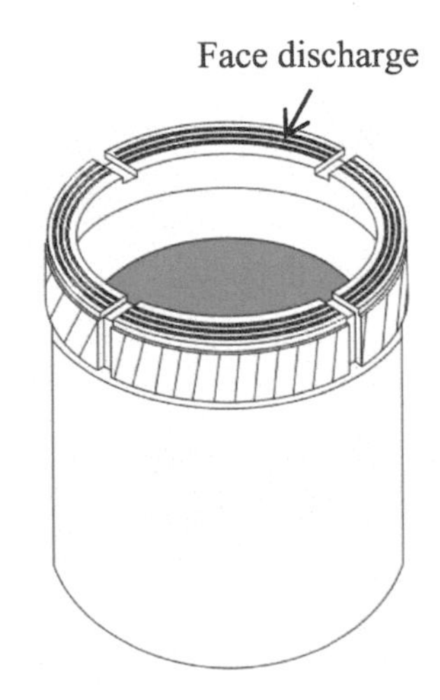

Figure 1.7 Diamond drill bits of the drill bit.

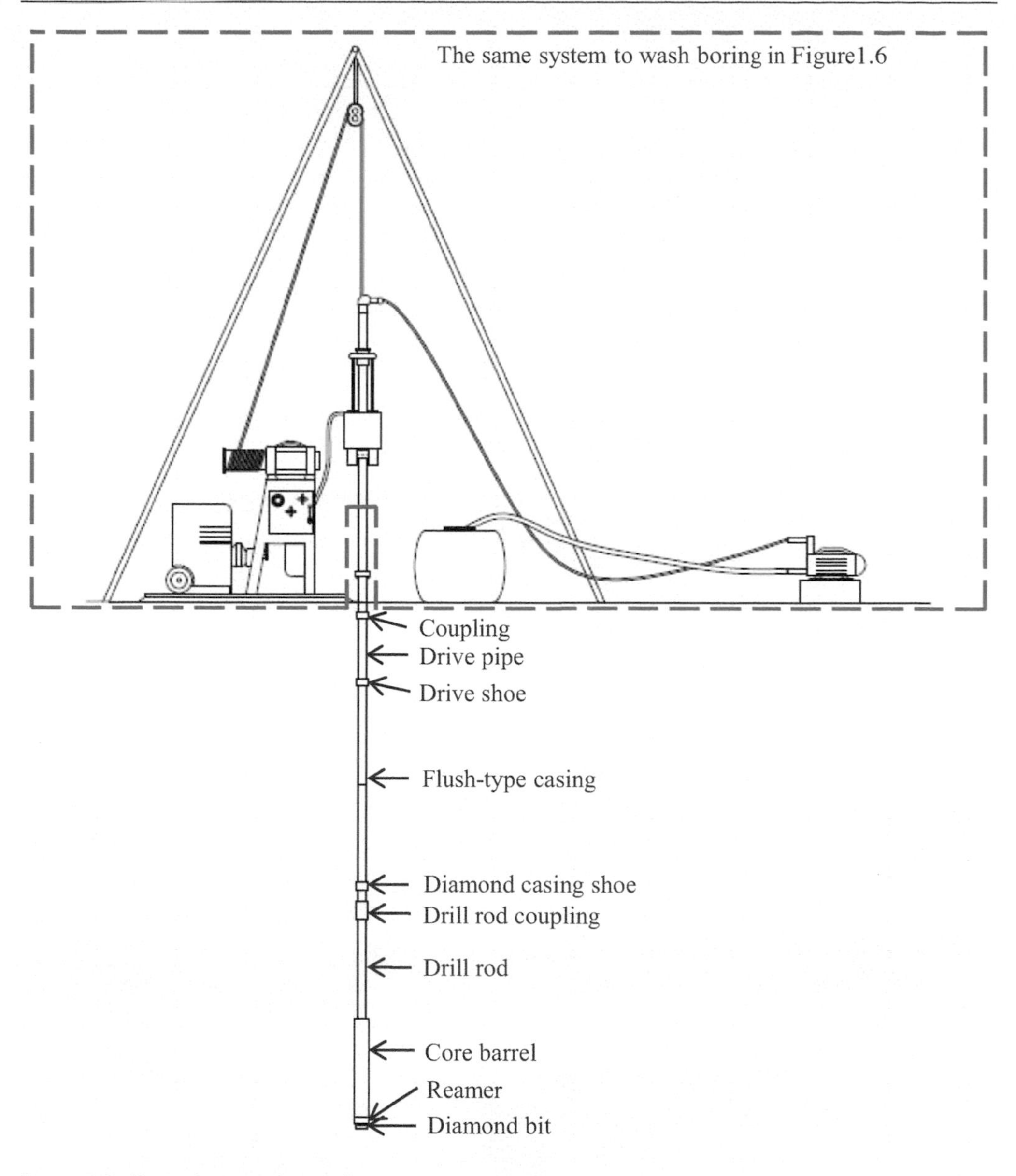

Figure 1.8 Typical rig with wireline coring equipment.

1.4 Soil sampling methods

For obtaining soils in the ground for different design parameters, the corresponding sampling methods are required. Generally speaking, for the properties of soil strength and compressibility, an undisturbed soil sample is preferred; for physical soil properties, a disturbed soil sample which is less costly is commonly used, so that three types of sampling methods and their disturbance to the soils are introduced herein.

1.4.1 Thin-walled tube sampler

The thin-walled tube sampler, made of mostly stainless steel, is commonly used to obtain relatively undisturbed samples of cohesive soils for strength, compressibility, and consolidation behavior. The sampler with a sharpened opening is commonly used, as shown in Figure 1.9, with a 76.2 mm outside diameter and a 72.9 mm inside diameter, and typically is 700 mm to 900 mm long. The pushing of the sampler should be gentle to keep the soils undisturbed and the samplers in a regular shape. Wax is used for sealing moisture, and caps are used to close up the tubes. The thin-walled tube should be delivered to the laboratory in very limited vibratory motion and preserved in a freezer for testing. ASTM D1587/D1587M-15 (2015) can be referred to herein.

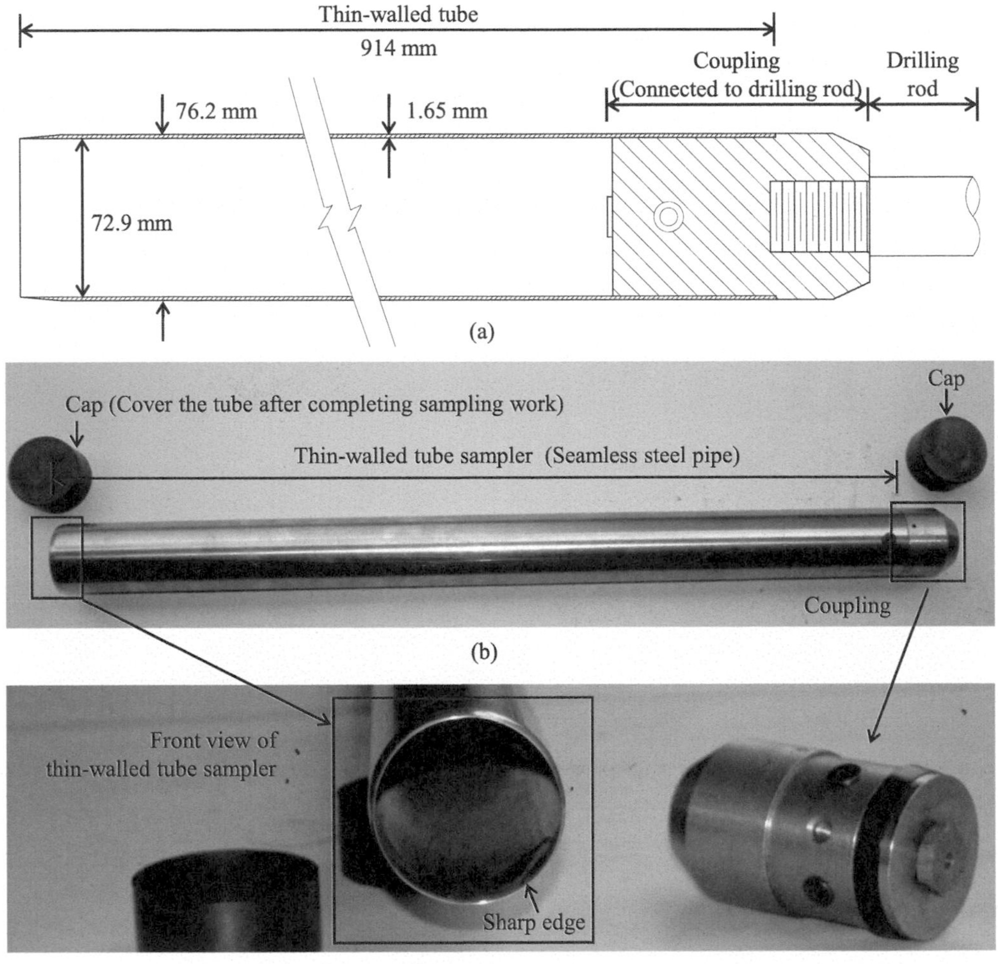

Figure 1.9 Thin-walled tube sampler: (a) illustration, (b) picture, (c) close-up picture.

1.4.2 Split-spoon sampler

Because thin-walled tube samplers have difficulty in penetrating hard soils, such as gravelly soils, sandy soil, and cemented soils, the thick-walled tube pushed into soils with hammer drops is often used to sample disturbed soils for general property tests. In the standard penetration test, the sampler usually has a 35 mm inside diameter and a 15 cm long thick-walled open-drive split barrel sampler (see Figure 1.10). It consists of a driving shoe, 560 mm split tube, and coupling part connected to a drill rod. There are 8 small tubes in the split tube, of which representative ones are shipped to the laboratory for analysis, observation, and test. The application of the split-spoon sampling method in the standard penetration test is discussed in Section 1.6. Readers can refer to the reference (ASTM D3550-01, 2001).

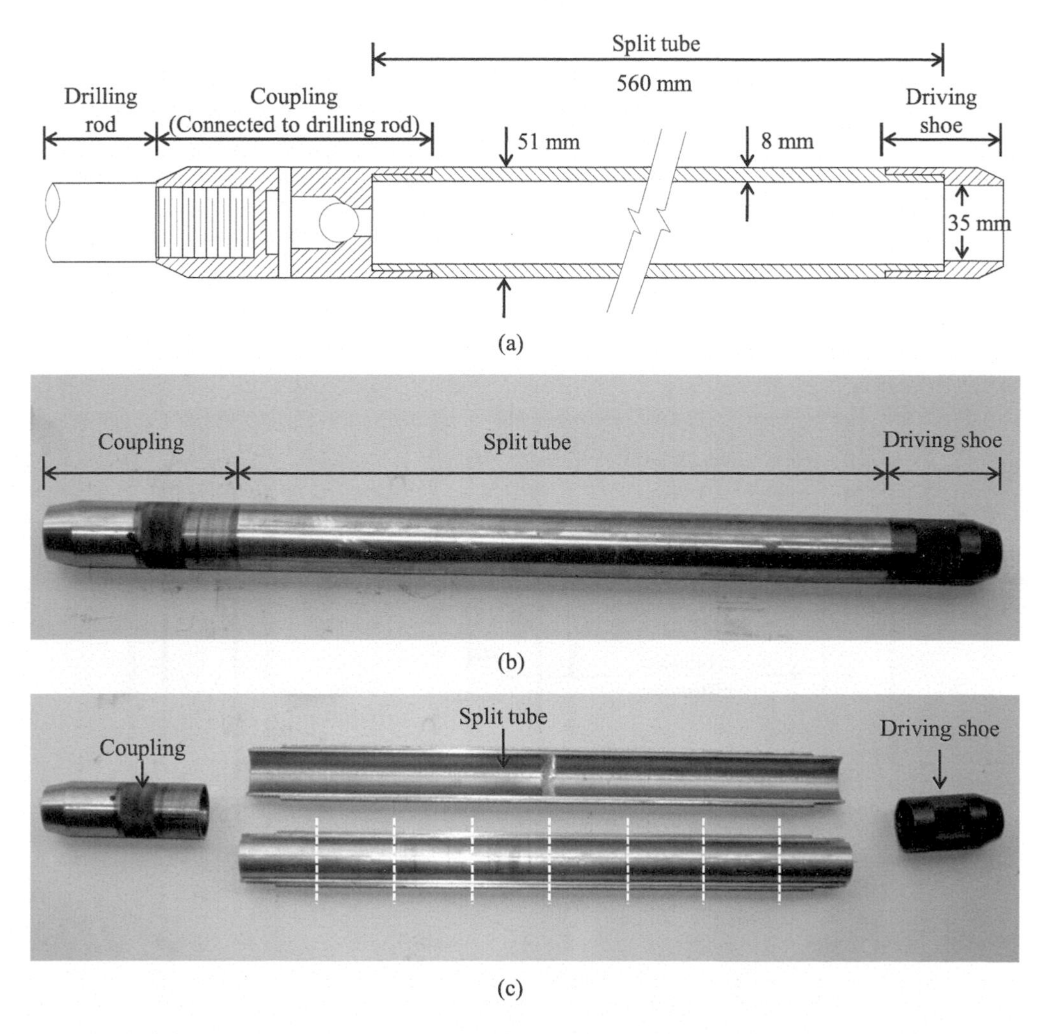

(c)

Figure 1.10 Split-spoon sampler: (a) illustration, (b) assembly photo, (c) disassembly photo.

1.4.3 Piston sampler

When sampling soils that are very sensitive or that consist of a high fines content, a piston sampler that consists of a thin-walled tube fitted with a piston can be used, in which the piston is attached to the length of the rod passing through the sampler head and running inside the hollow-boring rod. During the sampling process, the sampler is lowered into the borehole with a piston located at the lower end of the tube. The tube and piston are locked together with a clamping device at the top of the rods, as shown in Figure 1.11. Hence, the piston is able to prevent water or loose soils from entering the tube when sampling as well as retain the soil in the tube when being withdrawn with a vacuum. There are several types of piston samplers with different dimensions of the tubes; however, the operation principles are similar (Osterberg, 1952). Therefore, in Figure 1.11, the dimensions of the instruments are omitted.

1.4.4 Core sampler

The core sampler is usually used for sampling stiff to hard clays, silts, and sands with some cementation, as well as soft rock. The sampler is not suitable for gravelly soils and loose, cohesionless soils. The frequently employed rotary samplers are triple-tube core barrels with a removable thin-walled tube, known as a liner, inside an outer tube provided with a cutting bit, as shown in Figure 1.12. The rotary sampler has a diameter of 50.8 mm, namely, NX, which is often used for both plains and slopes, and 63.5 mm, namely, HQ, which is often used for slopes, with a length of 61 cm. The outer tube with the cutting bit is rotated and pushed into the soil to

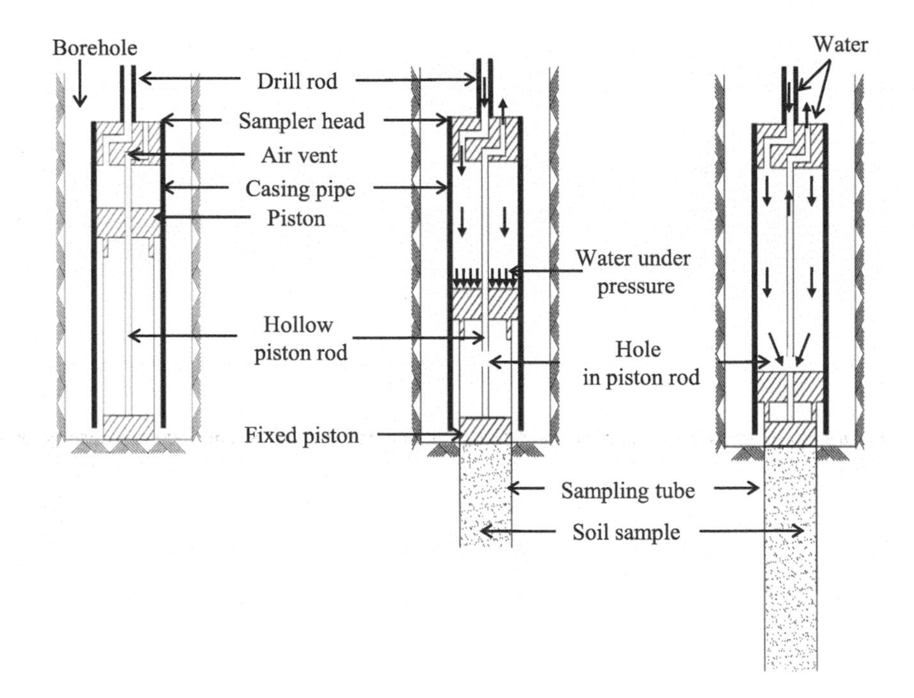

Figure 1.11 Sampling process by piston sampler.

Source: Redrawn from Osterberg (1952).

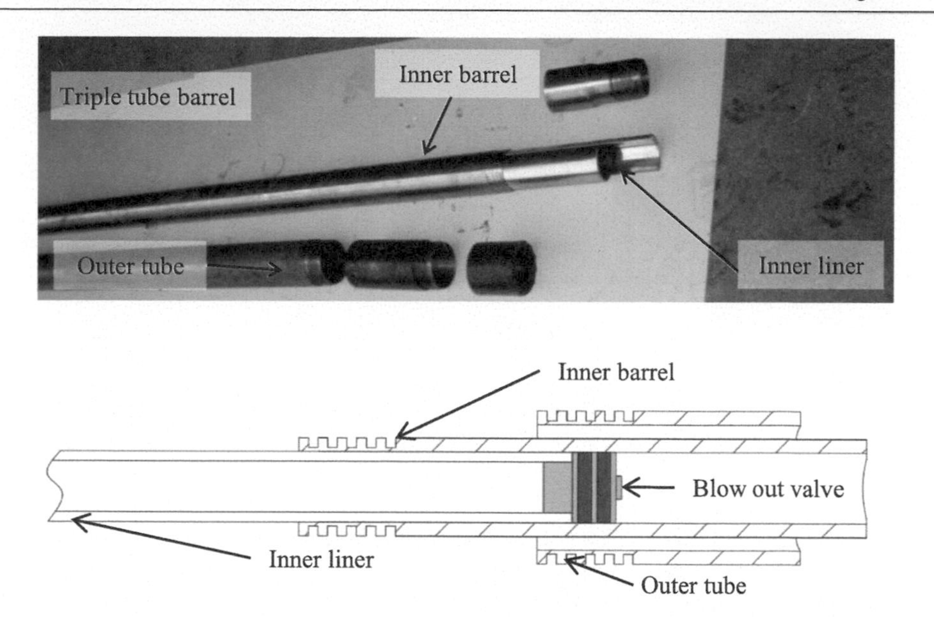

Figure 1.12 Triple-tube core barrel.

Table 1.2 Common borehole sizes

Core barrel type	*Hole diameter*	*Core diameter*	*Core sample* (*Sample preservation with labeled core box*)
NX	75.8 mm	50.8 mm	Upper one
HQ	96.3 mm	63.5 mm	Lower one

the required depth, and the sample enters the liner. The inner tube, that is, the liner, provided with a smooth cutting shoe, remains stationary, and the sample cut by the rotating outer tube slides into the liner. The rock core sample is thus collected in the liner. The rock core is reserved in the core box labeled with the project name, date, and sample depth for further geological investigation, as shown in Table 1.2.

Table 1.3 RQD classification index

RQD (%)	Rock mass quality
<25%	Very poor
25–50%	Poor
50–75%	Fair
75–90%	Good
90–100%	Excellent

After boring and sampling, a method known as rock quality designation (RQD), which uses the core sample to interpret the geological status of the area to measure the quality of a rock core taken from a borehole, is usually conducted. RQD signifies the degree of jointing or fracture in a rock mass measured as a percentage, where an RQD of 75% or more indicates good-quality hard rock and less than 50% indicates low-quality weathered rocks, as shown in Table 1.3. It should be reminded to the readers that *RQD* is defined as the percentage of intact drill core pieces longer than 10 cm recovered during a total core run.

1.5 Sampling disturbance

When pushing samplers introduced in Section 1.4 into the ground to withdraw soil, disturbances to the soil samples are unavoidable. When using the thin-walled tube method, inside wall friction is one of the principal causes of disturbance of the sample (Hvorslev, 1949). The thin-walled tube method is one way to reduce or eliminate wall friction between the soil and the sampler by providing inside clearance by making the diameter of the cutting edge slightly smaller than the inside diameter of the sampler tube. The area ratio can be expressed as:

$$A_r = \frac{d_0^2 - d_i^2}{d_i^2} \tag{1.1}$$

Where d_0 = outside diameter of the sampler tube.

d_i = inside diameter of the sampler tube.
A_r = area ratio (if A_r is smaller than or equal to 10%, the samples are considered undisturbed).

An example is shown to calculate the area ratio in Figure 1.12:

In the case of the thin-walled tube method, the outside diameter is 76.2 mm, and the inside diameter is 72.9 mm. According to eq. 1.1, $A_r = 9\% < 10\%$. (Undisturbed sample)

In the case of the split barrel method, the outside diameter is 51 mm, and the inside diameter is 35 mm. According to eq. 1.1, $A_r = 112\% >> 10\%$. (Disturbed sample)

Notably, although a thin-walled tube is able to withdraw undisturbed soil samples, the transportation of the samples to the laboratory is also a key factor to keep the samples undisturbed.

Table 1.4 lists the types of laboratory tests applicable to disturbed and undisturbed soil samples.

Table 1.4 Test for disturbed samples and undisturbed samples

Disturbed samples	*Undisturbed samples (Clayey soil and fine-grained soils)*
Sieve analysis	Shear strength
Liquid limit and plastic limit	Compressibility
Density	Permeability
Soil classification	–

Table 1.5 Common laboratory tests for soils

Laboratory tests	*Soil property*
General physical property tests	
Water (moisture) content test	ω, γ_{sat}, γ_d
Specific gravity test	G_s
Atterberg limits test	*LL*, *PL*, *PI*, *SL* Soil classification
Sieves and hydrometer analysis	Soil particle size distribution curve Soil classification
Relative density test	$\gamma_{d\max}$, $\gamma_{d\min}$, D_r
Permeability tests	
Constant head permeability test	*k* (for coarse-grained soil)
Falling head permeability test	*k* (for fine-grained soil)
Triaxial permeability test	*k* (for fine-grained soil)
Compressibility tests	
Oedometer test	σ'_c, *OCR*, C_c, C_{ur}
Shear strength tests	
Direct shear test	c', ϕ'
Triaxial compression test	
Triaxial consolidated drained (CD)	c', ϕ'
Triaxial consolidated undrained (CU)	c', ϕ' (with pore water pressure measurements)
Triaxial unconsolidated undrained (UU)	s_u
Triaxial unconfined compression test (UC)	q_u, s_u

1.6 Laboratory test

A laboratory test is essential to explore the mechanical and physical properties of soils in the field, and engineers are then able to provide a correct estimate of the engineering performance of the subsurface soil. Thus, practitioners should make good use of laboratory tests. Table 1.5 summarizes the laboratory tests and their applicability. Test procedures that are key to ensuring test quality should follow the corresponding specifications, which are beyond the scope of the book. The relevant reference (Casagrande and Fadum, 1940) can be referred to reader.

Symbols used in Table 1.5

ω (%)	water content	γ_{dmax} (kN/m³)	maximum dry unit weight	c' (kN/m²)	effective cohesion
γ_{sat} (kN/m³)	saturated unit weight	γ_{dmin} (kN/m³)	minimum dry unit weight	ϕ'	effective friction angle
γ_d (kN/m³)	dry unit weight	D_r (%)	relative density	q_u (kN/m²)	unconfined compression strength
Gs	specific gravity	k (m/s)	permeability	s_u (kN/m²)	undrained shear strength
LL	liquid limit	σ'_c (kN/m²)	preconsolidation stress		
PL	plastic limit	OCR	overconsolidation ratio		
PI	plasticity Index	C_c	compression index		
SL	shrinkage limit	C_{ur}	unload–reload index		

Table 1.6 Common laboratory tests for rocks

Laboratory tests	*Applicability*
Rock physical properties tests	Obtains unit weight, moisture content (γ_m)
Point load strength test	Provides an index of uniaxial compressive strength (q_u)
Unconfined compressive strength of intact rock core	Uniaxial compressive strength (q_u)
Direct shear test for weak rock	c', ϕ'

1.7 Field test

Field tests refer to all the in situ tests being executed during soil boring exploration. It includes the standard penetration test (SPT), cone penetration test (CPT), vane shear test, pressuremeter test, flat dilatometer test, and plate loading test. SPT is capable of withdrawing soil samples for laboratory tests, while CPT is capable of measuring continuous soil properties during penetration. In situ field tests have inherent advantages that are able to avoid disturbances of soil sampling. As mentioned in Section 1.5, undisturbed soil samples can be obtained through cautious sampling procedures; however, great attention must be paid to sample transportation and the test procedure in the laboratory. Much valuable information on soil behaviors can be provided, such as undrained and drained soil strength, stress path, and stress–strain responses of soils in the laboratory test. Therefore, both field tests and laboratory tests are required to explore the precise geological information of the site.

Although a field test can measure the properties of in situ soils, it is usually unable to derive the soil parameters directly that are required in a foundation design, such as the friction angle, undrained shear strength, Young's modulus, and so on. Therefore, some transformation equations or empirical formulas are necessary. Section 2.3 provides several useful transformation equations that have been published in literatures for readers' reference.

1.7.1 Standard penetration test (SPT)

The equipment is shown in Figure 1.14 and comprises a split tube with a driving head and either a heavy-duty cutting shoe or a solid cone point. The head of the tube is threaded for connection (via a series of drill rods) to a hammer. The device is driven into the ground at the base of the borehole with a free-falling 63.5 kg trip hammer dropping to a distance of 760 mm. The hammer and rods should be vertical, and therefore, the borehole should be plumb.

During the test, a count is made for the number of blows required to drive it though the initial 150 mm and a subsequent 300 mm test zone. The initial 150 mm is considered to be in distributed material. The recording of the blow count of the test zone is becoming widely accepted and is strongly recommended. The device is driven into the ground at the base of the borehole with a free-falling 63.5 kg trip hammer dropping a distance of 760 mm. The soils pushed into the split tube introduced in Figure 1.10 during the penetration are for general physical property soil test. Tables 1.7 and 1.8 are a general relationship between SPT-N value and the strength of clay or the relative density of sand.

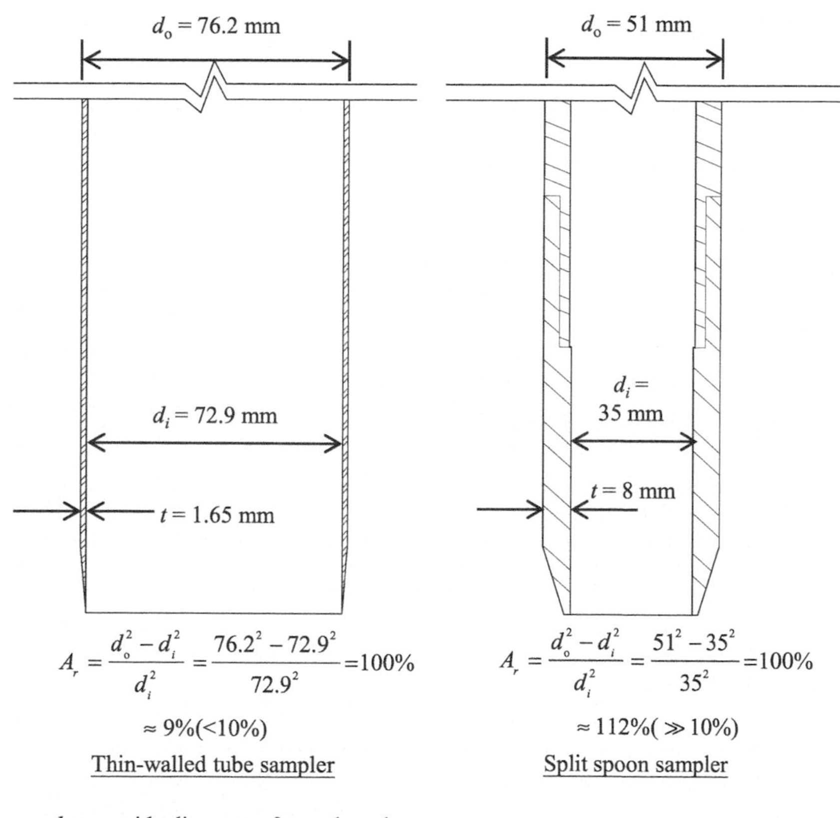

Figure 1.13 Calculation of the area ratio A_r.

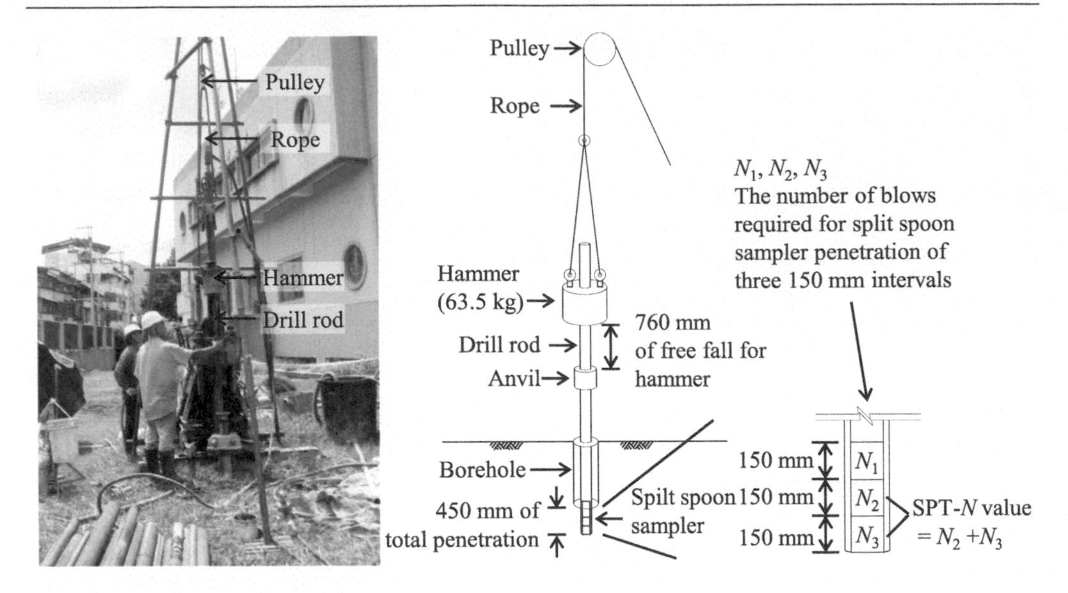

Figure 1.14 SPT in operation.

Table 1.7 Relationship of the consistency of clay, SPT-*N* value, and unconfined compressive strength q_u >.

SPT-N value (blows)	*Consistency*	*Unconfined compressive strength q_u (kPa)*
<2	Very soft	<25
2~4	Soft	25~50
4~8	Medium	50~100
8~15	Stiff	100~200
15~30	Very stiff	200~400
>30	Hard	>400

Source: Terzaghi et al. (1996).

Table 1.8 Relative density of sands according to SPT-*N* results

SPT-N value (blows)	*Relative density*
0~4	Very loose
4~10	Loose
10~30	Medium
30~50	Dense
>50	Very dense

Source: Terzaghi et al. (1996).

Due to variability in local practice and equipment, the energy efficiency of the drop hammer onto the drop rods varies from region to region. Moreover, the overburden stress also affects the SPT-*N* value significantly. The SPT-*N* values obtained should be corrected. The relevant correction and transformation equation for the parameters that are required for design can be found in Section 2.3.

1.7.2 Cone penetration test (CPT)

The cone penetration test (CPT) is also a commonly used field investigation method in engineering practice. A steel cone penetrates the ground for soil testing similar to the SPT, and data on tip resistance, side friction, and pore pressure during steady penetration is acquired for profiling continuous soil layers along the depth and estimating the soil strength, soil classification, and pile bearing capacity.

The configuration of the steel cone in the CPT, with an apex angle of 60° and cone bottom cross-sectional area of 10 cm^2, as shown in Figure 1.15, has been standardized in recent years

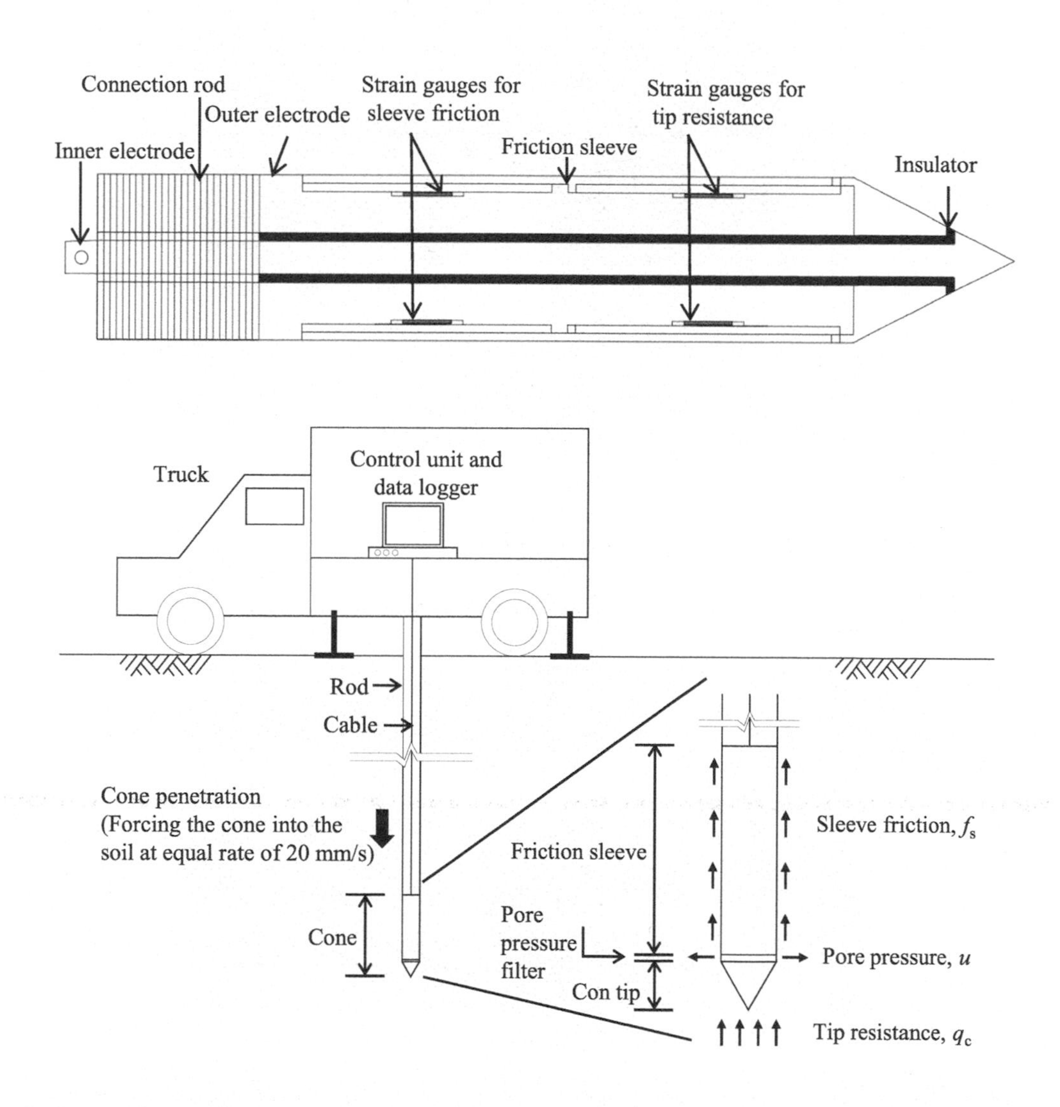

Figure 1.15 Illustration of CPT.

(David, 2018). Several empirical or transformation equations and figures to correlate design parameters have been developed, as discussed in Section 2.3.

Figure 1.16 demonstrates the work procedures of the CPT, in which the truck weight is employed to provide a reaction. Most commercially available CPT rigs operate electronic friction cone and piezocone penetrometers, whose testing procedures are outlined briefly in this book. These devices produce a computerized log of tip and sleeve resistance, the ratio between the two, induced pore pressure just behind the cone tip, pore pressure ratio (change in pore pressure divided by measured pressure with a penetration speed of 2 cm/s), and lithologic interpretation of each 2 cm interval, which are continuously logged and printed out. A guideline for evaluating soil types can be summarized: sands are identified when $q_c > 5$ MPa and $u_2 \approx u_o$, and

(a)

(b)

Figure 1.16 CPT in operation: (a) setting the CPT truck; (b) forcing the cone into the soil at equal rate (20 mm/s).

the presence of intact clays is prevalent when $q_c < 5$ MPa and $u_2 > u_o$ (hydro). The magnitude of pore water pressures can be helpful to indicate intact clays such as soft ($u_2 \approx 2 \cdot u_0$), firm ($u_2 \approx 4 \cdot u_0$), stiff ($u_2 \approx 8 \cdot u_0$), and hard ($u_2 \approx 20 \cdot u_o$). Fissured overconsolidated clays tend to have negative u_2 values such that $u_2 < 0$. As meaning of all the symbols can be listed as followings:

Tip resistance, q_c, is theoretically defined as the ratio of penetration force.
Sleeve friction, f_s, is measured by tension load cells embedded in the sleeve as the probe is advanced through the soil.
The hydrostatic pressure, u_0.
The pore pressure, u_2, is measured during penetration.

Compared with the SPT, the advantage of the CPT is that it is capable of acquiring continuous data on soils (see Figure 1.17), and the disadvantage is not being capable of sampling soils. In engineering practice in many countries, the SPT is still the main method employed in soil investigations, for two main reasons: SPT data are more abundantly collected, and the SPT labor cost is relatively lower than that of CPT trucks in many countries. In Figure 1.17, clayey soils are indicated by low tip resistance and high pore pressure, where $q_c < 5$ MPa and $u_2 > u_o$, such as the depths of 2–6 m and 10–17 m; on the other hand, high tip resistance and low pore pressure ($q_c > 5$ MPa and $u_2 \approx u_o$) indicate sandy soils or sand mixtures, with the depths of 8–10 m.

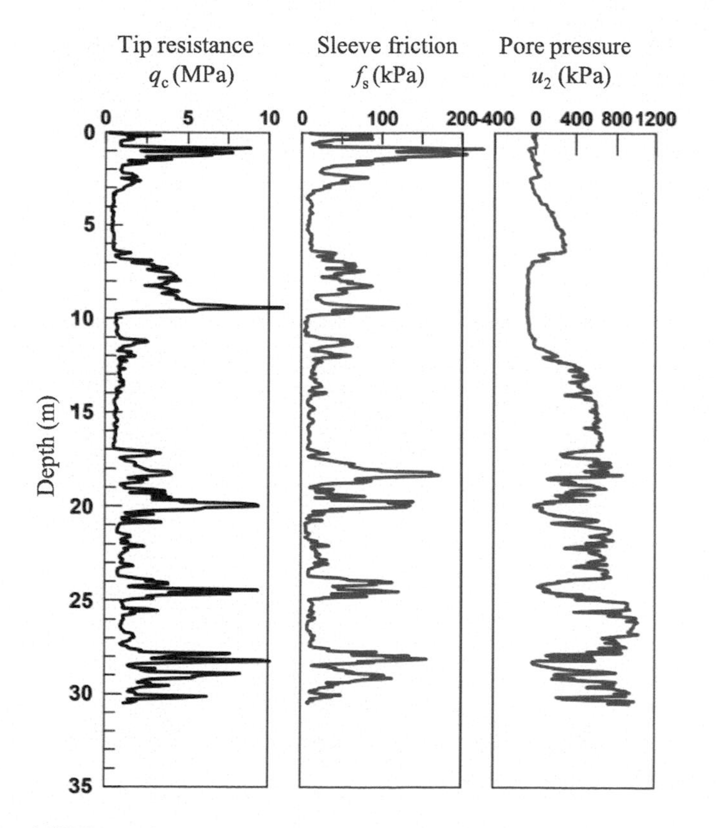

Figure 1.17 Typical CPT results.

1.7.3 Vane shear test (VST)

The vane shear test is a method for determining the undrained shear strength of soils, especially soft clays, in the field, while undisturbed soil sampling by the thin tube method is not easy to perform.

The vane shear test (VST) consists of a four-bladed vane, which is typically approximately 55 mm or 65 mm in diameter and 110 mm or 130 mm in height, with a height-to-diameter ratio of 2, depending on the test purpose, as shown in Figure 1.18. A rectangular bladed vane or tapered

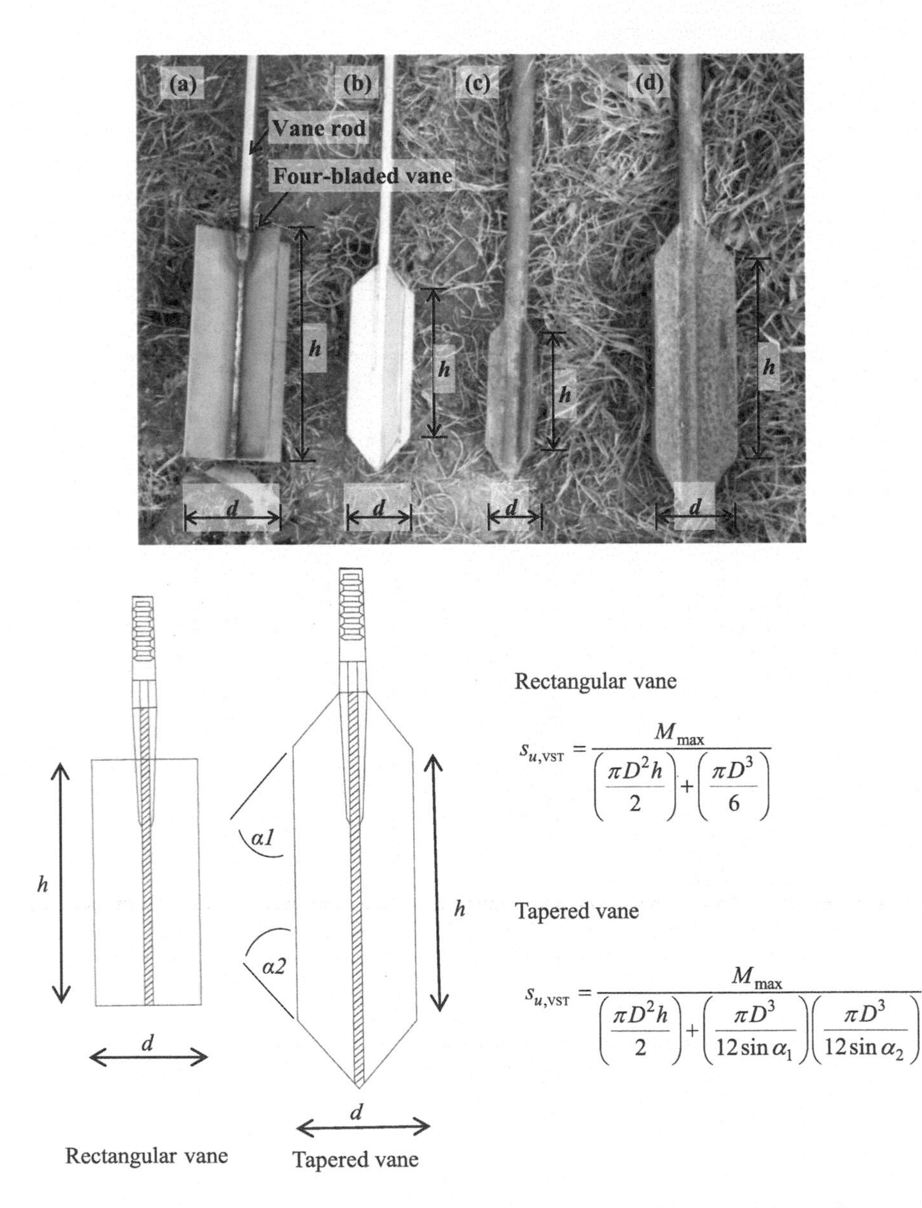

Figure 1.18 Variable types of four-bladed vane: (a) rectangular vane (b) to (d) tapered vane.

bladed vane with longer height is used in soft clay, while a tapered bladed vane with shorter height is used in harder clay. The vanes of the instrument are placed into a borehole using a suitably stiff rod until it reaches the desired depth. Then, the vane is gradually lowered into the undisturbed soils until the top of the vane is 50 cm into the soils. The vane is rotated using torque applied at the handle at a rate of 0.1° per second, until the soils reach the failure state. A picture of vane shear test performed in field is seen in Figure 1.19.

Eq. 1.2 assumes that the undrained shear strength ($s_{u,VST}$) of the soil is constant on the cylindrical sheared surface and at the top and bottom faces of the sheared cylinder, as shown in Figure 1.20. The torque applied (M_{max}) equals the sum of the resisting torque at the sides and

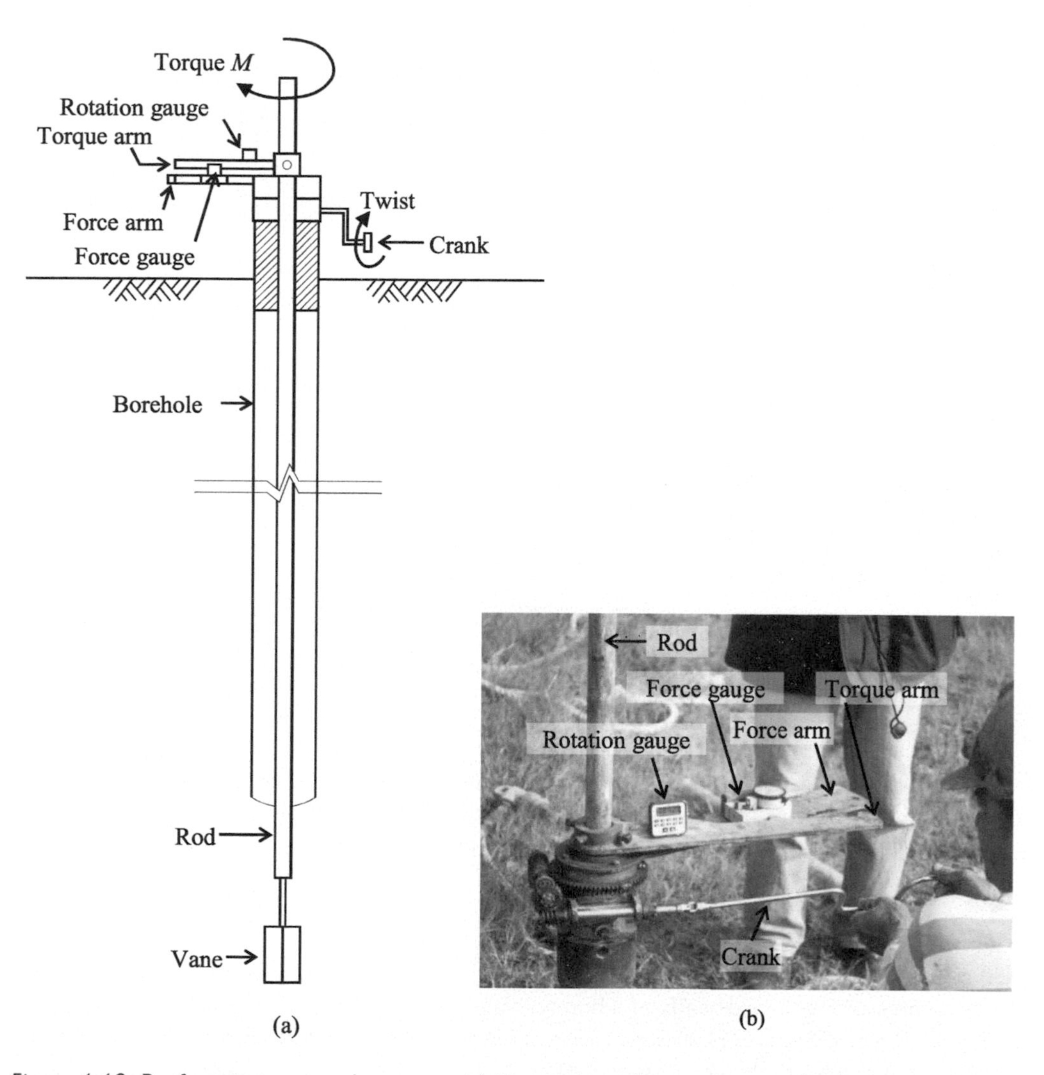

Figure 1.19 Performing a vane shear test: (a) illustration of vane shear test; (b) applying torque by twisting the crank at constant speed and recording the rotation angle θ and torque until the soils fail. Then, the max torque Mmax can be determined.

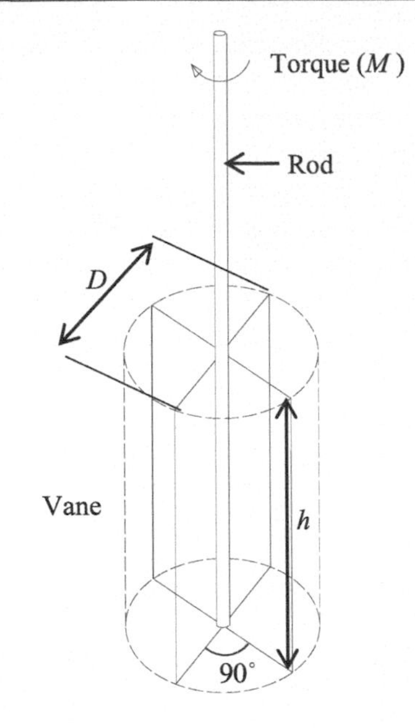

Figure 1.20 Definition of the parameters of the undrained shear strength formula.

that at the top and bottom. Thus, the shear strength of a given soil sample is calculated with the following formula.

$$s_{u,\mathrm{VST}} = \frac{M_{\max}}{\left(\frac{\pi D^2 h}{2}\right) + \left(\frac{\pi D^3}{6}\right)} \tag{1.2}$$

Where:

$s_{u,\mathrm{VST}}$ = undrained shear obtained by VST (kN/m^2).
$M_{\max}$ = max torque applied on the rod (kN-m).
D = diameter of the vane (m).
h = height of the vane (m).

The undrained shear strength obtained from a vane shear test depends on the rate of application of torque, which is, in turn, related to the plasticity of soils. Therefore, the obtained $s_{u,\mathrm{VST}}$ must be corrected. The relevant correction can be found in Section 2.3.

1.7.4 Pressuremeter test (PMT)

A pressuremeter, also named the borehole lateral load in situ test, is a device performed by applying pressure to the sidewalls of a borehole and recording the corresponding deformation. It was designed by Menard (1956) for determining the stress–strain relations of in situ soil at a certain depth in a borehole. By using these stress–strain relations, one can determine the elastic modulus, preconsolidated stress, friction angle, undrained strength, soil pressure of the remaining soil, etc.

Pressuremeters enter the ground by pushing into a pre-boring hole at a depth at which the probe should be placed. Then, increments of pressure by gas or water are applied to the inside of the membrane, forcing it to press against the material and loading a cylindrical two-dimensional state of stress. The probe of the pressuremeter, which is the major testing unit, consists of three inflatable cells, one of which is above the other, as shown in Figure 1.21. The middle cell is the measuring membrane cell, which is filled with water during the test and has a diameter of 58 or 74 mm and a height of 203 or 184 mm. The other two cells, which are at the top and bottom of the mid cell, are guard cells that protect the main cell from the end effects caused by the finite

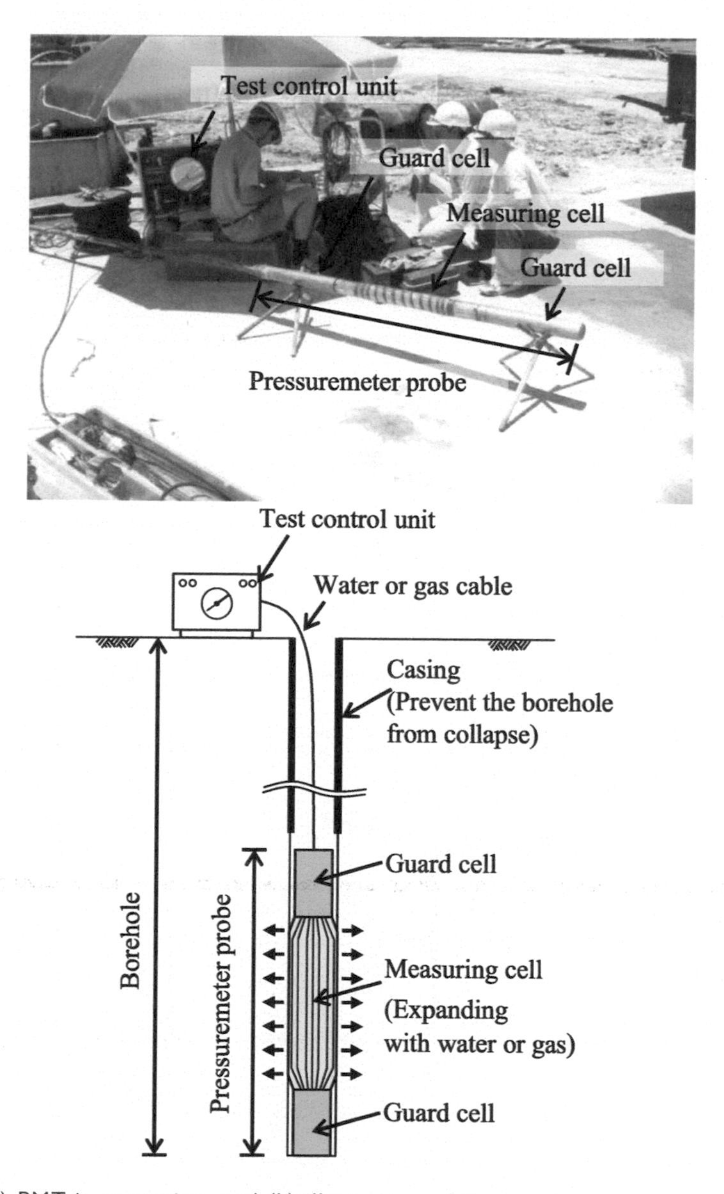

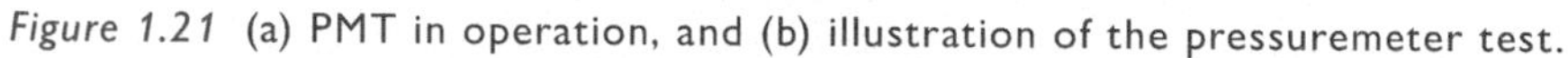

Figure 1.21 (a) PMT in operation, and (b) illustration of the pressuremeter test.

length of the cable (so-called end effects) to reduce testing errors. Relevant information can be referred to Baguelin et al. (1978) and Clarke (1994).

A self-boring pressuremeter test, namely, the SBPMT, is developed to overcome testing difficulties when experiencing frequent borehole collapse in soft soils or expansions in expansive soils before pressuremeter placement.

Figure 1.22 presents a typical pressuremeter curve to estimate Young's modulus, preconsolidation stress, internal friction angle, undrained shear strength, static earth pressure, etc. that can be found in Menard (1956). The formula used for determining Young's modulus is shown as an example in eq. 1.3.

$$E_P = 2 + (K_D / 1.5)^{0.47} - 0.6 \tag{1.3}$$

or

$$E_P = 2(1+\gamma)(V_0 + V_m)(\Delta P / \Delta V)$$

Where:

E_P = pressuremeter modulus.
K_D = horizontal stress index ($K_D = (P_0 - u_0)/\sigma'_{v0}$).
γ = Poisson's ratio (generally taken as 0.3 for pressuremeter applications).
$V_0 + \Delta V$ = volume of probe.
ΔV = volume increase in the straight-line portion of the test curve.
ΔP = pressure increase corresponding to ΔV volume increase.

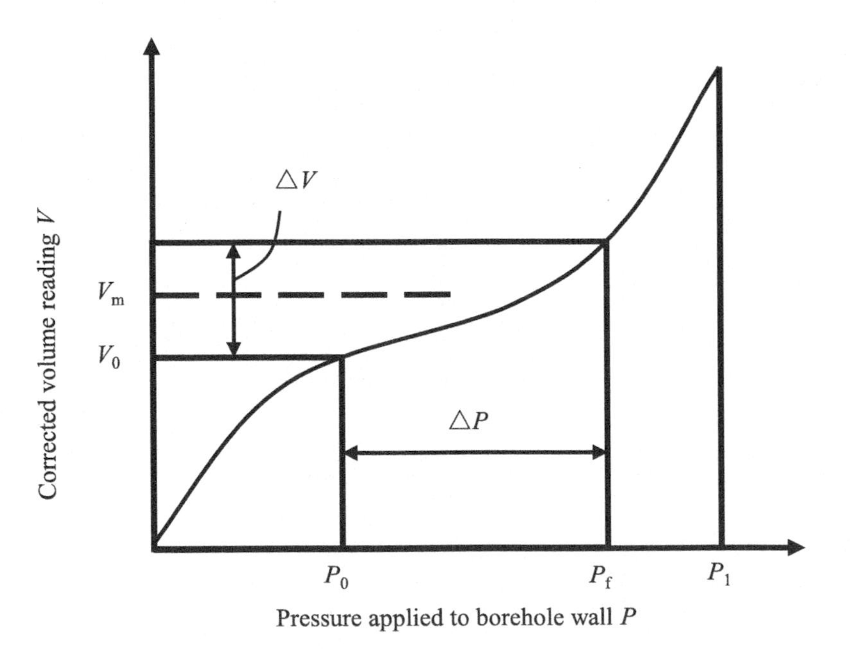

Figure 1.22 Typical pressuremeter test result.

Source: Redrawn from Rashed et al. (2012).

1.7.5 Dilatometer test (DMT)

The flat dilatometer test (DMT) was developed in Italy by Silvano Marchetti (1980) and is a push-in type in situ test that is quick, simple, economical, and highly reproducible for exploring ground profile, soil property, and modulus values by applied displacement and measured pressure. The device consists of an arrangement for lateral expansion after reaching the desired depth. This test uses a stainless steel blade with an 18° wedge, 240 mm length, 95 mm width, and 15 mm thickness. The blade has a 60 mm diameter circular steel membrane mounted flush on one side.

The blade in Figure 1.23(a) is connected to a control unit on the ground surface by a pneumatic-electrical tube (transmitting gas pressure and electrical continuity) running through the insertion rods. A gas tank, connected to the control unit by a pneumatic cable, supplies the gas pressure required to expand the membrane. The control unit is equipped with a pressure regulator, pressure gage, audiovisual signal, and vent valves.

First, pushing the blade into the ground against the reaction fame. Upon reaching the desired depth, the gas pressure was applied through a nylon tube connected to the surface assembly with control units. The lift-off pressure P_0 required to move the center of the membrane in line with the blade was measured approximately 15 sec after the blade reached the desired depth. Figure 1.23(b) shows that the pressure P_1 is the pressure required to move the center of the membrane by 1.1 mm into the soil and is recorded within 15 to 30 sec after the last reading. Upon reaching P_1, the membrane is quickly deflated, and the blade is pushed to the next test depth. When DMT is performed inside a borehole, the diameter of the borehole should be kept minimal (100–120 mm) to avoid buckling of the drill rod. The test is suitable for clays, sand, and silts.

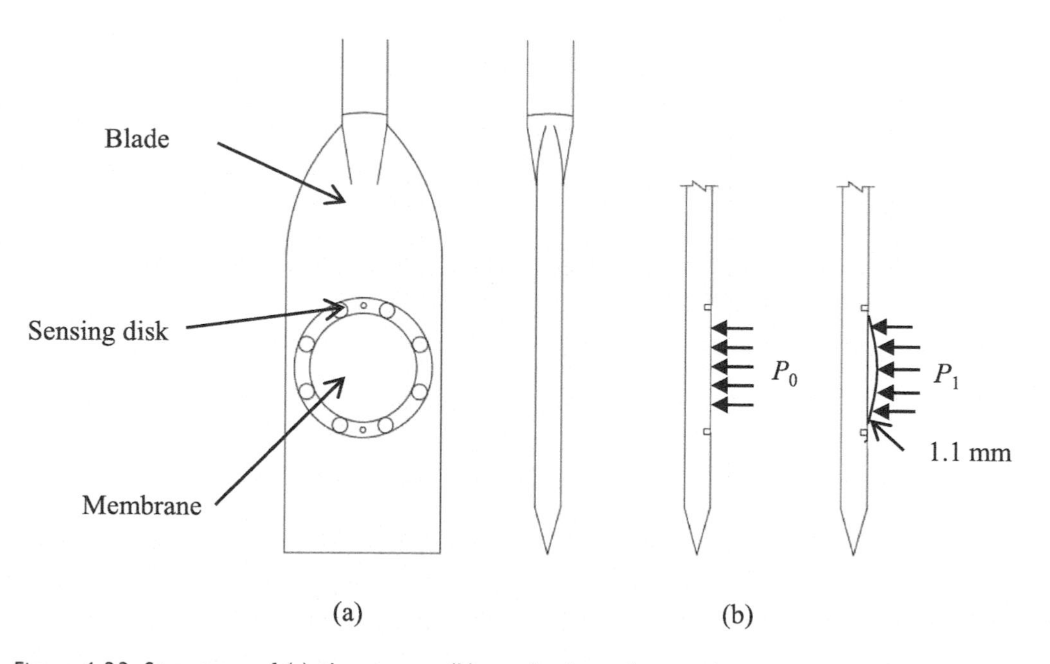

Figure 1.23 Structure of (a) the sensor, (b) mechanism of operation.

Source: Redrawn from Marchetti et al. (2001).

Although not suitable for gravels, the blade is strong enough to pass through a gravel layer of 0.5 m thickness. By using field record and data reduction P_0 (corrected first reading) and P_1 (corrected second reading), one can determine the coefficient of earth pressure at rest, as shown in the following (others can be found in Silvano Marchetti (1980)). Figure 1.24 presents a typical dilatometer test result with parameter K_D; furthermore, $K_{0,\,DMT}$ is able to be obtained by eq. 1.4.

The coefficient of earth pressure in situ can be determined by the formula:

$$K_{0,DMT} = (K_D/1.5)^{0.47}\text{-}0.6 \text{ For } I_D < 1.2 \tag{1.4}$$

Where:

$K_{0,DMT}$ = coefficient of earth pressure at rest, determined by the dilatometer test.
K_D = horizontal stress index ($K_D = (P_0 - u_0)/\sigma'_{v0}$).
σ'_{v0} = the initial effective stress and u_0 is the initial pore water pressure.
P_0 = the reading of the dilatometer test.
I_D = the material index ($I_D = (P_1 - P_0)/(P_0 - u_0)$).

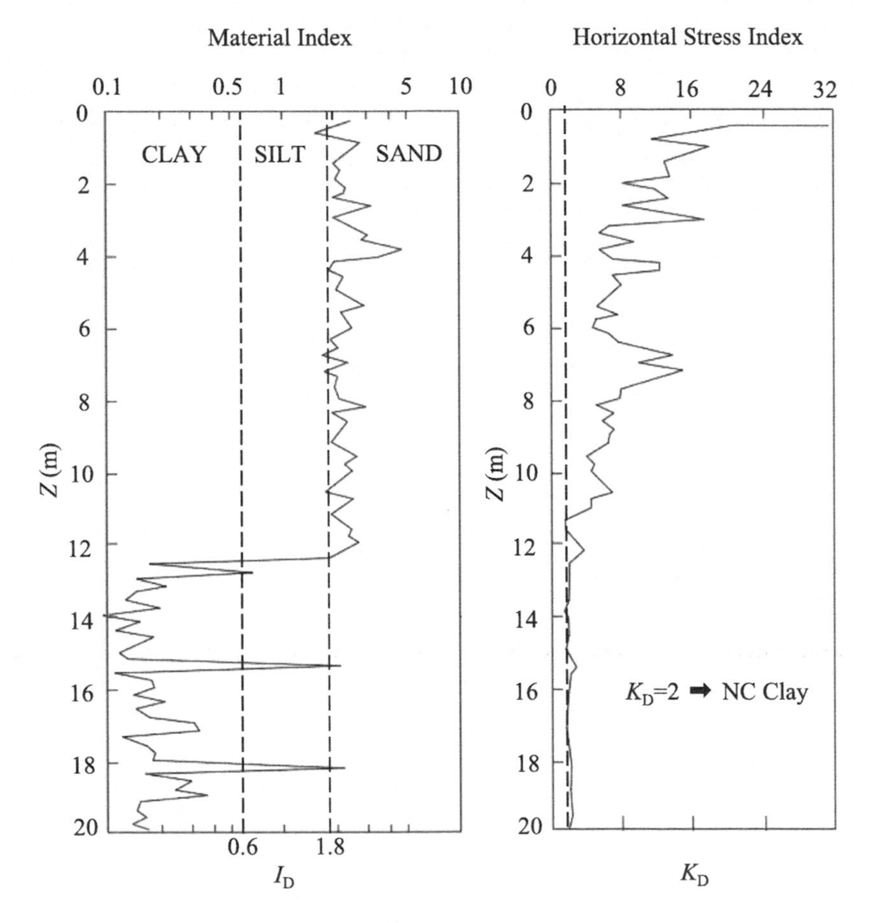

Figure 1.24 Typical dilatometer test result.
Source: Redrawn from Marchetti et al. (2015).

1.7.6 Plate loading test (PLT)

The plate load test (PLT) is designed to determine the vertical deformation and strength characteristics of soil by assessing the force with time when a rigid or flexible plate is made to act on the soil vertically. Then, the vertical plate load test can be used to evaluate the ultimate bearing capacity, shear strength, deformation parameters, and vertical subgrade reaction of the soil beneath the plate. If the plate is designed to act horizontally, the horizontal subgrade reaction of soils can be determined. Testing can be executed at the ground surface, in pits, or in trenches. The isotropic elastic continuum of the soils is hypothesized, and the stress and deformation information of the soil mass are approached in the plate load test.

The coefficient of the vertical subgrade reaction coefficient k_v is defined as the ratio of the uniform loading pressure exerted q to the plate displacement s, which is $k_v = q/s$. Importantly, since elastic theory is used to derive k_v, k_v is valid only under small levels of deformation, varies with the depth of the soils, and is affected by the size of the plate.

Figure 1.25 shows an example of a vertical plate loading test in operation. When the PLT is used to evaluate the k_v in the gravel layer, it is necessary to adjust the size of the plate to

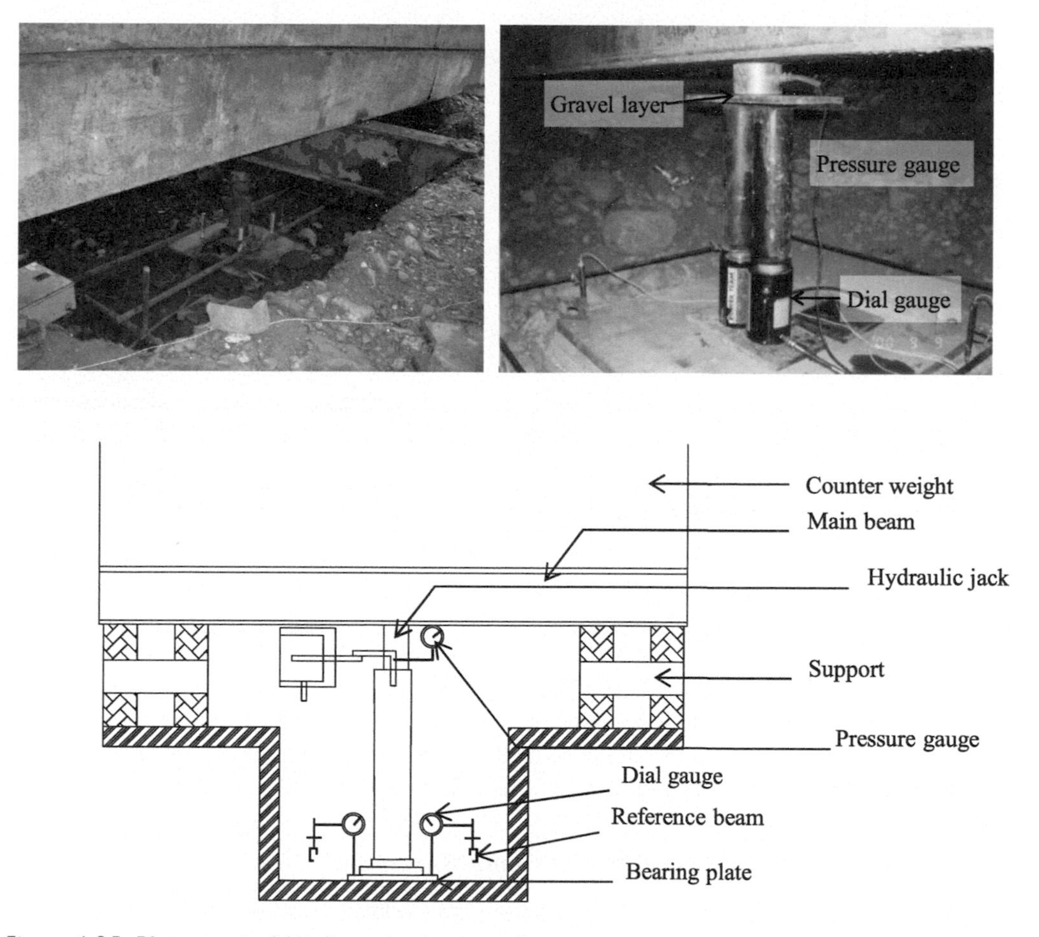

Figure 1.25 Plate vertical loading test in operation.

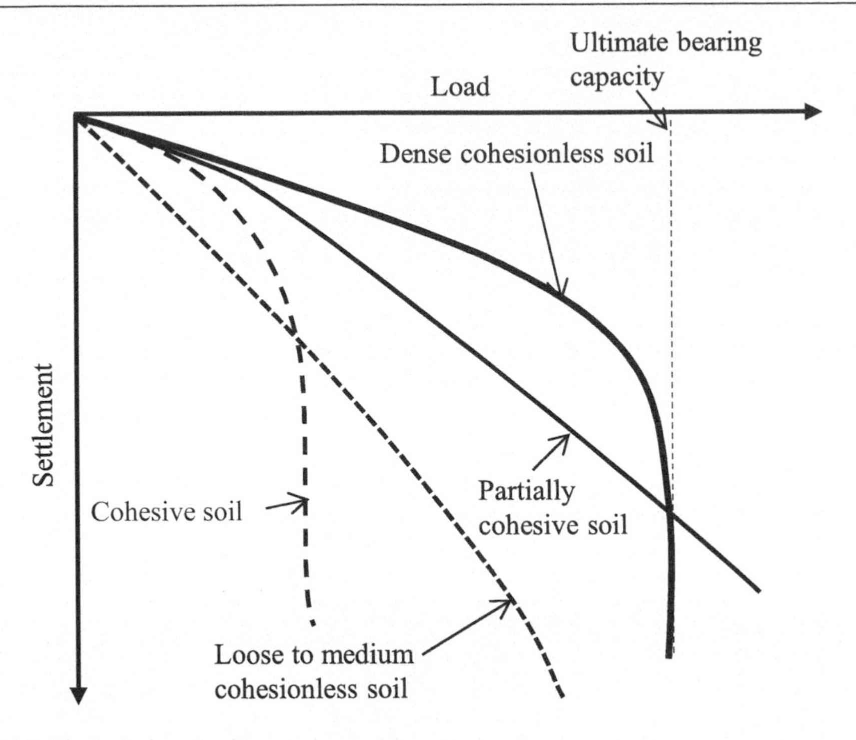

Figure 1.26 Typical plate loading test result.

Source: Redrawn from Anjan (2019).

approach the real value of k_v for the foundation design. Attention is also paid on the typical performance of plate load test on different soils in Figure 1.26 for finding the best test result in the plate load test.

The results from a plate load test can be used to estimate the bearing capacity and settlement of a foundation. Section 3.7 introduces the application of the plate load test to the foundation design.

1.8 Groundwater investigation

Groundwater, the major disaster-causing factor in geotechnical engineering, undoubtedly affects the stability of construction; therefore, it is a critical work factor in field investigations. Due to geo-structural inhomogeneity, discontinuities, or fissures, groundwater seepage is complex because of the characteristics of multiple aquifers (aquifers, confined aquifers, confining layers, subsurface perched water, etc.) that result in difficulties in field investigation. For a better understanding of the profile of groundwater variation, integrated investigation methods are necessary. This section introduces frequently employed methods, including groundwater table observation and water pressure monitoring and in situ permeability testing.

1.8.1 Groundwater table and pressure measurement

The groundwater table changes with the season and has a relevant relationship with rainfall. The groundwater table depth, water pressure profile, relationship between groundwater table and rainfall in the time domain, extreme conditions due to the groundwater table rise, groundwater influence on slope stability, and path of groundwater seepage are all critical in a field investigation of groundwater in geotechnical engineering.

The following paragraphs introduce two frequently employed methods in engineering practice: observation wells used for observing the groundwater table and piezometers used for monitoring water pressure in sandy or aquifer layers at different depths.

Observation wells are often constructed with a 50 mm diameter PVC opening with holes and screened with a filter pack or geotextile in completed boreholes, and the depth of groundwater table is measured in the PVC, as shown in Figure 1.27. Generally, the groundwater table measured in the PVC of observation wells represents a mixed groundwater table along its penetration depth; therefore, observation wells are more often employed in a single aquifer.

Open standpipe piezometers are typically installed with a solid casing down to the depth of interest and a slotted or screened casing within the zone where water pressure is being measured. Above and below the depth of interest, the water is sealed into the borehole with clay or bentonite to prevent the groundwater supply at other depths, as shown in Figure 1.28.

For monitoring groundwater pressure in the piezometer as well as the groundwater table in the observation well in modern days, data can be obtained manually as well as automatically,

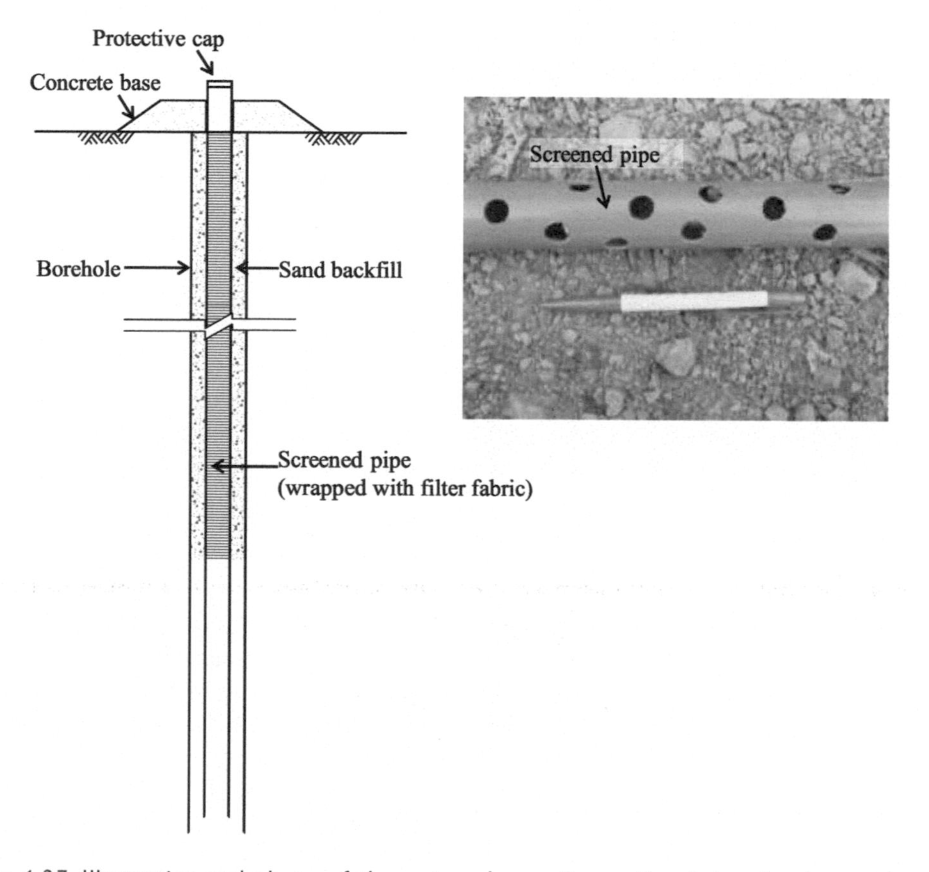

Figure 1.27 Illustration and photo of the water observation well and the pipe (measuring the water level of a single aquifer).

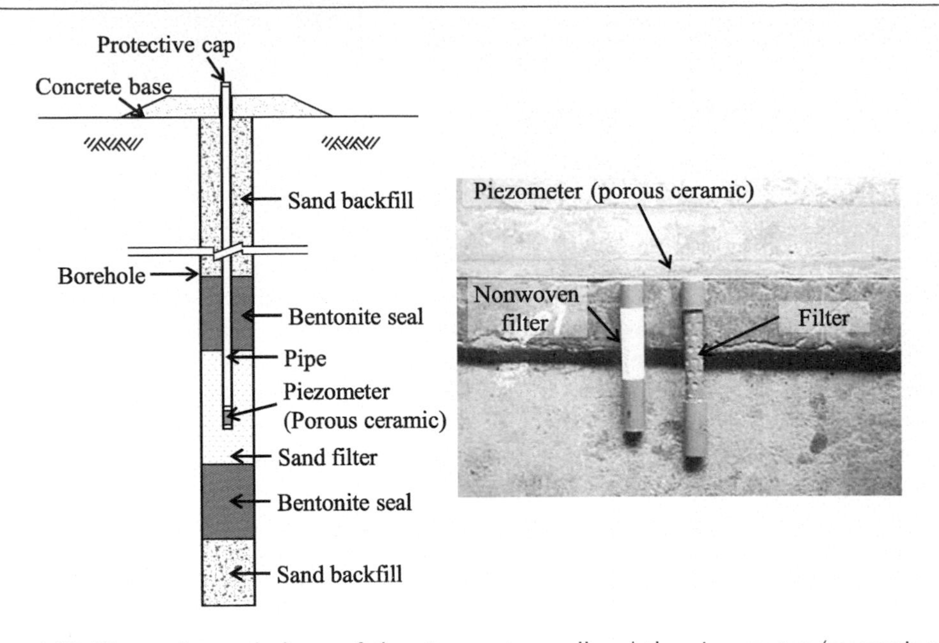

Figure 1.28 Illustration and photo of the piezometer well and the piezometer (measuring the water level of any aquifer).

and pressure gauges (transducer) can be vibrating-wire, pneumatic, or strain-gauge in operation, converting pressure into an electrical signal. These piezometers are cabled to the surface, where they can be read by data loggers or portable readout units, allowing for faster or more frequent reading. This capability can be a critical component in a warning system. Importantly, however, a manual reading of the groundwater table is necessary for confirming the accuracy of the installed automatic instrument.

Figure 1.28 presents an illustration of the piezometer wells and a picture of the piezometer. The bentonites in the borehole in Figure 1.28 were used for sealing water from other soil layers. Bentonites swell and form a semisolid, gel-like mass after mixing with water because of their special property, which can fill voids and limit water movement to create barriers between the soil and water.

1.8.2 Field permeability test

The soil permeability, also termed hydraulic conductivity, of soils in the field can be measured using several methods that include constant and falling head laboratory tests on undisturbed soil specimens sampled from the field in principle; however, in reality, the permeability of fine soils in laboratory testing is determined through triaxial test indirectly. Alternatively, permeability may be measured in the field by field constant head or falling head tests, or pumping tests. Also, the permeability may also be empirically obtained through the grain size distribution in engineering practice for a preliminary design purpose. A good knowledge of soil permeability is needed for estimating the quantity of seepage and designing drainage systems in construction site planning.

1. Permeameter test

The permeameter test measures soil permeability at the bottom of a borehole, as shown in Figure 1.29. The test measures the hydraulic conductivity by temporarily filling water into PVC in the borehole to create an unbalanced pressure head between water inside PVC and in soils at the bottom of the borehole with a constant head for sand, as in Figure 1.29(a), and gravel and a falling head for silt and clay, as in Figure 1.29(b). Based on the relationship of time and filling water in PVC, field permeability can be measured. The equations for the permeability of the constant head and falling head of the in-field test are similar to those of laboratory tests in soil mechanics books, but with horizontal seepage. In Figure 1.30, h_c shows the constant head, h_0 shows the initial head of time t_0, and h_1 shows the initial head of time t_1. Figure 1.30 also illustrates that the tests in boreholes that only involve a relatively small volume of soil or rock around the test section provide "small-scale" values of permeability in the field. If the soil/rock is heterogeneous or has significant fabric, such tests may not be representative of the mass permeability of the strata. Large-scale tests (such as pumping tests) may provide better results. Readers can also refer to the reference (Hvorslev, 1951).

2. Pumping test

A large-scale pumping test contains pumping wells and observation wells, in which a pumping well with a diameter larger than 30 cm and an observation well with a diameter larger than 15 cm are usually installed. The principle of a pumping test in Figure 1.31 is that if water from the well is pumped and the discharge q is recorded, the drawdown and in-observation wells h_1 and h_2 are measured, and the hydraulic conductivity (k) and other parameters can be estimated based on the measured drawdown curve and pumping discharge in the soil mechanics textbook.

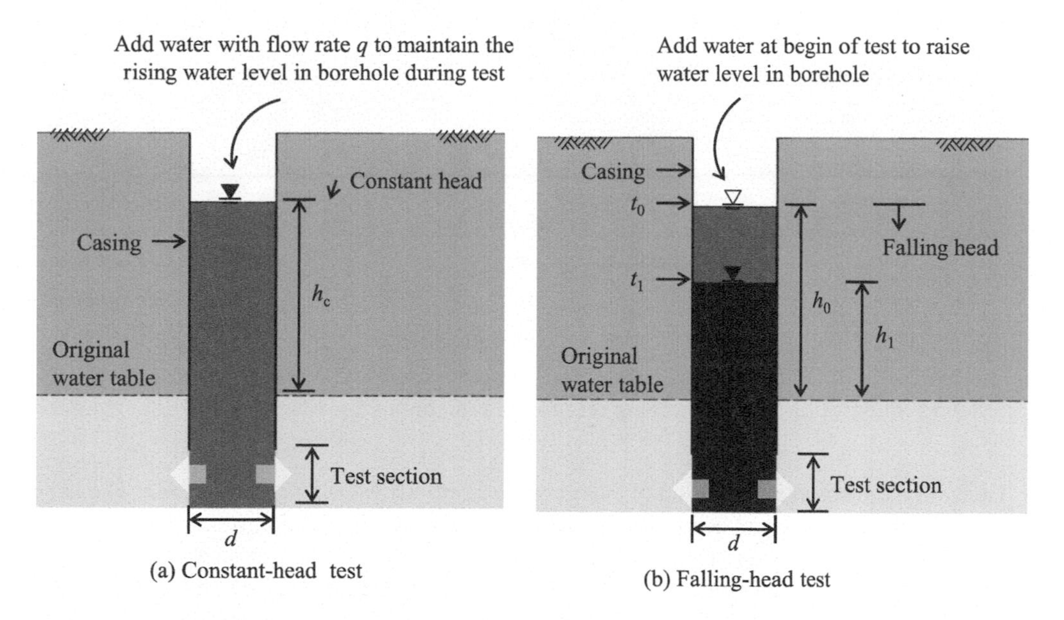

Figure 1.29 Illustration of borehole permeability tests: (a) constant head test, (b) falling head test.

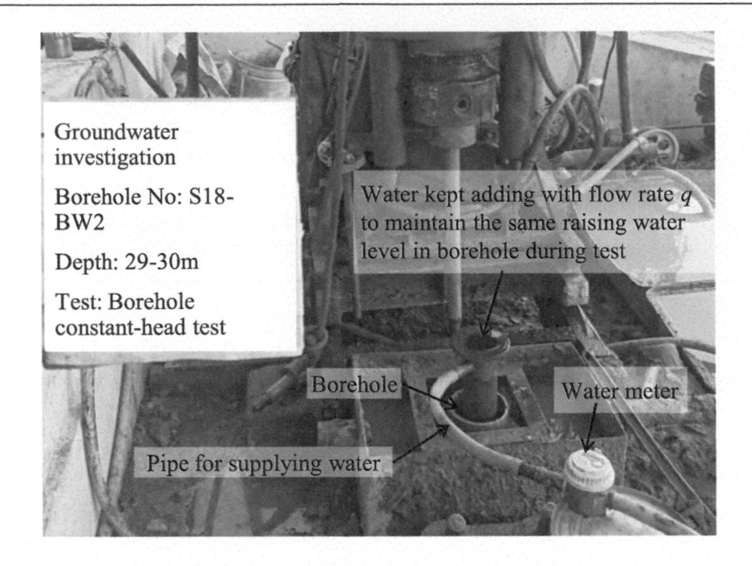

Figure 1.30 Borehole constant head test in operation.

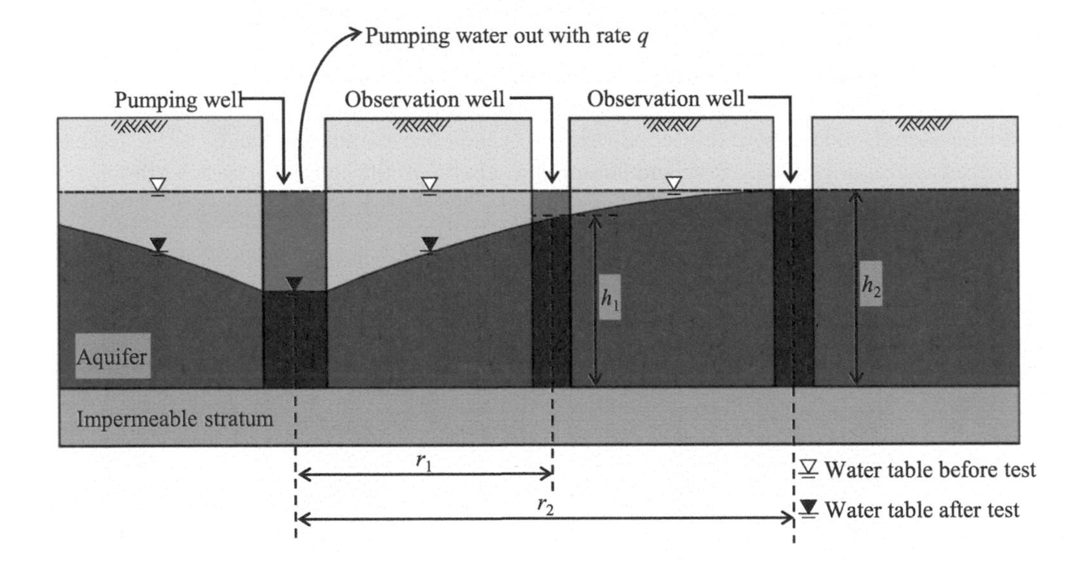

Figure 1.31 Illustration of the pumping test for the unconfined problem.

Generally, the observation period of a large-scale pumping test is longer than 24 hr. The large-scale pumping test result can be used for a workability evaluation of underground drainage work, such as horizontal drains, vertical pumping wells, or catchment wells. However, readers should keep in mind that overpumping at the site may cause adjacent buildings to tilt and settle, which is avoidable if the knowledge in the section is understood. Herein, a pumping test only in an unconfined layer is introduced, and the test in the confined layer may be interesting to some readers in relation to depressurization of confined aquifers, as referred to in the soil mechanics textbook.

1.9 Geophysical survey

Geophysical surveys are implementations of geophysical methods, which include electrical and electromagnetic surveys, seismic surveys, gravity surveys, magnetic methods, and radioactivity methods, to determine the geological, structural, physical, and mechanical characteristics of the foundation soil with airborne, marine, and land geophysical surveys and borehole surveys. These methods are used to explore the layout, thickness, and properties of individual layers below the terrain surface on which the construction of a specific structure is planned. Nondestructive testing methods have the merits of being quick, having lower cost, and offering continuous data logging and more comprehensive exploration of the site (Tokimatsu and Uchida, 1990; Griffiths and King, 1981; Gunn et al., 2015). This section will focus on electrical and electromagnetic survey and seismic survey methods.

1.9.1 Electrical and electromagnetic survey

Electrical resistivity, which measures the resistivity profile of layers, has been used widely in prospecting for groundwater, bedrock features, faults, buried objects based on conductivity differences of various cements, minerals, porosities, water contents, etc. of geological materials.

2D electrical resistivity tomography (ERT) can be used for mapping the distributions of the presence of groundwater and geological structures according to the resistivity of materials. Electrode array configurations are aimed at gathering data used to estimate lateral and vertical variations in ground resistivity values. An example of the results of implementing 2D electrical resistivity tomography (2D ERT) is shown in Figure 1.32. The ground resistivity detection line

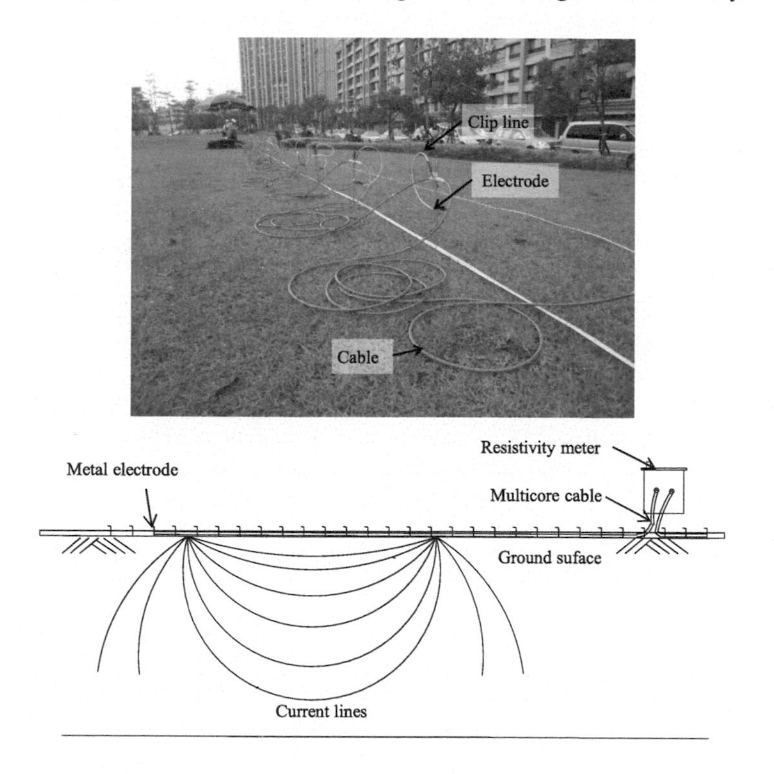

Figure 1.32 Electrical and electromagnetic survey operation.

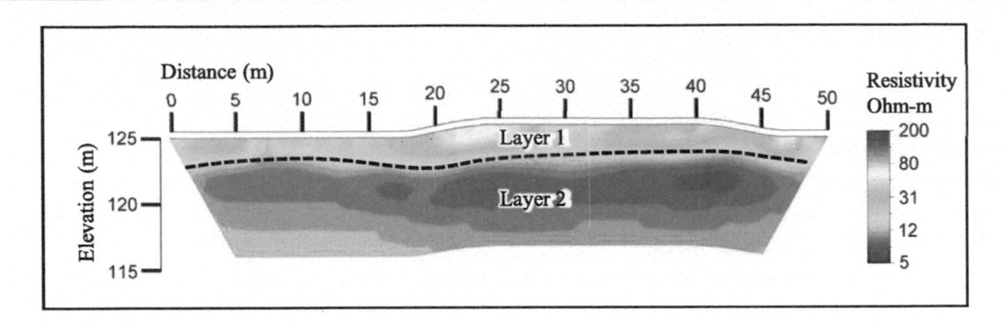

Figure 1.33 An example of a 2D ERT survey result.

is approximately 50 m long, and the detection depth range is approximately 20 m. The ground resistivity image profile detection data is presented to obtain the distribution of apparent resistivity (unit: Ohm-m) (see Figure 1.33). After reading the ground resistivity profile image, the survey area can be approximately divided into two layers. Layer 1: 0~3 m below the ground surface, with a higher resistivity distribution of approximately 10~80 ohm-m. The underground material has a larger particle or a higher void ratio, resulting in a higher resistivity. Layer 2: 3–20 m below the ground surface, generally showing a relatively low electrical resistivity of approximately 5~12 ohm-m. The particle size of the underground material is finer or the water content is higher, resulting in lower resistivity. Through ground resistivity detection, a wide range of survey data can be used to access the soil layer distribution. Importantly, because electrical properties are indirectly determined in this nondestructive testing method, the composition of the stratum, the size of the particles, and the water content and salinity of the stratum, which are factors affecting the resistivity responses, still require the core samples through boring to connect the resistivity and physical properties of those formation materials. The ERT provides more accurate investigation results when calibrated with other direct geological investigation methods, such as the SPT and CPT. Finally, readers should keep in mind that rainfall records must be involved in ERT projects because the resistivity of shallow layers is easily affected by rainfall and droughts.

1.9.2 Seismic survey

There are two major types of seismic surveys, refraction and reflection, depending on the specific type of waves being utilized. Active seismic waves are generated by using an impact source, such as a sledgehammer. The generated waves transmit differently according to densities, fractures, porosities, etc. in different geological media. The refraction and reflection recorded by an array of sensors or geophones are a topic of interest. After the measured wave velocity used for representing different geological materials is back-analyzed, the geological conditions of the area of interest can be obtained according to Table 1.9.

Figure 1.34 shows an example of the seismic refraction method implementation. In this case, the refraction seismic survey line is approximately 150 m long, and the detection depth range is approximately 50 m. The seismic wave is recorded at each measuring point on the survey line, the seismic wave velocity is calculated, and then the velocity layered profile distribution is explored. The refraction seismic survey shows that the ground can be approximately divided into five layers with the corresponding velocity wave. Layer 1: approximately 0~3 m below the

Table 1.9 Approximate range of P wave velocities Vp for some common geological materials

Medium	Velocity (km/s)	
	Min	Max
Air	0.31	0.36
Weathered soil horizon	0.10	0.50
Gravel, dry sand	0.10	0.60
Loam	0.30	0.90
Wet sand	0.20	1.80
Clay	1.20	2.50
Water	1.43	1.59
Friable sandstone	1.50	2.50
Dense sandstone	1.80	4.00
Chalk	1.80	3.50
Limestone	2.50	6.00
Marl	2.00	3.50
Gypsum	4.50	6.50
Ice	3.10	4.20
Granite	4.00	5.70
Metamorphosed rock	4.50	6.80

Source: Bettina Albers (2011).

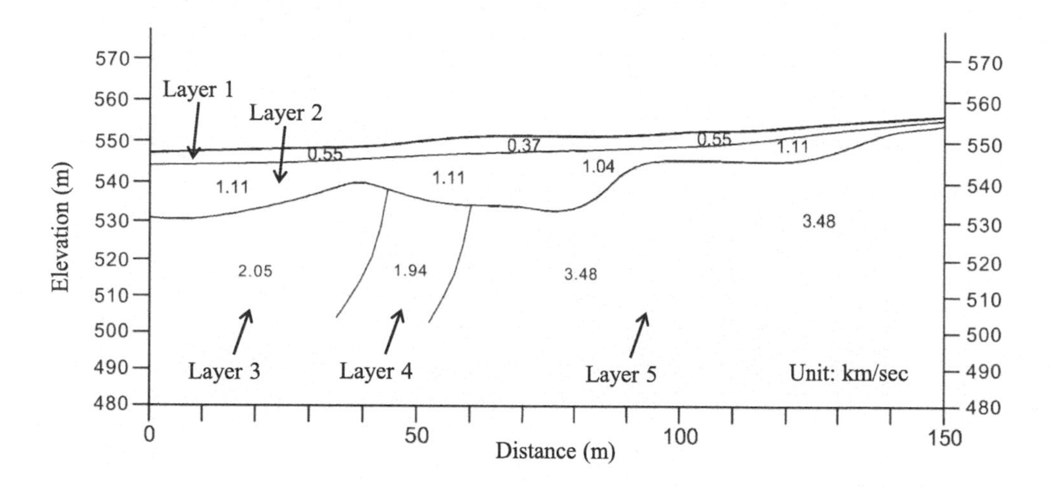

Figure 1.34 A seismic refraction survey result.

ground surface, and the seismic wave velocity is 0.4~0.6 km/sec. This surface soil layer is soft or porous due to the high energy loss during seismic wave transmission, so the wave velocity transmission is slow. Layer 2: approximately 1~11 m below the surface, and the seismic wave velocity is approximately 1.1 km/sec. This layer has a large-grained rock composition, or the formation composition is relatively compact; thus, the energy loss in the propagation is relatively low, and the wave velocity is fast. Layers 3 and 4: approximately 20–40 m below the ground surface, the seismic wave velocity is approximately 2.0 km/sec. It is a collapsed layer or a rock layer composed of many cracks or filled with muddy materials. The energy loss is relatively high during seismic wave propagation, so the wave velocity is slower than layer 5. However, this property requires further confirmation by geological survey core sampling. Layer 5: approximately 20~40 m below the ground surface, and the seismic wave velocity is approximately 3.50 km/sec. This layer is a rock formation. The rock mass is relatively intact, with few cracks and mud inclusions. The energy loss of wave transmission is low, and the wave velocity is therefore transmitted quickly. On the other hand, a reflection method, as shown in Figure 1.35, is also usually used for determining soil profile for shallow depth than seismic refraction method, of which V_1 and V_2, the shear velocities of the first and the second layer, are analyzed based on signals recorded by surface geophone for identifying the soil profile.

Essentially, soil boring methods can provide direct investigation results but are limited to explore a small scale of the geological materials in boreholes drilled in the area of interest. When they are used alongside geophysical surveys, a two- or three-dimensional geological

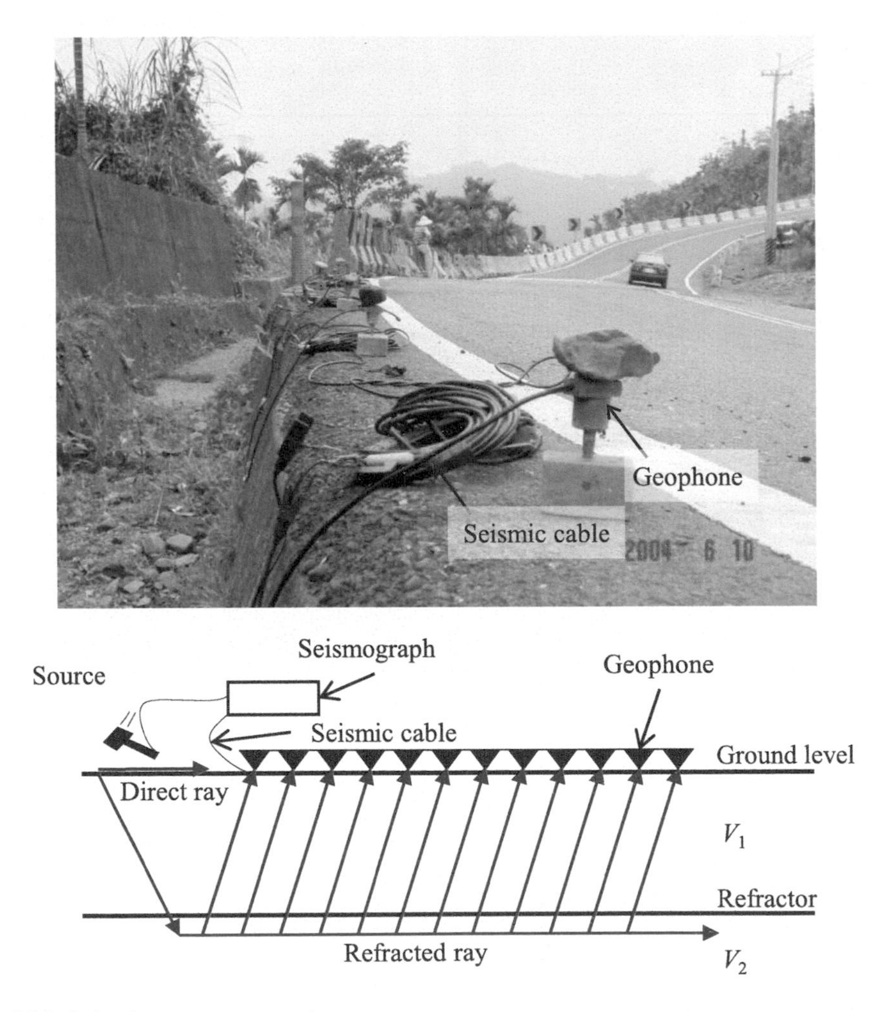

Figure 1.35 Seismic survey operation.

structure can be developed and provide overall information on subsurface geological conditions, especially useful for complex ground profile exploration.

1.10 Reports and interpretation

A site investigation report generally consists of two parts: a factual description of the investigation results and an engineering analysis with relevant recommendations.

A reliable report should precisely describe the features of the project site, the work performed, and the results obtained.

- The scope and purpose of the project and its importance.
- The basic conditions of the site, such as topographical and hydraulic features.

A concise description and summary of the amounts, procedures, and results of various field and laboratory tests should be prepared. Tables containing bore logs and other field and laboratory test results and a simplified geological profile should be presented.

Analyses of allowable bearing capacity, settlement, liquefaction potential, and/or recommended foundation type should also be provided in the report. Recommendations for further investigation should be reported specifically if there are still unclarified subjects due to the complexity of the geological conditions at the project site. Table 1.10 shows a list of items in geotechnical investigation report.

A building foundation construction, which is located in Taipei City, north Taiwan, is introduced here as an example to facilitate discussion on which foundation and excavation plan in that project are introduced. To integrate the aforementioned knowledge, the investigation works used to explore the geological conditions are introduced in what follows.

1.10.1 Abstract of investigation plan

The site plan of a building construction introduced in this chapter is shown in Figure 1.36. The excavation area is approximately 3,200 m^2, the length is approximately 60 to 105 m, the width is approximately 40 m, and the shape is similar to a trapezoid. This building is expected to have 18 stories above the ground surface and 5 stories underground. The design depth of excavation

Table 1.10 List of items in a geotechnical investigation report

General description	*Analysis*	*Other*
1. The purpose and importance of the project. 2. Topographical features and hydrological conditions of the site. 3. A brief description of the various field and laboratory tests carried out. 4. Boring log/CPT. 5. Geophysical surveying. 6. Table listing the testing results.	1. Analyze and discuss the test results. 2. Determine soil profiles and geotechnical and hydraulic parameters. 3. Determine allowable bearing pressures, pile loads, etc. 4. Evaluate soil liquefaction potential. 5. Analyze excavation- induced influences.	1. Recommendations, such as dewatering, ground improvement, construction procedures, and further investigation.

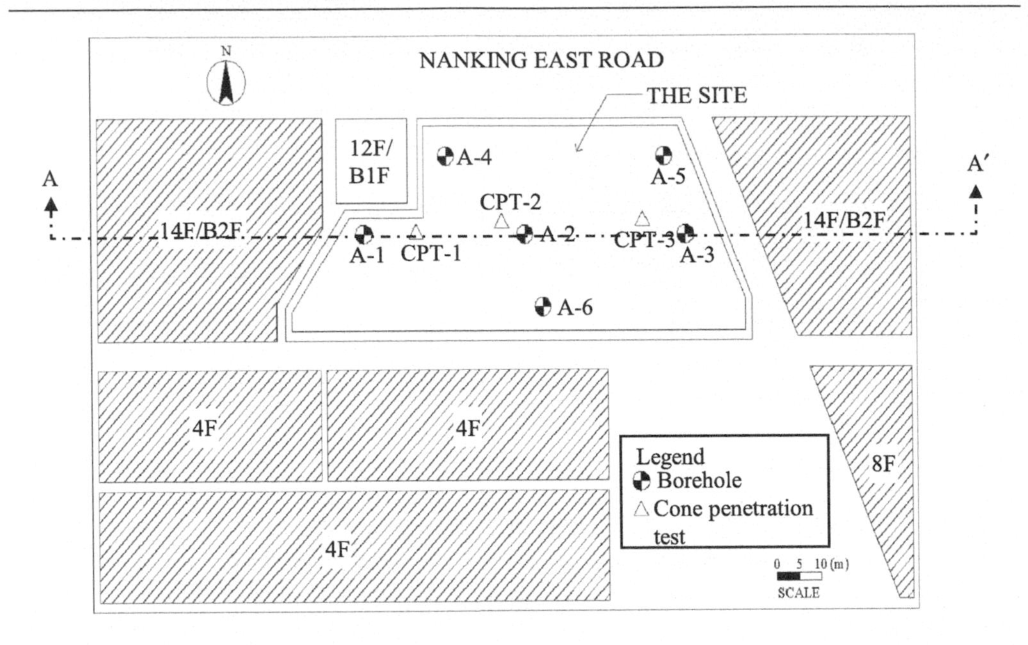

Figure 1.36 Site location plan and layout of borings.

Source: Redrawn from Ou et al. (1998).

is approximately 19.7 m below the ground surface. To plan and design an appropriate foundation, a series of investigations is performed to understand the geological conditions and obtain the soil properties (Liao, 1996).

1.10.2 Basic information collection

1.10.2.1 Location

Figure 1.36 shows that there are existing roads and 4-story and 12-story buildings around the plan site. Therefore, the safe protection and reduction of construction-induced vibration and noise for adjacent buildings and residents are important during the design and construction phases.

1.10.2.2 Regional geology

This plan site is located in the Taipei Basin, formed by a thick alluvium formation, known as the Sungshan Formation, which lies above the Chingmei Gravel Formation. The Sungshan Formation has six alternating sand and clay layers. Sungshan layers II, IV, and VI are silty clay (CL), and layers I, III, and V are silty sand (SM). The 25 m thick Sungshan layer IV, classified as soft to medium clay, is the main layer that affects excavation behavior in the study.

1.10.3 Borings and tests

The basic information collected indicates that the plan site will be situated on the clay deposits. Clay usually has lower shear strength and poor bearing capacity, and the settlement of consolidation must also be taken into consideration. The subsurface exploration items in this project site include geological borehole drillings and laboratory and field tests.

1.10.3.1 Layout and depth of borings

- Amount of geological borehole drilling: the base area of this project site is approximately 3,200 m^2, and the principle of borehole configuration is at least 1 borehole for each 600 m^2. Therefore, 3,200 m^2/600 m^2 = 5.3 boreholes, and 6 boreholes are required and averagely distributed in the plan site, as shown in Figure 1.36.
- Depth of geological borehole drilling: the excavation depth is 19.7 m, so the design of the borehole drilling depth must be deeper than the excavation depth to a hard stratum.
- Sampler: A thin-walled tube sampler is adopted to obtain undisturbed soil for laboratory testing in this project.

1.10.3.2 Field test

Field tests in this plan site investigation consist of a standard penetration test (SPT) and cone penetration test (CPT). Moreover, the vane shear test (VST) is used to directly determine the undrained shear strength of clay to ensure the quality of the test data. The shear strength of clay by VST will be very useful to obtain data for engineering design.

1.10.3.3 Laboratory tests

Laboratory soil Atterberg limit tests, triaxial permeability tests, oedometer tests, and shear strength tests are performed to determine the physical and mechanical soil properties in this project. Various shear strength tests, including the direct shear test, triaxial unconsolidated undrained test (UU), and triaxial consolidated undrained test (CU), are conducted to obtain necessary parameters that will be adopted in the design of the foundation.

1.10.4 Investigation results

After geological borehole drilling, laboratory and field testing are conducted in this site, a part of a boring log is obtained, as shown in Figure 1.37 as an example. According to the logs of Boreholes A-1, A-2, and A-3, designated as Section A-A in Figure 1.36, a soil profile can be determined, as Figure 1.38. Note that if inhomogeneous or non-uniform soil layers are justified, the soil profiles at other sections should be determined, for example, the section along boreholes A-4, A-2, and A-6.

The excavation depth for this excavation project is 19.7 m; the layer consists of silty clay (CL classified in USCS) influencing the behavior of this excavation with a SPT-N value of the layer 2 to 8 blows, which is assessed as soft to medium stiff soil. A distribution of the SPT-N values along the depth is shown in Figure 1.39.

Because this plan site consists mainly of clay, the main points of the foundation design will be considered bearing capacity, consolidation settlement, and heave; therefore, a detailed analysis needs to be carried out in the report. For obtaining a more detailed examination on the soil strength of the clay soil underlying the foundation, undisturbed soil is obtained for UU test in the laboratory and VS test is applied in the field. According to the shear strength test results, s_u is approximately 63 kPa at a depth of 19.7 m, as shown in Figure 1.40. Finally, a simplified subsoil profile judged by an experienced geotechnical engineer with the recommended parameters in Table 1.11 is to use for engineering design in the report.

Borehole Log															
Job title: Taipei survey						Location: Kuan Fu N. RD.						Borehole: A-3 Sheet 1 of 3			
Depth: 52.5 m				Elevation: 6.07 m		G.W.L: 5.58 m						Date: 1998/10/30~11/01			
Boring data					Testing										
Depth (m)	Sample no.	SPT-*N* blows	Log	Soil description	USCS	Particle size analysis(%)				Water content	Unit weight	Specific gravity	Void ratio	Liquid limit	Plasticity index
						G	S	M	C	ω (%)	(kN/m³)	G_s	*e*	*LL*	*PI*
				0.7											
1				Sand and gravel											
	S-1	4			CL	0	11	50	39	34.5	17.85	2.67	0.201	50.5	23.7
2															
3	S-2	2		Gray-black silty clay	CL					30	18.44				
4	T-1					0	1	60	39	40.8	17.66	2.68	0.214	29	8.5
	S-3	1			CL					39.5	17.17				
5															
				6.10	SM	0	3	68	29	42.2	17.56	2.68	0.217	30	8.8
6	S-4	2													
7															
	S-5	3		Sand mixed sandy silt	SM					25.6	18.93				
8															
				9.0											
9															
10	S-6	2			CL					30.9	18.54				
11	S-7	3			CL					27.8	19.62				
12	T-2				CL-ML	0	19	71	10	29.5	18.74	2.71	0.187	24.6	5.3
	S-8	2			CL	0	5	70	25	29.3	18.93	2.74	0.187	23.6	7.4
13															
14	S-9	3		Gray-black silty clay	CL					36.8	18.25				
15															
	S-10	3			CL	0	1	48	51	41.1	17.85	2.74	0.217	40.9	18.5
16															
17	S-11	3			CL					38.5	18.05				
18															
	T-3					0	0	52	48	40.8	17.66	2.7	0.215	38.3	15.6
19	S-12	4			CL					34.8	17.95				
20															

Figure 1.37 A part of one of the boring logs.

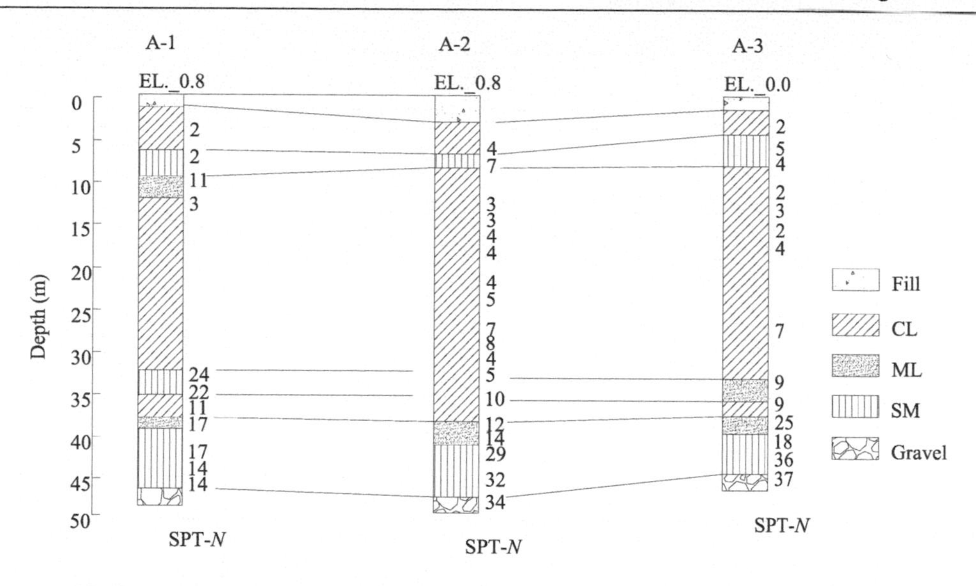

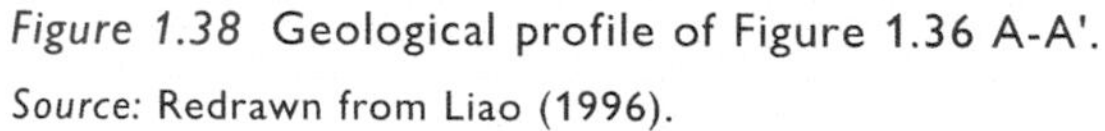

Figure 1.38 Geological profile of Figure 1.36 A-A'.

Source: Redrawn from Liao (1996).

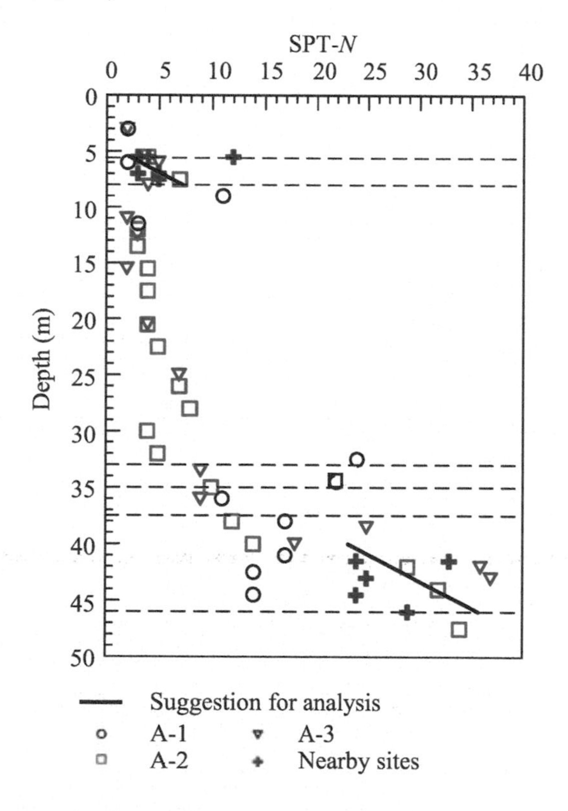

Figure 1.39 Distribution of the SPT-*N* values along the depth.

Source: Redrawn from Ou et al. (1998).

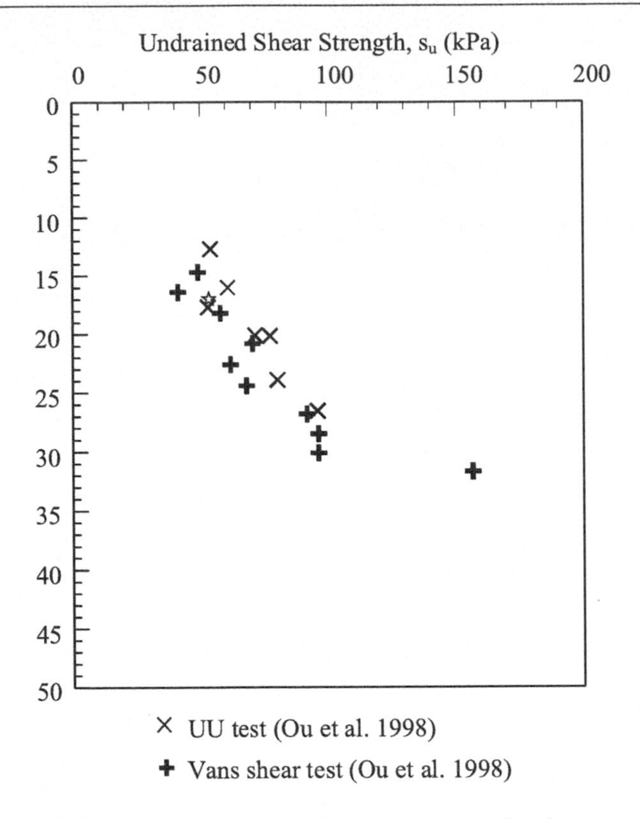

Figure 1.40 Undrained shear strength by variable tests versus depth.

Source: Redrawn from Ou et al. (1998).

Table 1.11 Simplified classification of strata and test results

Layer	*Depth (m)*	*USCS*	*SPT-N value*	*(kN/m³)*	*c' (kPa)*	*φ' (deg.)*	*ω (%)*	*LL*	*PI*	*s_u (kPa)*
1	0~5.6	CL	2~4	18.25	–	30	33~38	33~36	13~16	–
2	5.6~8	SM	4~11	18.93	0	31	21~25	–	–	–
3	8~33	CL	2~8	18.15	0	29	32~40	29~39	9~23	40~100
4	33~35	SM	22~24	19.62	0	31	25~26	–	–	–
5	35~37.5	CL	9~11	19.13	–	29	28~32	37~41	15~19	150
6	37.5~46	ML or SM	14~37	19.62	0	32	23~24	–	–	–
7	>46	GP	>100	–	–	–	1	–	–	–

Note: γ_t = unit weight; ω = water content; *LL* = liquid limit; *PI* = plasticity Index.

1.10.5 Foundation Analysis

For geotechnical reports, geotechnical engineers usually evaluate the foundation analysis according to the geotechnical investigations introduced earlier, which may include bearing capacity, settlement, and soil liquefaction potential assessments (Construction and planning agency ministry of the interior, R.O.C., 2000). The calculation method of bearing capacity and settlement can be found in Chapter 3.

Procedures for determining ultimate soil bearing capacity and structure loading are exhibited as follows:

- Ultimate bearing capacity q_u: The soil bearing capacity is calculated based on the general bearing capacity equation (Section 3.2). Because in this example, the bearing layer is located in clay with low permeability, the undrained behavior is considered here, that is, $c = s_u$ and $\phi = 0$. According to the shear strength test result shown in Figure 1.40, the s_u at a depth of 19.7 m is approximately 63.3 kPa. The foundation width B and length L are 40 m and 100 m, respectively. The depth of foundation D_f is 19.7 m. Based on these parameters, the ultimate bearing capacity q_u is calculated to be 779.8 kPa.
- Structure loading q_s (a combined load of foundation and building): It is preliminarily estimated that the dead and live load for each story is 14 to 16 kPa. The building contains 23 stories and a foundation; therefore, the structure loading is approximately 360 kPa. An accurate calculation for the structure result will be conducted in the detail design stage, and an approximate estimation in the site investigation report is usually carried out as follows:

Table 1.12 An approximate estimation of load

Story	*Total stories*	*Dead load (kPa)*	*Live load (kPa)*	*Total (kPa)*
2~18F	17	12	2	17 × (12 + 2) = 238
1F	1	12	5	1 × (12 + 5) = 17
B1~B5	5	11	5	5 × (11 + 5) = 80
Mat foundation	1	25	0	1 × 25 = 25
Sum (kPa)				360

1.10.5.1 Settlement

Because the foundation will be constructed on the clay layer, the consolidation settlement is necessarily taken into consideration:

- Preconsolidation stress σ'_c: The laboratory testing results show that $p'_c = 215$ kPa, and the in situ effective overburden pressure = 193 kPa.
- Overconsolidation ratio OCR: $OCR = 1.11 > 1$, which indicates overconsolidated clay.
- Compression index C_c: According to Section 2.3.3, an empirical correlation $C_c \approx PI/74$ for undisturbed clay, we therefore can obtain $C_c = 0.22$ as PI = 16.
- Swelling index C_{ur}: An empirical correlation for $C_{ur} \approx PI/370$ is helpful here so that $C_s =$ 0.043 can be determined.
- Consolidation settlement calculation: Hence, the clay layer is in an overconsolidation state and the stress increment of the clay layer is very small because the weight of the designed structure is close to the excavated soil; hence, the consolidation settlement can be ignored in this project.

1.10.5.2 Soil liquefaction potential assessment

Soil liquefaction occurs mainly in saturated sand or fine-grained soils that can cause the failure of buildings. The foundation of this building is to be situated on the clay layer, so the soil liquefaction risk of the foundation is evaluated to be very low. There are many soil liquefaction potential assessment methods for readers' further references: SPT-*N* result based (Seed and Idriss, 1971; Tokimatsu and Yoshimi, 1983; Seed et al., 1985, JRA, 1996), CPT-qc result based (Shibata and Teparaksa, 1988), and seismic test result based (Tokimatsu and Uchida, 1990).

1.11 Summary and general comments

Foundation engineering faces different geological hazards, such as a large settlement, bearing capacity failure, earthquake-induced liquefaction, soil piping, ground heave, and landslides during the construction or operation phase, which are related to the geological conditions and material properties. Therefore, the purposes of geotechnical site investigation are to survey subsurface conditions and to determine potential geological hazards by various investigation methods and technologies. Furthermore, geotechnical site investigation is also fundamental and essential to obtain correct parameters, including physical, hydraulic, and mechanical properties, to facilitate an appropriate design to avoid disasters and enhance the safety and economic performance of construction.

Each investigation method or technique has a distinct purpose, advantage, disadvantage, and applicability; moreover, no single method or technique can simultaneously obtain the various information of the ground for design and analysis, so in good engineering practice, a series of geotechnical site investigation methods is often applied during the investigation process. In addition, many investigation methods have been developed and continue to progress worldwide. This chapter lists some common and fundamental methods, such as borehole drilling and sampling, laboratory tests, standard penetration tests, groundwater investigations, electrical resistivity tomography, and seismic surveys. This section briefly integrates previous methods into a case study to provide readers with basic knowledge of geotechnical site investigation and its applications.

On the other hand, each investigation method or technique entails specific background knowledge (e.g., geology, geophysics). When those methods are being applied, practitioners sometimes face difficult cases with special and complex geological conditions, and it is not easy for a single individual to know all kinds of knowledge and different investigation techniques, so geotechnical site investigation work requires teamwork and group discussion to integrate the partial data obtained from various investigation methods and opinions from the team to build a reasonable and accurate geological model of the site for supporting design at the next stage.

Accurate and reliable investigation results are also based on reasonable investigation project planning and appropriate implementation, but determining the investigation plan and contents is often constrained by a limited timeframe and budget and the environmental conditions. The adoption of the most efficient and feasible method is always a challenge for development units and geotechnical engineers. Therefore, geotechnical engineers must first clearly understand the engineering project and the required parameters for design and analysis and based on thorough consideration and propose an appropriate investigation project. Moreover, investigation work is not a one-off task. A supplementary investigation can be conducted if there are still some doubts or unresolved questions. After the reports are carefully and comprehensively compiled, they can provide useful and important information to facilitate a viable engineering design.

Problems

1.1 Please indicate the purpose of each test.

Soil physical property tests

1. Triaxial unconsolidated undrained test (UU)
2. Triaxial consolidated undrained test (CU)
3. Oedometer test
4. Direct shear test
5. Permeability test
6. Dynamic triaxial test

1.2 Please describe the methods of downhole P-S logging and crosshole velocity logging, with an introduction to their application purpose and procedures for determining the soil parameters.

1.3 Please derive eq. 1.2 for determining the undrained shear strength with a vane shear test.

1.4 What is the difference between the blow count and SPT-*N* value of the standard penetration test (SPT)?

1.5 If soil permeability at depths of 5 m and 20 m in the boring log of Table 1.11 are needed, what types of test method would be used in the field and the laboratory?

1.6 Table P1.6 provides field standard penetration numbers' record data. Determine the N value at each depth.

Table P1.6

Depth (m)	N_1	N_2	N_3	N -value
1.5	2	3	2	
3.0	2	2	4	
4.5	1	3	3	
6.0	2	3	4	
7.5	3	4	4	

1.7 A dilatometer test was conducted in a clay deposit. The saturated unit weight of the soil is 19 kN/m^2, and the groundwater level is 2 m below the ground surface. Now, we have the P_0 = 250 kN/m^2 and P_1 = 300 kN/m^2 at a depth of 5 m. Determine the horizontal stress index (K_D) and the coefficient of earth pressure ($K_{0,DMT}$).

1.8 Compare the area ratios of the split-spoon sampler (outer diameter d_0 = 50.8 mm and inner diameter d_i = 34.9 mm) used in the standard penetration test and the thin-walled tube (outer diameter d_0 = 76.2 mm with wall thickness t = 1.55 mm).

References

Anjan, P. (2019), *Geotechnical Investigations and Improvement of Ground Conditions*, Duxford, UK and Cambridge, MA: Woodhead Publishing.

ASTM D1587/D1587M-15 (2015), *Standard Practice for Thin-Walled Tube Sampling of Fine-Grained Soils for Geotechnical Purposes*, ASTM International, West Conshohocken, PA.

ASTM D3550-01 (2001), *Standard Practice for Thick Wall, Ring-Lined, Split Barrel, Drive Sampling of Soils*, ASTM International, West Conshohocken, PA.

Baguelin, F., Jézéquel, J.F. and Shields, D.H. (1978), *The Pressuremeter and Foundation Engineering*, Clausthall: Trans Tech Publications.

Casagrande, A. and Fadum, R.E. (1940), *Notes on Soil Testing for Engineering Purposes*, Cambridge, MA: Harvard University Press.

Clarke, B.G. (1994), *Pressuremeters in Geotechnical Design*, London: CRC Press.

Construction and planning agency ministry of the interior, R.O.C. (2000), *Design Specifications of Foundation and Structure for Buildings* (In Chinese), Taiwan: R.O.C.

David, S. (2018), *Cone Penetration Test Design Guide for State Geotechnical Engineers*, Minnesota Department of Transportation, Research Services & Library, 395 John Ireland Boulevard, MS 330, St. Paul, Minnesota 55155–1899.

Geotechnical Engineering Office (2017), *Guide to Site Investigation*, FHWA NHI-16–072, Washington, DC, USA.

Griffiths, D.H. and King, R.F. (1981), *Applied Geophysics for Geologists and Engineers: The Elements of Geophysical Prospecting*, 2nd edition, Oxford: Pergamon Press.

Gunn, D.A., Chambers, J.E., Uhlemann, S., Wilkinson, P.B., Meldrum, P.I., Dijkstra, T.A., Haslam, E., Kirkham, M., Wragg, J., Holyoake, S., Hughes, P.N., Hen-Jones, R. and Glendinning, S. (2015), Moisture monitoring in clay embankments using electrical resistivity tomography. *Construction and Building Materials*, Vol. 92, pp. 82–94.

Hvorslev, M.J. (1949), *Subsurface Exploration and Sampling of Soils for Civil Engineering Purposes*, U.S. Army Corps of Engineers, Waterways Experiment Station, Vicksburg, Miss.

Hvorslev, M.J. (1951), *Time Lag and Soil Permeability in Ground-Water Observations*, Bull. No. 36, Waterways Exper. Sta. Corps of Engrs, U.S. Army, Vicksburg, Mississippi, pp. 1–50.

JRA (Japanese Road Association) (1996), *Specification for Highway Bridges*, Part V, Japan: Seismic Design, Japan.

Liao, J.T. (1996), *Performances of a Top Down Deep Excavation*, doctoral thesis (In Chinese), Taiwan.

Marchetti, S. (1980). In situ tests by flat dilatometer. *Journal of the geotechnical engineering division, 106*(3), pp. 299–321.

Marchetti, S., Marchetti, D., Monaco, P. (2015). Flat dilatometer (DMT). Applications and recent developments. In *Proceedings of the Indian Geotechnical Conference, Pune, India*, pp. 16–19.

Marchetti, S., Monaco, P., Totani, G. and Calabrese, M. (2001), *The Flat Dilatometer Test (DMT) in Soil Investigations*, ISSMGE Committee TC16, Italy: University of L'Aquila.

Osterberg, J.O. (1952), New piston type soil sampler. *Engineering News-Record*, Vol. 148, No. 17, pp. 77–78.

Ou, C.Y., Liao, J.T. and Lin, H.D. (1998), Performance of diaphragm wall constructed using top-down method. *Journal of Geotechnical and Geoenvironmental Engineering*, Vol. 124, No. 9, pp. 798–808.

Rashed, A., Bazaz, J. B., and Alavi, A. H. (2012). Nonlinear modeling of soil deformation modulus through LGP-based interpretation of pressuremeter test results. *Engineering Applications of Artificial Intelligence, 25*(7), pp. 1437–1449.

Seed, H.B., Yokimatsu, K., Harder, L.F. and Chung, R.M. (1985), Influence of SPT procedures in soil liquefaction resistance evaluation. *Journal of Geotechnical Engineering, ASCE*, Vol. 111, No. 12, pp. 1425–1445.

Shibata, T. and Teparaksa, W. (1988), Evaluation of liquefaction potentials of soil using cone penetration tests. *Soils and Foundations*, Vol. 28, No. 2, pp. 49–60.

Terzaghi, K., Peck, R.B. and Mesri, G. (1996), *Soil Mechanics in Engineering Practice*, 3rd edition, New York: John Wiley & Sons, Inc.

Tokimatsu, K. and Uchida, A. (1990, June), Correlation between liquefaction resistance and share wave velocity, soils and foundations. *Japanese Society of Soil Mechanics and Foundation Engineering*, Vol. 30, No. 2, pp. 33–42.

Tokimatsu, K. and Yoshimi, Y. (1983), Empirical correlation of soil liquefaction based on SPT N-value and fines content. *Soils and Foundations*, Vol. 23, No. 4, pp. 56–74.

Chapter 2

Principles of foundation design

2.1 Introduction

Foundation design is based on the properties of subsurface soil and loading on a foundation. Soil properties are highly related to the soil under drained or undrained conditions. This chapter will first explain the undrained and drained behavior of soil. Once we understand the drained and undrained behavior of soil, we can select appropriate soil parameters with drained or undrained analysis for target soils based on the effective stress or total stress analysis, which is also covered in this chapter.

Soil parameters, such as the cohesion and friction angle, Young's modulus, Poisson's ratio, compression index, unloading/reloading index, and coefficient of earth pressure at rest, are necessary for foundation design. In principle, these soil parameters should be obtained from related soil tests. However, some parameters can also be estimated by correlation with other soil properties, such as unit weight, density, N value from the standard penetration test (SPT), cone tip resistance from the cone penetration test (CPT), etc. This chapter provides reliable methods and correlations for obtaining design soil parameters for cohesive and cohesionless soils.

Three design methods are used for foundation design: working stress design, ultimate strength design, and limit state design methods. The first two methods are conventional. Working stress design involves designing a foundation system to sustain the design load (design stress) without exceeding its allowable load (allowable stress). The uncertainties of external load and soil resistance are lumped and addressed by a single factor of safety. Since the concept of working stress design is straightforward and simple, it is introduced to design various foundations in this book. Moreover, since both the external load and soil properties are subject to different degrees of uncertainties or variabilities, it is more reasonable to assign different factors to load and soil parameters. This design method is called ultimate strength design and has been adopted in some foundation design codes. This book also introduces the concept of ultimate strength design.

In fact, a foundation should be designed to avoid not only failure, that is, the ultimate limit state, but also excessive settlement, that is, the serviceability limit state. The design method that addresses both limits is often called limit state design. Recently, some geotechnical structural/foundation codes have adopted this method for design. The concept of limit state design will be introduced in this book. Finally, the determination of soil parameters that considers uncertainties and variabilities in soil properties is elaborated in the last section.

DOI: 10.1201/9781003350019-2

2.2 Geotechnical analysis method

2.2.1 *Drained behavior and undrained behavior*

When a soil is stressed, the ease of pore water flow out will significantly affect soil deformation behavior. Figure 2.1 shows a schematic diagram of saturated coarse-grained soil under normal pressure, $\Delta\sigma$. The spring represents the coarse-grained soil, which will deform as it is stressed. Therefore, the stress in the spring represents the effective stress in the soil. The initial pore water pressure is in the hydrostatic condition, that is, $u_w = \gamma_w z$. The initial effective stress and total stress are denoted by σ'_{v0} and σ_{v0}, respectively. As a pressure $\Delta\sigma$ acts on a piston, causing the piston to move down, the water would flow out rapidly and the volume thus decreases. The spring is therefore stressed. Under such a condition, the excess pore water pressure is nil and the pore water pressure remains the same as the hydrostatic pressure (Figure 2.1a). The applied pressure $\Delta\sigma$ directly transmits to the soil, causing the effective stress in the soil to be increased to $\sigma'_{v0} + \Delta\sigma$ (Figure 2.1b). The total pressure increases to $\sigma_{v0} + \Delta\sigma$ (Figure 2.1c). Such deformation behavior of the soil is called drained behavior.

Figure 2.2 shows a schematic diagram of saturated fine-grained soil, that is, clay, under normal pressure, $\Delta\sigma$. Similarly, $\gamma_w z$, σ'_{v0}, and σ_{v0} represent the initial pore water pressure, effective stress, and initial total stress, respectively. Under these conditions, when a pressure $\Delta\sigma$ acts on a piston, the pore water flows out very slowly. Therefore, at the initial stage of loading, the pressure $\Delta\sigma$ is borne entirely by the pore water, and the spring is unstressed. The stress in the spring or the effective stress in the soil remains unchanged (Figure 2.2b). This condition is called undrained behavior or short-term behavior. Pore water pressure is thus the sum of hydrostatic water pressure and excess pore water pressure, while excess pore water pressure is $\Delta\sigma$ (Figure 2.2a). If there is a shear stress-induced pore water pressure, then the excess pore water pressure will be generated by normal pressure ($\Delta\sigma$) as well as the shear stress.

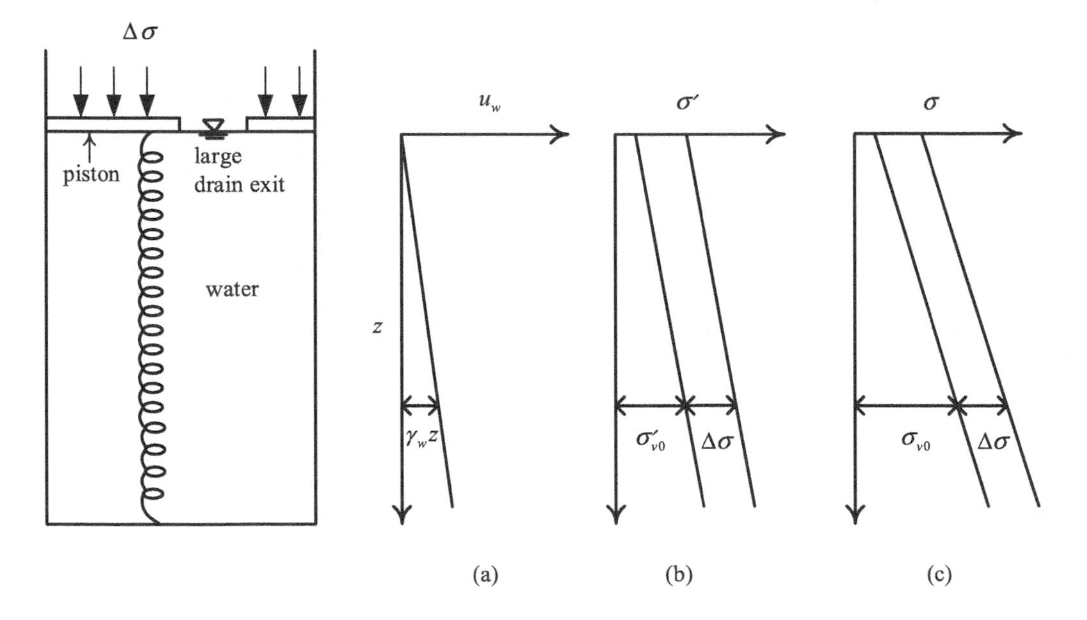

Figure 2.1 Schematic diagram of saturated coarse-grained soil under normal pressure: (a) pore water pressure, (b) effective stress, and (c) total stress.

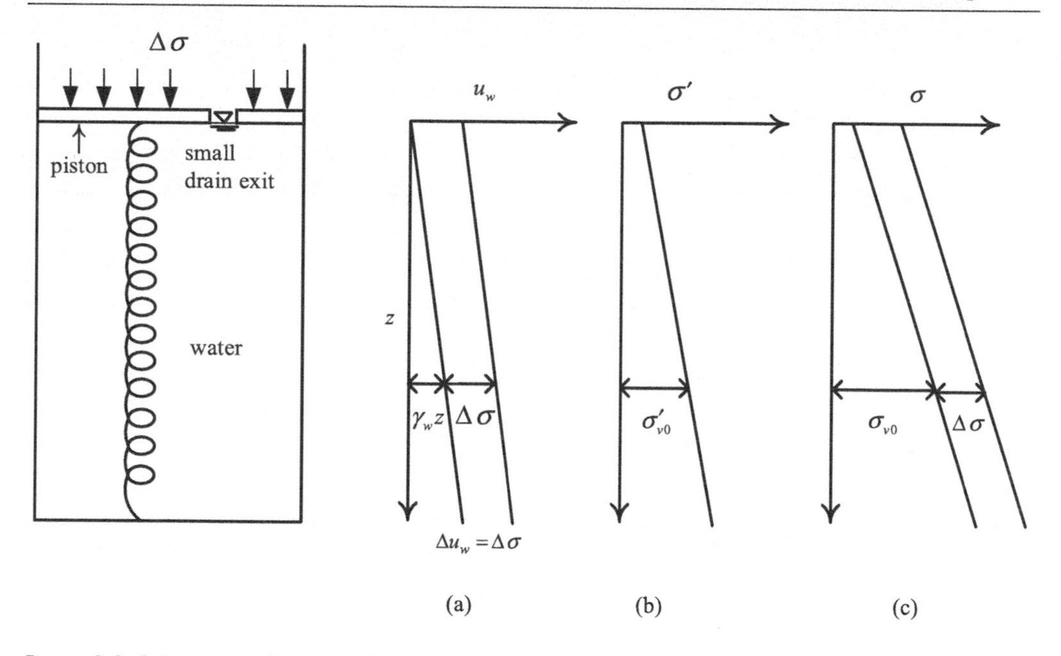

Figure 2.2 Schematic diagram of saturated fine-grained soil under normal pressure: (a) pore water pressure, (b) effective stress, and (c) total stress.

However, as shown in Figure 2.2, the pore water will eventually flow out. The piston eventually moves down, and the volume still changes. Excess pore water decreases or dissipates, and effective stress increases accordingly, but the total stress remains unchanged. When the excess pore water pressure is nil, the stress in the spring is equal to $\sigma'_{v0} + \Delta\sigma$. This condition is also called drained behavior or long-term behavior.

Materials with drained behavior are called drained material. Conversely, materials with undrained behavior are called undrained material. When conducting geotechnical analysis, one should first determine whether the target soil belongs to drained materials or undrained materials based on the soil classification test results of the target soil and the description of the site investigation and perform drained analysis or undrained analysis accordingly.

2.2.2 *Effective stress analysis and total stress analysis*

The effective stress analysis method and total stress analysis method are two types of analysis methods in geotechnical engineering.

The effective stress analysis method treats the soil as a two-phase material, that is, soil particles and pore water. When an external force acts on the soil, the stress and strain experienced by the soil depend on the contact stress between soil particles. The concept of effective stress represents the contact stress in soil mechanics. Due to the large void ratio of the soil, when the volume of the soil is changed by force, the pore water will flow out of (or in) the soil unit, so there is no excess pore water pressure (see Figure 2.1b). Basically, when soil particles are immersed in pore water, the shear strength or deformation of the soil is related only to the effective stress, not to the total stress. Therefore, the effective parameters, such as effective strength parameters (c', ϕ'), effective unit weight (γ'), effective Poisson's ratio (μ'), and effective Young's modulus (E'),

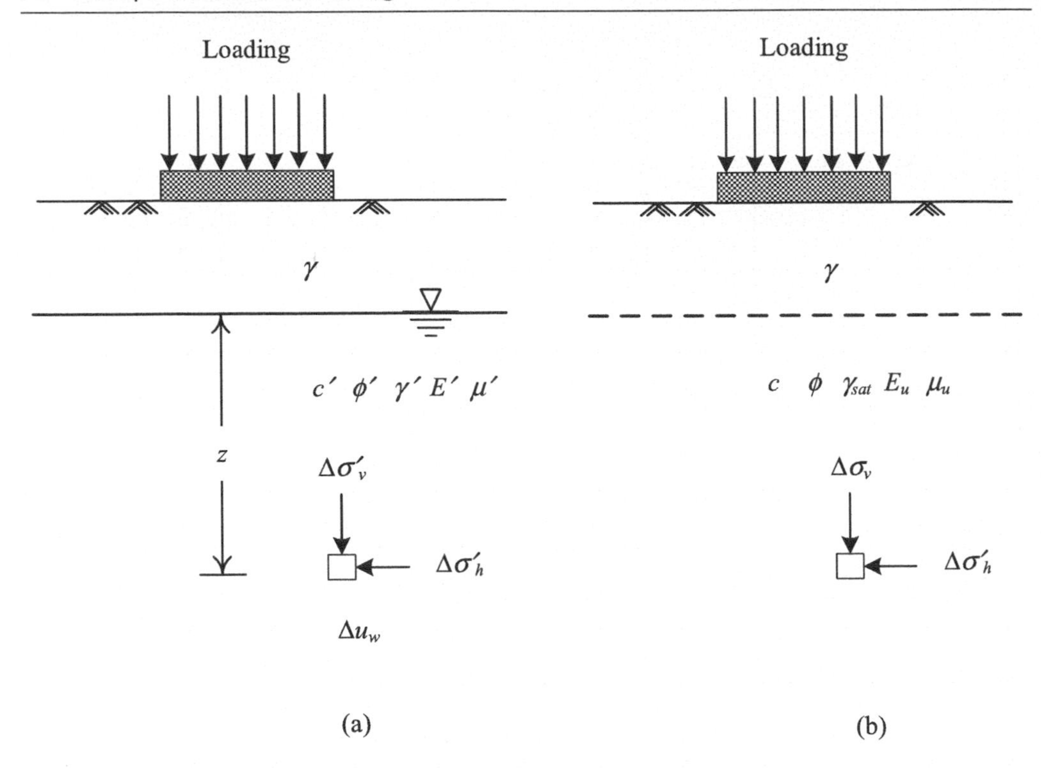

Figure 2.3 Types of analysis: (a) effective stress analysis and (b) total stress analysis.

should be adopted, and the groundwater level should be set in the analysis (Figure 2.3a). Effective stress analysis is applicable to coarse-grained soils with large void ratios, such as sand and gravel soil, or fine-grained soil under long-term conditions.

When fine-grained soil, such as clay, is stressed, due to very small clay voids, the pore water cannot be drained in a short period of time, and excessive pore water pressure is generated (see Figure 2.2a). Under the undrained condition, the amount of excess pore water pressure affects the effective stress of the soil, which, in turn, affects the stress–strain behavior of the soil. Therefore, total stress analysis is also called undrained analysis. This analysis method regards clay and pore water as a one-phase material or a soil–water mixture, with no pore water existing in the material (Figure 2.4). When the material is stressed, no excessive pore water pressure is generated because no pore water exists in this material. Therefore, even if the clay is below the groundwater level, the groundwater level should not be set in the analysis; otherwise, additional pore water pressure will be generated, that is, the groundwater level should be set outside the range of influence. The groundwater level is usually assumed at the bottom of the soil deposit or with no groundwater level. The total stress or undrained parameters, such as total or undrained strength parameters (c, ϕ), saturated unit weight (below groundwater level) (γ_{sat}), total or undrained Poisson's ratio (μ_u), and total or undrained Young's modulus (E_u), should be used in the analysis (Figure 2.3b). The total stress analysis method is applicable to the undrained behavior of fine-grained soils with a small void ratio, such as clay.

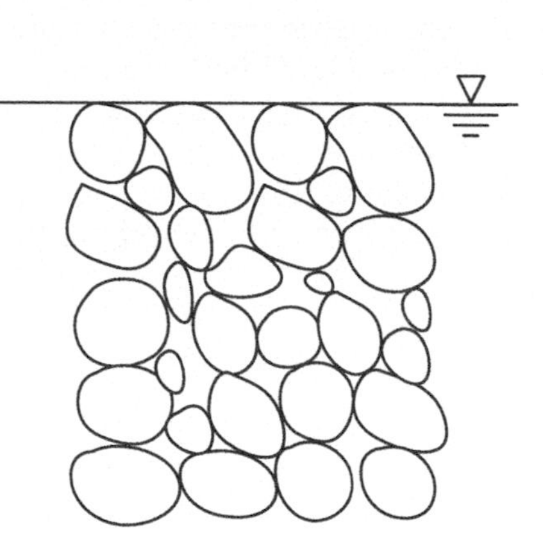

(a)

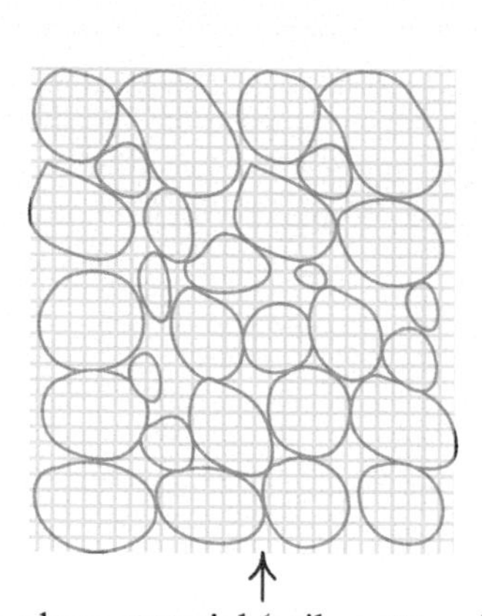

(b)

Figure 2.4 Concept of the total stress analysis for undrained material: (a) soil in water and (b) modeling as a soil-water mixture.

Table 2.1 Application of the effective stress analysis and total stress analysis

Method of analysis	*Parameters*	*Groundwater level*	*Pore water pressure*	*Application*
Effective stress analysis	c', ϕ', $\gamma'^{(1)}$, E', μ'	Setting the actual location	Existent	Granular soil, such as sand, silt, or gravel Fine soil, such as clay, drained under long-term conditions
Total stress analysis	c, ϕ, $\gamma_{sat}^{(2)}$, E_u, μ_u	Assumed nonexistent	Nonexistent	Fine soil, such as clay, under undrained conditions

Note:

(1): for the hydrostatic condition only, where $\gamma' = g_{sat} - \gamma_w$.
(2): for the soil below the groundwater level.

As shown in Figure 2.2, as the total pressure for saturated clay increases (or decreases), the increased (or decreased) pressure acts on the pore water, causing the excess pore water pressure to increase (or decrease), so the effective stress increment is zero. Based on the principle of effective stress, if the effective stress is not changed, the undrained shear strength will not change accordingly. This phenomenon is equivalent to the stress state of saturated clay in the triaxial unconsolidated undrained (UU) test. Therefore, the strength parameter can be represented by $\phi = 0$, c, where c is the cohesion. Notably, c is also called undrained shear strength, normally denoted s_u.

Table 2.1 summarizes the parameters used and the setting of the groundwater level in the effective stress and total stress analysis. The application of the analyses is also summarized in this table.

2.2.3 Stress paths

Figure 2.5a shows the Mohr's circle of a triaxial test specimen subject to horizontal pressure (σ_h) and vertical pressure (σ_v). The normal stress and shear stress at the top of Mohr's circle are equal to ($\sigma_v + \sigma_h$)/2 and ($\sigma_v - \sigma_h$)/2, respectively. According to the theory of Mohr's circle, these normal stress and shear stress are the stresses acting on a plane inclined at an angle of 45° with the horizontal (Figure 2.5b). Let p and q represent the normal and shear stresses at the top of Mohr's circle or on a 45° plane with the horizontal, which can be expressed by the following equations:

$$p = \frac{\sigma_v + \sigma_h}{2} \tag{2.1}$$

$$q = \frac{\sigma_v - \sigma_h}{2} \tag{2.2}$$

Where:

σ_v = vertical stress.
σ_h = horizontal stress.

Assuming that the initial pore water pressure is equal to 0 in the triaxial soil specimen after consolidation, the pore water pressure in the test specimen is equal to the excess pore water pressure (u_e) that is generated during undrained shear. The effective Mohr's circle can be obtained by shifting the total Mohr's circle to the left by an amount of u_e, as shown in Figure 2.6. The effective normal (p') and shear stresses (q) at the top of the effective Mohr's circle or a 45° plane with the horizontal can be derived as follows:

$$p' = \frac{\sigma_v' + \sigma_h'}{2} = \frac{(\sigma_v - u_e) + (\sigma_h - u_e)}{2} = \frac{\sigma_v + \sigma_h}{2} - u_e \tag{2.3}$$

$$q' = q = \frac{\sigma_v' - \sigma_h'}{2} = \frac{\sigma_v - \sigma_h}{2} \tag{2.4}$$

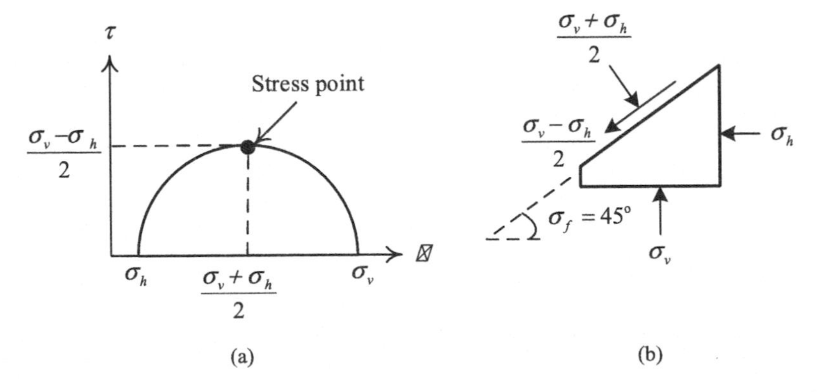

Figure 2.5 Mohr's circle for a soil specimen: (a) the stress state at the top of Mohr's circle and (b) the stresses on a 45° plane with the horizontal.

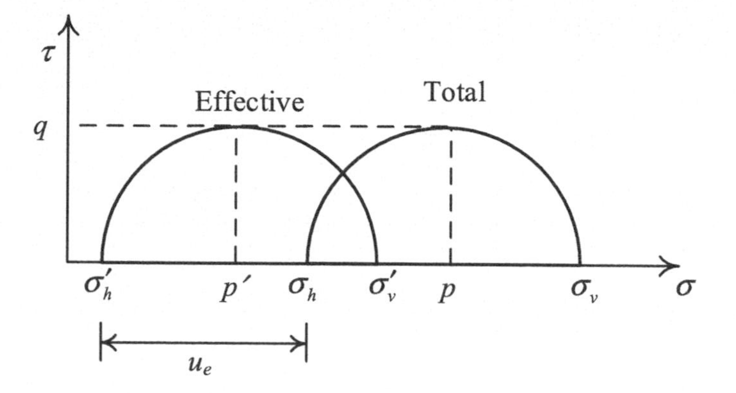

Figure 2.6 Total and effective Mohr's circles.

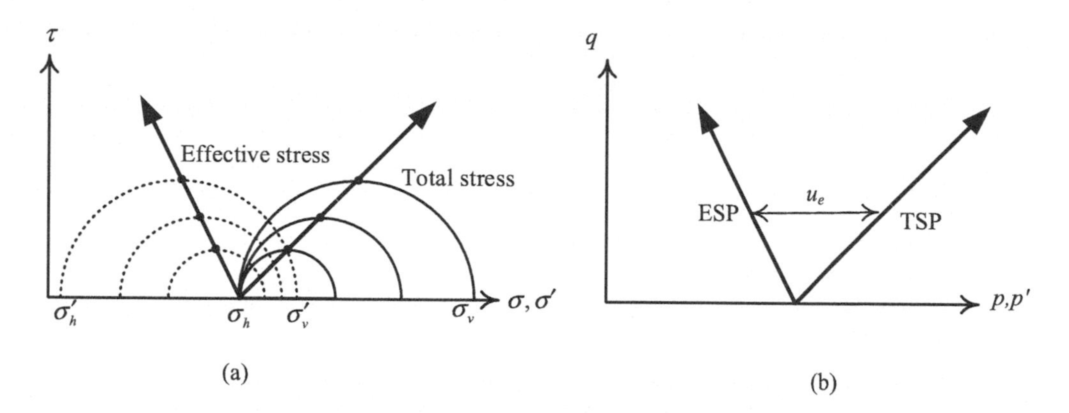

Figure 2.7 Stress path: (a) a series of Mohr's circles in the undrained test and (b) total and effective stress paths.

Connecting values of p and q on every Mohr's circle during shear on the p-q diagram results in the path of shearing, called the total stress path (TSP), as shown in Figure 2.7. Similarly, connecting the values of p' and q on each effective Mohr's circle, we obtain a path called the effective stress path (ESP) (Figure 2.7). In the triaxial consolidated drained (CD) test, the total Mohr's circles and the TSP are equal to the effective Mohr's circles because there is no excess pore water pressure generation in the triaxial CD test, $q = q'$, and $p = p'$. Therefore, the ESP and TSP would be the same curve. In the triaxial consolidated undrained (CU) test, if u_e is positive, the ESP shifts to the left of the TSP by an amount of u_e (Figure 2.7b).

As mentioned, the TSP represents the total normal and shear stresses on a plane inclined at an angle of 45° with the horizontal, while the ESP represents the effective normal and shear stresses on the plane. Therefore, the TSP and ESP can represent the changes in the normal stress, shear stress, and excess pore water pressure on the plane. By observing these changes, we are able to understand the shearing behavior of the soil specimen. We will use this concept to explain short-term strength and long-term strength in the next section. For more explanation of the stress path and application in geotechnical problems, readers can refer to the relevant references (Holtz et al., 2011).

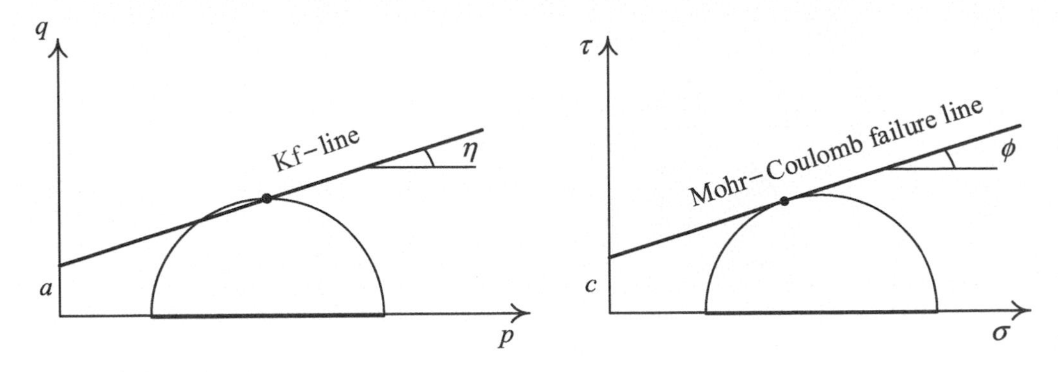

Figure 2.8 Relation between the K_f line and Mohr–Coulomb failure line.

The Mohr–Coulomb failure line, which is a line tangent to Mohr's circles, can be transformed into the *p-q* diagram, which is called the K_f line. The following geometric relationship exists between the K_f line and the Mohr–Coulomb failure line (Figure 2.8):

$$\sin\phi = \tan\eta \tag{2.5}$$

$$c = \frac{a}{\cos\phi} \tag{2.6}$$

Where a and η are the intercept on the q-axis and the angle of the K_f line with respect to the horizontal, respectively.

2.2.4 Short-term analysis and long-term analysis

Figure 2.9 shows the typical stress–strain behavior of dense and loose sand. The stress acting on loose sand increases up to the ultimate strength while its volume decreases, that is, the soil compresses. On the other hand, the stress acting on dense sand first increases up to the peak strength and then decreases with increasing strain to the ultimate strength. In turn, the soil first compresses and then dilates. With the same type of sand, the values of the ultimate strength for the dense and loose states are nevertheless close. The reason is that the strain at the ultimate strength is rather large, and the soil has been disturbed significantly. Under such conditions, soil strength relates largely to the characteristics of intergranular friction (i.e., roughness of the soil particles).

Normally consolidated (or lightly overconsolidated) clay in the drained test shows stress–strain behavior similar to that of loose sand. Because drained shearing tends to cause saturated normally consolidated (or lightly overconsolidated) clay to contract, positive pore water pressure is produced in the undrained test, where the volume of the clay remains unchanged. A representative stress–strain behavior along with the excess pore water pressure generated for normally consolidated (or lightly overconsolidated) clay during undrained shear is shown in Figure 2.10. The effective stress path (ESP) is therefore shifted to the left of the total stress path (TSP) by an amount of u_e, as shown in Figure 2.11.

Since the excess pore water pressure in the drained test is zero, curve ***ab*** in Figure 2.11 also represents the ESP and TSP. With the q value with regard to point ***b*** larger than that with regard

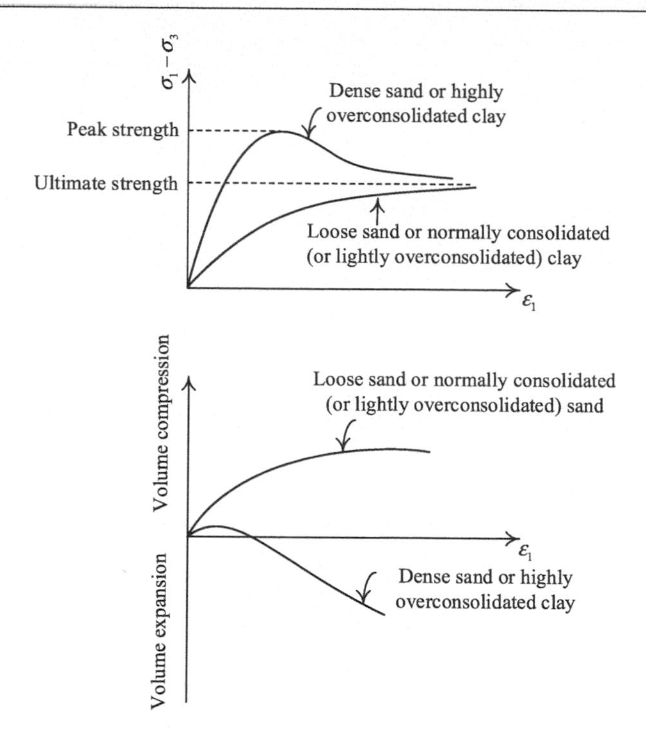

Figure 2.9 Typical stress-strain behavior and volume change in sand and clay under drained shear conditions.

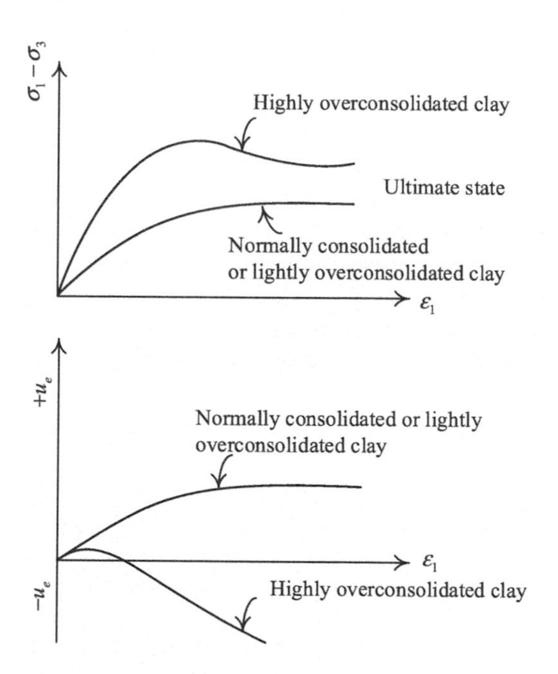

Figure 2.10 A typical stress-strain behavior and excess pore water pressure strain for clay under undrained shear conditions.

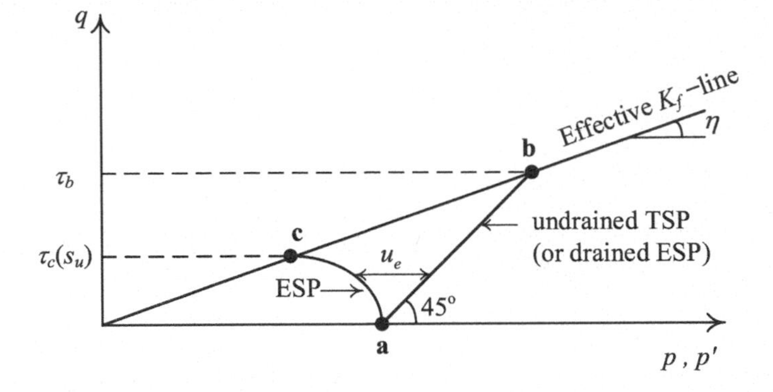

Figure 2.11 Possible effective and total stress paths for normally or lightly overconsolidated clay.

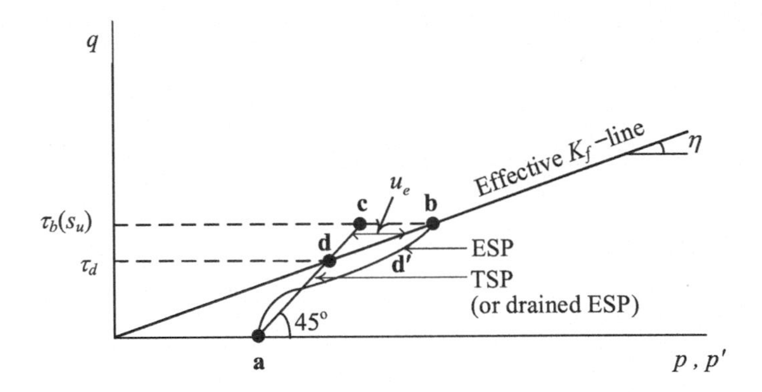

Figure 2.12 Possible effective and total stress paths for highly overconsolidated clay.

to point *c*, the undrained shear strength, τ_c or s_u, of normally consolidated (or lightly overconsolidated) clay is smaller than the drained shear strength, τ_b, of clay with the same consolidation pressure. That is, if positive pore water pressure is generated in an undrained test, the shear strength under undrained conditions will be smaller than that under drained conditions under the same consolidation pressure. The undrained shear strength should then be adopted for design. The undrained shear strength refers to the condition that the excess pore water pressure is not dissipated in a short time. Therefore, it is also called the short-term strength.

The behavior of highly overconsolidated clay is similar to that of dense sand. As shown in Figure 2.9, drained shearing will cause the saturated highly overconsolidated clay to compress first and then dilate, and therefore, undrained shearing will produce first positive and then negative pore water pressure (Figure 2.10). Figure 2.12 shows the TSP, as represented by curve ***adc***, and its corresponding ESP, as represented by curve ***ad'b***, for saturated clay sheared under the undrained condition. By definition, the undrained shear strength (s_u) is the shear strength (τ_b) corresponding to point ***b***, which is also called the short-term strength. In the long-term condition, excess pore water pressure would dissipate completely, and its TSP and ESP would be the same, as denoted by curve ***ad***. The drained shear strength corresponding to point ***d*** is also called the long-term strength. The q value with regard to point ***b*** is larger than that with regard to point ***d***.

Consequently, a highly overconsolidated clay normally has a larger undrained shear strength, τ_b or s_u, than the drained shear strength (long-term), τ_d, under the same consolidation pressure. In other words, if the undrained test produces negative pore water pressure (or the volume dilates in a drained test) at the ultimate state, the undrained shear strength will be larger than the drained shear strength. Therefore, the drained shear strength for a highly overconsolidated clay should be adopted for design.

2.3 Estimation of design soil parameters

Soils can be classified into two categories: cohesionless and cohesive. Cohesionless soils include gravels, sands, and silts. Cohesive soils include clays and organic soils.

Cohesive soils usually exhibit undrained behaviors during loading; hence, undrained (total stress) properties, such as the undrained shear strength (s_u), undrained Young's modulus (E_u), and undrained Poisson's ratio (μ_u), are important design parameters for foundation design. The undrained friction angle (ϕ_u) is zero whenever s_u is adopted to characterize the strength of an undrained soil. The consolidation deformation of cohesive soils is also an important foundation design problem, but it is a drained problem. The compression index (C_c) and unloading–reloading index (C_{ur}) are the relevant parameters for consolidation deformation.

Cohesionless soils usually exhibit drained behaviors during loading; hence, drained (effective stress) properties, such as the effective friction angle (ϕ'), effective Young's modulus (E'), and effective Poisson's ratio (μ'), are important design parameters for foundation design. The effective cohesion (c') for cohesionless soils is zero. Cohesionless soils also show consolidation deformation, but the deformation is considerably less than that for cohesive soils.

The at-rest lateral earth pressure coefficient (K_0) is also an important design soil parameter for both cohesive and cohesionless soils. Because K_0 is defined as the ratio of the in situ effective horizontal stress to the in situ effective vertical stress, K_0 is always effective by definition (even for cohesive soils).

2.3.1 *Soil strength*

The shear strength of soil can be expressed as the Mohr–Coulomb criterion:

$$\tau_f = c + \sigma \tan(\phi) \tag{2.7}$$

where:

τ_f = shear strength.
c = cohesion.
σ = normal stress.
ϕ = friction angle.

However, eq. 2.7 is rarely adopted in practice. In practical applications, eq. 2.7 is usually applied in two special forms. For the effective stress (drained) analysis, the following form is usually adopted for cohesionless soils:

$$\tau_f = \sigma' \tan(\phi') \tag{2.8}$$

where:

σ' = effective stress.
ϕ' = effective friction angle.

For the total stress (undrained) analysis, the following form is usually adopted for cohesive soils:

$$\tau_f = s_u \tag{2.9}$$

where:

s_u = undrained shear strength.

Cohesionless soils

For cohesionless soils, the main shear strength parameter is the effective friction angle (ϕ'). Depending on the soil density, the stress–strain curve of the cohesionless soil may exhibit different behaviors. For a dense sand, the (peak) ϕ' value is high and is developed under small strain (Figure 2.9). For very loose sand, ϕ' is relatively small and is developed under large strain (Figure 2.9). Laboratory tests, such as triaxial compression tests, can be adopted to determine ϕ' for undisturbed samples of cohesionless soils. However, it is challenging to extract such undisturbed samples for cohesionless soils. In practice, ϕ' for a cohesionless soil is typically estimated by correlation with other soil properties, such as unit weight, density, N value for the standard penetration test (SPT), cone tip resistance for the cone penetration test (CPT), etc.

Table 2.2 shows typical values of ϕ' for cohesionless soils with different densities (loose vs. dense). The effective friction angle of a cohesionless soil can also be estimated by its dry unit weight (γ_d) and relative density (D_r), as shown in Figure 2.13.

It is common to estimate ϕ' based on the in situ SPT N value, which is the number of blow counts for the standard sampler to advance 30 cm in its penetration depth. Two types of corrections for SPT N are necessary for better correlation to ϕ': (a) a correction for the energy efficiency and (b) a correction for the overburden stress. Due to variability in local practice and equipment, the energy efficiency of the drop hammer onto the drop rods varies by region. Skempton (1986) reviewed SPT data from Japan, China, the United States, and

Table 2.2 Typical values of the effective friction angles for cohesionless soils

Soil material	*Effective friction angle ϕ' (degrees)*	
	Loose	*Dense*
Sand, round grains, uniform	27.5	34
Sand, angular grains, well-graded	33	45
Sandy gravel	35	50
Silty sand	27 to 33	30 to 34
Inorganic silt	27 to 30	30 to 35

Source: Terzaghi and Peck (1967); reproduced from Kulhawy and Mayne (1990).

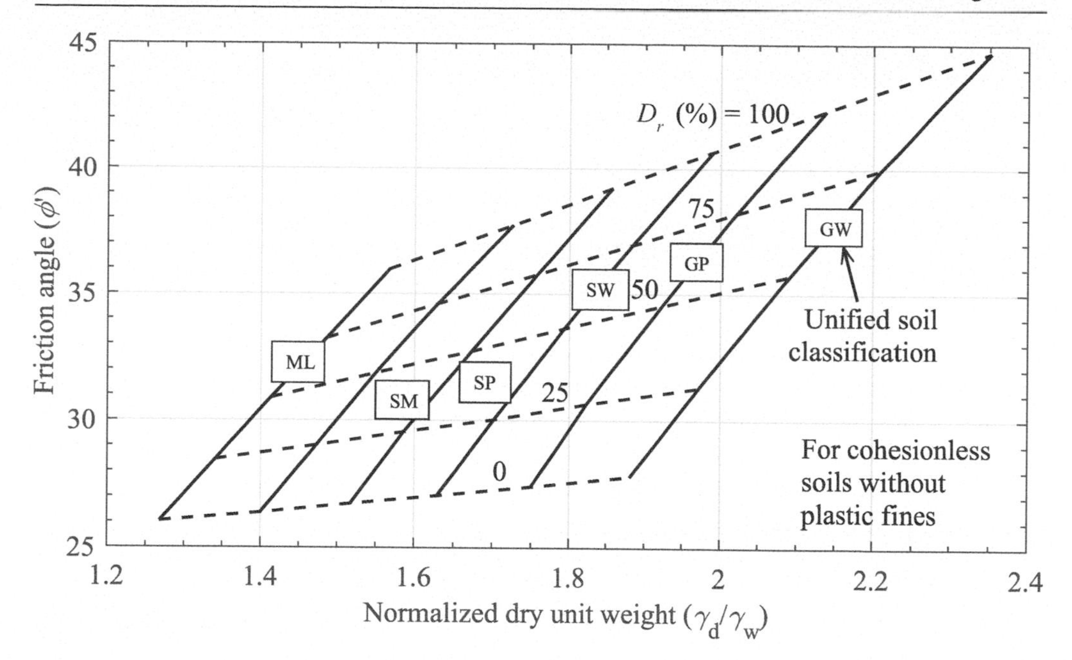

Figure 2.13 Effective friction angle versus dry unit weight and relative density.
Source: Redrawn from NAVFAC (1982).

the United Kingdom and suggested the following corrections for the SPT N value (Kulhawy and Mayne, 1990):

$$N_{60} = C_{ER}C_BC_SC_RN \tag{2.10}$$

where:

N_{60} = SPT N value corrected to a standard energy ratio of 60%.
C_{ER}, C_B, C_S, C_R = correction factors for the energy ratio, borehole diameter, sampling method, and rod length, respectively.
The values of C_{ER}, C_B, C_S, and C_R are given in Table 2.3.

In addition to the energy efficiency, the overburden stress also significantly affects the SPT N value. To obtain a better correlation to ϕ', it is necessary to apply the following correction for N_{60}:

$$(N_1)_{60} = C_N N_{60} \tag{2.11}$$

where:

$(N_1)_{60}$ = N_{60} value corrected to a reference overburden stress of one atmosphere pressure.
C_N = correction factor for overburden stress.

Table 2.3 Values for different SPT N correction factors

Factor	*Equipment variables*	*Correction*	
		Term	*Value*
Energy ratio	(Japan) Donut hammer with Tombi release	C_{ER}	1.3
	(Japan) Donut hammer with 2-turn rope release		1.1
	(China) Pilcon hammer		1.0
	(China) Donut hammer		0.9
	(US) Safe hammer		0.9
	(US) Donut hammer		0.75
	(UK) Various hammers with trip release		1.0
	(UK) Various hammers with 2-turn rope release		0.8
Borehole diameter	65–115 mm	C_B	1.0
	150 mm		1.05
	200 mm		1.15
Sampling method	Standard sampler	C_S	1.0
	Sampler without liner		1.2
Rod length	>10 m	C_R	1.0
	6–10 m		0.95
	4–6 m		0.85
	3–4 m		0.75

Source: Skempton (1986); Kulhawy and Mayne (1990).

The following overburden correction factor proposed by Liao and Whitman (1986) is well-known:

$$C_N = \left(p_a/\sigma'_{v0}\right)^{0.5} \tag{2.12}$$

where:

σ'_{v0} = in situ overburden (vertical effective) stress.
p_a = one atmosphere pressure = 101.3 kN/m^2.

The overburden correction factor calculated by eq. 2.12 is unbounded near the ground surface, where Σ'_{v0} is close to zero. Idriss and Boulanger (2008) recommended an upper bound of 1.7 for C_N. The effective friction angle ϕ' for a cohesionless soil can be estimated based on $(N_1)_{60}$ by the following transformation equation (Hatanaka and Uchida, 1996):

$$\phi' \approx \sqrt{15.4 \times (N_1)_{60}} + 20^o \tag{2.13}$$

This equation was developed based on ϕ' data of expensive undisturbed sand samples obtained by ground freezing. The coefficient multiplied by $(N_1)_{60}$ was not 15.4 in the original formula proposed by Hatanaka and Uchida (1996) but was revised to 15.4 by Mayne et al. (2001) to correct the SPT energy ratio in Japan. The equation in eq. 2.13 is plotted in Figure 2.14. Additional data

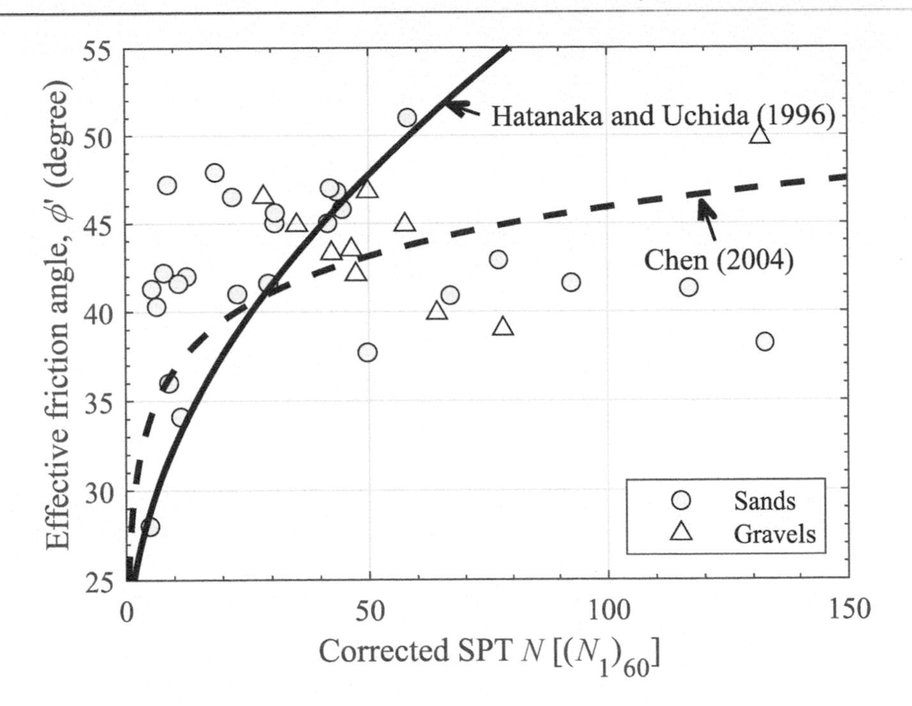

Figure 2.14 Correlation between ϕ' and $(N_1)_{60}$ for cohesionless soils.
Source: Ching et al. (2017).

for undisturbed sand/gravel samples extracted by ground freezing or block sampling are also plotted in Figure 2.14. The transformation equation proposed by Chen (2004) seems to provide a better fit to these additional data:

$$\phi' \approx 9.2 \times \log\left[(N_1)_{60}\right] + 27.5^o \tag{2.14}$$

It is also possible to estimate ϕ' based on the cone tip resistance (q_c) for a piezocone penetration test (CPTu). Similar to SPT N, some corrections for q_c are necessary: (a) a correction for the pore pressure acting on unequal areas of the cone geometry and (b) a correction for the overburden stress. The cone tip resistance corrected for the pore pressure acting on unequal cone areas is denoted by q_t:

$$q_t = q_c + (1 - a)u_{bt} \tag{2.15}$$

where:

q_t = cone tip resistance corrected for the pore pressure acting on unequal cone areas.
a = net area ratio due to unequal areas of the cone geometry (typically, a = 0.7–0.85)
u_{bt} = pore water pressure measured behind the cone tip.

In addition to the correction for unequal cone areas, an overburden correction is necessary:

$$Q_{tn} = C_N \times (q_t / p_a) \tag{2.16}$$

where:

Q_{tn} = q/p_a value corrected to a reference overburden stress of p_a = 101.3 kN/m².
C_N = correction factor for overburden stress in eq. 2.12.

The effective friction angle ϕ' for a cohesionless soil can be estimated based on CPTu (Kulhawy and Mayne, 1990):

$$\phi' \approx 11.0 \times \log(Q_{tn}) + 17.6^\circ \tag{2.17}$$

This transformation equation is plotted in Figure 2.15. The data of reconstituted sands that are used to develop this equation are shown in the figure (Q_{tn} is based on calibration chamber tests).

Cohesive soils

For cohesive soils (clays), the main shear strength parameter for foundation design is the undrained shear strength (s_u). However, s_u is not a fundamental soil property. It is the response of a clay during undrained loading. It depends on the initial condition, effective stress, stress history,

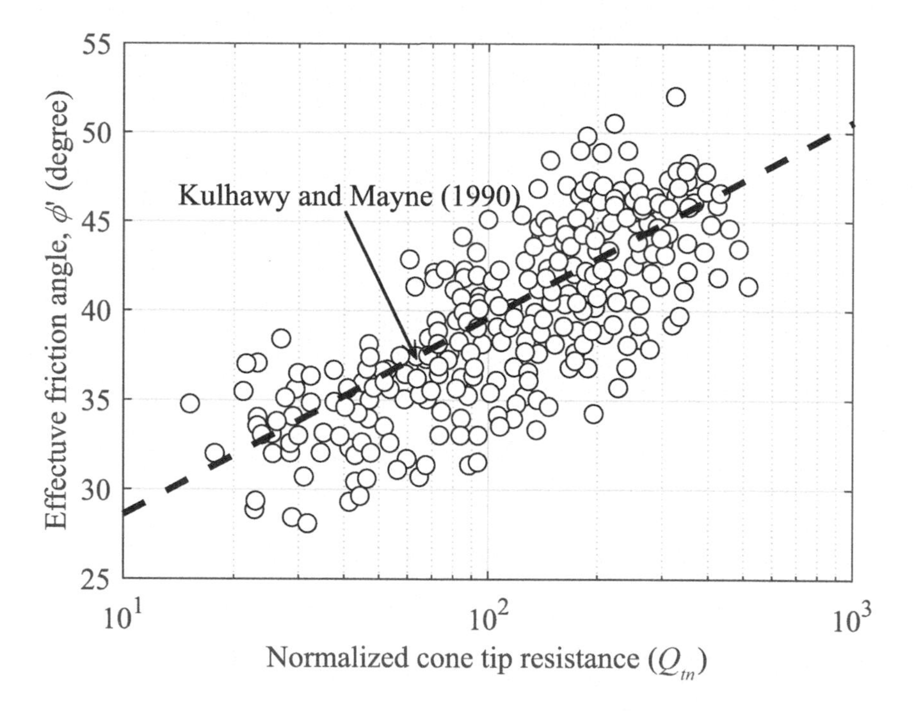

Figure 2.15 Correlation between ϕ' and Q_{tn} for cohesionless soils.
Source: Kulhawy and Mayne (1990).

mode of loading, rate of loading, etc. For a remolded clay (e.g., reconstituted clay prepared in the laboratory), the undrained shear strength can be estimated by the liquidity index (Locat and Demers, 1988):

$$s_u^{re}/p_a \approx 0.0144 \times \mathrm{LI}^{-2.44} \quad (2.18)$$

where:

s_u^{re} = undrained shear strength of a remolded clay.
LI = liquidity index.

Figure 2.16 illustrates this equation together with data for remolded clays.

For an intact (in situ) clay, the s_u value is affected by the effective stress, stress history, loading mode, etc. For the loading mode of embankment failures, Bjerrum (1972), Jamiolkowski et al. (1985), and Mesri and Huvaj (2007) proposed the following transformation equations to estimate s_u based on the vane shear test (VST), overconsolidation ratio, and CPTu:

$$s_u \approx \mu_B \times s_{u,VST} \quad \text{Bjerrum (1972)} \quad (2.19)$$

$$s_u/\sigma'_{v0} \approx 0.23 \times \mathrm{OCR}^{0.8} \quad \text{Jamiolkowski et al. (1985)} \quad (2.20)$$

$$s_u \approx (q_t - \sigma_{v0})/16 \quad \text{Mesri and Huvaj (2007)} \quad (2.21)$$

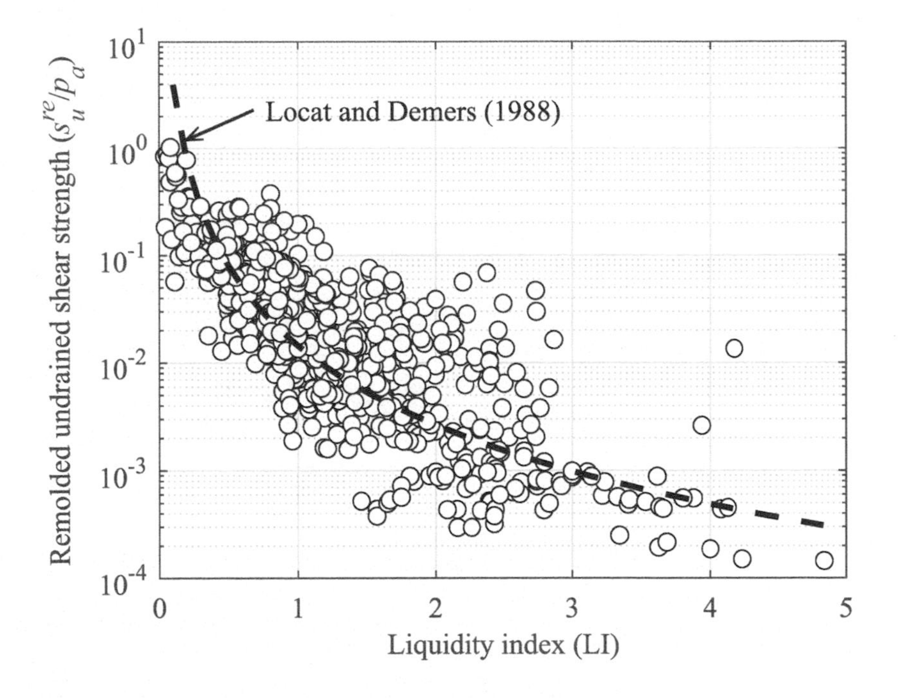

Figure 2.16 Correlation between s_u^{re}/p_a and *LI* for remolded clays.
Data source: Ching and Phoon (2014).

where:

$s_{u,VST}$ = undrained shear strength by VST.
μ_B = VST correction factor proposed by Bjerrum (1972).
OCR = overconsolidation ratio.
σ_{v0} = total overburden stress.

Figure 2.17 shows how the VST correction factor (μ_B) changes with the plasticity index (PI). The data shown in the figure are back-calculated by real failure cases of embankments, footings, and excavations. Figures 2.18 and 2.19 illustrate the transformation equations proposed by Jamiolkowski et al. (1985) and Mesri and Huvaj (2007), respectively, together with some data from in situ clays.

Although the correlation is relatively weak, the s_u value for in situ clay can also be estimated by the SPT N value (Kulhawy and Mayne, 1990):

$$s_u / p_a \approx 0.06 \times N_{60} \tag{2.22}$$

Figure 2.20 illustrates this transformation equation together with data from in situ clays.

2.3.2 *Soil modulus*

For cohesionless soils, the main parameters for immediate deformation are the effective (or drained) Young's modulus (E') and effective Poisson's ratio (μ'). For cohesive soils, they are the undrained Young's modulus (E_u) and undrained Poisson's ratio (μ_u). The Young's modulus of a soil varies over a wide range, depending on the initial condition, effective stress, loading mode, etc. Due to its dependence on effective stress, the Young's modulus is not a constant. As illustrated in Figure 2.21, for a particular stress–strain curve, the modulus can be defined as

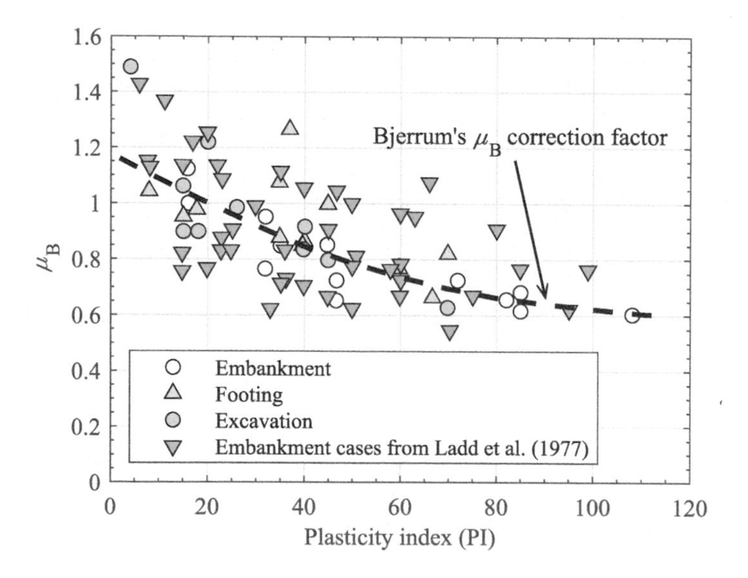

Figure 2.17 VST correction factor (μ_B) versus PI.

Source: Bjerrum (1972) and Ladd et al. (1977).

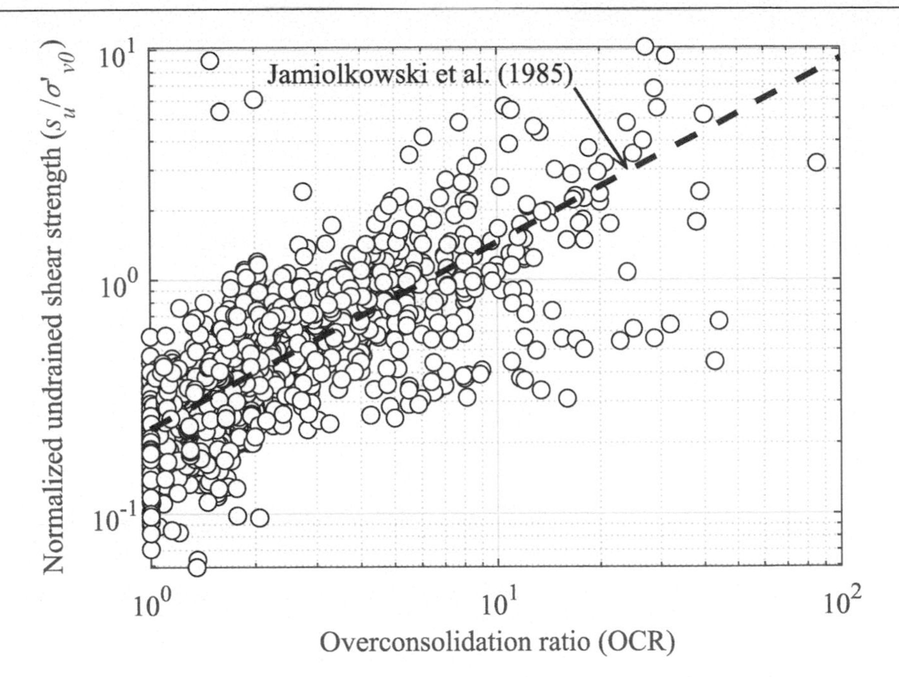

Figure 2.18 Correlation between s_u/σ'_{v0} and OCR for in situ clays.

Data source: Ching and Phoon (2014).

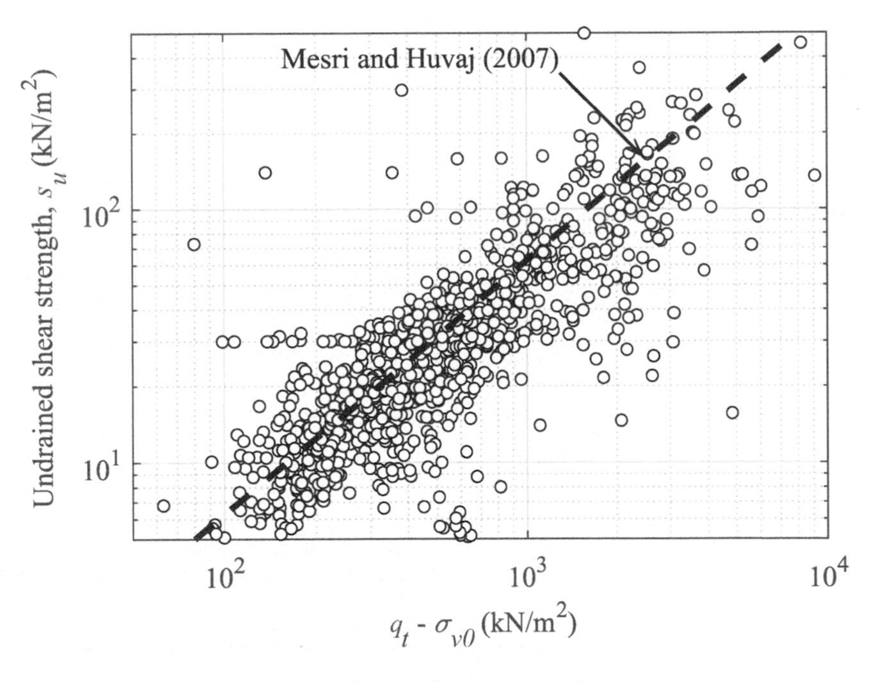

Figure 2.19 Correlation between s_u and $(q_t - \sigma_{v0})$ for in situ clays.

Data source: Ching and Phoon (2014).

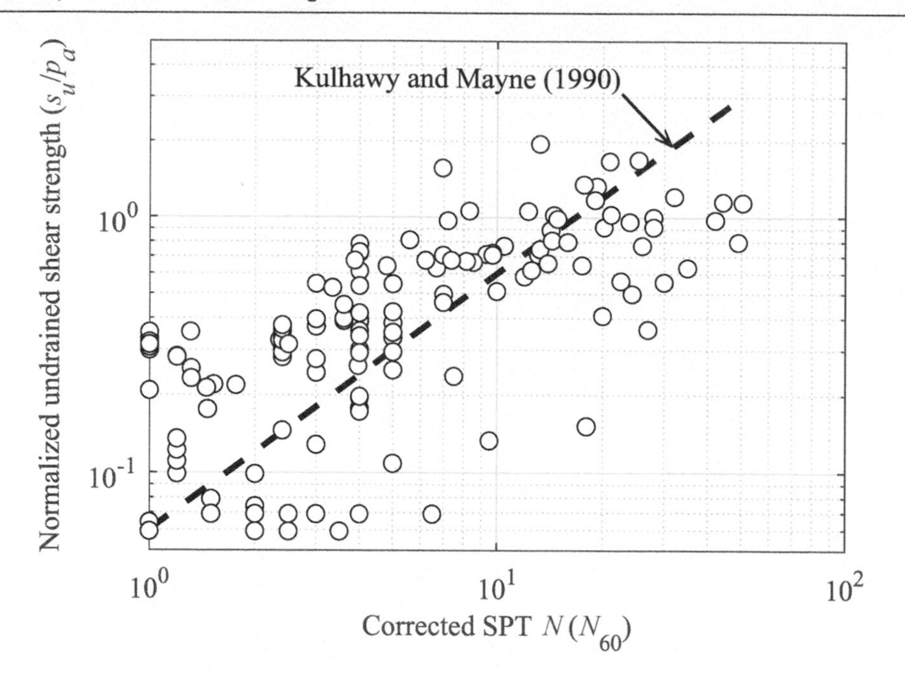

Figure 2.20 Correlation between s_u/p_a and N_{60} for in situ clays.
Data source: Ching and Phoon (2014).

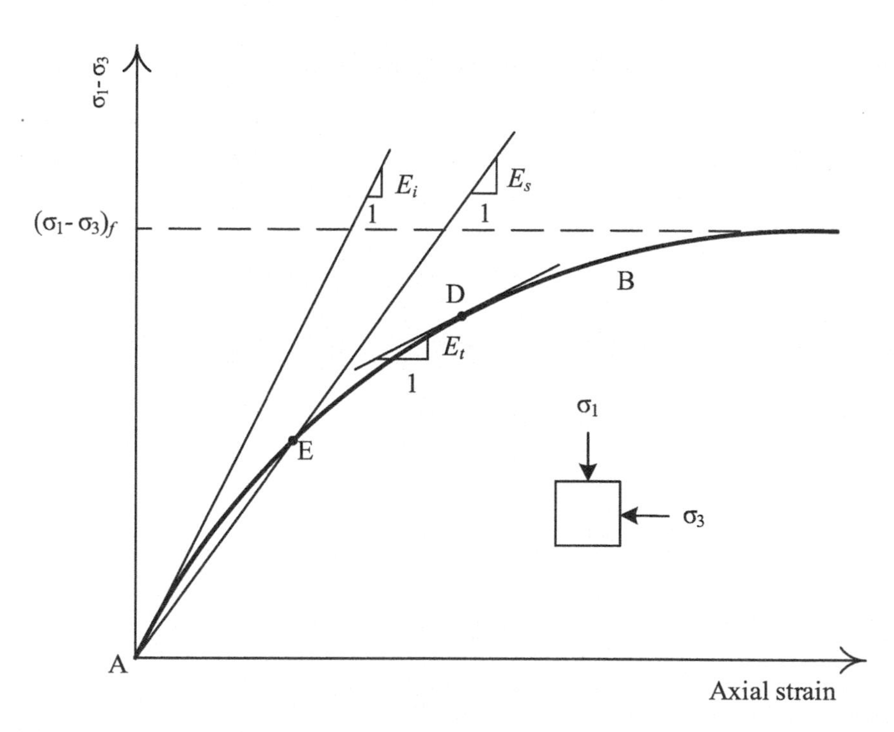

Figure 2.21 Different definitions for the soil modulus.

the initial modulus (E_i), tangent modulus at a certain stress level (E_t), and secant modulus at a certain stress level (E_s). In the following, the notation E is used as the generic Young's modulus, whereas the notations E_i, E_t, and E_s are used in special cases. For instance, E_u denotes a generic undrained modulus, whereas E_{us} denotes the secant undrained modulus. For the secant modulus, it is common to adopt a stress level of one-half of the peak strength, as illustrated in Figure 2.21.

Poisson's ratio

The effective Poisson's ratio (μ') varies over a relatively narrow range. Table 2.4 shows typical values of μ' for various soils. A clay can also behave in a drained manner under long-term loading. The typical value of μ' for clays is also shown in the table. The undrained Poisson's ratio (μ_u) is equal to 0.5 because undrained soils are incompressible.

Cohesionless soils

The typical range for the effective Young's modulus (E') of a cohesionless soil is shown in Table 2.5. E' can also be estimated by correlation with the SPT N value. Many equations for E' versus SPT N have been proposed, and Kulhawy and Mayne (1990) summarized them by the following transformation equations:

$$\begin{aligned} &E'/p_a \approx 5 \times N_{60} && \text{sand with fines} \\ &E'/p_a \approx 10 \times N_{60} && \text{clean nornally consolidated sand} \\ &E'/p_a \approx 15 \times N_{60} && \text{clean overconsolidated sand} \end{aligned} \tag{2.23}$$

The correlations of these equations are usually relatively weak.

The pressuremeter test (PMT) provides a direct measurement of the in situ horizontal modulus (denoted by E_{PMT}). Figure 2.22 illustrates the correlation between E_{PMT} and SPT N for cohesionless soils. The data points from Ohya et al. (1982) in the figure are Japanese cases, and

Table 2.4 Typical ranges of the effective Poisson's ratio (μ')

Soil	*Range of μ'*
Clay	0.2 to 0.4
Dense sand	0.3 to 0.4
Loose sand	0.1 to 0.3

Source: Kulhawy and Mayne (1990).

Table 2.5 Typical ranges of E'/p_a

Density	*Range of E'/p_a*
Loose	100 to 200
Medium	200 to 500
Dense	500 to 1000

Source: Poulos (1975); Kulhawy and Mayne (1990).

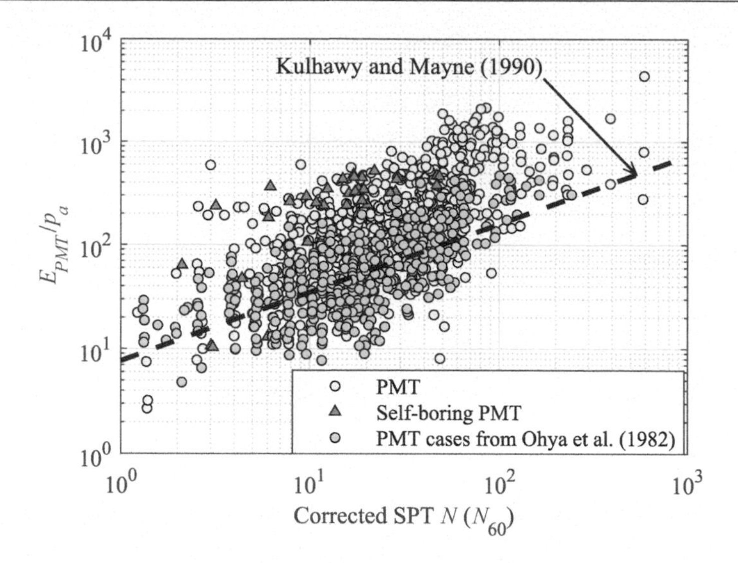

Figure 2.22 Correlation between E_{PMT}/p_a and N_{60} for cohesionless soils

Data source: Ohya et al. (1982) and Ching (2020).

Kulhawy and Mayne (1990) proposed the equation $E_{PMT}/p_a = 9.08 \times N^{0.66}$ to fit them. Because Japanese SPT has a higher energy ratio, the coefficient of 9.08 is modified to convert to N_{60}:

$$E_{PMT}/p_a \approx 7.64 \times N_{60}^{0.66} \tag{2.24}$$

In practice, eq. 2.24 can be adopted to estimate E' because $E' \approx E_{PMT}$ (Kulhawy and Mayne, 1990).

The oedometer test provides a measurement of the tangent effective constrained modulus (M'_t), the ratio of the consolidation stress increment to the vertical strain increment. This M'_t value is often correlated to the cone tip resistance for CPTu. Mayne (2007) proposed the following transformation equation:

$$M'_t \approx 5 \times (q_t - \sigma_{v0}) \tag{2.25}$$

Figure 2.23 illustrates the correlation between (q_t–σ_{v0}) and M'_t together with some data for cohesive and cohesionless soils. Once M'_t is estimated by eq. 2.25, the tangent effective Young's modulus (E'_t) can be computed:

$$E'_t = M'_t \times \frac{(1+\mu')(1-2\mu')}{1-\mu'} \tag{2.26}$$

Cohesive soils

The typical range for the undrained Young's modulus (E_u) of cohesive soils is shown in Table 2.6. E_u is commonly estimated by undrained shear strength (s_u). Ladd and Edgers (1972) conducted a series of undrained direct simple shear (DSS) tests for clays. Based on the E_{us}/s_u, OCR, and PI

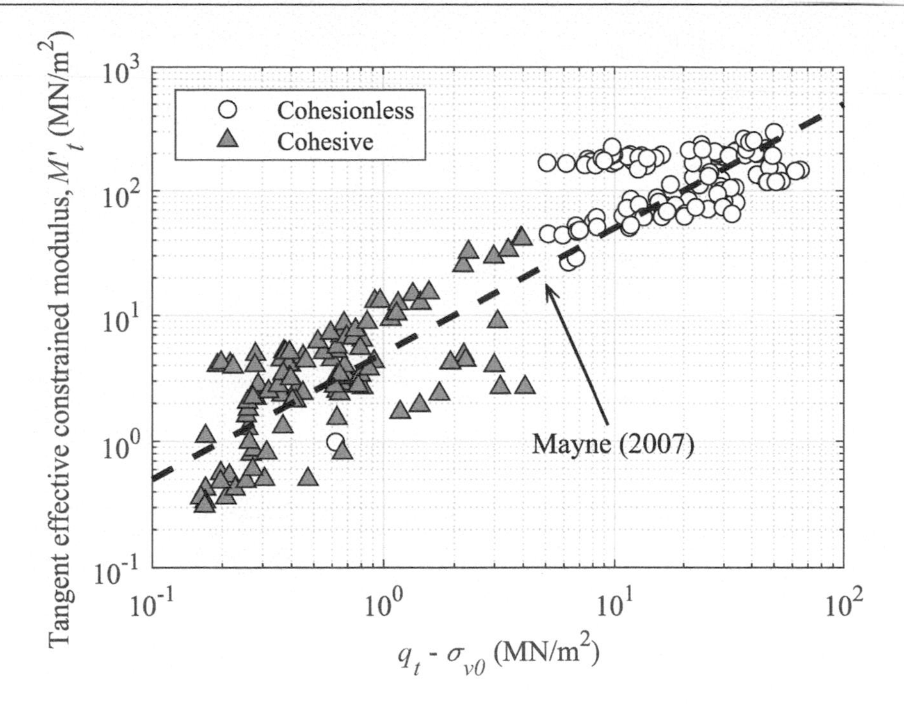

Figure 2.23 Correlation between $(q_t - \sigma_{v0})$ and M'_t.

Data source: Ching (2020).

Table 2.6 Typical ranges of E_u/p_a

Consistency	*Range of* E_u/p_a
Soft	15 to 40
Medium	40 to 80
Stiff	80 to 200

Source: Kulhawy and Mayne (1990).

data of these DSS tests, Duncan and Buchignani (1976) proposed the E_{us}/s_u–OCR relationship in Figure 2.24. The DSS data from Ladd and Edgers (1972) are shown in the figure, along with DSS data from other studies.

The PMT provides a direct measurement of the in situ horizontal modulus (E_{PMT}). Figure 2.25 illustrates the correlation between E_{PMT} and SPT N for cohesive soils. Again, based on the Japanese cases from Ohya et al. (1982), Kulhawy and Mayne (1990) proposed $E_{PMT}/p_a = 19.3 \times N^{0.63}$. To convert to N_{60}, the coefficient of 19.3 is modified:

$$E_{PMT} / p_a \approx 16.36 \times N_{60}^{0.63} \tag{2.27}$$

Eq. 2.27 can be adopted to estimate E_u because $E_u \approx E_{PMT}$ for clays (Kulhawy and Mayne, 1990).

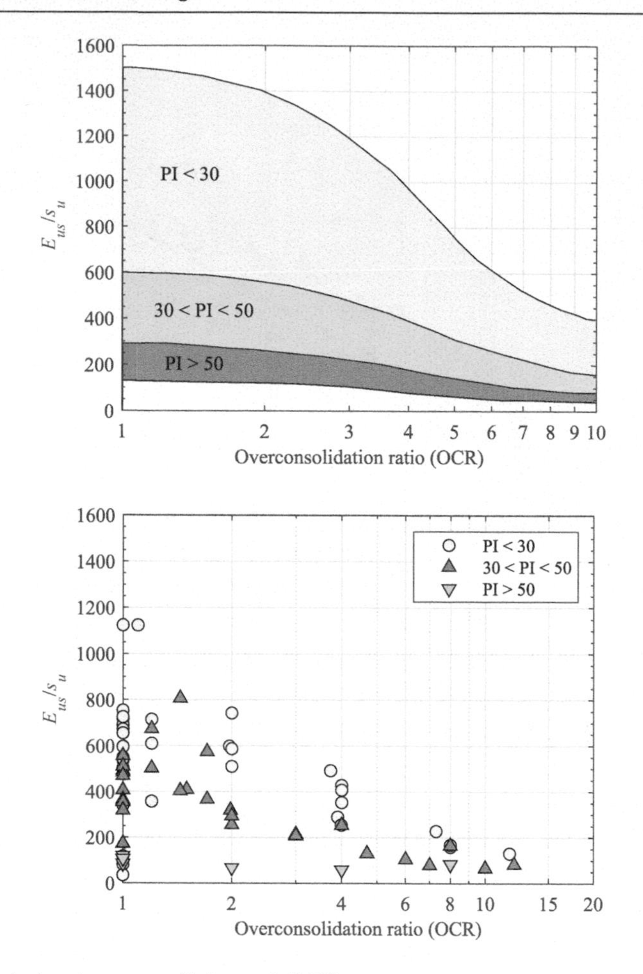

Figure 2.24 Correlation between E_{us}/s_u and OCR.

Data source: Ladd and Edgers (1972) and Ching (2020).

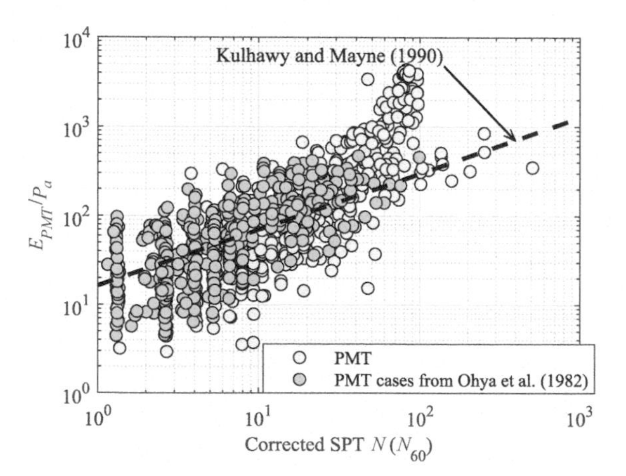

Figure 2.25 Correlation between E_{PMT}/p_a and N_{60} for clays.

Data source: Ohya et al. (1982) and Ching (2020).

2.3.3 Soil consolidation

The consolidation deformation of cohesive soils is an important foundation design problem, and the compression index (C_c) and unload–reload index (C_{ur}) are the two main design parameters. Their definitions are shown in Figure 2.26. Kulhawy and Mayne (1990) proposed the following transformation equations to estimate C_c and C_{ur} based on PI:

$$C_c \approx PI/74 \tag{2.28}$$

$$C_{ur} \approx PI/370 \tag{2.29}$$

Figure 2.27 illustrates these equations with data for cohesive soils. Note that eq. 2.28 significantly underestimates the C_c values for clays with LI > 1 (sensitive clays).

2.3.4 At-rest lateral earth pressure coefficient

The at-rest lateral earth pressure coefficient (K_0) is defined as the ratio of the in situ horizontal effective stress to the in situ vertical effective stress. Mayne and Kulhawy (1982) proposed the following equation to estimate K_0 by OCR for both cohesive and cohesionless soils:

$$K_0 \approx (1-\sin\phi')OCR^{\sin\phi'} \tag{2.30}$$

Figure 2.28 illustrates this equation for selected ϕ' values, together with comparison data. Marchetti (1980) proposed the following equation to estimate K_0 by the horizontal stress index (K_D) for the dilatometer test (DMT):

$$K_0 \approx (K_D/1.5)^{0.47} - 0.6 \tag{2.31}$$

Figure 2.29 illustrates this equation along with comparison data.

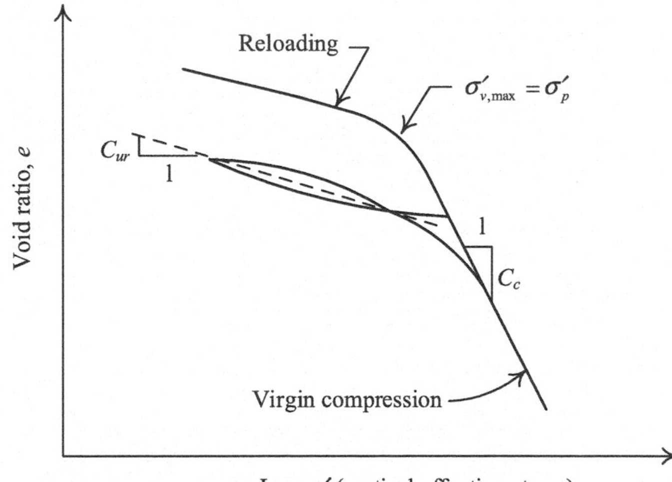

Figure 2.26 Definitions of C_c and C_{ur}.

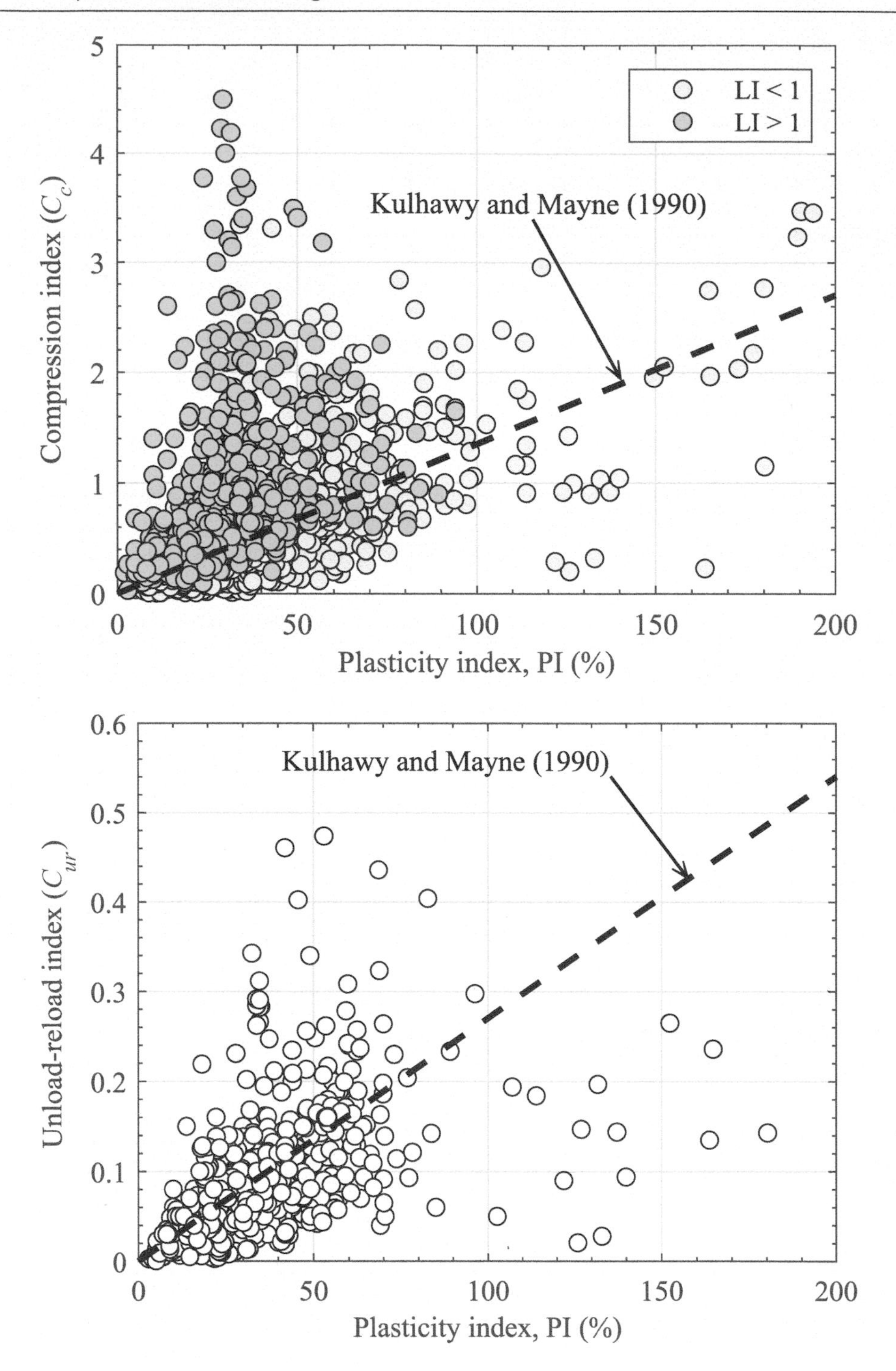

Figure 2.27 Correlation between (C_c, C_{ur}) and PI.

Data source: Ching et al. (2022).

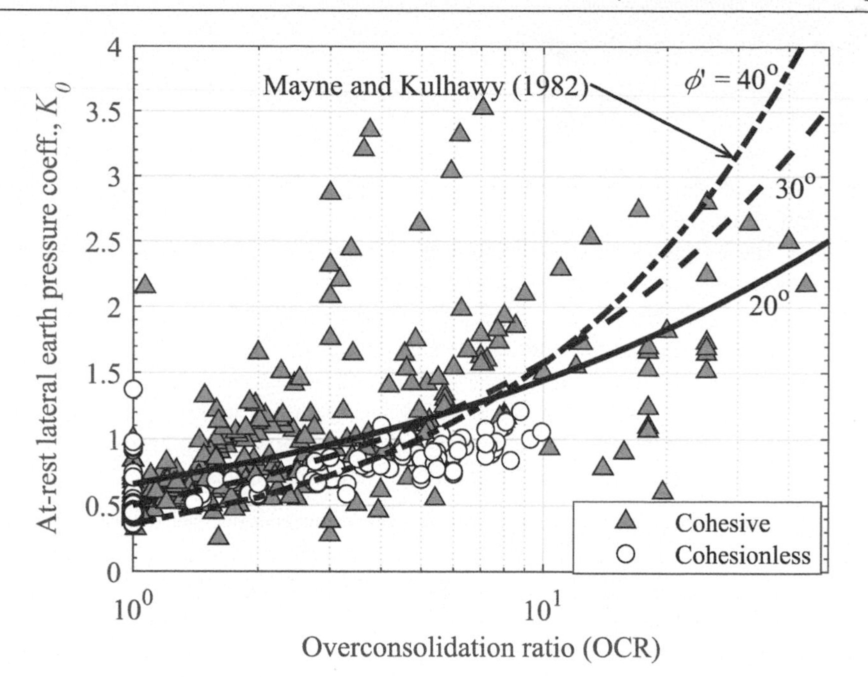

Figure 2.28 Correlation between K_0 and OCR.
Data source: Ching (2020).

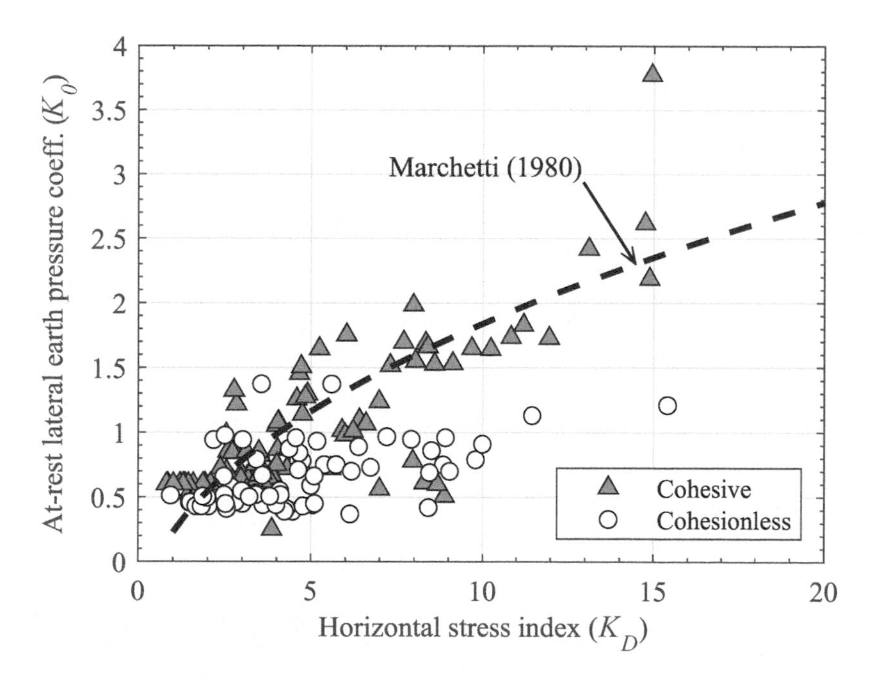

Figure 2.29 Correlation between K_0 and K_D.
Data source: Ching (2020).

2.4 Type of loads and design loads

Loads supported by a foundation need to be estimated in foundation design. The sources of loads generally include the following:

1. Dead loads: loads due to the weight of the structure (i.e., the load associated with all weights permanently attached to the structure).
2. Live loads: the weight of people, traffic surcharge, and objects that are on the structure but are not permanently attached to it.
3. Environmental loads: the loads caused by environmental factors, such as water and earth pressures, wind, snow, wave action, and seismic loads.

Loads adopted in design are called design loads. Design loads are calculated according to the design methods specified in applicable codes by considering load combinations. Load combinations typically include dead loads, live loads, and environmental loads.

2.5 Foundation design methods

There are three design methods adopted in, but not limited to, foundation design: working stress design, ultimate strength design, and limit state design. The first two methods are conventional and force-based. Design criteria are used to evaluate whether the design load exceeds the load capacity of a foundation system (or component). In contrast, the limit state design method is nonconventional and is both "force-based" and "displacement-based" (generally called "performance-based"). In addition to checking for the load capacity, this method evaluates whether the displacement of a foundation system (or component) under the design load exceeds its displacement capacity. The design concepts for the three methods are introduced in the following sections.

2.5.1 *Working stress design (WSD) method*

The concept of the working stress design method is straightforward. It is to design a system or component to sustain the design load (or design stress) without exceeding its allowable load (or allowable stress). Therefore, this method is also called the allowable stress design (ASD) method. The design criterion is expressed as follows:

$$Q_d \leq Q_a \tag{2.32}$$

where Q_d is the design load (or working load or service load) and Q_a is the allowable load.

In this design method, a factor of safety is commonly adopted to compute the allowable load Q_a, which is expressed as:

$$Q_a = \frac{Q_u}{FS} \tag{2.33}$$

where Q_u is the ultimate capacity and FS is the factor of safety. A large FS implies a small allowable load. The magnitude of FS varies with the importance of structures, design load conditions, target failure probability, consequence of failure, uncertainty in design load and capacity, etc.

Different values are adopted in the long-term load condition and short-term load conditions. A larger *FS* is adopted for the long-term condition than for the short-term condition. For foundation structures, such as shallow foundations under vertical loading, an F_S of 3 is commonly used in long-term conditions, under which the elastic response of the foundation system is expected. Because the concept of this method is straightforward and simple, it is introduced for designing various foundations in this book.

2.5.2 Ultimate strength design (USD) method

Contrary to the WSD method, which generally assumes elastic structural behavior, the ultimate strength design (USD) method considers the ultimate strength of structures to resist the largest loads anticipated during their design lives. The design criterion of this method is commonly expressed as follows:

$$\sum (LF)_i Q_{ni} \leq (RF) R_n \tag{2.34}$$

where LF_i and Q_{ni} are the load factor and nominal load, respectively, for load type *i*. Different *i* indices represent different load sources, such as the dead load and live load. *RF* and R_n are the resistance factor and nominal resistance, respectively.

In eq. 2.34, the left-hand-side term of the inequality represents the design load, which is a combination of nominal loads; the latter are multiplied by load factors that are typically greater than 1. The design load should be less than or equal to the design resistance. As defined by the right-hand-side term of the inequality in eq. 2.34, the design resistance is the nominal resistance multiplied by a resistance factor, which is typically less than 1. The word "nominal" means that both the loads and resistances are determined according to certain guidelines.

Because the load and resistance factors are used to modify the nominal loads and resistance, respectively, the ultimate strength design method is also called the load and resistance factor design (LRFD) method. Different values of the load and resistance factors are adopted to account for uncertainties in load and resistance, the target probability of failure, and the consequence of failure. The LRFD method was initially applied to structural design practice for reinforced concrete structures in the United States. It has been widely applied in the design of bridge foundations in North America (AASHTO, 2012; CSA, 2014). For example, in the AASHTO (American Association of State Highway Transportation Officials) bridge design specifications, load factors of 1.25–1.5 (or 0.9 when the effect of load is favorable) for dead loads and of 1.35–1.75 for live loads are adopted for the ultimate state. The resistance factors depend on foundation types, soil types, analysis methods, and determination methods of soil parameters. The suggested resistance factors for the bearing capacity evaluation of shallow foundations for the ultimate state are given in Table 2.7.

Notably, the design load for the WSD method is a combination of unfactored loads, that is, the load factor LF_i in eq. 2.34 is not adopted. Uncertainties in loads and resistances are lumped in *FS*. However, for the USD method, the design load is a combination of factored loads. Uncertainties in loads and resistances are reflected in load and resistance factors, respectively. Because both the USD and WSD methods are force-based, additional checking for foundation settlement and displacement is needed.

Table 2.7 Resistance factors for shallow foundations for the ultimate state suggested in AASHTO bridge design specifications

Method/soil/condition		*Resistance factor*
Bearing resistance	Theoretical method (Munfakh et al., 2001), in clay	0.50
	Theoretical method (Munfakh et al., 2001), in sand, using CPT	0.50
	Theoretical method (Munfakh et al., 2001), in sand, using SPT	0.45
	Semiempirical method (Meyerhof, 1957), all soils	0.45
	Footings on rock	0.45
	Plate load test	0.55
Sliding resistance (base)	Precast concrete placed on sand	0.9
	Cast-in-place concrete on sand	0.8
	Precast or cast-in-place concrete on clay	0.85
	Soil on soil	0.9
Sliding resistance (lateral)	Passive earth pressure	0.5

Source: AASHTO (2012).

2.5.3 Limit state design (LSD) method

The limit state design (LSD) method introduces a more complete view to design a system considering different limit states. A *limit state* is a set of critical conditions to avoid. Each state shall have a required performance. Two basic limit states, serviceability and ultimate limit states, are commonly used for design. For the serviceability limit state, a system fails to perform its intended function due to excessive settlement or displacement. For the ultimate limit state, a system is no longer safe due to overall instability, bearing failure, failure of foundation components, etc.

The AASHTO bridge design specifications have implemented the LSD method. In addition to the ultimate limit state, the format of the LRFD method eq. 2.34 is applied to other limit states. In the AASHTO bridge design specifications, different load factors are specified for different limit states. For the serviceability limit state, the load factors are generally set to 1, and the deformations and displacements of structures and their components are checked to ensure they are below the tolerable limits. For the ultimate limit state, the aforementioned USD method is adopted to check the strength and stability of structures and their components.

The Eurocode 7 (Geotechnical Design, EC7) (CEN, 2004) has also implemented the LSD method. A distinct feature of EC7 is the use of the characteristic soil parameter, defined as a cautious estimate of the value affecting the occurrence of the limit state. More discussions of the characteristic soil parameters are given in Section 2.6. The characteristic value of a soil parameter is usually more conservative than the nominal value. For instance, given soil shear strength data, the nominal shear strength may be calculated as the average value of the test data (called the sample mean). However, the characteristic shear strength is usually less than the sample mean. The sample mean is not a cautious estimate of the value affecting the limit state (the true mean) because there is an approximately 50% probability that the sample mean underestimates the true mean. A cautious estimate requires a smaller probability (e.g., 5%) of underestimating the true mean, so the characteristic shear strength is usually less than the sample mean. In EC7, a conservative design is achieved on two levels. At the first level, the use of the characteristic soil parameter leads to a conservative design. At the second level, the characteristic soil parameters are further divided by partial (safety) factors for additional conservatism. The first level of conservatism concerns soil parameter variabilities and uncertainties, whereas the second level concerns the required safety. Different from the AASHTO bridge design specifications, EC7 does not adopt load and resistance factors in eq. 2.34 but adopts partial factors. Soil strengths (such as friction angle and undrained shear strength) are divided by partial factors, whereas loads are multiplied by partial factors:

$$\sum(\gamma_Q)_i Q_{ki} \leq R(X_k/\gamma_X) \tag{2.35}$$

where Q_{ki} is the characteristic load for load type i; $(\gamma_Q)_i$ is its partial factor; R is the resistance, which depends on soil parameter X; X_k is the characteristic soil parameter; and γ_X is its partial factor. The partial factors $(\gamma_Q)_i$ and γ_X play a similar role as the traditional safety factor, but each load or soil parameter now has its own partial safety factor. For the serviceability limit state, partial factors are not adopted (set to one). For the ultimate limit state, the characteristic loads are multiplied by their partial factors, whereas the characteristic soil parameters are divided by their partial factors.

Recently, the performance-based design (PBD) method was proposed based on the concept of the limit state design (LSD) method by considering the types of actions and importance of structures. The design procedure is commonly expressed in terms of a matrix (often called the performance matrix), as displayed in Figure 2.30. In this matrix, the structural performances (limit states) and the magnitude of external actions are taken as two axes. The required performance for the magnitude of each external action is indicated in the matrix depending on the importance of the structure (Honjo, 2003). The magnitude of external actions is divided into three levels based on the occurrence frequency and impact of the actions. Three limit states, serviceability, reparability, and ultimate, are considered to account for different levels of structural damage. According to this matrix, for example, an important structure is required to be in the serviceability limit state under high frequency and low impact and medium frequency and medium impact of loading and in the reparability limit state under low frequency and high impact loading. The railway foundation design code of Japan (Railway Technical Research Institute, RTRI, 2012) has shifted toward PBD. In the code, both force and displacement are checked to ensure they meet the requirements. To perform PBD, a complete capacity curve of a foundation system is needed, as presented in Figure 2.31. The capacity curve displays the relationship between the foundation load capacity and displacement. The limit states in the code consider general and earthquake conditions. The general condition includes the serviceability and safety limit states. The foundation is not allowed to exceed the yield point of the foundation system (Δ_y, Q_y). The long-term and short-term allowable bearing capacities are set to 1/3 and 2/3 of the yield bearing capacity of the foundation (Q_y), respectively. The earthquake condition includes the reparability limit state for checking the residual displacement and the safety limit state. The foundation can be allowed to exceed the yield point of the foundation system, but the reparability and safety of the foundation must be considered. Table 2.8 presents an example of detailed performance requirements for shallow foundations under different limit states provided in the code. In this table, the performance requirements for limit states under general and earthquake conditions are completely specified.

2.6 Characteristic soil parameters

Eurocode 7 (EC7) requires the determination of characteristic soil parameters. This section discusses the concept of the characteristic soil parameter as well as its determination.

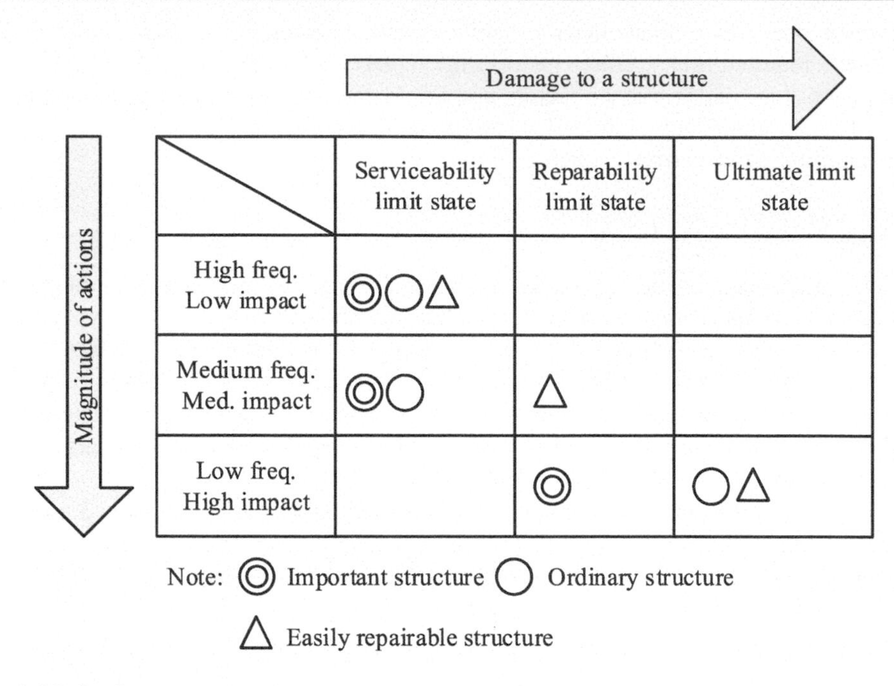

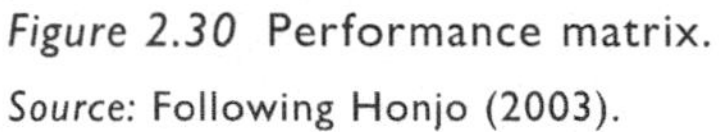

Figure 2.30 Performance matrix.

Source: Following Honjo (2003).

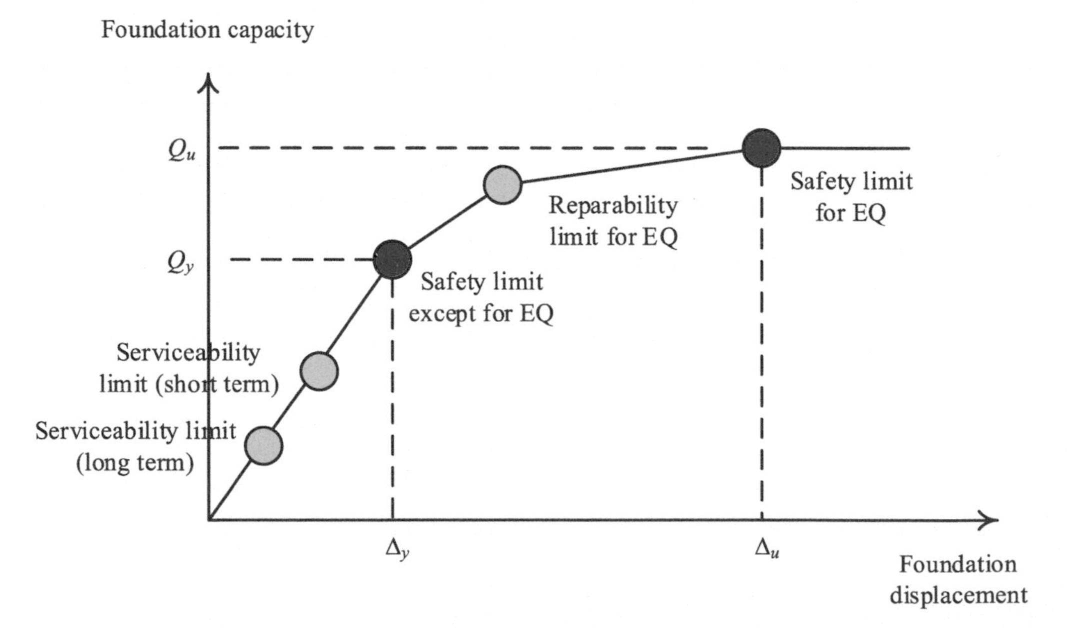

Figure 2.31 Capacity curve of a foundation system.

Table 2.8 Example of performance requirements for shallow foundations

Required performance	*Performance item*		*Performance requirements*
Safety limit	General	Ground failure	Vertical load ≤ allowable bearing capacity (f_r = 0.67)
		Horizontal stability	Horizontal load ≤ allowable lateral capacity (f_r = 0.67)
		Overturning	Moment ≤ allowable moment capacity (f_r = 0.67)
	Earthquake	Ground failure	Plastic zone at base ≤ B/4
		Horizontal stability	Horizontal load ≤ allowable lateral capacity (f_r = 1.0)
		Overturning	Max. rotation ≤ 30/1,000 rad
Serviceability limit	Long-term performance	Vertical	Vertical load ≤ allowable bearing capacity (f_r = 0.33)
		Horizontal	Horizontal load ≤ allowable lateral capacity (f_r = 0.33)
		Rotational	Eccentricity ≤ B/6
	Short-term performance	Vertical	Vertical load ≤ allowable bearing capacity (f_r = 0.5)
		Horizontal	Horizontal load ≤ allowable lateral capacity (f_r = 0.5)
		Rotational	Eccentricity ≤ B/4
Reparability limit (residual displacement)	Earthquake level I	Vertical	Vertical load ≤ allowable bearing capacity (f_r = 0.67)
		Horizontal	Horizontal load ≤ allowable lateral capacity (f_r = 0.67)
		Rotational	Moment ≤ allowable moment capacity (f_r = 0.67)
	Earthquake level II	Vertical	Plastic zone at base ≤ B/6
		Horizontal	Horizontal load ≤ allowable lateral capacity (f_r = 1.0)
		Rotational	Max. rotation ≤ 20/1,000 rad

Source: RTRI (2012).

2.6.1 Concept of the characteristic soil parameter

Eurocode 7 defines the characteristic soil parameter as "a cautious estimate of the value affecting the occurrence of the limit state" (EC7 Clause 2.4.5.2(2)). There are two main aspects in the definition of characteristic values: (a) a cautious estimate and (b) the value affecting the occurrence of the limit state. These two aspects are elaborated next.

For the first aspect, a cautious estimate is needed because soil parameters show variabilities and uncertainties. For instance, soils are inherently variable in space, so there are spatial variabilities. Various laboratory and field tests are also subject to measurement errors. The transformation equations introduced in Section 2.3 are not exact, so there are transformation uncertainties. There is only a finite amount of site investigation data, and the sample mean of the finite data is different from the true mean, so there are statistical uncertainties.

For the second aspect, the value affecting the occurrence of the limit state is related to the failure mechanism. EC7 provides some guidelines on the value affecting the occurrence of the limit state. For instance, EC7 Clause 2.4.5.2(4) lists factors to consider when selecting the characteristic value, which include, among other factors, "the extent of the zone of ground governing the behavior of the geotechnical structure at the limit state being considered." In the literature, the zone governing the limit state is sometimes called the "influence zone." EC7 Clause 2.4.5.2(7) further states that "the value of the governing parameter is often the mean of a range of values covering a large surface or volume of the ground." Consequently, the value affecting the occurrence of the limit state is often the mean value over the influence zone.

For instance, the slip circle for the base shear failure of an excavation in clay is shown in Figure 6.10. The design soil parameter relevant to this ultimate limit state is the undrained shear strength (s_u). When the characteristic s_u is determined, the value affecting the occurrence of the ultimate limit state may be taken as the mean s_u value over the influence zone defined by slip arcs a-g-f-b in the figure.

2.6.2 Determination of the characteristic soil parameter

The characteristic soil parameter can be determined based on statistical methods, as EC7 Clause 2.4.5.2(11) specifies: "If statistical methods are used, the characteristic value should be derived such that the calculated probability of a worse value governing the occurrence of the limit state under consideration is not greater than 5%." In other words, the probability that the characteristic soil parameter underestimates the true mean over the influence zone should not be greater than 5%. If the probability distribution of a soil parameter is lognormal, the characteristic soil parameter can be determined as the 5% fractile of the lognormal distribution:

$$X_k = m \times e^{-1.645 \times \sqrt{\ln\left(1+COV_m^2\right)}} \Big/ \sqrt{1+COV_m^2} \qquad (2.36)$$

where X_k is the characteristic soil parameter, m is the sample mean over the influence zone, COV_m is the coefficient of variation of this sample mean, and −1.645 is the 5% fractile of the standard normal random variable (namely, the probability that the standard normal random variable is less than −1.645 is 5%). The COV_m quantifies all variabilities and uncertainties, including the spatial variability, measurement error, transformation uncertainty, and statistical uncertainty (Phoon and Kulhawy, 1999). For the slip circle example earlier, m in eq. 2.36 can be taken as the sample mean of the s_u data over the influence zone (slip arc a-g-f-b in Figure 6.10). For COV_m in eq. 2.36, Table 2.9 shows the range of COV_m for the mean shear strength (s_u or ϕ') over the influence zone with a size (depth) of 5 m. If the shear strength data are based on laboratory triaxial

Table 2.9 Range of COV_m for soil shear strength parameters (s_u and ϕ')

Soil parameter	*Test*	*Soil type*	COV_m *(%)*	*Transformation equation*
s_u	Direct (lab UC)	Clay	10–40	–
s_u	Direct (lab UU)	Clay	7–25	–
s_u	Direct (lab CIUC)	Clay	10–30	–
s_u	VST	Clay	15–50	Eq. 2.19
s_u	q_t	Clay	30–40	Eq. 2.21
s_u	SPT N	Clay	40–55	Eq. 2.22
ϕ'	Direct (lab CD)	Clay, sand	6–20	–
ϕ'	q_t	Sand	10	Eq. 2.17
ϕ'	SPT N	Sand	11*	Eq. 2.13

* Also based on the transformation uncertainty *COV* summarized in Table 1 of Ching et al. (2017).

Source: Phoon and Kulhawy (1999), assuming the size of the influence zone = 5 m.

Table 2.10 Range of COV_x for other soil parameters (E and K_0)

Soil parameter	*Test*	*Soil type*	COV_m(%)	*Transformation equation*
E_{PMT}	Direct (PMT)	Sand	15–70	–
E_D	Direct (DMT)	Sand	10–70	–
E_{PMT}	SPT N	Clay	85–95	Eq. 2.27
K_0	Direct (PMT)	Clay	15–45	–
K_0	Direct (PMT)	Sand	20–55	–
K_0	K_D	Clay	35–50	Eq. 2.31

Source: Phoon and Kulhawy (1999), assuming the size of the influence zone = 5 m.

tests (direct tests), s_u or ϕ' data are directly obtained. Transformation equations, such as those in Section 2.3, are not necessary, so COV_m is typically smaller because there is no transformation uncertainty. If the shear strength is estimated by a transformation equation, COV_m is typically larger. Table 2.9 shows the results for the shear strength. Table 2.10 further shows the range of COV_m for the modulus and at-rest lateral earth pressure coefficient (E and K_0). The increase in shear strength and modulus usually has a positive effect on resistance or capacity. Their characteristic values can be determined by eq. 2.36. However, the increase in K_0, for example, sometimes has an adverse effect on resistance or capacity. In this case, eq. 2.36 should not be adopted. Instead, the following equation should be adopted:

$$X_k = m \times e^{1.645 \times \sqrt{\ln\left(1+COV_m^2\right)}} \Big/ \sqrt{1+COV_m^2} \tag{2.37}$$

where 1.645 rather than –1.645 is adopted in eq. 2.36.

Example 2.1

Consider the base shear failure (an ultimate limit state) of an excavation in clay. The influence zone is defined by the slip arc a-g-f-b in Figure 6.10. Suppose the influence zone extends from depths of 20 m to 25 m. The site investigation data include the undrained shear strength data from the unconsolidated undrained test (UU) and SPT N data shown in Figures EX 2.1a and

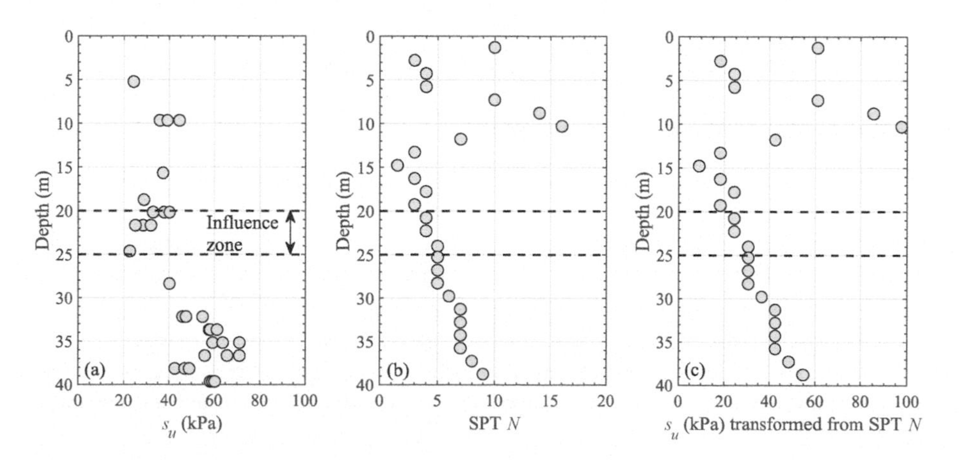

Figure EX 2.1 Site investigation data: (a) s_u data from UU, (b) SPT N data, and (c) s_u data transformed from SPT N.

EX 2.1b, respectively. Determine the characteristic undrained shear strength for the excavation by considering two scenarios: (a) only the undrained shear strength (s_u) data are available, and (b) only the SPT N data are available.

Solution

Let the characteristic undrained shear strength be denoted by $s_{u,k}$.

(a)

There are 7 s_u data points from depths of 20 m to 25 m (22.6, 32.9, 37.8, 40.2, 28.4, 31.9, and 25.0 kPa). The sample mean (m) of the s_u data is:

$$m = (22.6 + 32.9 + 37.8 + 40.2 + 28.4 + 31.9 + 25.0)/7 = 31.2 \text{ kPa}$$

The influence zone has a size of 25 m – 20 m = 5 m, consistent with that assumed in Table 2.9, so the table is applicable. The COV_m checked from the table (soil parameter = s_u; test = lab UU) is COV_m = 7–25%. An intermediate value of COV_m = 15% is adopted. Inserting the m and COV_m values into eq. 2.36 yields $s_{u,k}$ = 24.1 kPa.

(b)

First, the SPT N data are transformed to s_u by eq. 2.22 (Figure EX 2.1c). There are three transformed s_u data from depths of 20 m to 25 m (24.4, 24.4, 30.5 kPa). The sample mean of the transformed s_u data is:

$$m = (24.4 + 24.4 + 30.5)/3 = 26.4 \text{ kPa}$$

The COV_m checked from Table 2.9 (soil parameter = s_u; test = SPT N) is COV_m = 40–55%. An intermediate value of COV_m = 45% is adopted. Inserting the m and COV_m values into eq. 2.36 yields $s_{u,k}$ = 11.9 kPa.

For this particular example, the characteristic value $s_{u,k}$ obtained by SPT N is relatively low because of the significant uncertainty (COV_m = 45%). This significant uncertainty is mainly due to the uncertainty of the transformation equation (i.e., eq. 2.22). The characteristic value $s_{u,k}$ obtained by lab UU does not require such a transformation equation.

2.7 Summary and general comments

This chapter introduces various phenomena of drained and undrained behaviors as well as the concept of total stress and effective stress analyses. The correlation equations (or transformation equations) that can estimate design soil parameters are provided. The working stress design (WSD), ultimate strength design (USD), and limit state design (LSD) methods are explained in this chapter. A summary and general comments regarding this chapter are as follows:

1. In foundation design, the target soils should be classified as either drained or undrained material, and effective stress analysis or total stress analysis should be carried out accordingly.

2. The effective stress analysis method treats the soil as a two-phase material, that is, soil particles and pore water, during the analysis process. The groundwater level and pore water pressure in the soil should be taken into account. All soil parameters are expressed in terms of effective stress. The total stress analysis method regards fine-grained soil and pore water as a one-phase material or a soil–water mixture, as if pore water does not exist in the material. The groundwater level is not considered. All soil parameters are expressed in terms of total stress.
3. As the soil is stressed, under undrained or short-term conditions, a positive or negative excess pore water pressure is generated, which is dissipated under long-term conditions. The critical shear strength of the soil should be smaller between short-term and long-term conditions. According to the characteristics of the stress–strain behavior of clays, the critical shear strength for normally consolidated or slightly overconsolidated clay is the undrained shear strength, where the undrained soil parameters s_u and $\phi = 0$ should be adopted for analysis or design. The critical shear strength for highly overconsolidated clay is the drained shear strength, where the effective strength parameters c' and ϕ' are adopted for analysis and design.
4. Direct determination of design soil parameters (such as shear strengths, moduli, at-rest lateral earth pressure coefficient, etc.) usually requires undisturbed soil specimens and sophisticated tests. In practice, transformation equations are often adopted to estimate these design soil parameters based on the results of convenient tests, such as SPT N values, CPT cone tip resistance, and Atterberg limits. These transformation equations are not exact, and there is usually a significant amount of transformation uncertainty.
5. The WSD and USD methods are force-based. The design criteria are used to evaluate whether the design load exceeds the load capacity of a foundation system (or component). In contrast, the limit state design method is both "force-based" and "displacement-based." In addition to checking for the load capacity, the LSD method also evaluates whether the displacement of a foundation system (or component) under the design load exceeds its displacement capacity.
6. The design load for the WSD method is a combination of unfactored loads. Uncertainties in loads and resistances are lumped in FS. For the USD method, the design load is a combination of factored loads. Uncertainties in loads and resistances are reflected in load and resistance factors, respectively. Because both methods are force-based, additional checking for foundation settlement and displacement is needed.
7. The LSD method introduces a more complete view to design a system considering different limit states. A limit state is a set of critical conditions to be avoided. Each state has corresponding performance requirements. Two basic limit states are the serviceability and ultimate limit states. Recently, the performance-based design (PBD) method was proposed based on the concept of the LSD method by considering the types of actions and importance of structures. To perform PBD, a complete capacity curve of a foundation system for the relationship between the foundation load capacity and displacement is needed.
8. In Eurocode 7, the characteristic value is defined as a cautious estimate of the value affecting the occurrence of the limit state. The characteristic value explicitly addresses various uncertainties related to the estimation of design soil parameters (such as spatial variability, transformation uncertainty, and statistical uncertainty). In Eurocode 7, the required safety level is separately addressed for partial factors.

References

AASHTO. (2012), *LRFD Bridge Design Specifications*, 6th edition, Washington, DC: American Association of State Highway and Transportation Officials (AASHTO).

Bjerrum, L. (1972), *Embankments on Soft Ground.* ASCE Conf. on Performance of Earth and Earth-Supported Structures, ASCE, New York, NY, 2, 1–54.
CEN. (2004), *Eurocode 7: Geotechnical Design. Part 1: General Rules, EN 1997–1*, Brussels: European Committee for Standardisation.
Chen, J.R. (2004), *Axial Behavior of Drilled Shafts in Gravelly Soils*, Ph.D. dissertation, Cornell University, Ithaca, New York.
Ching, J. (2020), *Personal Database.*
Ching, J., Lin, G.H., Chen, J.R. and Phoon, K.K. (2017), Transformation models for effective friction angle and relative density calibrated based on a multivariate database of coarse-grained soils. *Canadian Geotechnical Journal*, Vol. 54, No. 4, pp. 481–501.
Ching, J. and Phoon, K.K. (2014), Transformations and correlations among some clay parameters—the global database. *Canadian Geotechnical Journal*, Vol. 51, No. 6, pp. 663–685.
Ching, J., Phoon, K.K. and Wu, C.T. (2022), Data-centric quasi-site-specific prediction for compressibility of clays. *Canadian Geotechnical Journal*. Vol. 59, No. 12, pp. 2033–2049.
CSA (Canadian Standards Association). (2014), *Canadian Highway Bridge Design Code*. CAN/CSA-S6-14, Mississauga, Ontario, Canada: CSA.
Duncan, J.M. and Buchignani, A.L. (1976), *An Engineering Manual for Settlement Studies*. Department of Civil Engineering, Berkeley: University of California at Berkeley.
Hatanaka, M. and Uchida, A. (1996), Empirical correlation between penetration resistance and internal friction angle of sandy soils. *Soils and Foundations*, Vol. 36, No. 4, pp. 1–9.
Holtz, R.D., Kovacs, W.D. and Sheahan, T.C. (2011), *An Introduction to Geotechnical Engineering*, New York: Pearson Education, Inc.
Honjo, Y. (2003), *Comprehensive Design Codes Development in Japan: Geo-code 21 ver. 3 and Code PLATFORM ver. 1*, LSD2003: International Workshop on Limit State Design in Geotechnical Engineering Practice.
Idriss, I.M. and Boulanger, R.W. (2008), *Soil Liquefaction during Earthquakes*. Monograph MNO-12, Oakland, CA: Earthquake Engineering Research Institute.
Jamiolkowski, M., Ladd, C.C., Germain, J.T. and Lancellotta, R. (1985), New developments in field and laboratory testing of soils. In *Proceedings of the 11th International Conference on Soil Mechanics and Foundation Engineering*, AA Balkema, Rotterdam, Netherlands, Vol. 1, pp. 57–153.
Kulhawy, F.H. and Mayne, P.W. (1990), *Manual on Estimating Soil Properties for Foundation Design*, Report EL-6800, Palo Alto: Electric Power Research Institute.
Ladd, C.C. and Edgers, L. (1972), *Consolidated-Undrained Direct-simple Shear Tests on Saturated Clays*. MIT Research Report R72-82, Cambridge, MA: Massachusetts Institute of Technology.
Ladd, C.C., Foott, R., Ishihara, K., Schlosser, F. and Poulos, H.G. (1977), Stress-deformation and strength characteristics. In *Proceedings of 9th International Conference on Soil Mechanics and Foundation Engineering*, Japanese Society of Soil Mechanics and Foundation Engineering, Tokyo, pp. 421–494.
Liao, S.S. and Whitman, R.V. (1986), Overburden correction factors for SPT in sand. *Journal of Geotechnical Engineering, ASCE*, Vol. 112, No. 3, pp. 373–377.
Locat, J. and Demers, D. (1988), Viscosity, yield stress, remolded strength, and liquidity index relationships for sensitive clays. *Canadian Geotechnical Journal*, Vol. 25, No. 4, pp. 799–806.
Marchetti, S. (1980), In-situ tests for flat dilatometer. *Journal of the Geotechnical Engineering Division, ASCE*, Vol. 106, No. GT3, pp. 299–321.
Mayne, P.W. (2007), *NCHRP Synthesis 368: Cone Penetration Test*. Transportation Research Board, Washington, DC: National Academies Press.
Mayne, P.W., Christopher, B.R. and DeJong, J. (2001), *Manual on Subsurface Investigations*. National Highway Institute Publication No. FHWA NHI-01-031, Federal Highway Administration, Washington, DC.
Mayne, P.W. and Kulhawy, F.H. (1982), K0-OCR relationships in soil. *Journal of the Geotechnical Engineering Division, ASCE*, Vol. 108, No. GT6, pp. 851–872.

Mesri, G. and Huvaj, N. (2007), Shear strength mobilized in undrained failure of soft clay and silt deposits. In *Advances in Measurement and Modeling of Soil Behaviour (GSP 173)*, Ed., D.J. DeGroot et al., Reston, VA: ASCE, pp. 1–22.

Meyerhof, G.G. (1957), The Ultimate Bearing Capacity of Foundations on Slopes. In *Proc., 4th International Conference on Soil Mechanics and Foundation Engineering*, Butterworths, London.

Munfakh, G., Arman, A., Collin, J.G., Hung, J.C.-J. and Brouillette, R.P. (2001), *Shallow Foundations Reference Manual*, FHWA-NHI-01-023. Federal Highway Administration, U.S. Department of Transportation, Washington, DC.

NAVFAC (1982), *Soil Mechanics (DM 7.1)*, Naval Facilities Engineering Command, Alexandra.

Ohya, S., Imai, T. and Matsubara, M. (1982), Relationships between N value by SPT and LLT pressuremeter results. In *Proceedings, 2nd European Symposium on Penetration Testing*, CRC Press, Amsterdam, Vol. 1, pp. 125–130.

Phoon, K.K. and Kulhawy, F.H. (1999), Evaluation of geotechnical property variability. *Canadian Geotechnical Journal*, Vol. 36, No. 4, pp. 625–639.

Poulos, H.G. (1975), Settlement of isolated foundations. In *Soil Mechanics—Recent Developments*, Eds., Valliappan et al., Australia, NSW: Unisearch, 181–212.

Railway Technical Research Institute (RTRI). (2012), *Design Standards for Railway Structures and Commentary (Foundation Structures)*. (in Japanese). Tokyo: Railway Technical Research Institute.

Skempton, A.W. (1986), Standard penetration test procedures and the effect in sands of overburden pressure, relative density, particle size, aging, and overconsolidation. *Geotechnique*, Vol. 36, No. 3, pp. 425–447.

Terzaghi, K. and Peck, R.B. (1967), *Soil Mechanics in Engineering Practice*, 2nd edition, New York: John Wiley and Sons.

Chapter 3

Shallow foundations

3.1 Introduction

All structures, such as buildings and bridges, need foundations to transfer structure loads to soil. According to the embedment depth of foundations, structure foundations are classified into two main categories: shallow and deep. Shallow foundations transfer structure loads to the near-surface soils, as shown in Figure 3.1. They are the simplest and oldest foundation type because of their easy construction and low construction cost.

In general, shallow foundations are situated on sturdy ground. However, for sites with soft soil, because of the low stiffness and strength of the soil, shallow foundations under loading may undergo unfavorable settlement and even cause ground failure. Under these conditions, some approaches, such as enlargement of the foundation base, inclusion of a basement, and ground improvement, can be used to improve the performance of shallow foundations. Alternatively, deep foundations may be more appropriate to transfer loading to deep stronger soil.

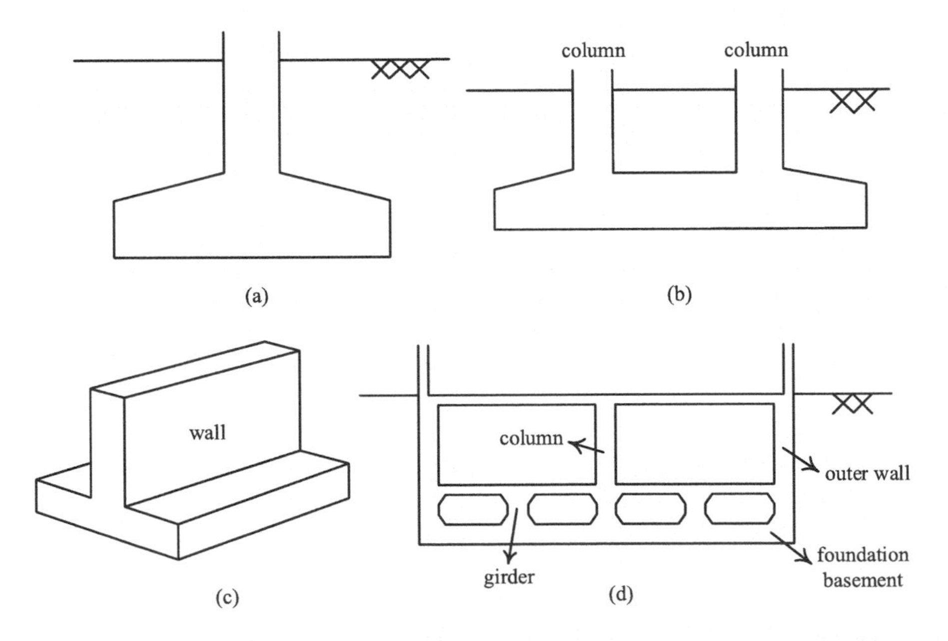

Figure 3.1 Types of shallow foundations: (a) isolated footing, (b) combined footing, (c) continuous footing, and (d) mat foundation.

DOI: 10.1201/9781003350019-3

To ensure that a structure achieves its anticipated performance, foundations should be well designed and constructed. In this chapter, we will introduce the design principles and basic design methods of shallow foundations.

3.2 Types of shallow foundations

Shallow foundations include spread footings and mat foundations, as shown in Figure 3.1.

A spread footing is an enlargement at the bottom of a column or bearing wall that is used to spread structure loads over a large soil area. Spread footings are often used in small- to medium-sized structures on sites with moderate to good soil conditions. A footing that supports a single column is called an isolated footing (Figure 3.1a). In some cases, when columns are close, a combined footing (Figure 3.1b) may be used to support two or more columns. When a footing supports a load-bearing wall, it is referred to as a continuous footing (Figure 3.1c).

For very weak and highly compressible soils, it is more appropriate and economical to use mat foundations (also called raft foundations) (Figure 3.1d). A mat foundation is a very large spread footing that supports an entire structure. The large size of the footing base can help reduce the contact pressure between the bottom of the footing and the soil below, and the rigidity of the footing slab can limit excessive differential settlement among columns.

3.3 Components of shallow foundation design

Designing a shallow foundation involves the following four main items:

1. Evaluation of vertical bearing capacity
2. Evaluation of eccentricity
3. Evaluation of settlement
4. Structural design of foundation components

The first two items concern the capacity issue to avoid bearing failure. Generally, a foundation is designed for centric loading. However, in many cases, the foundation may also be subjected to eccentric loading, such as slabs of retaining walls and footings under moment loading. The eccentric loading will cause nonuniform contact pressure and reduce the bearing capacity. The third item concerns the foundation settlement, which is essential for the functionality of structures. Finally, the fourth item is to perform a detailed structural design of foundation components, including foundation slabs and slab–column connections, to ensure sufficient strength for transmitting structure loads to the ground.

In the following sections, we will introduce the details of items 1–3. Item 4 is not addressed since it concerns reinforced concrete design.

3.4 Vertical bearing capacity

3.4.1 *Ultimate bearing capacity*

The bearing pressure q_b is the contact pressure between the foundation and soil, which is defined as:

$$q_b = \frac{Q}{A} \tag{3.1}$$

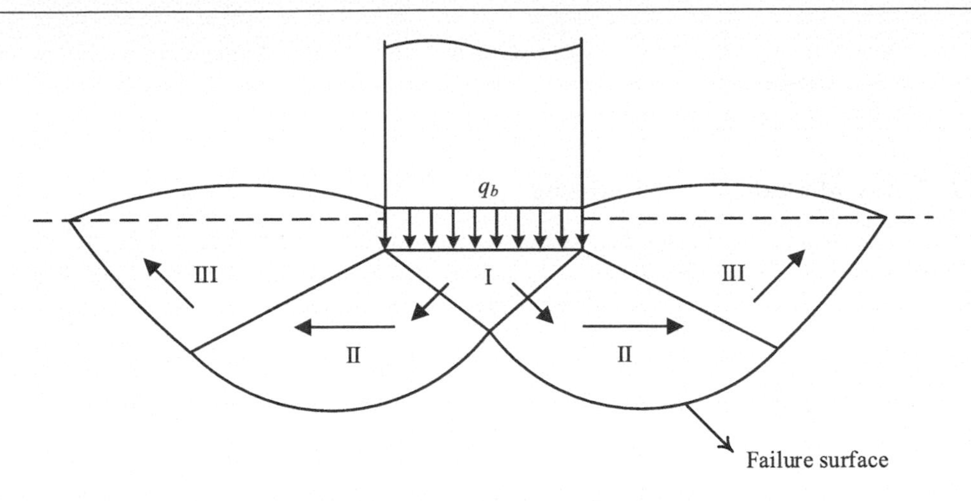

Figure 3.2 Failure mechanism of a footing under vertical loading

where Q is the load carried by the footing and A is the bottom area of the footing.

As the load increases, the bearing pressure of the footing may reach its ultimate capacity, called the ultimate bearing capacity q_u, leading to bearing failure. Shear failure occurs within the surrounding soil, as shown in Figure 3.2, generally accompanied by excessive foundation settlement. Therefore, to prevent shear failure, a footing should be well designed to prevent the bearing pressure from reaching its ultimate bearing capacity.

3.4.2 Bearing failure modes

A complete shear failure for a footing normally has three failure zones, as shown in Figure 3.2. As the bearing pressure increases, zones I to III gradually develop. Zone I is initiated first, and the soil in this zone is displaced downward. Then, zone II is formed, and the soil in this zone is pushed laterally. Eventually, zone III is generated and pushed upward. Depending on the development of the three failure zones, there are three main bearing failure modes: general shear failure, local shear failure, and punching shear failure, as shown in Figure 3.3. The bearing failure mode is dependent on the soil type and density and the embedment depth.

General shear failure, as shown in Figure 3.3a, occurs in strong soil, such as dense sand or stiff clay. The aforementioned three failure zones are fully mobilized. The failure surface is well-defined and global. Failure occurs suddenly (brittle failure) when the bearing pressure reaches a peak, called the ultimate bearing capacity in the bearing pressure–settlement plot, as displayed in Figure 3.4. The settlement for the ultimate bearing capacity is approximately 4–10% of the width of the footing. For this type of failure, a clear bulge appears on the ground surface.

Local shear failure, as shown in Figure 3.3b, occurs in medium strong soil, such as medium dense sand. The failure surface in the soil will gradually extend outward from the foundation. Unlike general shear failure, only zones I and II (solid lines) are generated. The failure surface is well-defined only below the footing but is not obvious near the ground surface. The bearing pressure–settlement plot is gentler than that under general bearing failure, as displayed in Figure 3.4. A sudden failure does not occur (ductile failure) without producing a peak in the bearing pressure–settlement plot. When zones I and II are mobilized, the bearing pressure reaches q_y,

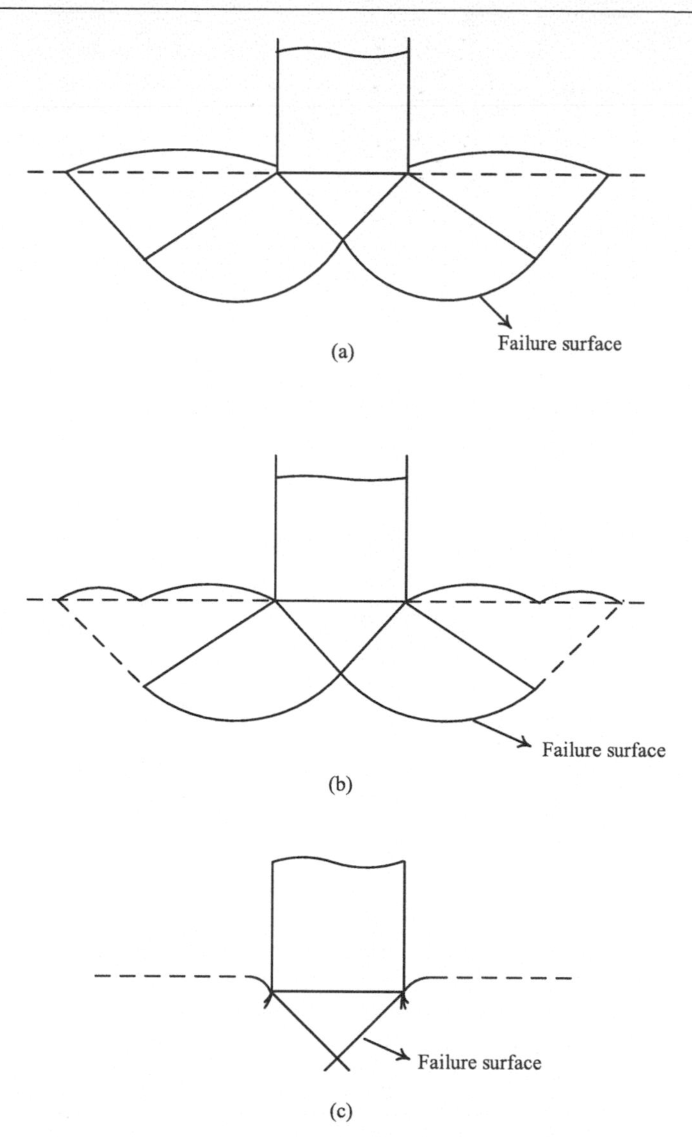

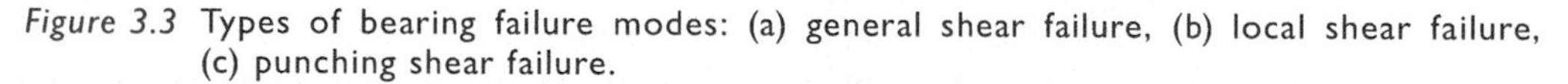

Figure 3.3 Types of bearing failure modes: (a) general shear failure, (b) local shear failure, (c) punching shear failure.

Source: Adapted from Vesic (1963).

generally referred to as the yield bearing capacity. Beyond this point, settlement significantly increases as the bearing pressure increases. Considerable movement is required for the failure surface in the soil to extend to the ground surface (dashed lines, zone III). The bearing pressure at which this development happens is referred to as the ultimate bearing capacity q_u. The load–settlement plot after this point is steep and linear. The settlement for the ultimate bearing capacity is approximately 12–22% of the width of the footing.

The punching shear failure, as shown in Figure 3.3c, occurs in loose sand or soft clay. The shear failure develops downward gradually with increasing settlement, with little or no bulging

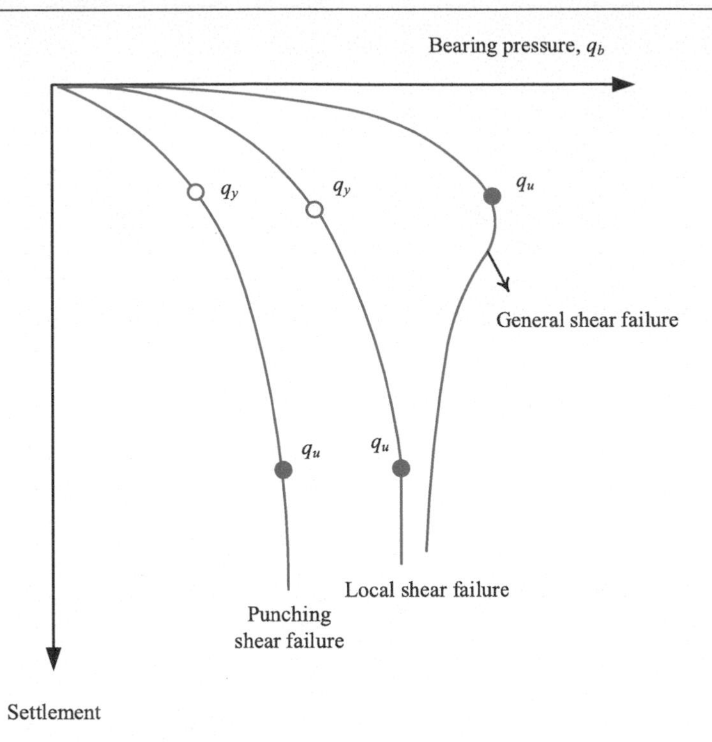

Figure 3.4 Bearing pressure–settlement plots for the three bearing failure modes.

at the ground surface. The failure surface in the soil does not extend to the ground surface (only zone I and/or zone II can be mobilized). The bearing pressure–settlement plot is similar to that under local shear failure (no peak is observed); however, it is softer, as displayed in Figure 3.4. The bearing pressure reaches q_y as zone I or a part of zone II is mobilized. Beyond the yield point, settlement significantly increases as the bearing pressure increases. Generally, when the settlement reaches 15–25%, the width of the footing under bearing pressure reaches its ultimate value. Beyond the ultimate bearing capacity q_u, the pressure–settlement plot is steep and linear. The difference between q_y and q_u is small since zone III (even zone II) is not mobilized.

For footings on sand, Vesic (1963) found from model tests that the failure modes are dependent on the density of sand and the embedment depth of the footing, as shown in Figure 3.5. In this figure, B indicates the footing width, and D_f indicates the embedment depth. For footings with shallow embedment, general shear failure occurs on dense sand, local shear failure occurs on medium dense sand, and punching shear failure occurs on loose sand. With increasing embedment depth, global shear failure is unlikely to occur. For embedment depths greater than five times the width of the footing, punching shear failure always occurs.

3.4.3 *Terzaghi's bearing capacity theory*

Terzaghi (1943) developed a theory to estimate the ultimate bearing capacity of footings. Terzaghi defined foundations with embedment depths less than or equal to the foundation width as shallow foundations. The theory considers a rough continuous or strip foundation (width-to-length ratio approaches 0) and the general failure mode. The following failure zones developed below and around the footing are assumed, as shown in Figure 3.6.

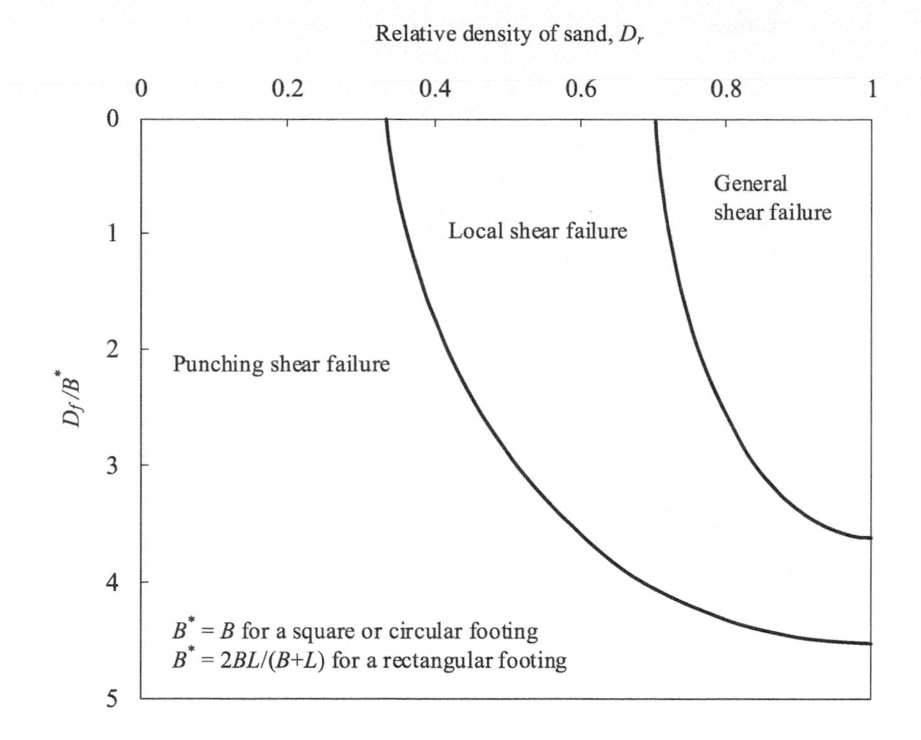

Figure 3.5 Types of shear failure for various embedment depths and relative densities.
Source: Vesic (1973).

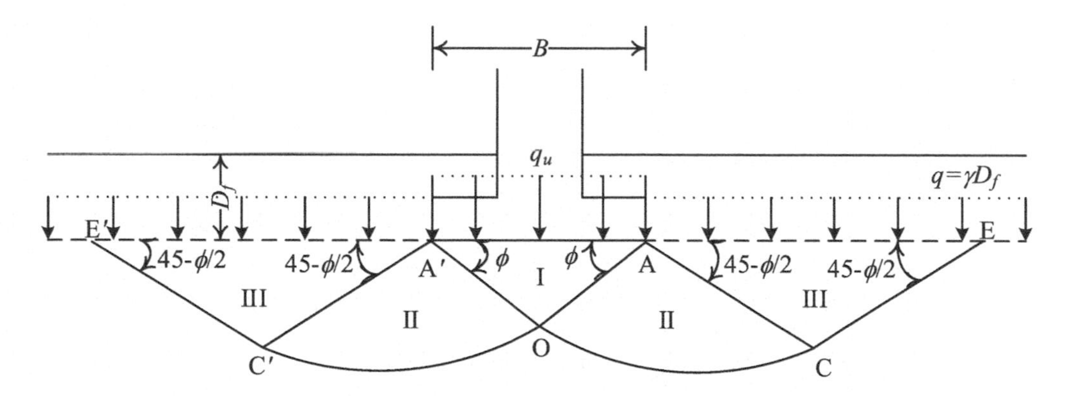

Figure 3.6 Shear failure mechanism of Terzaghi's bearing capacity theory.

Zone I: The triangular (elastic) zone A′OA immediately under the foundation. The angles A′AO and AA′O are assumed to be equal to the soil friction angle ϕ. Zone II: The radial shear zones AOC and A′OC′, with the curves OC and OC′, respectively, being arcs of logarithmic spirals. Zone III: Two triangular Rankine passive zones ACE and A′C′E′.

The shear resistance offered by the soil and footing sides within depth D_f is considered not significant and is ignored. The overburden effect of the soil above the bottom of the foundation is replaced by an equivalent surcharge $q - \gamma D_f$ (γ = unit weight of soil).

The ultimate bearing capacity of the foundation can be deduced by considering the equilibrium of the triangular wedge A′OA. Consider that AO and A′O are two walls that push the soil wedges AOCE and A′OC′E′, respectively, to cause passive failure. Passive force P_p is inclined at an angle ϕ to the perpendicular drawn to the wedge faces (AO and A′O), assuming the angle of wall friction equals the angle of friction of the soil. If the bearing pressure is applied to the foundation to cause general shear failure, the passive force will act on each face of the soil wedge. According to the free body diagram of zone I shown in Figure 3.7, considering the equilibrium of vertical forces:

$$q_u \cdot 2b \cdot 1 = -W + 2C\sin\phi + 2P_p \tag{3.2}$$

where q_u is the ultimate bearing capacity, $b = B/2$, W is the weight of the wedge A′OA carried by the footing, and C is the cohesion on the faces of the wedge A′OA.

The passive force P_p is the sum of three components: $P_{p\gamma}$ produced by the weight of the shear zone AOCE, P_{pc} produced by the soil cohesion, and P_{pq} produced by the surcharge, as expressed in eq. 3.3.

$$P_p = P_{pq} + P_{pc} + P_{p\gamma} \tag{3.3}$$

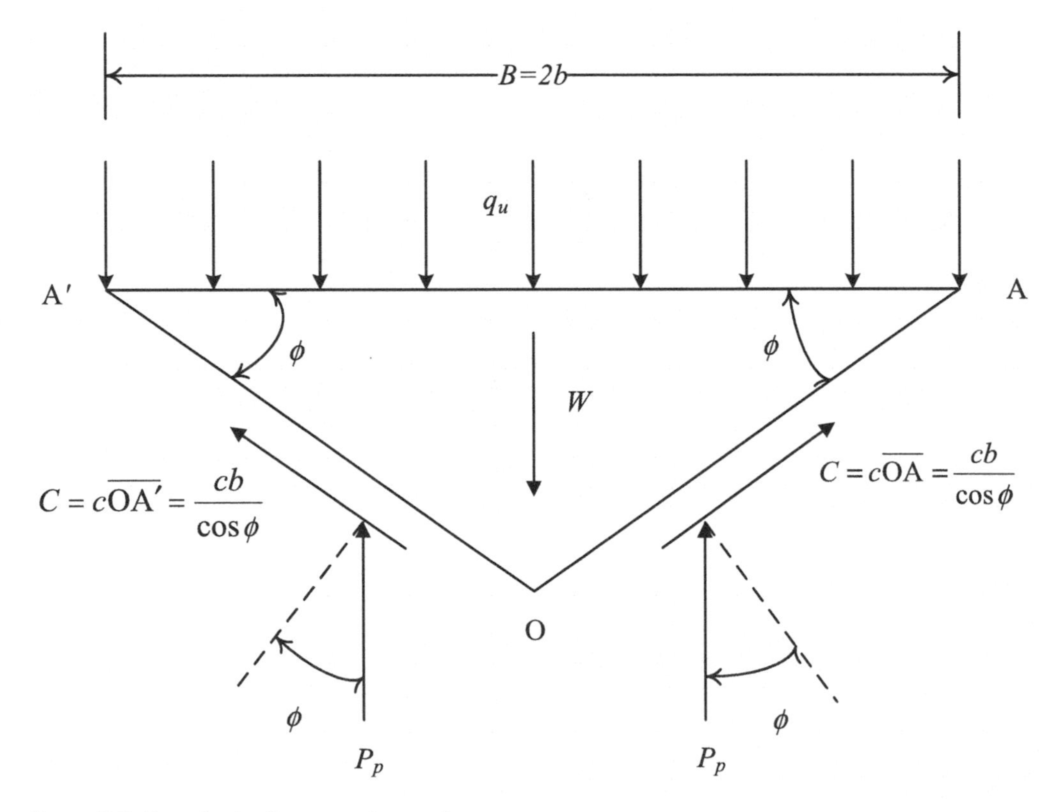

Figure 3.7 Free body diagram of zone I.

Combining eqs. 3.2 and 3.3 gives:

$$q_u = \frac{P_{pq}}{b} + \left(\frac{P_{pc}}{b} + c\tan\phi\right) + \left(\frac{P_{p\gamma}}{b} - \frac{\gamma b}{2}\tan\phi\right) \quad (3.4)$$

The three terms of eq. 3.4 can be further expressed in the form:

$$q_u = qN_q + cN_c + \frac{1}{2}\gamma BN_\gamma \quad (3.5)$$

where N_q, N_c, and N_γ are the bearing capacity factors.

Comparing eqs. 3.4 and 3.5, N_q, N_c, and N_γ are the bearing capacity factors representing the contributions of surcharge, cohesion, and soil weight to the ultimate bearing capacity, respectively. Once P_{pq}, P_{pc}, *and* $P_{p\gamma}$ are given, the factors N_q, N_c, and N_γ can be determined accordingly. Considering the force equilibrium of zones 2 and 3, P_{pq}, P_{pc}, and $P_{p\gamma}$ are derived individually by ignoring the contributions of the other two components. Therefore, N_q, N_c, and N_γ are determined as functions of ϕ, as shown in Table. 3.1 and Figure 3.8, in which N_q and N_c have analytical forms as follows:

$$N_q = \frac{e^{2(3\pi/4-\phi/2)\tan\phi}}{2\cos^2(45+\phi/2)} \quad (3.6)$$

$$N_c = \cot\phi\left(\frac{e^{2(3\pi/4-\phi/2)\tan\phi}}{2\cos^2(45+\phi/2)} - 1\right) = \cot\phi(N_q - 1) \quad (3.7)$$

As the angle of friction influences the size of the failure zones, a larger value of the angle of friction induces a larger failure surface and therefore results in larger bearing capacity factors.

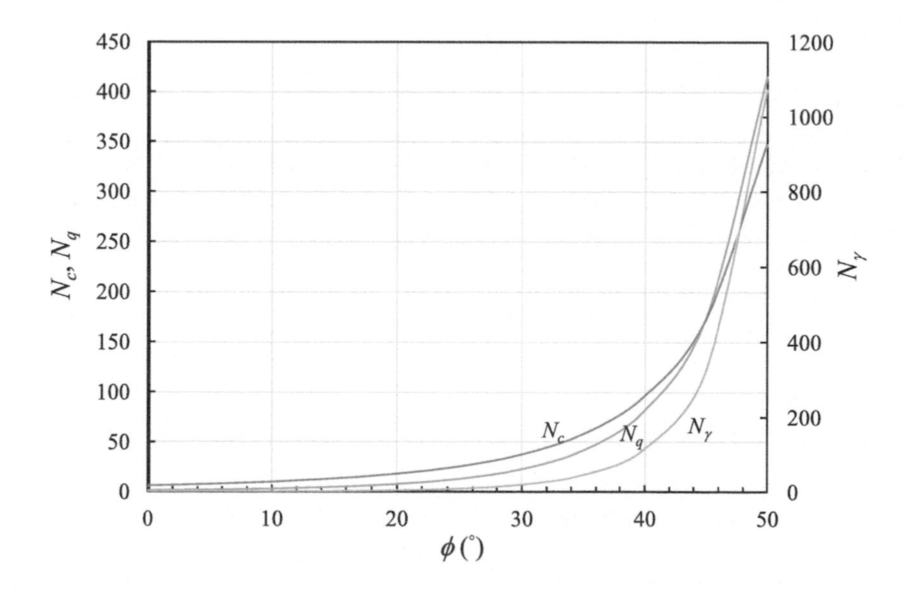

Figure 3.8 Terzaghi's bearing capacity factors.

Table 3.1 Values of Terzaghi's bearing capacity factors

ϕ	N_c	N_q	N_γ	ϕ	N_c	N_q	N_γ
0	5.70	1.00	0.00	26	27.09	14.21	9.84
1	6.00	1.10	0.01	27	29.24	15.90	11.60
2	6.30	1.22	0.04	28	31.61	17.81	13.70
3	6.62	1.35	0.06	29	34.24	19.98	16.18
4	6.97	1.49	0.10	30	37.16	22.46	19.13
5	7.34	1.64	0.14	31	40.41	25.28	22.65
6	7.73	1.81	0.20	32	44.04	28.52	26.87
7	8.15	2.00	0.27	33	48.09	32.23	31.94
8	8.60	2.21	0.35	34	52.64	36.50	38.04
9	9.09	2.44	0.44	35	57.75	41.44	45.41
10	9.60	2.69	0.56	36	63.53	47.16	54.36
11	10.16	2.98	0.69	37	70.07	53.80	65.27
12	10.76	3.29	0.85	38	77.50	61.55	78.61
13	11.41	3.63	1.04	39	85.97	70.61	95.03
14	12.11	4.02	1.26	40	95.66	81.27	115.31
15	12.86	4.45	1.52	41	106.81	93.85	140.51
16	13.68	4.92	1.82	42	119.67	108.75	171.99
17	14.56	5.45	2.18	43	134.58	126.50	211.56
18	15.52	6.04	2.59	44	151.95	147.74	261.60
19	16.56	6.70	3.07	45	172.29	173.29	325.34
20	17.69	7.44	3.64	46	196.22	204.19	407.11
21	18.92	8.26	4.31	47	224.55	241.80	512.84
22	20.27	9.19	5.09	48	258.29	287.85	650.67
23	21.75	10.23	6.00	49	298.72	344.64	831.99
24	23.36	11.40	7.08	50	347.51	415.15	1072.80
25	25.13	12.72	8.34				

*From Kumbhojkar (1993).

In addition, when the angle of friction becomes larger than 30°, there is a rapid increase in the values of the factors.

The preceding bearing capacity factors are derived in general form for a c-ϕ soil. For the drained, long-term condition, effective stress analysis is applied using $c = c'$, $\phi = \phi'$, $\gamma = \gamma_{moist}$ (above groundwater table) or γ' (below groundwater table) in eq. 3.5. However, for saturated clay under the undrained condition (short-term condition), total stress analysis based on the "$\phi = 0$ concept" is applied, and the values $c = c_u$, $\phi = \phi_u = 0$, and $\gamma = \gamma_{sat}$ are used in eq. 3.5. For this condition, $N_q = 1$, $N_c = 5.7$, and $N_\gamma = 0$.

Eq. 3.5 is used for strip foundations. For square and circular foundations, empirical modifications are made from model tests to consider the effect of footing shape as follows:

For square foundations,

$$q_u = 1.3cN_c + qN_q + 0.4\gamma BN_\gamma \tag{3.8}$$

For circular foundations with diameter D:

$$q_u = 1.3cN_c + qN_q + 0.3\gamma DN_\gamma \tag{3.9}$$

Example 3.1

A square footing is 2 m × 2 m in plan view, situated on a dry sandy stratum with an embedment depth of 1 m. The angle of friction of sand is 32°. The unit weight of sand is 18 kN/m³. Apply Terzaghi's bearing capacity equation to estimate the ultimate bearing capacity of the footing.

Solution

Eq. 3.8 is used.

For this sandy layer:

$$c = c' = 0$$
$$\phi = \phi' = 32°, N_c = 44.04, N_q = 28.52, N_\gamma = 26.87 \text{ (see Table 3.1)}$$
$$q_u = (1.3)(0)(44.04) + (1)(18)(28.52) + (0.4)(18)(2)(26.87) = 900.29 \text{ kN/m}^2$$

Example 3.2

A square footing is 2 m × 2 m in plan view, situated on a saturated clay stratum with an embedment depth of 1 m. The groundwater table is located at the ground surface. The saturated unit weight of the soil is 18 kN/m³. The effective angle of friction and cohesion of the soil are 25° and 20 kN/m², respectively. The undrained shear strength of the clay is 100 kN/m². Apply Terzaghi's bearing capacity equation to estimate the ultimate bearing capacity of the footing for long-term and short-term conditions.

Solution

Eq. 3.8 is used.

For the long-term condition, effective analysis is conducted.

$$c = c' = 20 \text{ kN/m}^2, \phi = \phi' = 25°, \gamma = \gamma' = 18\text{-}9.81 = 8.19 \text{ kN/m}^3$$
$$N_c = 25.13, N_q = 12.72, N_\gamma = 8.34 \text{ (see Table 3.1)}$$
$$q_{u\ (\text{long-term})}$$
$$= (1.3)(20)(25.13) + (1)(8.19)(12.72) + (0.4)(8.19)(2)(8.34) = 812.2 \text{ kN/m}^2$$

For the short-term condition, total stress analysis is conducted under the $\phi = 0$ condition.

$$c = c_u = 100 \text{ kN/m}^2, \phi = 0°, \gamma = \gamma_{sat} = 18 \text{ kN/m}^3$$
$$N_c = 5.7, N_q = 1, N_\gamma = 0$$
$$q_{u\ (\text{short-term})}$$
$$= (1.3)(100)(5.7) + (1)(18)(1) + (0.4)(18)(2)(0) = 759 \text{ kN/m}^2$$

Because $q_{u\ (\text{short-term})} < q_{u\ (\text{long-term})}$, the short-term bearing capacity governs the design.

3.4.4 General bearing capacity equation

There are three major limitations when using Terzaghi's bearing capacity equation mentioned in Section 3.4.3. Terzaghi's equation does not address cases of rectangular footings of $B/L \neq 1$. It does not consider the shear resistance of the soil above the foundation base. The load on the foundation is vertical without any inclination. To improve the aforementioned limitations, Meyerhof (1963) proposed a general bearing capacity equation by introducing shape, depth, and load inclination factors.

$$q_u = cN_c F_{cs} F_{cd} F_{ci} + qN_q F_{qs} F_{qd} F_{qi} + \frac{1}{2}\gamma B N_\gamma F_{\gamma s} F_{\gamma d} F_{\gamma i} \qquad (3.10)$$

where:

c = cohesion.
q = overburden pressure at the bottom of the foundation.
γ = unit weight of soil.
B = width of foundation (= diameter for a circular footing).
$F_{cs}, F_{qs}, F_{\gamma s}$ = shape factors.
$F_{cd}, F_{qd}, F_{\gamma d}$ = depth factors.
$F_{ci}, F_{qi}, F_{\gamma i}$ = load inclination factors.
N_q, N_c, N_γ = bearing capacity factors.

The shape factors F_{qs}, F_{cs}, and $F_{\gamma s}$ are for rectangular footings of $B/L \neq 1$ and square and circular footings ($B/L = 1$). The depth factors F_{qd}, F_{cd}, and $F_{\gamma d}$ reflect the shear resistance above the base of the embedded footings. The load inclination factors F_{qi}, F_{ci}, and $F_{\gamma i}$ are for inclined loading on the footing. These factors are listed in Table 3.2.

Table 3.2 Meyerhof's shape, depth, and inclination factors

Shape factors
For $\phi = 0$
$F_{cs} = 1 + 0.2(B/L)$
$F_{qs} = F_{\gamma s} = 1$
For $\phi \geq 10°$
$F_{cs} = 1 + 0.2(B/L)\tan^2(45 + \phi/2)$
$F_{qs} = F_{\gamma s} = 1 + 0.1(B/L)\tan^2(45 + \phi/2)$
Depth factors
For $\phi = 0$
$F_{cd} = 1 + 0.2(D_f/B)$
$F_{qd} = F_{\gamma d} = 1$
For $\phi \geq 10°$
$F_{cd} = 1 + 0.2(D_f/B)\tan(45 + \phi/2)$
$F_{qd} = F_{\gamma d} = 1 + 0.1(D_f/B)\tan(45 + \phi/2)$ D_f: embedment depth of footing
Inclination factors
$F_{ci} = F_{qi} = 1 - (\beta/90°)^2$
$F_{\gamma i} = 1 - (\beta/\phi)^2$ β: inclination of the load on the footing with respect to the vertical

Source: Meyerhof (1963).

Different from Terzaghi's assumption, the angle of triangular zones A′AO and AA′O (zone I) of the failure surface with respect to the ground surface is assumed to be 45 + ϕ/2, as shown in Figure 3.9, considering that the footing base is smooth. This zone is usually referred to as the Rankine active zone. Upon this modified failure surface, another set of bearing capacity factors is derived as follows, with the values presented in Table 3.3.

$$N_q = \tan^2(45 + \phi/2)e^{\pi \tan\phi} \tag{3.11}$$

$$N_c = \cot\phi(N_q - 1) \tag{3.12}$$

$$N_\gamma = \tan 1.4\phi(N_q - 1) \tag{3.13}$$

Similar to Terzaghi's equation, Meyerhof's equation is presented in general form for a c-ϕ soil. For the drained long-term condition, effective stress analysis is applied using the values

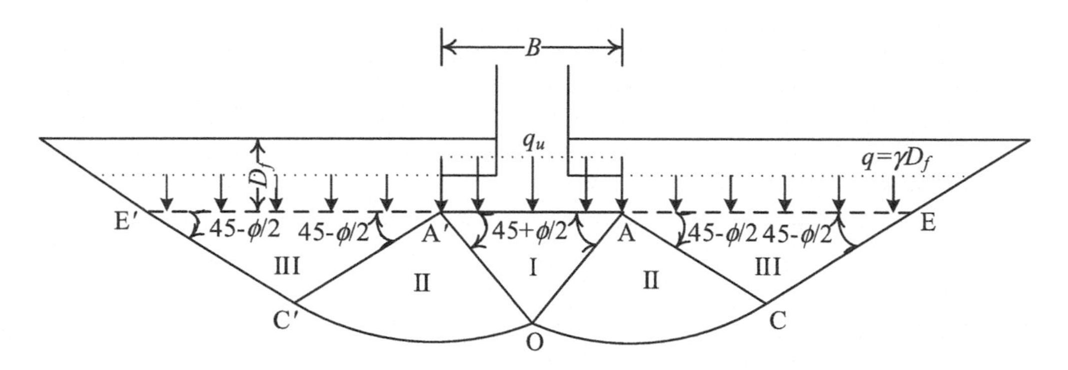

Figure 3.9 Shear failure mechanism for Meyerhof's bearing capacity equation.

Table 3.3 Values of Meyerhof's bearing capacity factors

ϕ	N_c	N_q	N_γ	ϕ	N_c	N_q	N_γ
0	5.14	1.00	0.00	31	32.67	20.63	18.56
1	5.38	1.09	0.00	32	35.49	23.18	22.02
2	5.63	1.20	0.01	33	38.64	26.09	26.17
3	5.90	1.31	0.02	34	42.16	29.44	31.15
4	6.19	1.43	0.04	35	46.12	33.30	37.15
5	6.49	1.57	0.07	36	50.59	37.75	44.43
6	6.81	1.72	0.11	37	55.63	42.92	53.27
7	7.16	1.88	0.15	38	61.35	48.93	64.07
8	7.53	2.06	0.21	39	67.87	55.96	77.33
9	7.92	2.25	0.28	40	75.31	64.20	93.69
10	8.34	2.47	0.37	41	83.86	73.90	113.99
11	8.80	2.71	0.47	42	93.71	85.37	139.32
12	9.28	2.97	0.60	43	105.11	99.01	171.14
13	9.81	3.26	0.74	44	118.37	115.31	211.41
14	10.37	3.59	0.92	45	133.87	134.87	262.74
15	10.98	3.94	1.13	46	152.10	158.50	328.73

ϕ	N_c	N_q	N_γ	ϕ	N_c	N_q	N_γ
16	11.63	4.34	1.37	47	173.64	187.21	414.33
17	12.34	4.77	1.66	48	199.26	222.30	526.45
18	13.10	5.26	2.00	49	229.92	265.50	674.92
19	13.93	5.80	2.40	50	266.88	319.06	873.86
20	14.83	6.40	2.87	51	311.75	385.98	1,143.95
21	15.81	7.07	3.42	52	366.66	470.30	1,516.08
22	16.88	7.82	4.07	53	434.42	577.50	2,037.29
23	18.05	8.66	4.82	54	518.80	715.07	2,781.13
24	19.32	9.60	5.72	55	624.92	893.48	3,865.77
25	20.72	10.66	6.77	56	759.79	1,127.44	5,487.58
26	22.25	11.85	8.00	57	933.17	1,437.96	7,986.27
27	23.94	13.20	9.46	58	1,158.83	1,855.52	11,979.47
28	25.80	14.72	11.19	59	1,456.54	2,425.08	18,664.42
29	27.86	16.44	13.24	60	1,855.10	3,214.14	30,570.95
30	30.14	18.40	15.67				

Source: Meyerhof (1963).

$c = c'$, $\phi = \phi'$, and $\gamma = \gamma_{moist}$ (above groundwater table) or γ' (below groundwater table) in eq. 3.10. However, for saturated clay under the undrained condition (short-term condition), total stress analysis based on the $\phi = 0$ condition is applied, and the values $c = c_u$, $\phi = \phi_u = 0$, and $\gamma = \gamma_{sat}$ are used in eq. 3.10. For this condition, $N_q = 1$, $N_c = 5.14$, and $N_\gamma = 0$.

Example 3.3

A rectangular footing is 2 m × 4 m in plan view, situated on a dry sandy stratum with an embedment depth of 1 m. The angle of friction of sand is 32°. The unit weight of sand is 18 kN/m³. Apply Meyerhof's bearing capacity equation to estimate the ultimate bearing capacity of the footing.

Solution

Eq. 3.10 is used.

For this sandy layer:

$c = c' = 0$
$\phi = \phi' = 32°, N_c = 35.49, N_q = 23.18, N_\gamma = 22.02$
$B = 2\text{ m}, L = 4\text{ m}$
$F_{cs} = 1 + (0.2)(2/4)\tan^2(45 + 32/2) = 1.325$
$F_{qs} = F_{\gamma s} = 1 + (0.1)(2/4)\tan^2(45 + 32/2) = 1.163$
$F_{cd} = 1 + (0.2)(1/2)\tan(45 + 32/2) = 1.180$
$F_{qd} = F_{\gamma d} = 1 + (0.1)(1/2)\tan(45 + 32/2) = 1.090$
$F_{ci} = F_{qi} = 1$
$F_{\gamma i} = 1$
$q_u = (0)(35.49)(1.325)(1.180)(1) + (1)(18)(23.18)(1.163)(1.09)(1) + (1/2)(18)(2)(22.02)(1.163)(1.09)(1) = 1031.38\text{ kN/m}^2$

3.4.5 Factor of safety and allowable bearing capacity

Calculating the allowable load-bearing capacity of shallow foundations requires applying a factor of safety (F_S) to the ultimate bearing capacity. The magnitude of the factor of safety depends on the design load conditions. An $F_S \geq 3$ is usually adopted for the long-term load condition. The following two approaches for the design inequality and associated allowable bearing capacity are commonly adopted in design codes:

1. In terms of the gross ultimate bearing capacity q_u

$$q_b \leq q_{all} \; (q_{all} = q_u/F_S) \tag{3.14}$$

where q_{all} = gross allowable bearing capacity

2. In terms of the net ultimate bearing capacity $q_{u(net)}$

In general, the embedment depth of a footing is a given parameter in design, and therefore, the surcharge q is regarded as a certain value without uncertainty. This approach does not apply an FS to the surcharge. To address the net contribution of ultimate bearing capacity and bearing pressure exceeding the surcharge above the foundation level before the foundation is constructed, the net ultimate bearing capacity $q_{u(net)}$ and net bearing pressure $q_{b(net)}$ are defined as follows.

$$q_{u(net)} = q_u - q \tag{3.15}$$

$$q_{b(net)} = q_b - q \tag{3.16}$$

Therefore, the design inequality is:

$$q_{b(net)} \leq q_{all(net)} \; (q_{all(net)} = q_{u(net)}/F_S) \tag{3.17}$$

where $q_{all(net)}$ = net allowable bearing capacity.

Example 3.4

A rectangular footing is 2 m × 4 m in plan view, situated on a dry sandy stratum with an embedment depth of 1 m. The angle of friction of sand is 32°. The unit weight of sand is 18 kN/m^3.

1. Apply Meyerhof's bearing capacity equation to estimate the net ultimate bearing capacity of the footing.
2. Compute a factor of safety in terms of the net ultimate bearing capacity, given that the footing is subjected to a total centric vertical load of 3,200 kN on its base.

Solution

1. According to example 3.3, q_u = 1031.38 kN/m^2.

   The surcharge $q = \gamma D_f = 18 \times 1 = 18$
   $q_{u(net)} = 1031.38 - 18 = 1013.38$ kN/m^2

2. $q_{b(net)} = 3200/(2\times4) - 18 = 382$ kN/m^2

$F_s = 1013.38/382 = 2.65$

3.4.6 Modification of bearing capacity equations for the groundwater table

This section describes the considerations for the groundwater table when using the bearing capacity equation in effective stress and total stress analyses.

1. Effective stress analysis

The position of the groundwater table may alter the effective stress in the soil and further the ultimate bearing capacity of footings. Different groundwater levels may yield different values of the parameter q in the second term and parameter γ in the third term of the bearing capacity equations.

Figure 3.10 illustrates the three groundwater table conditions. Case 1 is for the groundwater table located above the bottom of the footing (the depth of groundwater table $D_w <$ the embedment depth D_f). Cases 2 and 3 are for the groundwater table below the bottom of the footing. Assuming that the main influence depth of bearing failure of the footing is the same as the width of footing B, case 2 indicates that the groundwater table is within the limit of the influence zone ($D_w < D_f + B$), whereas case 3 indicates that the groundwater table is on or outside the limit of the influence zone ($D_w \geq D_f + B$).

Assume γ_2 is the unit weight of the soil above the bottom of the footing, which is used to compute q in the second term of the bearing capacity equation, that is, $q = \int_0^{D_f} \gamma_2 dz$, and γ_1 is the unit weight of the soil below the bottom of the footing, which is used in the third term of the bearing capacity equation. γ_1 and γ_2 are computed as follows for these three cases.

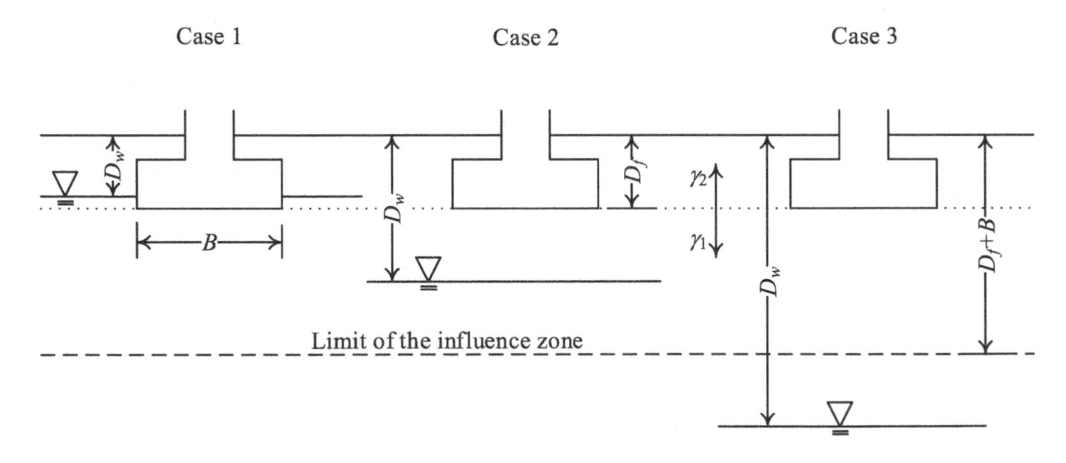

Figure 3.10 Three cases for groundwater table depth.

For case 1, γ_1 uses the submerged unit weight of soil, and γ_2 is dependent on the location of the groundwater table:

$$\gamma_1 = \gamma_1' = \gamma_{1,sat} - \gamma_w$$

$$\gamma_2 = \gamma_{2,moist} \text{ (above groundwater table)}$$

$$\gamma_2 = \gamma_2' = \gamma_{2,sat} - \gamma_w \text{ (below groundwater table)}$$

where γ_1', $\gamma_{1,sat}$ are the effective and saturated unit weights of the soil below the bottom of the footing, respectively, and γ_2', $\gamma_{2,moist}$, $\gamma_{2,sat}$ are the effective, moist, and saturated unit weights of the soil above the bottom of the footing, respectively.

For case 2, γ_2 is not influenced by the location of the groundwater table. However, since the groundwater table is located within the influence zone of bearing failure of the footing, a depth-weighted average γ_1 as what follows is used for partially considering the effect of water.

$$\gamma_1 = \left(\frac{D_w - D_f}{B}\right)\gamma_{1,moist} + (\gamma_{1,sat} - \gamma_w)\left(\frac{B-(D_w - D_f)}{B}\right) = \gamma_1' + \left(\frac{D_w - D_f}{B}\right)(\gamma_{1,moist} - \gamma_1')$$

$$\gamma_2 = \gamma_{2,moist}$$

For case 3, neither q nor γ is influenced by the location of the groundwater table.

$$\gamma_1 = \gamma_{1,moist}$$

$$\gamma_2 = \gamma_{2,moist}$$

where $\gamma_{1,moist}$, $\gamma_{2,moist}$ are the moist unit weight of the soil below and above the bottom of the footing, respectively.

2. Total stress analysis

This analysis is applied for the saturated clay condition and is applicable only to case 1 in Figure 3.10. γ_1 is the saturated unit weight, but its value has no effect on the bearing capacity as $N_\gamma = 0$; γ_2 needs to consider the position of the groundwater table. That is:

$$\gamma_1 = \gamma_{1,sat}$$

$$\gamma_2 \text{ (above water table)} = \gamma_{2,moist}$$

$$\gamma_2 \text{ (below water table)} = \gamma_{2,sat}$$

Example 3.5

A rectangular footing is 2 m × 4 m in plan view, situated on a sandy stratum with an embedment depth of 1 m. The angle of friction of sand is 32°. The groundwater table is located at a depth of 0.5 m. The unit weights of sand above and below the groundwater table are 18 and 20 kN/m^3, respectively. Apply Meyerhof's bearing capacity equation to estimate the ultimate bearing capacity of the footing.

Solution

Eq. 3.10 is used.
For this sandy layer:

$c = c' = 0$
$\phi = \phi' = 32°, N_q = 23.18, N_c = 35.49, N_\gamma = 22.02$
$B = 2$ m, $L = 4$ m
$F_{cs} = 1 + (0.2)(2/4)\tan^2(45 + 32/2) = 1.325$
$F_{qs} = F_{\gamma s} = 1 + (0.1)(2/4)\tan^2(45 + 32/2) = 1.163$
$F_{cd} = 1 + (0.2)(1/2)\tan(45 + 32/2) = 1.180$
$F_{qd} = F_{\gamma d} = 1 + (0.1)(1/2)\tan(45 + 32/2) = 1.09$
$F_{ci} = F_{qi} = 1$
$F_{\gamma i} = 1$
$q = 18 \times 0.5 + (20 - 9.81) \times 0.5 = 14.095$
$\gamma_1 = \gamma_1' = (20 - 9.81) = 10.19$
$q_u = (0)(35.49)(1.325)(1.180)(1) + (14.095)(23.18)(1.163)(1.09)(1) + (1/2)(10.19)(2)$
$(22.02)(1.163)(1.09)(1) = 698.62$ kN/m^2

3.5 Eccentrically loaded foundations

In several instances, as with the base of a retaining wall or structures under lateral loading, foundations are subjected to moments in addition to vertical loading. In this situation, the distribution of bearing pressure beneath the foundation is no longer uniform. As shown in Figure 3.11, the distribution bearing pressure along the footing base varies with different eccentricities e. When $e < B/6$, the footing is fully in contact with the soil, and the shape of the bearing pressure distribution is trapezoidal. The bearing pressure at one edge is larger than that at the other edge. However, when $e = B/6$, the shape of the bearing pressure distribution is triangular, with 0 bearing pressure at one edge. At this time, the footing begins to lift off. When $e > B/6$, the shape of the bearing pressure distribution is also triangular; however, because the soil cannot sustain tension, the footing lifts off without full contact with the soil below. To avoid a large amount of eccentricity from affecting the serviceability and stability of the footing, a common design requires that the maximum eccentricity should not be larger than 1/6 the width of the foundation slab under long-term load conditions and should not be larger than 1/3 the width of the foundation slab under short-term load conditions.

3.5.1 Bearing pressure distributions of eccentrically loaded foundations

According to the equilibria of the vertical force and moment, the bearing pressure distribution can be determined for different eccentricities.

1. When the eccentricity $e < B/6$ (Figure 3.11), the maximum and minimum of the trapezoidal bearing pressure distribution are:

$$q_{b,\max} = \frac{Q}{BL}\left(1 + \frac{6e}{B}\right) \tag{3.18.1}$$

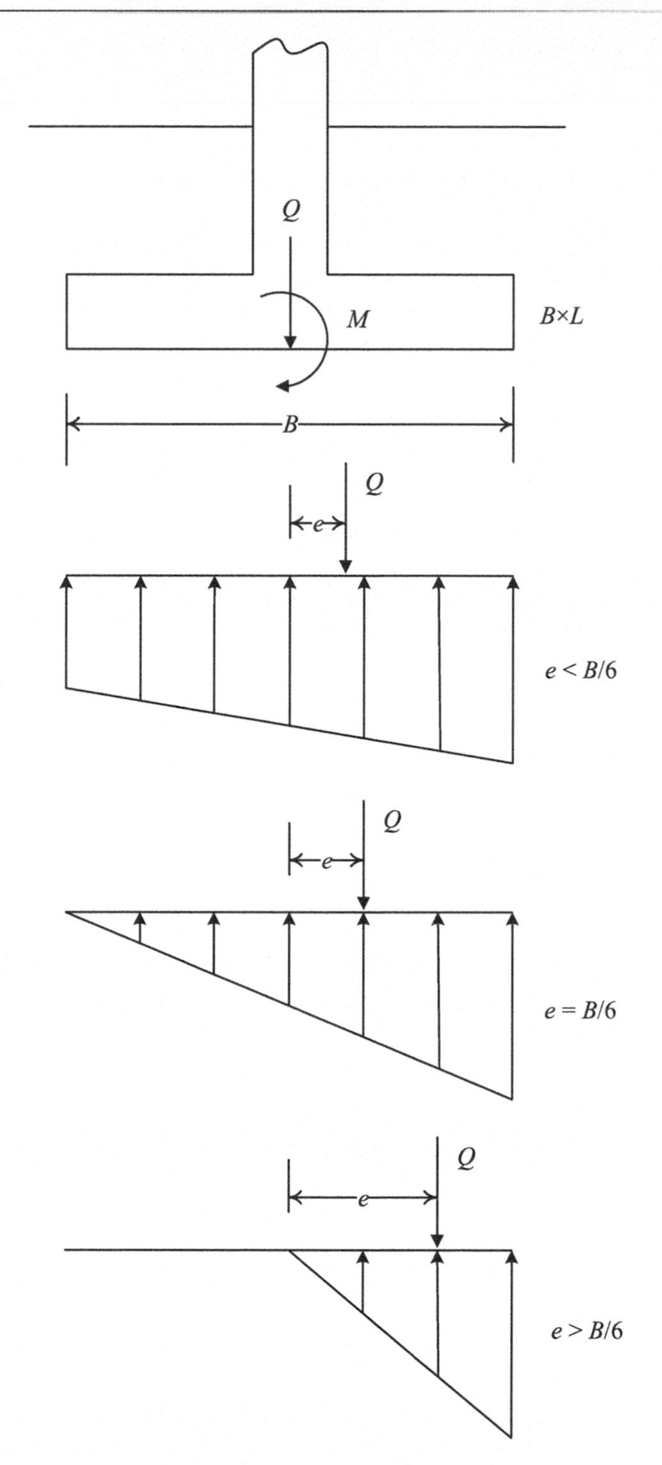

Figure 3.11 Bearing pressure of eccentrically loaded foundations.

$$q_{b,\min} = \frac{Q}{BL}\left(1 - \frac{6e}{B}\right) \tag{3.18.2}$$

The preceding equations are derived based on the equilibria of the vertical force and moment at the footing base considering a trapezoidal distribution of bearing pressure.

2. When the eccentricity $e = B/6$ (Figure 3.11), the footing begins to lift off, and the maximum and minimum of the bearing pressure distribution become:

$$q_{b,\max} = \frac{2Q}{BL} \tag{3.19.1}$$

$$q_{b,\min} = 0 \tag{3.19.2}$$

The preceding equations are derived by substituting $e = B/6$ into eqs. 3.18.1 and 3.18.2. At this state, the trapezoidal distribution of the bearing pressure becomes triangular.

3. When $e > B/6$ (Figure 3.11), because the soil cannot sustain any tension, a separation between the foundation and underlying soil will occur.

$$q_{b,\max} = \frac{4Q}{3L(B-2e)} \tag{3.20.1}$$

$$q_{b,\min} = 0 \tag{3.20.2}$$

The contact area of the footing B_c is reduced to $B_c = 3\left(\frac{B}{2} - e\right)$.

The preceding equations are derived based on the equilibria of the vertical force and moment at the footing base, considering a triangular distribution of bearing pressure within the contact area.

3.5.2 Ultimate bearing capacity of eccentrically loaded foundations

Compared with a centric footing, an eccentric footing has a nonuniform bearing pressure distribution and even a reduced contact area, which may influence the ultimate bearing capacity of the footing. To account for this influence, the effective area concept is applied to compute the ultimate bearing capacity of a footing under eccentric loading. In this approach, the ultimate bearing capacity equation is modified using an effective foundation size as follows.

$$q_u = cN_c F_{cs} F_{cd} F_{ci} + qN_q F_{qs} F_{qd} F_{qi} + \frac{1}{2}\gamma B' N_\gamma F_{\gamma s} F_{\gamma d} F_{\gamma i} \tag{3.21}$$

where B' is the effective width of the footing (short side), and the shape factors F_{qs}, F_{cs}, and $F_{\gamma s}$ are determined based on the effective width B' and length L' of the footing. Note that the determination of F_{qd}, F_{cd}, and $F_{\gamma d}$ uses the original footing width for B, without using B'.

Figure 3.12 demonstrates the determination of the effective size of a footing for one-way and two-way eccentric loading. For one-way eccentric loading, assuming the eccentricity is in the direction of the footing width:

$$B' = B - 2e$$

$$L' = L$$

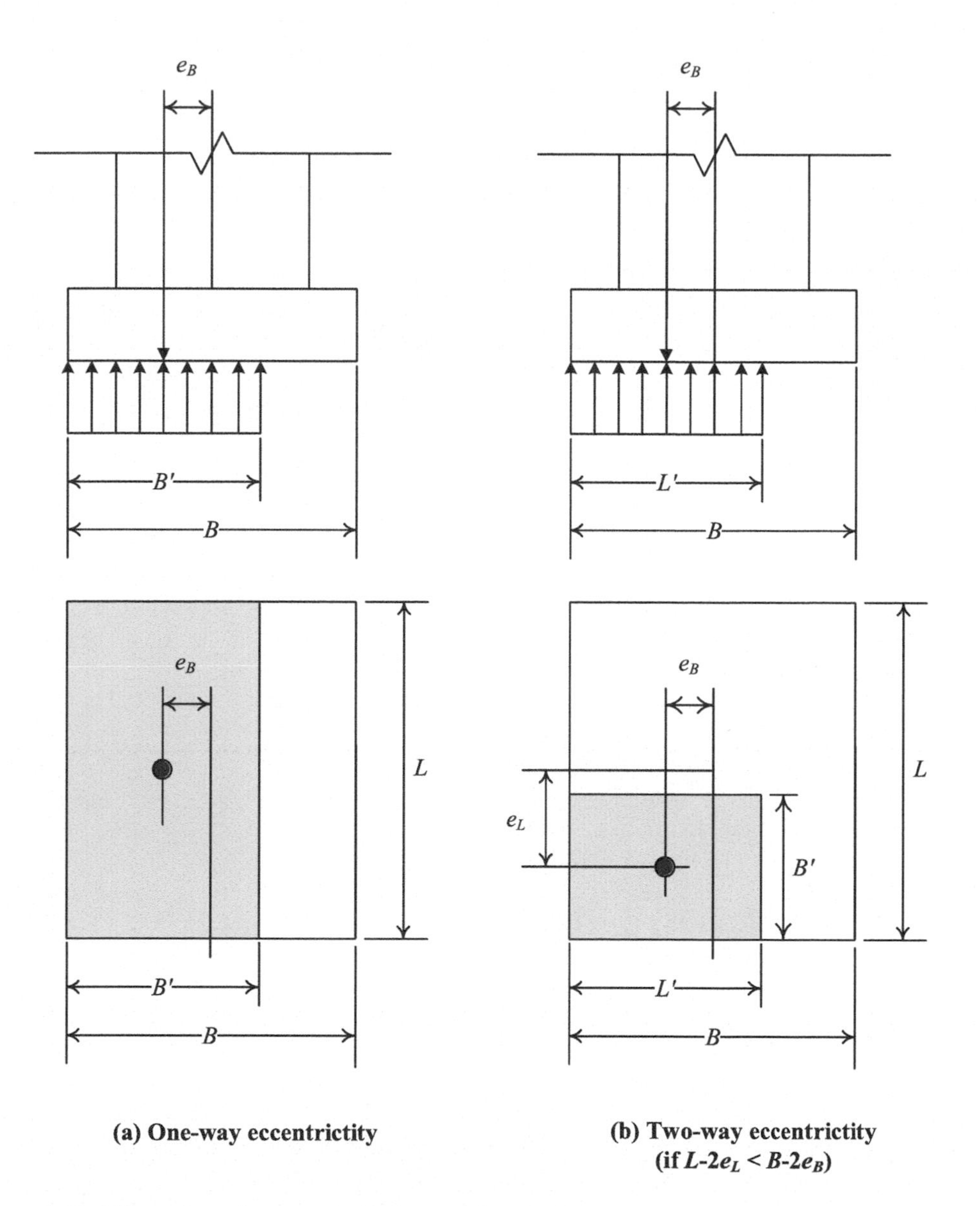

Figure 3.12 Effect size of eccentrically loaded foundations.

If the eccentricity is in the longitudinal direction, B' always refers to the short side, and therefore:

$$B' = \min(B, L-2e)$$
$$L' = \max(B, L-2e)$$

For two-way eccentric loading, eccentricities e_B and e_L occur in the B and L directions, respectively. The effective width and length of the footing are:

$$B' = \min(B-2e_B, L-2e_L)$$
$$L' = \max(B-2e_B, L-2e_L)$$

Once the effective size of the footing is determined, q_u is computed based on eq. 3.21. In the same way, the factor of safety of q_u/q_{max} is checked if its value satisfies the design requirement.

Example 3.6

A rectangular footing is 2 m × 4 m in plan view, situated on a dry sandy stratum with an embedment depth of 1 m. The footing is subjected to a total vertical load of 3,200 kN and a longitudinal moment 640 kN-m (toward the transverse direction) on its base at the centerline. The angle of friction of the sand is 32°. The unit weight of sand is 18 kN/m³. Compute $q_{b,max}/q_{b,\min}$ and the ultimate bearing capacity.

Solution

$B = 2$ m, $L = 4$ m, $Q = 3{,}200$ kN, $M_L = 640$ kN-m, $e_L = 640/3{,}200 = 0.2$ m ($< L/6$ $(= 4/6)$)
$q_{b,max/min} = [(3200)/(2 \times 4)](1 \pm (6)(0.2)/4) = 520 \text{ kN/m}^2/280 \text{ kN/m}^2$
Using eq. 3.21:
$B' = \min(2, 4 - 2 \times 0.2) = 2$ m, $L' = \max(2, 4 - 2 \times 0.2) = 3.6$ m
$c = c' = 0$
$\phi = \phi' = 32°$, $N_q = 23.18$, $N_c = 35.49$, $N_\gamma = 22.02$
$F_{cs} = 1 + (0.2)(2/3.6)\tan^2(45 + 32/2) = 1.362$
$F_{qs} = F_{\gamma s} = 1 + (0.1)(2/3.6)\tan^2(45 + 32/2) = 1.181$
$F_{cd} = 1 + (0.2)(1/2)\tan(45 + 32/2) = 1.180$
$F_{qd} = F_{\gamma d} = 1 + (0.1)(1/2)\tan(45 + 32/2) = 1.09$
$F_{ci} = F_{qi} = F_{\gamma i} = 1$
$q_u = (0)(35.49)(1.362)(1.18)(1) + (1)(18)(23.18)(1.181)(1.09)(1) + (1/2)(18)(2)(22.02)$
$(1.181)(1.09)(1) = 1047.34 \text{ kN/m}^2$

Example 3.7

A rectangular footing is 2 m × 4 m in plan view, situated on a dry sandy stratum with an embedment depth of 1 m. The footing is subjected to a total vertical load of 3,200 kN and a transverse moment 640 kN-m (toward longitudinal direction) on its base at the centerline. The angle of friction of the sand is 32°. The unit weight of sand is 18 kN/m³. Compute $q_{b,max}/q_{b,\min}$ and the ultimate bearing capacity.

Solution

$B = 2$ m, $L = 4$ m, $Q = 3{,}200$ kN, $M_T = 640$ kN-m, $e_B = 640/3{,}200 = 0.2$ m ($< B/6$ $(= 2/6)$)
$q_{b,max/min} = [(3{,}200)/(2 \times 4)](1 \pm (6)(0.2)/2) = 640$ kN/m²/160 kN/m²
Using eq. 3.21:
$B' = \min(2 - 2 \times 0.2, 4) = 1.6$ m, $L' = \max(2 - 2 \times 0.2, 4) = 4$ m
$c = c' = 0$
$\phi = \phi' = 32°$, $N_q = 23.18$, $N_c = 35.49$, $N_\gamma = 22.02$
$F_{cs} = 1 + (0.2)(1.6/4)\tan^2(45 + 32/2) = 1.26$
$F_{qs} = F_{\gamma s} = 1 + (0.1)(1.6/4)\tan^2(45 + 32/2) = 1.13$
$F_{cd} = 1 + (0.2)(1/2)\tan(45 + 32/2) = 1.18$
$F_{qd} = F_{\gamma d} = 1 + (0.1)(1/2)\tan(45 + 32/2) = 1.09$
$F_{ci} = F_{qi} = F_{\gamma i} = 1$
$q_u = (0)(35.49)(1.362)(1.18)(1) + (1)(18)(23.18)(1.13)(1.09)(1) + (1/2)(18)(1.6)(22.02)$
$(1.13)(1.09)(1) = 904.47$ kN/m²

3.6 Foundation settlement

Checking for the bearing capacity of a footing cannot ensure the actual performance of the footing. It is also necessary to check the settlement of the footing upon loading. When the settlements of footings at different locations are different, differential settlement between footings may occur and cause additional stresses in the structure members.

There are two major types of foundation settlement: immediate and delayed settlement. Immediate settlement occurs during or immediately after the construction of the structure. Delayed settlement occurs over time. The total settlement of a foundation is the sum of the immediate settlement and delayed settlement. For footings on sand upon loading, the settlement is an immediate settlement. For footings on saturated clay upon loading, the settlement comprises both an immediate settlement and a delayed settlement. Generally, the delayed settlement is contributed by a consolidation of clay from the dissipation of excess pore water pressure. Consolidation settlement occurs in two phases: primary and secondary. Primary consolidation settlement is more significant than secondary settlement in inorganic clays and silty soils. Secondary consolidation settlement occurs after the primary consolidation caused by the slippage and reorientation of soil particles under a sustained load (σ' is unchanged), and it is significant in organic soils.

The following sections discuss how to compute immediate and delayed settlements.

3.6.1 *Immediate settlement*

The immediate settlement of a shallow foundation can be estimated using the mechanics of elasticity. For example, as shown in Figure 3.13, the settlement at a position immediately beneath the foundation can be computed by integrating the vertical strains induced by stress increments from the net bearing pressure of the foundation along the depth below the position.

$$S_i = \int_0^\infty \varepsilon_z dz = \frac{1}{E_s}\int_0^\infty (\Delta\sigma_z - \mu_s\Delta\sigma_x - \mu_s\Delta\sigma_y) \quad (3.22)$$

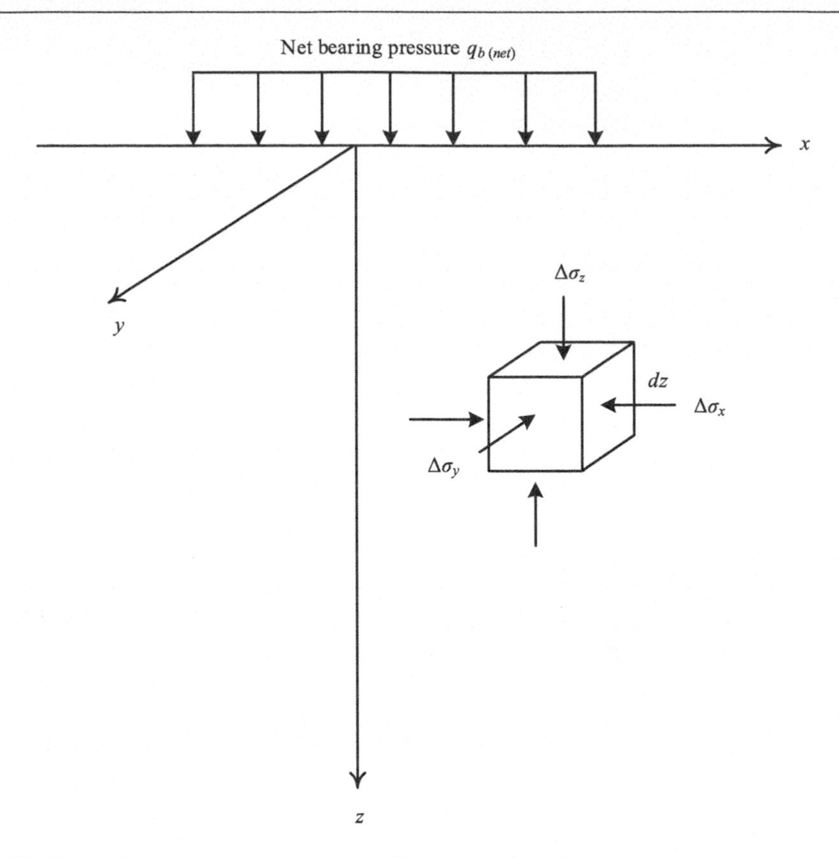

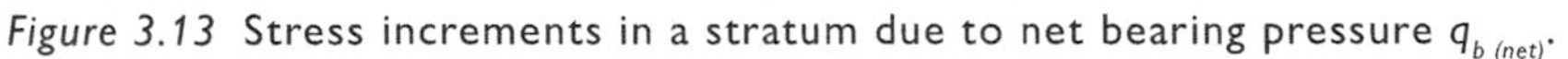
Figure 3.13 Stress increments in a stratum due to net bearing pressure $q_{b\,(net)}$.

where:

S_i = immediate settlement.
E_s = secant modulus of elasticity of the soil (refer to Section 2.3.2).
μ_s = Poisson's ratio of soil (refer to Section 2.3.2).
$\Delta\sigma_x$, $\Delta\sigma_y$, and $\Delta\sigma_z$ = stress increments in the *x, y,* and *z* directions, respectively, due to the net bearing pressure.

Based on this theory, Steinbrenner (1934) proposed the following equation to calculate the settlement of a perfectly flexible foundation under the net bearing pressure $q_{b(net)}$. Referring to Figure 3.14:

$$S_i = q_{b(net)}(\alpha B')\left(\frac{1-\mu_s^2}{E_s}\right)I_s I_f \tag{3.23}$$

Where E_s is the average secant modulus of elasticity of the soil under the foundation, generally measured from z = 0 to z = 5*B*, considering the major influence range of stress.

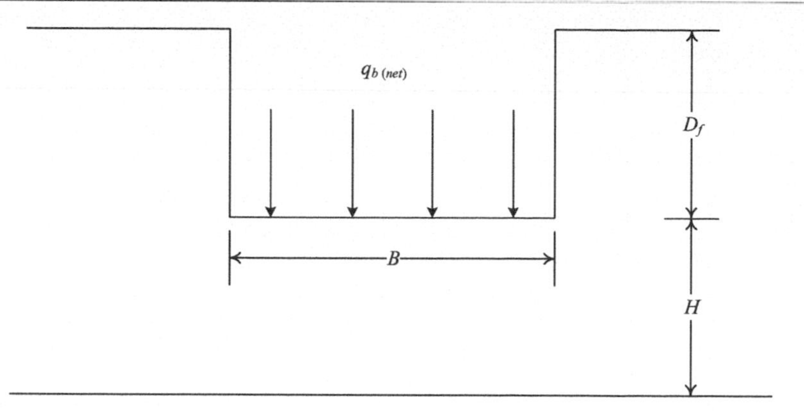

Figure 3.14 A flexible foundation under net bearing pressure $q_{b(net)}$.

Table 3.4 The depth factor I_f

μ_s	D_f/B	B / L		
		0.2	0.5	1
0.3	0.2	0.95	0.93	0.9
	0.4	0.9	0.86	0.81
	0.6	0.85	0.8	0.74
	1.0	0.78	0.71	0.65
0.4	0.2	0.97	0.96	0.93
	0.4	0.93	0.89	0.85
	0.6	0.89	0.84	0.78
	1.0	0.82	0.75	0.69
0.5	0.2	0.99	0.98	0.96
	0.4	0.95	0.93	0.89
	0.6	0.92	0.87	0.82
	1.0	0.85	0.79	0.72

$B' = B$ and $\alpha = 1$ at the corner of the foundation.
$B' = B/2$ and $\alpha = 4$ at the center of the foundation.
I_s is the shape factor and is a function of width B and length L.
I_f is the depth factor and is a function of the depth of the foundation base D_f (refer to Table 3.4).

For the shape factor I_s, define $m = L/B$ and $n = H/B'$, where H is the thickness of the soil below the foundation:

$$I_s = F_1 + \frac{1-2\mu_s}{1-\mu_s} F_2 \tag{3.24}$$

$$F_1 = \frac{1}{\pi}(A_0 + A_1) \tag{3.25}$$

$$F_2 = \frac{n}{2\pi} \tan^{-1} A_2 \tag{3.26}$$

$$A_0 = m \ln \frac{(1+\sqrt{m^2+1})\sqrt{m^2+n^2}}{m(1+\sqrt{m^2+n^2+1})} \tag{3.27}$$

$$A_1 = \ln \frac{(m+\sqrt{m^2+1})\sqrt{1+n^2}}{m+\sqrt{m^2+n^2+1}} \tag{3.28}$$

$$A_2 = \frac{m}{n\sqrt{m^2+n^2+1}} \tag{3.29}$$

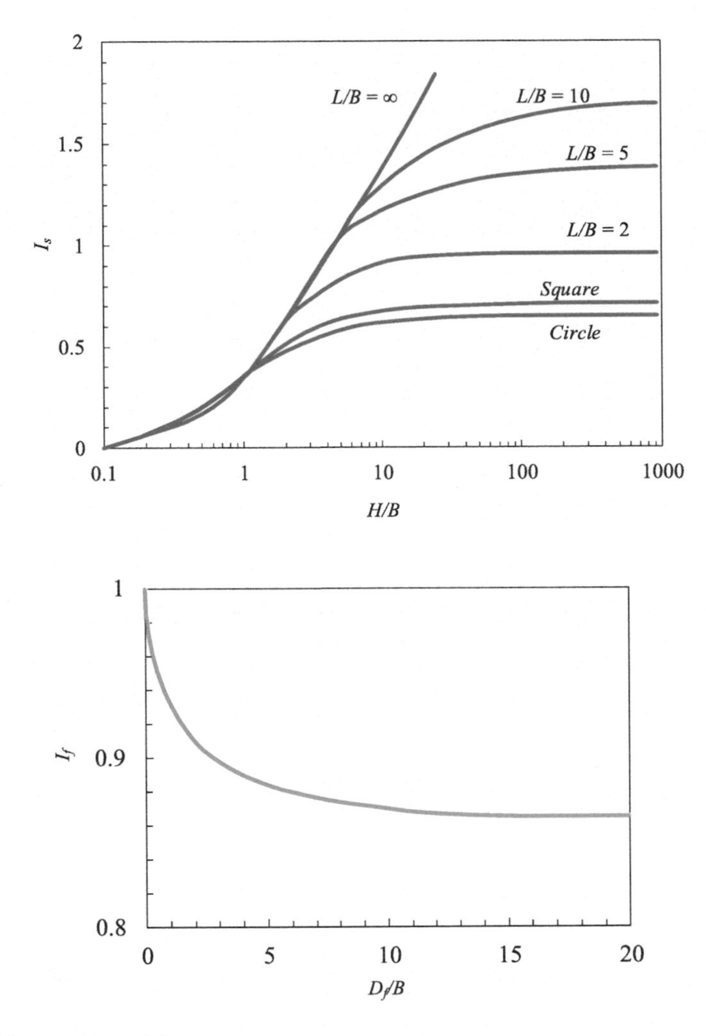

Figure 3.15 Values of I_s and I_f.

Source: After Christian and Carrier (1978).

The immediate settlement of a rigid footing can be approximated as:

$$S_{i\ (\text{rigid})} \approx 0.93 S_{i\ (\text{flexible, center})} \tag{3.30}$$

where $S_{i\ (\text{rigid})}$ and $S_{i\ (\text{flexible, center})}$ are the immediate settlement of a rigid footing and the immediate settlement of a flexible footing at the center, respectively.

For saturated clay under loading in the short-term condition, the soil is regarded as in the undrained condition without volume change. For this state, Poisson's ratio is set to 0.5 for the no volume strain condition. Similar to Steinbrenner's solution, Janbu et al. (1956) derived an equation for the average settlement of flexible foundations on saturated clay soils under net bearing pressure $q_{b(net)}$, as follows. Referring to Figure 3.14:

$$S_i = I_s I_f \left(\frac{q_{b(net)} B}{E_s} \right) \tag{3.31}$$

where I_s the shape factor and is a function of width B and length L (refer to Figure 3.15), and I_f is the depth factor and is a function of the depth of the foundation base D_f (refer to Figure 3.15).

Note that the modulus of elasticity for saturated clays is for the undrained condition. Section 2.3.2 describes the secant undrained Young's modulus as follows.

$$E_s = E_{us} \tag{3.32}$$

E_{us}/s_u is a function of the plasticity index and overconsolidation ratio (OCR), as shown in Figure 2.24.

Example 3.8

A rigid footing of 2 m × 4 m in plan view is situated on a dry sandy stratum with a thickness of 6 m. The embedment depth of the footing is 1 m. The average modulus of elasticity is 10,000 kN/m^2, and Poisson's ratio is 0.3. Calculate the immediate settlement of the footing under a net bearing pressure of 200 kN/m^2.

Solution

Eq. 3.23 is used.

For the center of the footing, $\alpha = 4$:

$m = L/B = 4/2 = 2$, $n = H/(B/2) = 5/(2/2) = 5 \Rightarrow F_1 = 0.526$, $F_2 = 0.058$ ($A_0 = 0.593$, $A_1 = 1.061$, $A_2 = 0.073$)

$I_s = 0.526 + [(1 - 2(0.3))/(1 - 0.3)](0.058) = 0.559$

$B/L = 2/4 = 0.5$, $D_f/B = 1/2 = 0.5 \Rightarrow I_f = (0.86 + 0.80)/2 = 0.83$

$S_{i\ (flexible)} = (200)(4)(2/2)[(1 - 0.3^2)/10000](0.559)(0.83) = 0.0338\ \text{m} = 33.8\ \text{mm}$

$S_{i\ (rigid)} = (0.93)\ (33.8) = 31.4\ \text{mm}$

Example 3.9

A footing of 2 m × 4 m in plan view is situated on a saturated clay stratum with a thickness of 6 m. The embedment depth of the footing is 1 m. The undrained shear strength of the clay is

150 kN/m², OCR = 2, PI = 30. Calculate the average immediate settlement of the footing under a net bearing pressure of 200 kN/m².

Solution

Eq. 3.31 is used.

For OCR = 2, PI = 30 => $E_{us}/s_u = 550$ (according to Figure 2.24).

$E_s = E_{us} = (550)(150) = 82{,}500$ kN/m²
$L/B = 2,\ H/B = 5/2 = 2.5 \Rightarrow I_s = 0.7$
$D_f/B = 1/2 = 0.5 \Rightarrow I_f = 0.95$
$S_{i\,(av)} = (0.7)(0.95)[(200)(2)/82{,}500] = 0.003224$ m = 3.22 mm

3.6.2 Consolidation settlement

Consolidation settlement occurs over time in saturated clay subjected to an increased load caused by the construction of the foundation and the supported structure. According to the one-dimensional consolidation theory, the following approach can be applied to compute the consolidation settlement, considering the initial soil stress condition, stress increment, and overconsolidation ratio (OCR) of clay. For example, a clay with a thickness of H_c is subjected to the net bearing pressure $q_{b(net)}$, as shown in Figure 3.16. First, the simplified 2 (vertical): 1 (horizontal) rule can be used to approximately estimate the average stress increment due to $q_{b(net)}$. The average stress increment $\Delta\sigma'_{av}$ at the midpoint of the clay layer is computed as follows:

$$\Delta\sigma'_{av} = \frac{q_{b(net)}B^2}{(B+z_0)^2} \quad \text{or} \quad \frac{q_{b(net)}BL}{(B+z_0)(L+z_0)} \tag{3.33}$$

where z_0 is the depth from the foundation base to the midpoint of the clay layer.

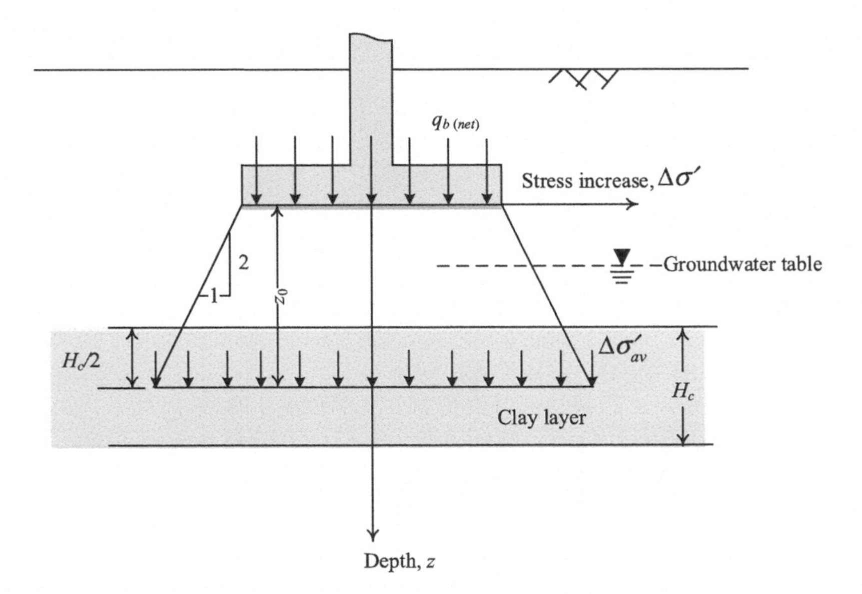

Figure 3.16 Stress increment in saturated clay due to the net bearing pressure $q_{b(net)}$.

The settlement of the clay layer S_c upon the stress increment can be computed for different soil conditions.

For normally consolidated clay:

$$S_c = \frac{C_c H_c}{1+e_0} \log \frac{\sigma_0' + \Delta\sigma_{av}'}{\sigma_0'} \tag{3.34}$$

where e_0 and C_c are the initial void ratio and compression index of the clay, respectively, and σ_0' is the initial overburden pressure.

For overconsolidated clay, two scenarios are considered:

1. When the stress in the clay layer upon the stress increment is less than the preconsolidation stress p_c' (i.e., $\Delta\sigma_0' + \Delta\sigma_{av}' < p_c'$), then:

$$S_c = \frac{C_s H_c}{1+e_0} \log \frac{\sigma_0' + \Delta\sigma_{av}'}{\sigma_0'} \tag{3.35}$$

where C_s is the swell index of the clay.

2. When the stress in the clay layer upon the stress increment is larger than the preconsolidation stress p_c' (i.e., $\Delta\sigma_0' + \Delta\sigma_{av}' > p_c'$), then:

$$S_c = \frac{C_s H_c}{1+e_0} \log \frac{p_c'}{\sigma_0'} + \frac{C_c H_c}{1+e_0} \log \frac{\sigma_0' + \Delta\sigma_{av}'}{p_c'} \tag{3.36}$$

Example 3.10

A footing of 2 m × 4 m in plan view is situated on a saturated clay stratum with a thickness of 6 m. The embedment depth of the footing is 1 m. Assume that $\gamma = 16$ kN/m³, $e_0 = 0.8$, $C_c = 0.3$, $C_s = 0.1$, and $p_c' = 100$ kN/m². Calculate the average consolidated settlement of the footing under a net bearing pressure of 200 kN/m².

Solution

$H_c = 5$ m
σ_0' (at the middle of the clay layer below the footing) = (16 − 9.81)(1 + 5/2) = 21.67 kN/m²
$< p_c' = 100$ kN/m² => OC clay
$q_{b\,(net)} = 200$ kN/m²
$\Delta\sigma'_{av}$ = (200)(4)(2)/[(4 + 5/2)(2 + 5/2)] = 54.7 kN/m²
$\sigma_0' + \Delta\sigma'_{av}$ = 21.67 + 54.7 = 76.37 $< p_c'$

By using eq. 3.35:

$S_{c\,(av)}$ = [5/(1 + 0.8)](0.1)log(76.37/21.67) = 0.1520 m = 152 mm

Much larger than the immediate settlement!

3.6.3 *Allowable bearing capacity considering allowable settlement (example of saturated clay)*

The allowable bearing capacity determined from the factor of safety does not provide information about the settlement of footings. To consider the serviceability of a footing, the allowable bearing capacity also needs to account for the foundation settlement. This section illustrates an example to determine the allowable bearing capacity of a surface footing on saturated clay considering the bearing capacity and foundation settlement. As displayed in Figure 3.17, two lines that describe the relationships between the allowable net bearing pressure $q_{all\,(net)}$ and the footing width B are built based on the criteria of ultimate bearing capacity and settlement individually.

Line 1 is the relationship of the allowable bearing capacity based on the ultimate bearing capacity $q_{u\,(net)}$ divided by a factor of safety F_S. That is:

$$q_{u(net)} = cN_c \tag{3.37}$$

$$q_{all(net),1} = \frac{q_{u(net)}}{F_S} = \frac{cN_c}{F_S} \tag{3.38}$$

From eq. 3.37, the curve is a horizontal line, implying that the allowable bearing capacity is independent of the footing width B.

Line 2 (blue curve) is the relationship of the allowable bearing capacity based on the allowable foundation settlement S_{all}. Consider the immediate settlement, according to eq. 3.31:

$$q_{all(net),2} = \frac{S_{all}E_s}{I_s I_f B} \tag{3.39}$$

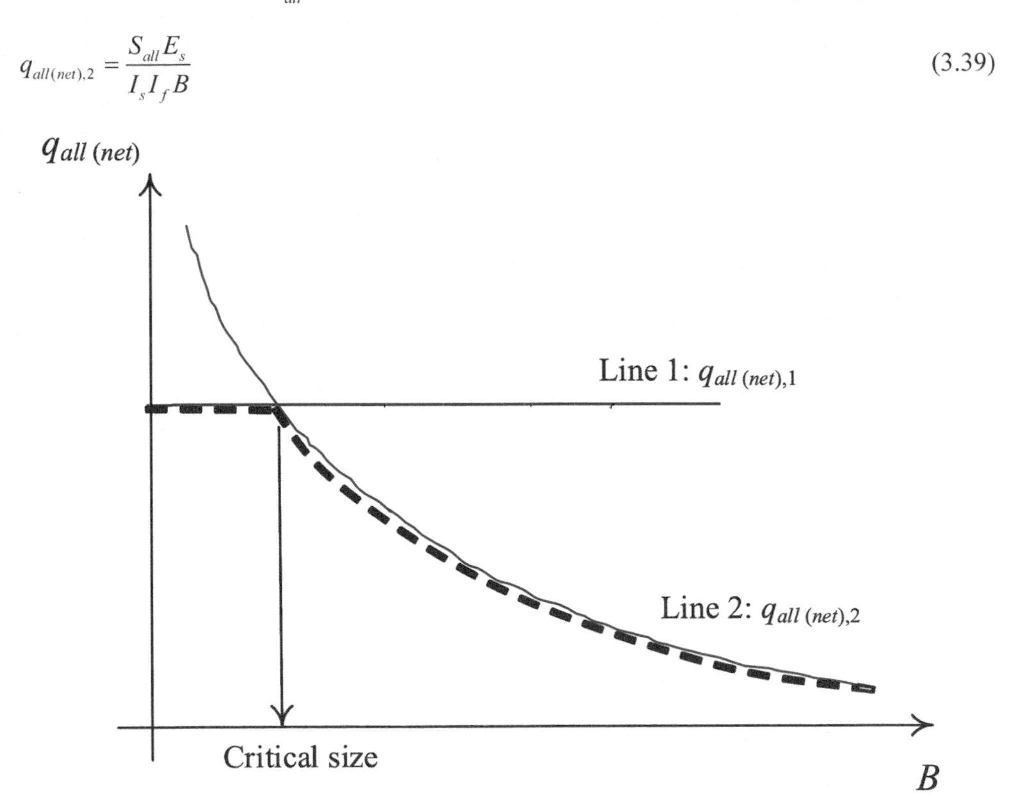

Figure 3.17 Schematic diagram of the interaction curve for allowable bearing capacity in saturated clay.

From eq. 3.39, for a given allowable settlement, the allowable bearing capacity decreases with an increase in the footing width. More precisely, it is inversely proportional to the footing width.

The final relationship of the allowable bearing capacity versus the footing size is the lesser of $q_{all\ (net),1}$ and $q_{all\ (net),2}$, as shown by the dashed line in Figure 3.17. The intersection of the two lines indicates the critical size of the footing. Below this critical size, the allowable bearing capacity is governed by the bearing capacity, but for a footing larger than the critical size, the allowable bearing capacity is governed by the allowable footing settlement.

Example 3.11

A rigid footing of 2 m × 4 m in plan view is situated on a dry sandy stratum with a thickness of 6 m. The embedment depth of the footing is 1 m. The average modulus of elasticity is 10,000 kN/m², Poisson's ratio is 0.3, the angle of friction of sand is 32°, and the unit weight of sand is 18 kN/m³. Calculate the allowable net bearing pressure of the footing, considering an allowable settlement of 40 mm and $F_S = 3$.

Solution

1. With the settlement controlled:

Use eq. 3.23.

$$S_i = q_{b(net)}(\alpha B')\frac{1-\mu_s^2}{E_s} I_s I_f$$

For the center of the footing, $\alpha = 4$:

$m = L/B = 4/2 = 2,\ n = H/(B/2) = 5/(2/2) = 5 \Rightarrow F_1 = 0.526,\ F_2 = 0.058$
$I_s = 0.526 + [(1 - 2(0.3))/(1 - 0.3)](0.058) = 0.559$
$B/L = 2/4 = 0.5,\ D_f/B = 1/2 = 0.5 \Rightarrow I_f = (0.86 + 0.80)/2 = 0.83$
$S_{i\ (rigid)} = (0.93)q_{all(net)}(4)(2/2)[(1 - 0.3^2)/10{,}000](0.559)(0.83) = 0.04 \text{ m} = 40 \text{ mm}$
$q_{all(net)} = 254.6 \text{ kN/m}^2$

2. With the bearing capacity controlled:

Referring to example 3.3, $q_u = 1031.38$ kN/m²
$q_{u\ (net)} = 1031.38 - 18(1) = 1013.38 \text{ kN/m}^2$
$q_{all(net)} = 1013.38/3 = 337.79 \text{ kN/m}^2$

Considering conditions 1 and 2, choose the smaller one:

$q_{all(net)} = \min(254.6, 337.79) = 254.6$ kN/m² (Condition 1 governs!)

3.6.4 *Tolerable settlement of buildings*

In most conditions, the subsoil is not homogeneous, and the load carried by foundations of a structure may vary at different locations. Varying degrees of settlement may occur in different parts of a building. The differential settlement of the parts of a building can lead to damage to the structure. Figure 3.18 presents schematic settlement profiles at the base of a building with length L. Figure 3.18a shows the building settlement without tilt, whereas Figure 3.18b shows the building settlement with tilt. In the figure, points A, B, C, D, and E indicate points on the building base before settlement, and points A', B', C', D', and E' indicate their positions after settlement. According to this figure, the following parameters are defined to quantify the settlement condition of the building:

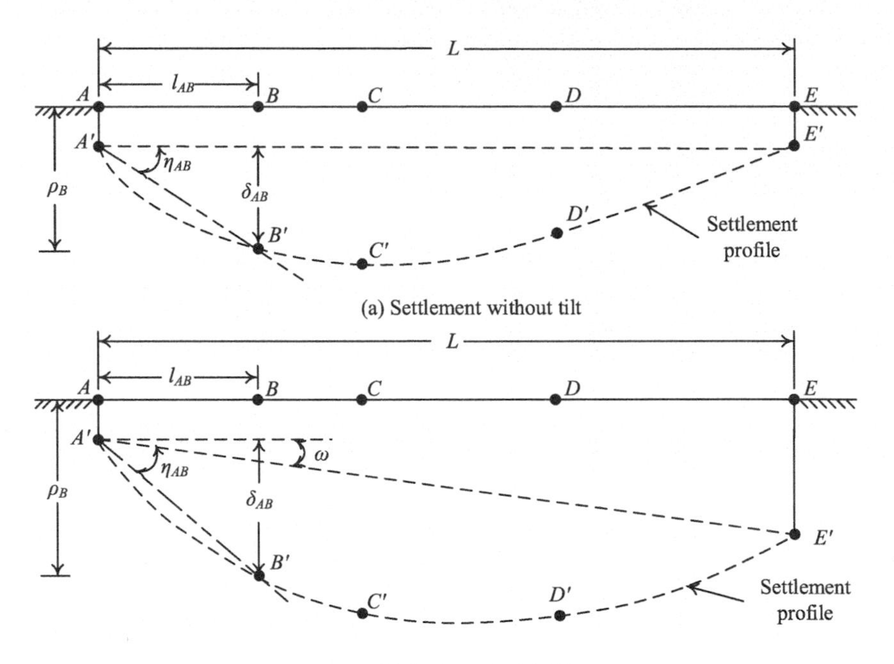

Figure 3.18 Definitions of settlement parameters.
Source: Redrawn following Wahls (1981).

ρ_i = total settlement of point i.
δ_{ij} = differential settlement between points i and j (= $\rho_i - \rho_j$).
ω = rigid body rotation.
η_{ij} = angular distortion between points i and j (= δ_{ij}/l_{ij} for the building settlement without tilt, = $\delta_{ij}/l_{ij} - \omega$ for the building settlement with tilt, where l_{ij} is the distance between points i and j).

In the preceding parameters, the angular distortion is directly related to the degree of damage to a building. As defined, when the foundation settlement exhibits tilt (refer to Figure 3.18b), the tilt ω is removed to deduce an actual angular distortion as the tilt is a rigid-body rotation that does not induce additional stress in the structure. Table 3.5 relates the potential structural damage to the degree of angular distortion. As angular distortion increases, more damage is anticipated to occur in structures. According to this table, if the first cracks in structures are not allowed, the allowable angular distortion is limited to less than 1/300.

Alternatively, differential settlement and total settlement are usually used to determine the allowable settlement. Similar to angular distortion, differential settlement can also be related to structural damage. However, it is not dimensionless. An angular distortion may correspond

Table 3.5 Limiting angular distortion to potential damage of structures

Category of potential damage	η
Danger to machinery sensitive to settlement	1/750
Danger to frame with diagonals	1/600
Safe limit for no cracking of buildings	1/500
First cracking of panel walls	1/300
Difficulties with overhead cranes	1/300
Tilting of high rigid buildings becoming visible	1/250
Considerable cracking of panel and brick walls	1/150
Danger of structural damage to general buildings	1/150
Safe limit for flexible brick walls, L/H > 4[b]	1/150

Source: Bjerrum[a] (1963).

a Following Wahls (1981).
b Safe limits include a factor of safety, H = height of building.

Table 3.6 Allowable total and differential settlements

Foundation type	*Soil type*	*Total settlement (cm)*	*Differential settlement (cm)*	*Note*
Isolated footing	Sand	2.5	2.0	T
		5.0	3.0	S
		3.0	–	J
Isolated footing	Clay	7.5	–	S
		10.0	–	J
Mat foundation	Sand	5.0-7.5	3.0	S
		6.0-8.0	–	J
		–	3.0	G
Mat foundation	Clay	7.5-12.5	4.5	S
		20.0-30.0	–	J
			5.6	G

Note: T: Terzaghi and Peck (1967); S: Skempton and MacDonald (1957) for 1/300 angular distortion; J: Architectural Institute of Japan (1988); G: Grant et al. (1974) for 1/300 angular distortion.

to different values of differential settlement for different column spans. For a general building, the distance of two neighboring columns is approximately 6 m. According to this distance and Table 3.5 ($\eta = 1/300$ for the first cracking of panel walls), the differential settlement between two columns can be limited to be less than 2 cm to avoid cracking in a building. Differential settlement is generally not easy to measure. A large differential settlement is usually accompanied by a large total settlement of the footing. Based on past experience, the differential settlement for a building on sand is approximately 3/4 of the total settlement; thus, the allowable total settlement of an isolated footing on sand is set to 2.5 cm. However, footings on clay or buildings with mat foundations generally settle more uniformly, and the allowable total settlement can be larger. Table 3.6 lists some allowable total and settlement values suggested in the literature. Note that these values are applicable for the column spacing of approximately 6 m; however, they may not be appropriate for span distances larger or less than 6 m.

Considering uncertainties in settlement evaluation and measurement, in practical applications, two parameters, angular distortion and total settlement, or differential settlement and total settlement, are used to simultaneously evaluate the settlement condition of a building.

3.7 Applications of plate loading tests

The estimation of bearing capacity and foundation settlement using the aforementioned theories may lead to uncertainties in the theoretical models and the determination and selection of material parameters. To reduce uncertainties and confirm results in design, plate load testing performed in the field is beneficial to estimate the bearing capacity and settlement of footings. Figure 1.25 demonstrates a general setup and a field picture of vertical plate load testing. First, a test pit is excavated to the foundation bottom. The width of the pit is generally larger than four times the width of the plate to ignore the effect of the overburden above the foundation base. Reaction beams are set to react to the jack when applying vertical loading to the plate. With this testing, the pressure–settlement relationship of the plate can be recorded, as presented in Figure 1.26. Based on this relationship, the settlement and ultimate pressure of the plate can be used to estimate the settlement and ultimate bearing capacity of the footing for design reference.

3.7.1 *Estimation of q_u*

For tests in clays:

$$q_{u(F)} = q_{u(P)} \tag{3.40}$$

where $q_{u(F)}$ is the ultimate bearing capacity of the proposed foundation, and $q_{u(P)}$ is the ultimate bearing capacity of the test plate.

This equation implies that the ultimate bearing capacity in clay is practically independent of the plate size.

For tests in sandy soils:

$$q_{u(F)} = q_{u(P)} \frac{B_F}{B_P} \tag{3.41}$$

where B_F is the width of the foundation, and B_P is the width of the test plate.

This equation implies that the ultimate bearing capacity in sand is linearly proportional to the plate size.

3.7.2 *Estimation of settlement*

A footing generally has a larger size than the plate. Because of their different influence depths of stress D_i and nD_i, the footing and plate may have different settlements under the same bearing pressure. As shown in Figure 3.19, considering the difference in size between the footing and plate:

For tests in clayey soils (overconsolidated clay), assume that the elastic modulus of the soil is constant with depth:

$$S_{i(F)} = S_{i(P)} \frac{B_F}{B_P} \tag{3.42}$$

This equation is derived based on eq. 3.31, which implies that the settlement of a footing in clay linearly increases with the plate size.

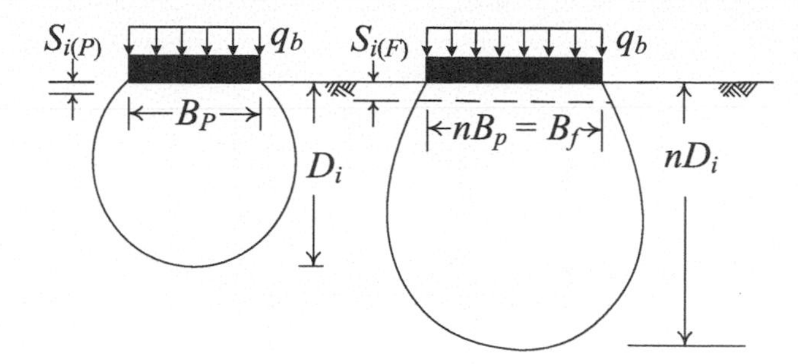

Figure 3.19 Influence zone of stress for the plate and foundation.

For tests in sandy soils:

$$S_{i(F)} = S_{i(P)} \left(\frac{B_F}{B_P} \right)^2 \left(\frac{B_P + 0.3}{B_F + 0.3} \right)^2 \tag{3.43}$$

where B_F and B_P are in meters.

Because the elastic modulus of sandy soil is dependent on the confining stress, this equation was empirically formulated by Terzaghi and Peck (1967) based on several plate load tests with different plate sizes.

3.8 Mat foundations

A mat foundation (raft foundation) is a combined footing that covers the entire area under a structure, supporting several columns, and walls. It is sometimes applied to soil that has a low bearing capacity but supports large column or wall loads. Mat foundations might be more economical when spread footings need to cover a large area of the building. Figure 3.20 displays three common types of mat foundations: (a) the foundation is a flat plate with uniform thickness, (b) the foundation is composed of a slab with girder beams, in which the columns are located at the intersections of the beams, and (c) the foundation is the basement with slabs and walls. The foundation of type (c) is often referred to as a compensated foundation because the weight of soil in the basement area is excavated to counterbalance a part of or all the structure loads.

3.8.1 Bearing capacity of mat foundations

The net ultimate bearing capacities of a mat foundation can be determined by eq. 3.15. A factor of safety is used to calculate the net allowable bearing capacity. A factor of safety of 3 is normally used.

3.8.2 Compensated foundations

The net bearing pressure under a mat foundation can be reduced by increasing the excavation depth of the foundation. As shown in Figure 3.21, for a mat foundation subjected to the dead

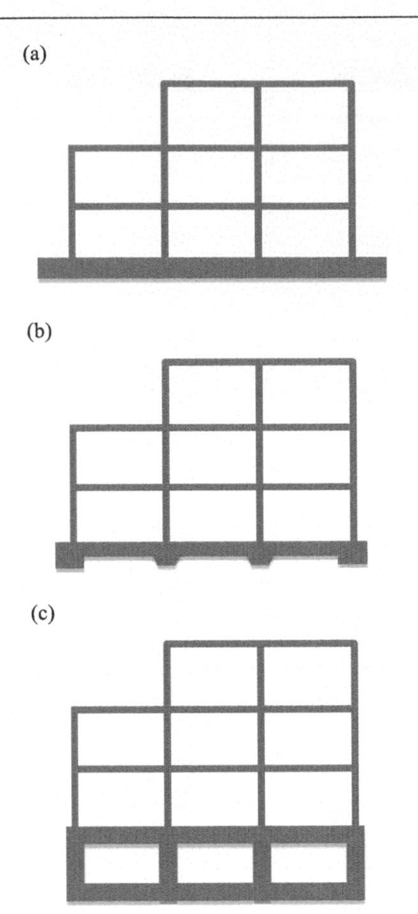

Figure 3.20 Common types of mat foundations: (a) flat plate, (b) beams and slab, and (c) slab with basement walls.

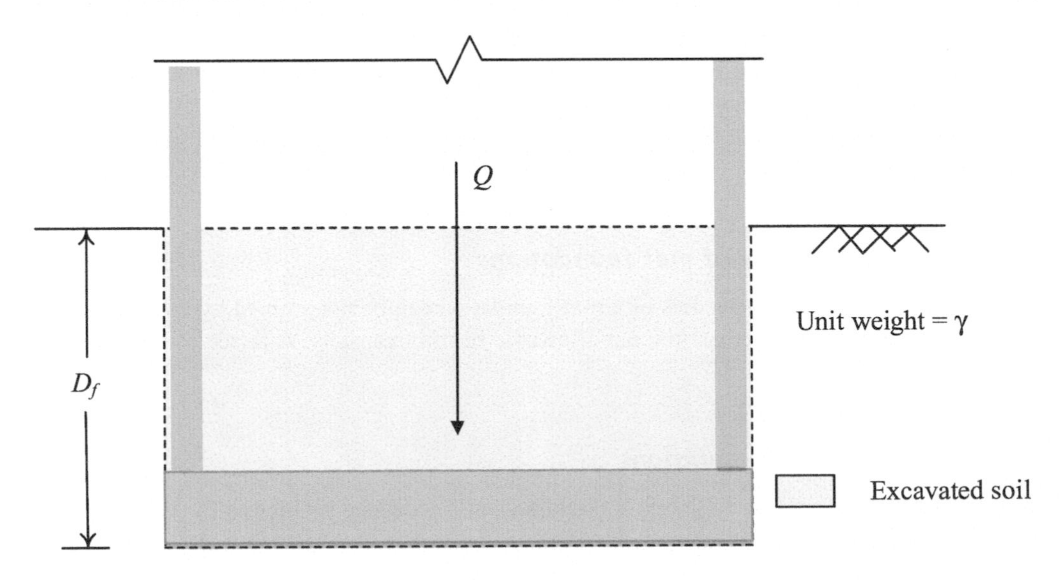

Figure 3.21 Compensated foundation design concept.

weight of the structure and the live load Q, when its base is embedded at depth D_f and the soil within this depth (highlighted zone) is removed, the net bearing pressure is reduced as follows:

$$q_{b(net)} = \frac{Q}{A} - \gamma D_f \tag{3.44}$$

where A is the area of the foundation.

According to the earlier equation, increasing the embedment depth of the foundation implies that the more that is removed, the smaller the net bearing pressure becomes. Because of the decrease in the net bearing pressure, the factor of safety against bearing capacity failure increases, and the foundation settlement decreases. This approach is known as compensated foundation design and is useful when structures are built on soft clays. For this design, a basement is made below the structure, as shown in Figure 3.20(c).

Furthermore, for no increase in the pressure in the soil below a foundation, $q_{b(net)}$ should be zero. The foundation for this condition is called a fully compensated foundation, and its excavation depth is:

$$D_f = \frac{Q}{\gamma A} \tag{3.45}$$

Example 3.12

A mat foundation has dimensions of 20 m × 30 m in plan view. The total load on the mat is 100 MN. The mat is situated on a saturated clay with a unit weight of 18 kN/m^2. Determine the excavation depth for this mat to be a fully compensated foundation.

Solution

Use eq. 3.45, D_f = 100,000/[(18)(20)(30)] = 9.26 m.

3.8.3 Stress analysis of mat foundations

Compared with spread footings, mat foundations are generally more flexible because of a large foundation base. Figure 3.22 displays the distribution of bearing pressure for a mat foundation for different soil consistencies. When the mat is on soft soil, it may behave like a rigid footing, and the bearing pressure is linearly distributed because of a linear footing displacement profile, as presented in Figure 3.11. Therefore, for a rigid mat, a linear bearing pressure profile is assumed to calculate the stress in the mat. However, when the same mat is on much stiffer soil, the mat may behave flexibly, the bearing pressure profile is nonlinear, and a larger pressure occurs around the location of the load. For this condition, the analysis method for a rigid footing cannot be applied. To simulate a more accurate distribution of bearing pressure and stress in the foundation for design, the slab on the Winkler foundation model is commonly used in practice, as presented in Figure 3.23. In this model, the mat is no longer rigid, and the flexibility of the mat is simulated in the model; therefore, the method is called the flexible method. In addition, this model simulates the soil as a bed of springs (Winkler springs), each with a stiffness k_v. k_v is called the coefficient of vertical

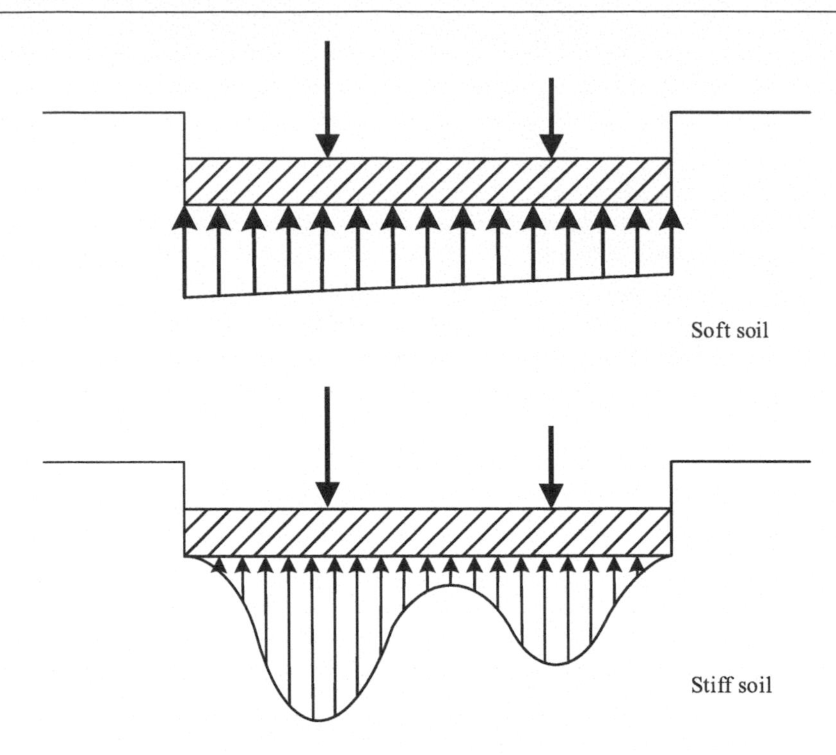

Figure 3.22 Distributions of bearing pressure for a footing on soft and stiff soil.

subgrade reaction, which defines the relationship between bearing pressure and settlement. k_v is expressed as follows:

$$k_v = \frac{p}{\rho} \tag{3.46}$$

where p is the bearing pressure and ρ is the settlement.

Since discrete springs are adopted in numerical modeling, the stiffness of the springs is:

$$K_v = k_v A \tag{3.47}$$

where A is the tributary area of the spring, as shown in Figure 3.23.

In the slab on the Winkler foundation model, the determination of k_v is an important and difficult task. Theoretical analysis and empirical judgment are usually needed to determine an appropriate value. In essence, the coefficient of subgrade reaction is not a fundamental soil property but a fictitious property that relates the bearing pressure to the settlement. It varies with the properties of the subgrade and the beam or slab and even the size of the footing. In practice, k_v can be estimated by field load testing, the results of boring and laboratory tests, or past experience. Terzaghi (1955) proposed an approach to determining k_v by using field plate load testing with a plate size of 0.3 m (1 ft). Considering the settlement characteristics of a footing on clay and sandy soils and the difference in size between the foundation and the plate, Terzaghi (1955)

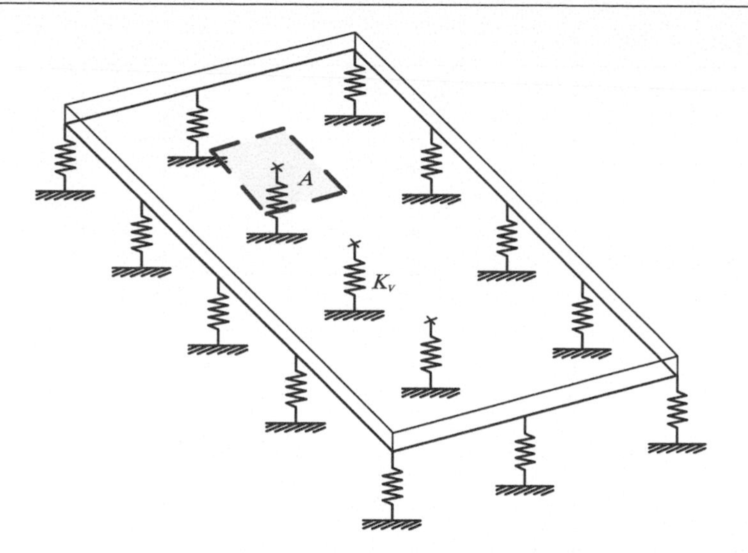

Figure 3.23 Slab on Winkler foundation model.

further related the coefficient of subgrade reaction of a rigid footing with that of a rigid plate as follows, based on eqs. 3.48 and 3.49 for clayey and sandy soils, respectively. These two equations clearly show that k_v decreases as the footing size increases.

For clayey soils:

$$k_{vB} = \frac{B_1}{B} k_{v1} \tag{3.48}$$

For sandy soils:

$$k_{vB} = \left(\frac{B + B_1}{2B} \right)^2 k_{v1} \tag{3.49}$$

where k_{v1} and k_{vB} are the coefficients of the subgrade reaction of the plate of size 0.3 m (1 ft) and the rigid foundation, respectively, and B_1 and B are the sizes of the rigid plate and rigid foundation, respectively. In addition, Terzaghi (1955) suggested empirical values of k_{v1} for different types of soils, as listed in Table 3.7.

Vesic (1961) compared the solutions of an infinite beam on a continuum model and a beam on an elastic foundation and proposed:

$$k_v = \frac{0.65 E_s}{1 - \mu_s^2} \sqrt[12]{\frac{E_s B^4}{EI}} \tag{3.50}$$

where B is the width of the footing, E is the modulus of elasticity of the foundation, and I is the moment of inertia of the foundation's cross section.

Table 3.7 Suggested values for k_{v1}

Clay	*Stiff*	*Very stiff*	*Hard*
c_u (kN/m^2)	47.5–95	95–190.5	>190.5
k_{v1} (MN/m^3)	15.6–31.3	31.3–62.7	>62.7
Sand	*Loose*	*Medium*	*Dense*
Unsaturated k_{v1} (MN/m^3)	6.3–18.8	18.8–94	94–313.4
Saturated k_{v1} (MN/m^3)	7.8	25.1	94

Source: Terzaghi (1955).

To evaluate the rigidity of a foundation system, a parameter λ that compares the rigidities of the foundation and soil can be used. It is defined as follows:

$$\lambda = \sqrt[4]{\frac{k_v B}{4EI}} \tag{3.51}$$

With this parameter, when $\lambda L < 1.0$ (where L is the foundation length), the foundation can be regarded as rigid.

The actual distribution of k_v is not uniform because of the nonuniform distributions of bearing pressure and foundation displacement. To consider a detailed variation of k_v, a more delicate approach considering the actual interaction between the bearing pressure and foundation displacement can be performed. This approach requires iterations. First, assuming an initial distribution of k_v, the structure model is built, and the bearing pressure is calculated after the structure is loaded. Second, according to the obtained bearing pressure, the ground surface settlement is analyzed with numerical analysis, in which the adopted soil model can consider the nonlinear behavior of soil and complex soil and boundary conditions. Fourth, based on the bearing pressure and ground surface settlement, a variation in k_v can be formulated and used to update the original k_v distribution. With the updated k_v distribution, the aforementioned procedure is repeated until the distribution of k_v converges.

3.8.4 Piled mat foundations

A mat foundation may have a sufficient bearing capacity to resist structure loads because of a large foundation size, but it may still be subjected to excessive settlement when it supports a high-rise building on soft soil. To overcome this excessive settlement issue, a piled mat foundation can be an option. A piled mat is a hybrid foundation in which piles are used to support the mat to reduce the total and differential settlements of the mat; in this foundation type, the piles serve as settlement reducers. This foundation type differs from a conventional capped group pile foundation (which is introduced in Chapter 7). For a piled mat, most of the bearing capacity of the foundation is provided by the mat, but for a capped group pile foundation, its bearing capacity is supposed to be provided entirely by the piles, considering that the pile cap is rigid and the cap is not in contact with the soil.

Designing a piled mat foundation requires more comprehensive analyses than those when mat or pile foundations are used alone. The interaction among the piles, soil, and mat needs to

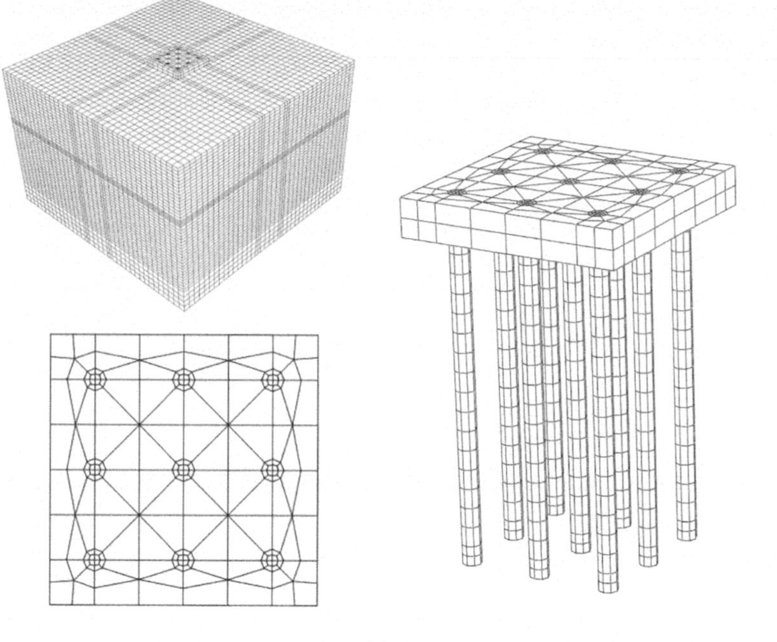

(a)

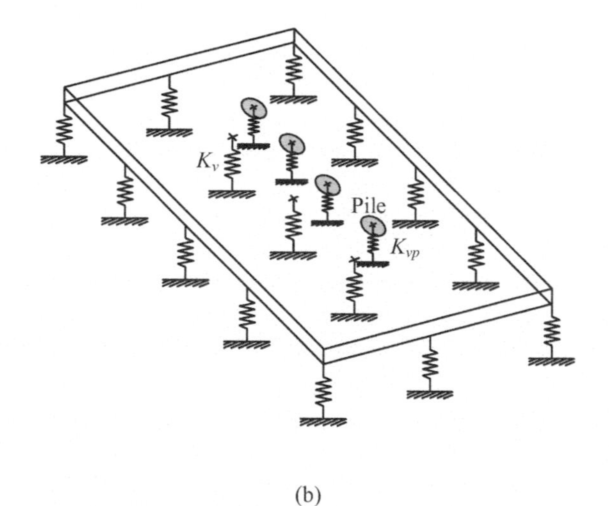

(b)

Figure 3.24 Numerical models for piled mat foundations: (a) finite element model and (b) slab on spring foundation model.

be simulated in the analyses. Issues such as the proportions of loads carried by the piles and mat, how the total and differential settlements are reduced by the inclusion of the piles, and how the internal forces (shear forces and moments) in the mat are reduced by the piles need to be addressed. The answers to these issues determine the final design parameters, for example, the mat thickness and reinforcement, the pile arrangement, and the pile diameter and length. The piles in the mat are not necessarily uniformly distributed; an efficient design can be achieved by placing the piles at the locations where the settlement is considerable to reduce the differential settlement and therefore the internal forces in and the thickness of the mat. No simple design charts for addressing the preceding concerns are available. The response of piled rafts is often analyzed numerically either by using finite element simulation (Figure 3.24a) or by the slab on the spring foundation model (Figure 3.24b). The finite element analyses that use solid elements to model the mat and soil are the most common because they can directly simulate the effect of pile–soil–mat interactions; however, the computational cost is large. Similar to mat foundations, the slab on the spring foundation model, in which the soil reactions and piles are modeled by springs K_v and K_{vp}, respectively, can be applied for piled mats and is more efficient. However, the determination of spring properties of the pile and soil reactions should consider the interaction of the piles and the surrounding soil, which is not an easy task.

3.9 Summary and general comments

This chapter introduces shallow foundation design methods. The summary of and general comments on this chapter are as follows:

1. Foundations with embedment depths less than or equal to the foundation width are called shallow foundations. Designing a shallow foundation involves four main items: evaluation of vertical bearing capacity, evaluation of eccentricity, evaluation of settlement, and structural design of foundation components.
2. The bearing failure modes of a footing include general, local, and punching modes. The bearing failure mode is dependent on the soil type and density and the embedment depth. For the general bearing failure mode, Terzaghi's and Meyerhof's bearing capacity formulas are introduced, which show that the ultimate bearing capacity of a footing is influenced by three parts: cohesion, surcharge, and soil weight. The bearing capacity is sensitive to the angle of friction of the soil. For $\phi > 30°$, the bearing capacity increases rapidly. For footing on saturated clay, effective stress and total stress analyses with appropriate soil parameters can be applied to evaluate the long-term and short-term bearing capacities, respectively.
3. The ultimate bearing capacity is also influenced by the foundation eccentricity and location of the groundwater table. The parameters in the bearing capacity formulas need to be modified to consider their influences. In addition to the stability of the footing, the eccentricity not only increases the bearing pressure but also reduces the bearing capacity because of reduced effective contact area between the footing and soil. The location of the groundwater table may alter the effective stress in the soil and thus the ultimate bearing capacity of footings.
4. The total settlement of a foundation is the sum of the immediate settlement and delayed settlement. Immediate settlement occurs during or immediately after the construction of a structure. Delayed settlement occurs over time. For footings on sand upon loading, the settlement is an immediate settlement. For footings on saturated clay upon loading, the settlement comprises both an immediate settlement and a delayed settlement. The immediate settlement can be computed based on the theory of elasticity. The delayed settlement of saturated clay is due to the consolidation of clay, which can be computed based on the theory of consolidation. The angular distortion (differential settlement divided by the distance of two points) is directly related to the degree of damage to a building. Both total settlement and differential settlement (or angular distortion) are used in the design to evaluate the settlement condition of the building.

5. The allowable bearing capacity determined from the factor of safety does not provide information about the settlement of a footing. To determine the serviceability of a footing, the allowable bearing capacity also needs to account for the foundation settlement. Interactive curves can be built to address the relationship between the allowable bearing capacity and footing size, considering the settlement and bearing failure. Generally, the allowable bearing capacity for a small footing is governed by the bearing failure, but that for a large footing is governed by the foundation settlement. Field plate load testing can be used to estimate the bearing capacity and settlement of a footing for modifying and confirming foundation design parameters and results.
6. A mat foundation is commonly applied to soil that has a low bearing capacity but supports large structure loading. To increase the factor of safety against vertical bearing capacity and reduce settlement, a type of mat foundation called a compensated foundation can be applied. A compensated foundation uses basements as the foundation body and excavates the soil within the basement to reduce the net bearing pressure. Compared with spread footings, mat foundations are more flexible because of their large foundation base, especially when the mat is on stiff soil. To simulate the actual distribution of bearing pressure and stress in the mat foundation for design, the beam-on-elastic foundation model is commonly used in practice.
7. A piled mat is a hybrid foundation in which piles are used to support a mat to reduce the total and differential settlement of the mat. This design can reduce excessive settlement in the mat foundation when supporting a high-rise building. The main role of the piles in this foundation type is a settlement reducer. The piles in the mat can be efficiently arranged by allocating them at the locations where the settlement is significant. The response of piled rafts is often analyzed numerically either by using finite-element simulation or by the slab-on-a-spring foundation model. The finite element analyses that use solid elements to model the mat and soil are the most common because they can directly simulate the pile–soil–mat interaction.

Problems

3.1 A footing is situated on a dry sandy stratum with an embedment depth of 1 m. The angle of friction of sand is 34°. The unit weight of sand is 18.5 kN/m^3. Apply Terzaghi's bearing capacity equation to estimate the ultimate bearing capacity of the footing in the following cases:

a. 3 m wide strip footing
b. 3 m × 3 m square footing
c. circular footing 3 m in diameter

3.2 A square footing is 3 m × 3 m in plan view, situated on a saturated clay stratum with an embedment depth of 1 m. The groundwater table is located at the ground surface. The saturated unit weight of the soil is 18 kN/m^3. The effective angle of friction and cohesion of the soil are 24° and 10 kN/m^2, respectively. The undrained shear strength of the clay is 80 kN/m^2. Apply Terzaghi's bearing capacity equation to estimate the ultimate bearing capacity of the footing for long-term and short-term conditions.

3.3 Same as problem 3.1. Use Meyerhof's bearing capacity to estimate the ultimate bearing capacity of the footing.

3.4 Same as problem 3.2. Use Meyerhof's bearing capacity to estimate the ultimate bearing capacity of the footing for long-term and short-term conditions.

3.5 A rectangular footing is 3 m × 4 m in plan view, situated on a dry sandy stratum with an embedment depth of 1 m. The angle of friction of sand is 34°. The unit weight of sand is 18.5 kN/m^3.

a. Apply Meyerhof's bearing capacity equation to estimate the gross and net ultimate bearing capacity of the footing.
b. Compute a factor of safety in terms of the net ultimate bearing capacity, given that the footing is subjected to a total centric vertical load of 6,000 kN on its base.

3.6 Redo problem 3.5 when the groundwater table is located at a depth of 0.5 m. The unit weights of sand above and below the groundwater table are 16.5 and 19.5 kN/m^3, respectively.

3.7 A rectangular footing is 3 m × 4 m in plan view, situated on a dry sandy stratum with an embedment depth of 1 m. The footing is subjected to a total vertical load of 6,000 kN and a moment of 1,200 kN-m in the transverse direction of the footing on its base at the centerline. The angle of friction of the sand is 34°. The unit weight of sand is 18.5 kN/m^3. Compute $q_{b,max}/q_{b,min}$ and the ultimate bearing capacity.

3.8 Redo problem 3.7 when the moment is applied in the longitudinal direction of the footing.

3.9 A rigid footing of 3 m × 4 m in plan view is situated on a dry sandy stratum with a thickness of 5 m. The embedment depth of the footing is 1 m. The average modulus of elasticity is 12,000 kN/m^2, and Poisson's ratio is 0.3. Calculate the immediate settlement of the footing under a net bearing pressure of 250 kN/m^2.

3.10 A footing of 3 m × 4 m in plan view is situated on a saturated NC clay stratum with a thickness of 5 m. The embedment depth of the footing is 1 m. The undrained shear strength of the clay is 20 kN/m^2, PI = 30. Calculate the average immediate settlement of the footing under a net bearing pressure of 200 kN/m^2.

3.11 A footing of 3 m × 4 m in plan view is situated on a saturated clay stratum with a thickness of 5 m. The embedment depth of the footing is 1 m. Assume that γ = 18 kN/m^3, e_0 = 0.8, C_c = 0.25, C_s = 0.05, and p_c' = 80 kN/m^2. Calculate the average consolidated settlement of the footing under a net bearing pressure of 200 kN/m^2.

3.12 A rigid footing of 3 m × 4 m in plan view is situated on a dry sandy stratum with a thickness of 5 m. The embedment depth of the footing is 1 m. The average modulus of elasticity is 12,000 kN/m^2, Poisson's ratio is 0.3, the angle of friction of sand is 34°, and the unit weight of sand is 18.5 kN/m^3. Calculate the allowable net bearing pressure of the footing, considering an allowable settlement of 25 mm and F_S = 3.

3.13 Redo problem 3.12 for an allowable settlement of 40 mm and F_S = 2.

3.14 A mat foundation has dimensions of 25 m × 30 m in plan view. The total load on the mat is 120 MN. The mat is situated on a saturated clay having a unit weight of 17.5 kN/m^2. Determine the excavation depth for this mat to be a fully compensated foundation.

References

Architectural Institute of Japan (AIJ). (1988), *Recommendation for the Design of Building Foundations* (in Japanese), Tokyo: Architectural Institute of Japan.

Bjerrum, L. (1963), Allowable settlement of structures. In *Proceedings, 3rd European Conference on Soil Mechanics and Foundation Engineering*, Wiesbaden, Germany, Vol. 3, pp. 135–137.

Christian, J.T. and Carrier, W.D. (1978), Janbu, Bjerrum and Kjaernsli's chart reinterpreted. *Canadian Geotechnical Journal*, Vol. 15, pp. 123–128.

Grant, R., Christian, J.T. and Vanmarcke, E.H. (1974), Differential settlement of buildings of buildings. *Journal of the Geotechnical Engineering Division, ASCE*, Vol. 100, No. 9, pp. 973–991.

Janbu, N., Bjerrum, L. and Kjaernsli, B. (1956), Veuledning ved losning av fundamenteringsoppgaver. *Norwegian Geotechnical Institute Publication*, Vol. 16, pp. 30–32 (in Norwegian).

Kumbhojkar, A.S. (1993), Numerical evaluation of Terzaghi's N_γ. *Journal of the Geotechnical Engineering Division, ASCE*, Vol. 119, No. 3, pp. 598–607.

Meyerhof, G.G. (1963), Some recent research on the bearing capacity of foundations. *Canadian Geotechnical Journal*, Vol. 1, No. 1, pp. 16–26.

Skempton, A.W. and McDonald, D.H. (1957), Allowable settlement of buildings. *Proceedings, Institute of Civil Engineers, Part III*, Vol. 5, pp. 727–768.

Steinbrenner, W. (1934), Tafeln zur setzungsberechnung. *Die Strasse*, Vol. 1, pp. 121–124.

Terzaghi, K. (1943), *Theoretical Soil Mechanics*, New York: John Wiley & Sons, Inc.

Terzaghi, K. (1955), Evaluation of coefficient of subgrade reaction. *Geotechnique*, Vol. 5, No. 4, pp. 297–326.

Terzaghi, K. and Peck, R.B. (1967), *Soil Mechanics in Engineering Practice*, New York: John Wiley and Sons.

Vesic, A.S. (1961), Bending of beam resting on isotropic elastic solid. *Journal of the Engineering Mechanics Division, ASCE*, Vol. 87, No. EM2, pp. 35–53.

Vesic, A.S. (1963), Bearing capacity of deep foundations in sand. *Highway Research Record*, No. 39, pp. 112–153.

Vesic, A.S. (1973), Analysis of ultimate loads of shallow foundations. *Journal of Soil Mechanics and Foundations Division, ASCE*, Vol. 99, No. SM1, pp. 45–73.

Wahls, H.E. (1981), Tolerable settlement of buildings. *Journal of the Geotechnical Engineering Division, ASCE*, Vol. 107, No. 11, pp. 1489–1504.

Chapter 4

Lateral earth pressure

4.1 Introduction

Lateral earth pressure is an important design element in several foundation engineering problems, such as retaining walls, excavations, slopes, and even pile foundations. Thus, it is necessary to estimate the lateral earth pressure acting on structural members, which is a function of several factors, including the unit weight of soil, shear strength parameters, and the types and amount of structural member displacement.

Figure 4.1 shows three types of movement of a retaining wall. When the wall is restrained from movement, the lateral earth pressure is called the at-rest earth pressure, σ_{h0} (Figure 4.1a). Under the action of lateral earth pressure on the wall, the wall rotates with respect to the wall toe. When the rotation is sufficiently large, a failure surface forms behind the wall, and the lateral earth pressure acting on the wall is called the active earth pressure, σ_a (Figure 4.1b). The soil behind the wall settles or moves downward. On the other hand, if the wall is pushed backward, it induces the formation of a failure surface (Figure 4.1c). Under such conditions, the lateral earth pressure is referred to as passive earth pressure, σ_p. The soil behind the wall heaves or moves upward.

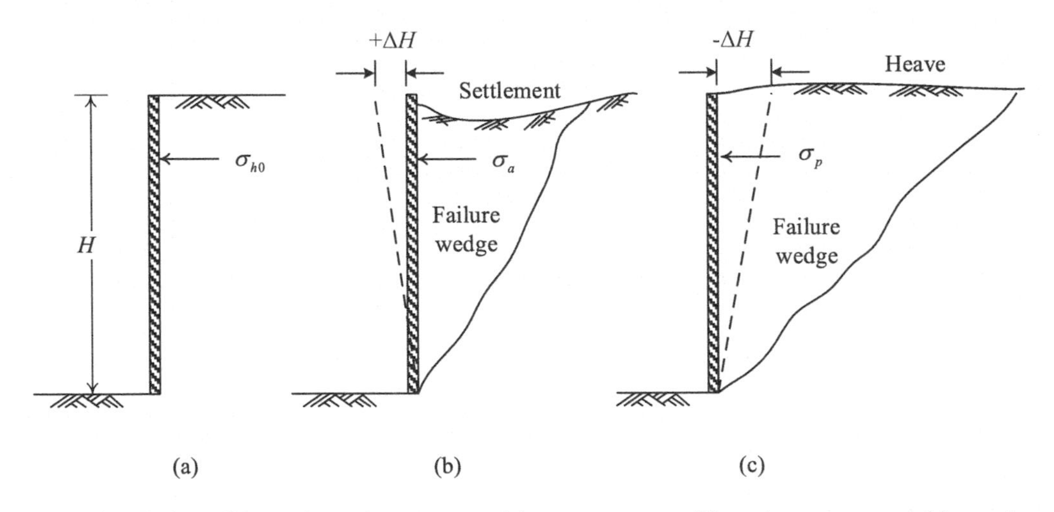

Figure 4.1 Types of lateral earth pressure: (a) at-rest state, (b) active state, and (c) passive state.

DOI: 10.1201/9781003350019-4

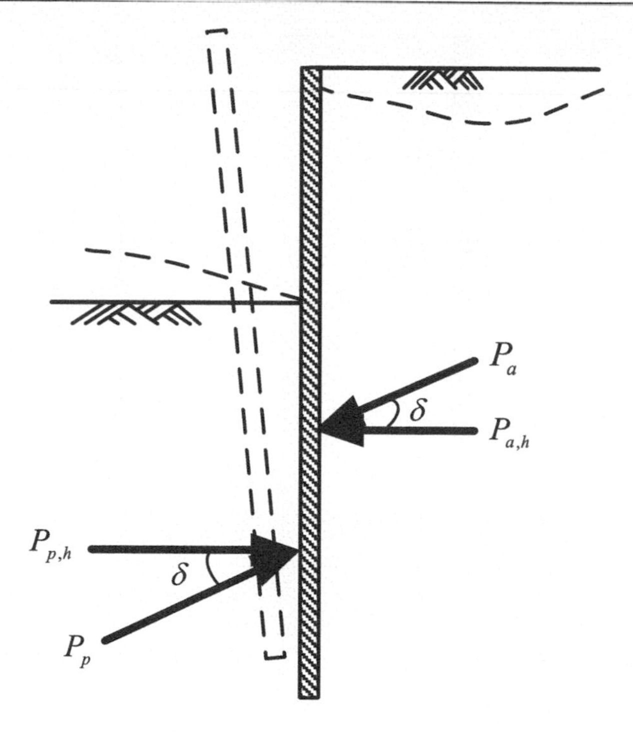

Figure 4.2 Direction of active and passive forces on a vertical wall.

As shown in Figure 4.2, in the problem of retaining walls, the soil behind the wall is initially in the at-rest state. As the wall moves, the earth pressure evolves into active earth pressure. Assuming the friction angle between the wall and soil is δ, the active earth pressure, P_a, inclines at an angle δ to the normal to the wall surface as the soil moves forward and downward with the forward movement of the wall. Conversely, the forward movement of the wall compresses the soil in front of the wall, and the earth pressure gradually transitions into a passive state. The passive earth pressure P_p inclines at an angle δ to the normal to the wall surface as the soil is compressed and heaves with the forward movement of the wall.

This chapter addresses the estimation of those lateral earth pressures based on various earth pressure theories.

4.2 Lateral earth pressure at rest

Figure 4.3a shows a vertical retaining wall with height H. Assume that friction does not exist between the retaining wall and that the soil and the wall are infinitely long. When the wall is restrained from movement, the stresses at depth z below the ground surface are under elastic equilibrium with no shear stress. Supposing σ'_{v0} represents the effective vertical overburden pressure, the effective lateral pressure σ'_{h0} at rest is:

$$\sigma'_{h0} = K_0 \sigma'_{v0} \tag{4.1}$$

where K_0 is the coefficient of lateral earth pressure at rest.

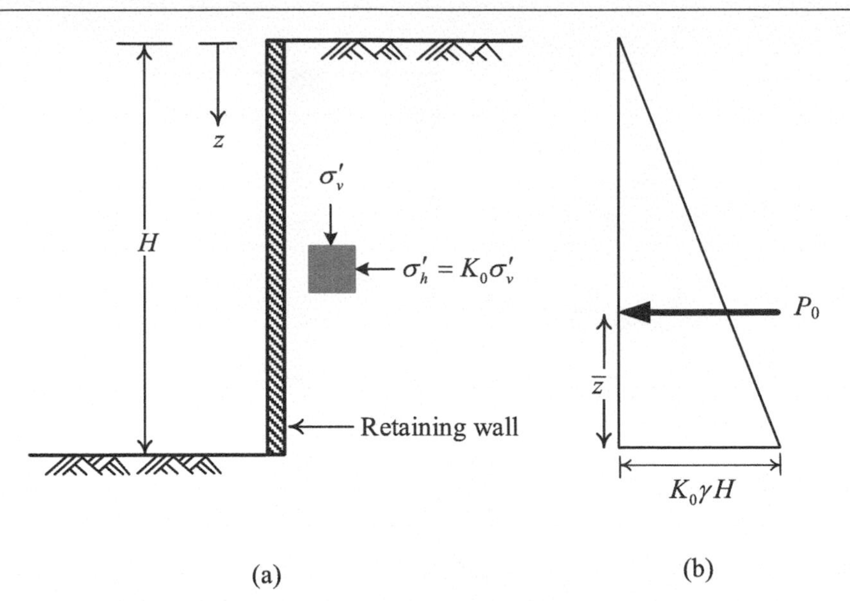

Figure 4.3 Lateral earth pressure at rest: (a) stress of soil at depth z; (b) distribution of the at-rest earth pressure.

The total lateral earth pressure at rest is:

$$\sigma_{h0} = \sigma'_{h0} + u_w \tag{4.2}$$

where u_w is the pore water pressure, which is the summation of the hydrostatic water pressure and excess pore water pressure.

For cohesionless soil, K_0 can be estimated by Jaky's (1944) equation:

$$K_0 = 1 - \sin\phi' \tag{4.3}$$

where ϕ' is the effective internal angle of friction, also called the drained angle of friction.

When cohesionless soil is in the preconsolidated state, that is, overconsolidated, K_0 can be estimated by the following equation (Schmidt, 1967, Alpan, 1967):

$$K_{0,\text{OC}} = K_{0,\text{NC}}(\text{OCR})^{\alpha} \tag{4.4}$$

where:

$K_{0,\text{OC}}$ = coefficient of lateral earth pressure at rest for overconsolidated soil with the overconsolidation ratio, OCR

$K_{0,\text{NC}}$ = coefficient of lateral earth pressure at rest for normally consolidated soil.
α = empirical coefficient, $\alpha \approx \sin\phi'$.

Ladd et al. (1977) suggested that K_0 can also be estimated by eq. 4.3 for normally consolidated cohesive soil and by eq. 4.4 for overconsolidated cohesive soil.

Eqs 4.1 and 4.2 can be used to determine the variation in the at-rest earth pressure with depth, as shown in Figure 4.3b. The resultant force, P_0, per unit length of the wall can be obtained from the area of the pressure diagram in Figure 4.3b as:

$$P_0 = K_0 \gamma H \times H \times \frac{1}{2} = \frac{1}{2} K_0 \gamma H^2 \tag{4.5}$$

The location of the line of action of the resultant force can be obtained by computing the moment about the bottom of the wall, which is $\overline{z} = H/3$ for this case.

In addition to estimating the lateral earth pressure on the wall that is restrained from movement, the at-rest earth pressure or K_0 value is treated as the initial stress in the numerical analysis.

Example 4.1

A retaining wall 6 m high in sand is shown in Figure EX 4.1. The groundwater level is located 2 m below the surface. A uniformly distributed load $q = 30$ kN/m² is applied on the surface. The unit weight of the soil above and below the groundwater level is 17 and 20 kN/m³, respectively. The effective strength parameters (c', ϕ') are shown in the figure. Compute the at-rest earth pressure acting on the wall, including the pore water pressure.

Solution

According to eq. 4.3, $K_0 = 1 - \sin\phi' = 1 - \sin 34^\circ = 0.44$.

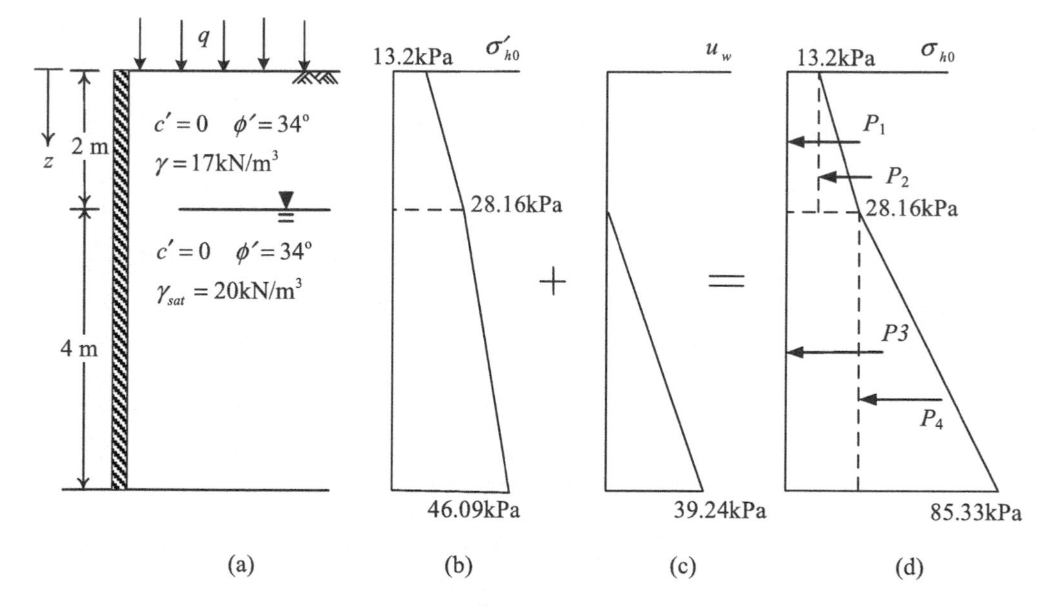

Figure EX 4.1 (a) A 6-m high retaining wall in sand, (b) variation in the effective lateral earth pressure with depth, (c) variation in the pore water pressure with depth, and (d) variation in the total lateral earth pressure with depth.

As stated in Section 4.7, as a uniformly distributed load, q, is applied on the entire surface, the increase in the vertical stress at any depth below the surface is equal to q, and the increase in the lateral stress is $K_0 q$.

At $z = 0$ m:

$\sigma'_{v0} = \sigma_{v0} = q = 30\ \text{kN/m}^2$

$\sigma'_{h0} = \sigma_{h0} = K_0 \times \sigma'_{v0} = 30 \times 0.44 = 13.2\ \text{kN/m}^2$

At $z = 2$ m:

$\sigma'_{v0} = \sigma_{v0} = q + \gamma z = 30 + 17 \times 2 = 64\ \text{kN/m}^2$

$\sigma'_{h0} = K_0 \times \sigma'_{v0} = 64 \times 0.44 = 28.16\ \text{kN/m}^2$

$u_w = 0$

$\sigma_{h0} = \sigma'_{h0} + u_w = 28.16\ \text{kN/m}^2$

At $z = 6$ m:

$\sigma_{v0} = 64 + 20 \times 4 = 144\ \text{kN/m}^2$

$u_w = 9.81 \times 4 = 39.24\ \text{kN/m}^2$

$\sigma'_{h0} = K_0(\sigma_{v0} - u_w) = 46.09\ \text{kN/m}^2$

$\sigma_{h0} = \sigma'_{h0} + u_w = 85.33\ \text{kN/m}^2$

Figures EX 4.1b, c, and d show the variation in effective lateral earth pressure, pore water pressure, and total lateral earth pressure at a given depth, respectively. The total resultant force per unit length of the wall (Figure EX 4.1d) can then be calculated as:

$$P_1 = 13.2 \times 2 = 26.4\ \text{kN/m},\ P_2 = (28.16 - 13.2) \times 2/2 = 14.96\ \text{kN/m},$$
$$P_3 = 28.16 \times 4 = 112.64\ \text{kN/m},\ P_4 = (85.33 - 28.16) \times 4/2 = 114.34\ \text{kN/m}$$
$$P_0 = P_1 + P_2 + P_3 + P_4 = 268.34\ \text{kN/m}$$

Location of the line of action of the total resultant force measured from the bottom of the wall:

$$\bar{z} = \frac{P_1(2/2+4) + P_2(2 \times 1/3 + 4) + P_3(4 \times 1/2) + P_4(4 \times 1/3)}{P_1 + P_2 + P_3 + P_4} = 2.16\ \text{m}$$

Example 4.2

Figure EX. 4.2a shows a retaining wall 6 m high in normally consolidated clay. The groundwater level is located 2 m below the surface. The unit weight of the soil above and below the groundwater level is 15 and 18 kN/m^3, respectively. Both the effective strength parameters (c', ϕ') and total strength undrained parameters (c_u, ϕ_u) are also shown in the figure. Compute the at-rest earth pressure acting on the wall, including the pore water pressure.

Solution

According to eq. 4.3, $K_0 = 1 - \sin\phi' = 1 - \sin 30^\circ = 0.5$.

Since the soil is in the at-rest state, the wall does not move, and no shear stress is induced in the soil. Therefore, no excess pore water pressure is generated in the soil, and the effective

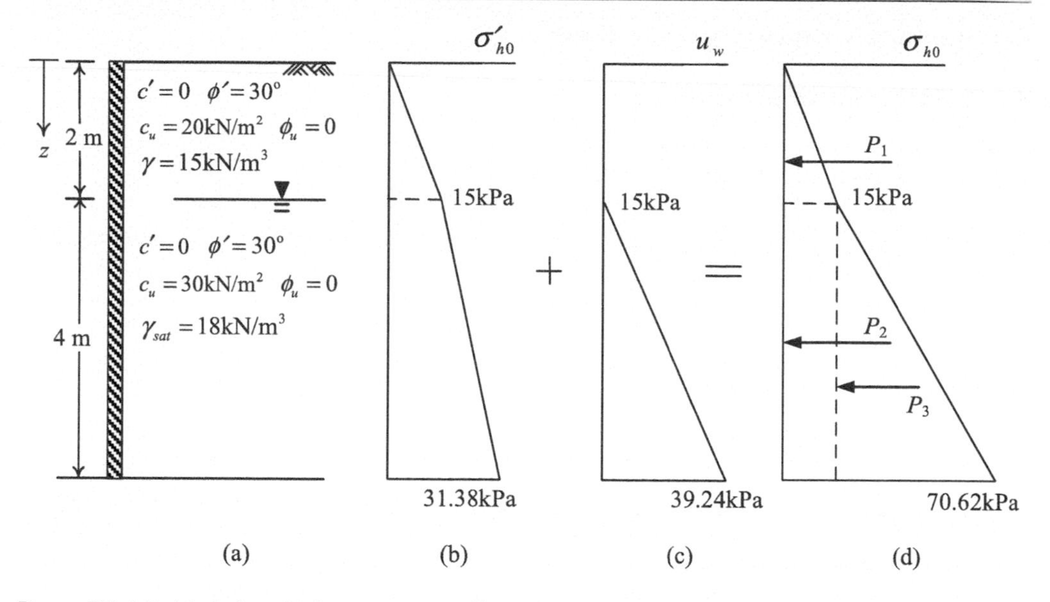

Figure EX 4.2 (a) A 6-m high retaining wall in clay, (b) variation in the effective lateral earth pressure with depth, (c) variation in the pore water pressure with depth, (d) variation in the total lateral earth pressure with depth.

stress analysis should be conducted. The effective strength parameters will be used in the analysis.

At $z = 2$ m:

$$\sigma_{v0} = 15 \times 2 = 30 \text{ kN/m}^2$$

$$\sigma'_{h0} = K_0 \times \sigma'_{v0} = 30 \times 0.5 = 15 \text{ kN/m}^2, \; u_w = 0$$

$$\sigma_{h0} = \sigma'_{h0} + u_w = 15 \text{ kN/m}^2$$

At $z = 6$ m:

$$\sigma_{v0} = 15 \times 2 + 18 \times 4 = 102 \text{ kN/m}^2, \; \sigma'_{h0} = K_0(\sigma_{v0} - u_w) = 31.38 \text{ kN/m}^2,$$

$$u_w = 9.81 \times 4 = 39.24 \text{ kN/m}^2$$

$$\sigma_{h0} = \sigma'_{h0} + u_w = 70.62 \text{ kN/m}^2$$

The variations in the effective lateral earth pressure, pore water pressure, and total lateral earth pressure with depth are shown in Figures EX 4.2b, c, and d, respectively. The total resultant force per unit length of the wall (Figure EX 4.2d) can then be calculated as:

$$P_1 = 15 \times 2/2 = 15 \text{ kN/m}, \; P_2 = 15 \times 4 = 60 \text{ kN/m}, \; P_3 = (70.62 - 15) \times 4/2 = 111.24 \text{ kN/m}$$
$$P_0 = P_1 + P_2 + P_3 = 15 + 60 + 111.24 = 186.24 \text{ kN/m}$$

Location of the line of action of the total resultant force measured from the bottom of the wall:

$$\bar{z} = \frac{P_1(2 \times 1/3 + 4) + P_2(4 \times 1/2) + P_3(4 \times 1/3)}{P_1 + P_2 + P_3} = 1.82 \text{ m}$$

4.3 Rankine's earth pressure theory

4.3.1 Active earth pressure

Rankine (1857) developed a theory of lateral earth pressure in front of and behind a retaining wall on the basis of the concept of plastic equilibrium.

Assume that the retaining wall is infinitely long. As shown in Figure 4.4a, the parameters for the Mohr–Coulomb failure line in the back of the retaining wall are c and ϕ. Assuming that there is no friction between the retaining wall and the soil in the back of the retaining wall, the stress state is in the K_0 state before the retaining wall moves. The stress state of the soil at depth z below the ground surface can be represented by Mohr's circle **a** in Figure 4.4b. Due to the earth pressure acting on the back of the retaining wall, the wall rotates with respect to the wall toe and moves to AB' from AB. The horizontal stress thus decreases with the movement (ΔH) at the top of the wall, while the vertical stress remains unchanged, as shown by Mohr's circle **b**. As the movement (ΔH) or Mohr's circle grows sufficiently large, causing Mohr's circle to be tangential to the Mohr–Coulomb failure line, the soil is in failure. This type of failure is called active failure, and the lateral earth pressure on the retaining wall is called the active earth pressure, which is represented by σ_a on circle **c**. Therefore:

$$\sin\phi = \frac{\text{AB}}{\text{O'A}} = \frac{\text{AB}}{\text{O'O+OA}} = \frac{(\sigma_v - \sigma_a)/2}{c\cot\phi + (\sigma_v + \sigma_a)/2}$$

Which can be simplified as follows:

$$\sigma_a = \sigma_v \frac{1-\sin\phi}{1+\sin\phi} - 2c\frac{\cos\phi}{1+\sin\phi} \tag{4.6}$$

$$= \sigma_v \tan^2(45^\circ - \frac{\phi}{2}) - 2c\tan(45^\circ - \frac{\phi}{2})$$

$$= \sigma_v K_a - 2c\sqrt{K_a} \tag{4.6a}$$

where K_a is the coefficient of Rankine's active earth pressure, $K_a = \tan^2(45^\circ - \phi/2)$.

According to Mohr's failure theory, a failure zone forms behind the retaining wall that is called the active failure zone. The soil in the failure zone is all in failure. The failure surfaces all form an angle of $(45^\circ + \phi/2)$ with the horizontal (see Figure 4.4c).

For effective stress analysis or drained analysis, the parameters of the Mohr–Coulomb failure line should be expressed in terms of the effective cohesion (c') and the effective internal angle of friction (ϕ'). Eqs. 4.6 and 4.6a should be rewritten in terms of effective stress as follows:

$$\sigma'_a = \sigma'_v \tan^2(45^\circ - \frac{\phi'}{2}) - 2c'\tan(45^\circ - \frac{\phi'}{2}) \tag{4.7}$$

$$= \sigma'_v K_a - 2c'\sqrt{K_a} \tag{4.7a}$$

where $K_a = \tan^2(45^\circ - \phi'/2)$.

According to Mohr–Coulomb theory, the angle between the active failure surface and the horizontal plane is $45^\circ + \phi'/2$.

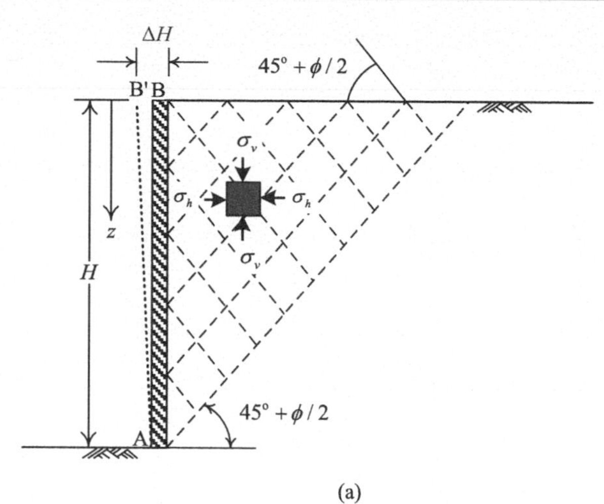

(a)

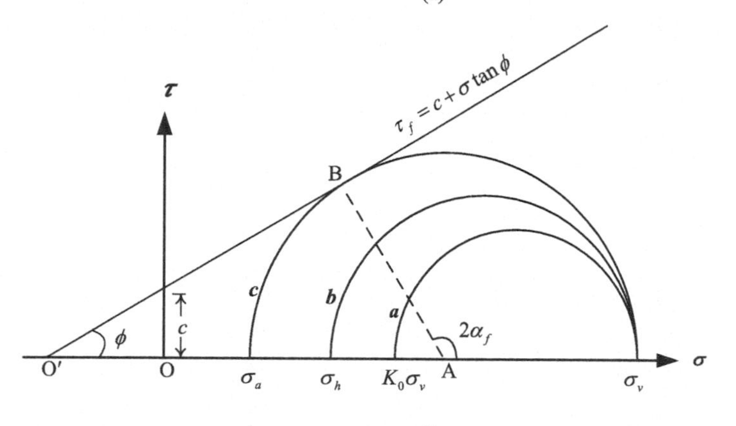

(b)

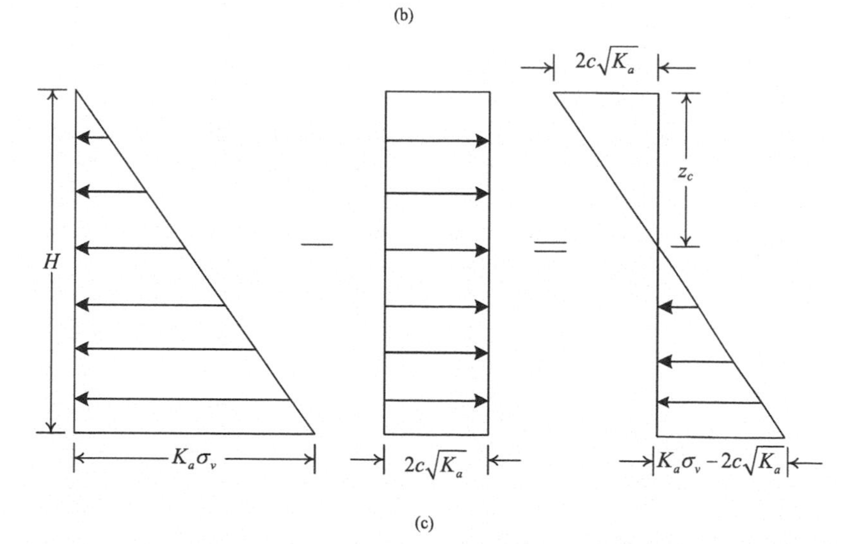

(c)

Figure 4.4 Rankine's active earth pressure: (a) stress state at depth z, (b) evolution of Mohr's circle, (c) variation in the active earth pressure with depth.

For total stress undrained analysis, the pressure (or stress) and strength parameters should be expressed in terms of total stress parameters. For example, σ_a, σ_v, σ_h, c, and ϕ. The equation has exactly the same form as that shown in eq. 4.6.

For saturated clay under the short-term or undrained condition, the "$\phi = 0$" concept is applied. Assuming the undrained shear strength is s_u, the total active earth pressure (σ_a) on the wall is:

$$\sigma_a = \sigma_v - 2s_u \tag{4.8}$$

where $K_a = \tan^2(45^\circ - \phi/2) = 1$.

According to eq. 4.6, we can obtain the theoretical distribution of earth pressure as shown in Figure 4.4c. As shown in this figure, there is a tension zone in the cohesive soil behind the wall. Soil is seldom subjected to tension stress, and therefore, tension cracks often occur. Assuming the soil is unable to bear tension stress, let the lateral earth pressure be 0 in the tension crack zone. According to eq. 4.6a, we have:

$$\sigma_a = \gamma z K_a - 2c\sqrt{K_a} = 0 \tag{4.9}$$

The depth of a tension crack is:

$$z_c = \frac{2s_u}{\gamma\sqrt{K_a}} \tag{4.10}$$

where z_c is the depth of tension cracks.

Once tension cracks occur, rain and other environmental factors will further fill the cracks with water, which will increase the total pressure on the wall.

The long-term behavior of cohesive soil should be analyzed on the basis of the complete dissipation of excess pore water pressure. The distribution of earth pressure is estimated to be similar to that of cohesionless soil, as discussed in the following section.

4.3.2 *Passive earth pressure*

Similarly, Mohr's circle **a** in Figure 4.5a shows the initial stress state of the soil at depth z below the ground surface before the movement of the wall. When the wall rotates backward and moves to AB' from AB, the horizontal stress increases with the movement (ΔH), while the vertical stress remains unchanged. The corresponding Mohr circle decreases, as represented by Mohr's circle **b**. When the horizontal stress is greater than the vertical stress, Mohr's circle begins to enlarge with the increase in ΔH until Mohr's circle is tangent to the Mohr–Coulomb failure line, which indicates the soil in failure. This type of failure is called passive failure, and the lateral earth pressure acting on the retaining wall is called passive earth pressure, as represented by σ_p on Mohr's circle **c**.

Again, as shown in Figure 4.5b, we have:

$$\sin\phi = \frac{\text{QD}}{\text{O'D}} = \frac{\text{QD}}{\text{O'O+OD}} = \frac{(\sigma_p - \sigma_v)/2}{c\cot\phi + (\sigma_v + \sigma_p)/2}$$

The preceding equation can be simplified as:

$$\sigma_p = \sigma_v \tan^2(45^\circ + \frac{\phi}{2}) + 2c\tan(45^\circ + \frac{\phi}{2}) \tag{4.11}$$

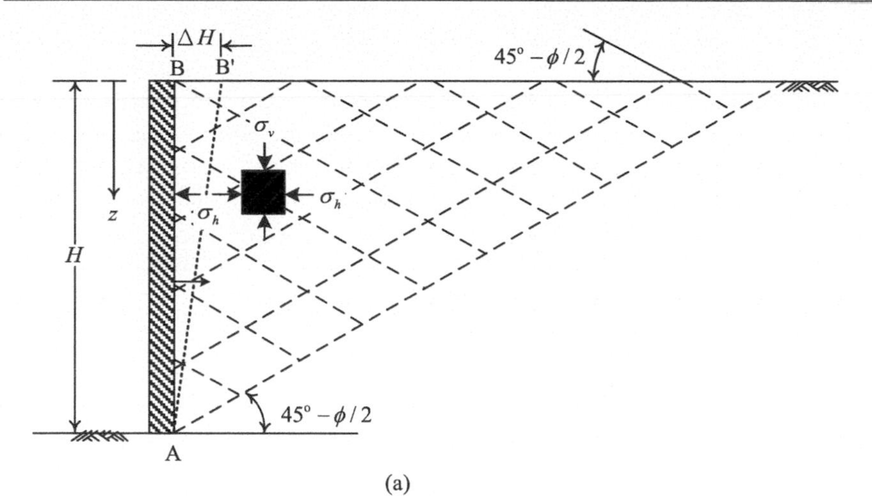

(a)

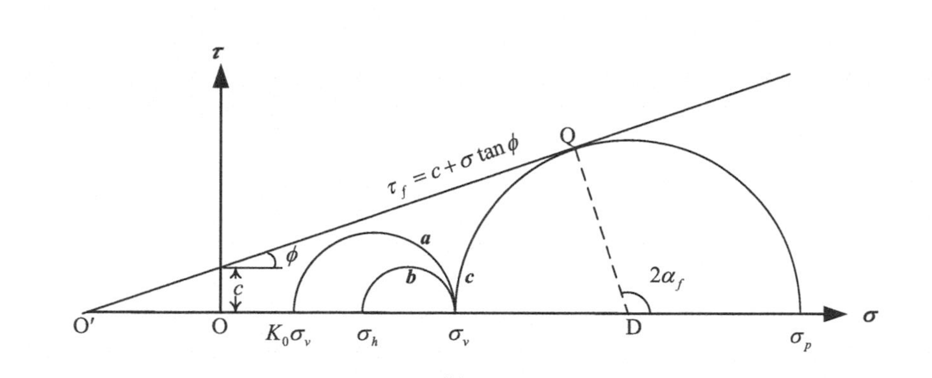

(b)

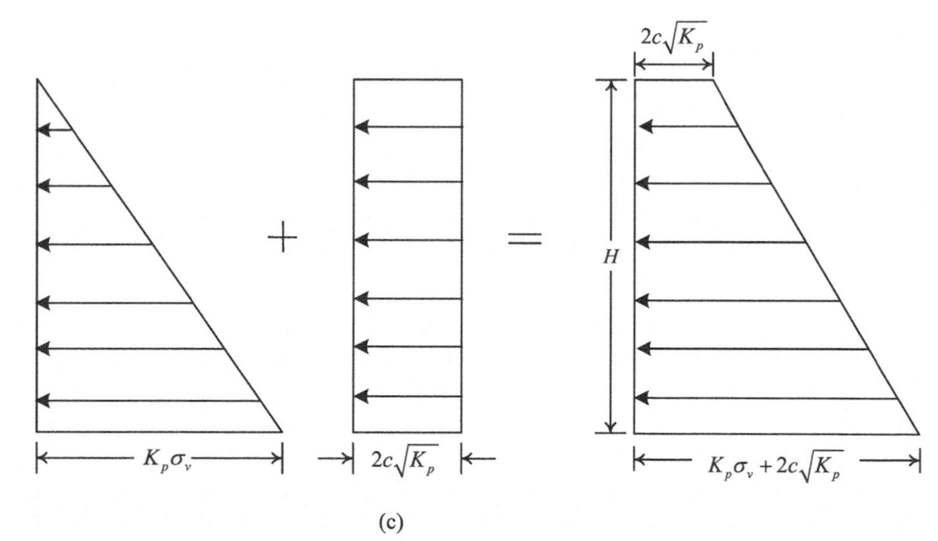

(c)

Figure 4.5 Rankine's passive earth pressure: (a) stress state at depth z, (b) evolution of Mohr's circle, and (c) variation in the passive earth pressure with depth.

$$= \sigma_v K_p + 2c\sqrt{K_p} \tag{4.11a}$$

where K_p is the coefficient of Rankine's passive earth pressure, $K_p = \tan^2(45^\circ + \phi/2)$.

Similarly, according to Mohr's failure theory, the soil within the failure zone where passive failures occur is called the passive failure zone. The soil in the area is all in failure, whose failure surfaces form angles of $45^\circ - \phi/2$ with the horizontal plane, as shown in Figure 4.5c.

Similar to the active earth pressure, for effective stress analysis or drained analysis, eqs. 4.11 and 4.11a should be rewritten in terms of effective stress as:

$$\sigma'_p = \sigma'_v \tan^2(45^\circ + \frac{\phi'}{2}) + 2c' \tan(45^\circ + \frac{\phi'}{2}) \tag{4.12}$$

$$= \sigma'_v K_p + 2c'\sqrt{K_p} \tag{4.12a}$$

where $K_p = \tan^2(45^\circ + \phi'/2)$.

For saturated clay under the short-term or undrained condition, the "$\phi = 0$" concept is applied. Assuming the undrained shear strength is s_u, the total passive earth pressure (σ_p) acting on the wall is:

$$\sigma_p = \sigma_v + 2s_u \tag{4.13}$$

4.3.3 *Earth pressure with sloping ground*

According to Rankine's earth pressure theory, the coefficients of the earth pressure for sloping ground (Bowles, 1988) can be computed as follows:

$$K_a = \frac{\cos\beta - \sqrt{\cos^2\beta - \cos^2\phi}}{\cos\beta + \sqrt{\cos^2\beta - \cos^2\phi}} \tag{4.14}$$

$$K_p = \frac{\cos\beta + \sqrt{\cos^2\beta - \cos^2\phi}}{\cos\beta - \sqrt{\cos^2\beta - \cos^2\phi}} \tag{4.15}$$

where β = slope of the ground.

Thus, the active and passive earth pressures and their resultants acting on the retaining wall, as shown in Figure 4.6, are:

$$\sigma_a = \gamma z K_a \tag{4.16}$$

$$P_a = \frac{1}{2}\gamma H^2 K_a \tag{4.17}$$

$$\sigma_p = \gamma z K_p \tag{4.18}$$

$$P_p = \frac{1}{2}\gamma H^2 K_p \tag{4.19}$$

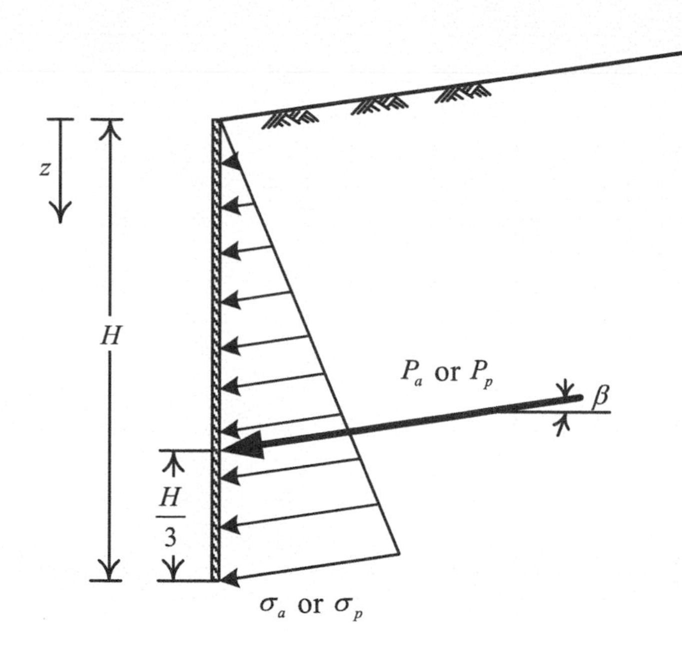

Figure 4.6 Rankine's earth pressures with sloping ground.

where:
γ = unit weight of soil.
H = height of the retaining wall.
P_a = resultant of active earth pressure.
P_p = resultant of passive earth pressure.

The horizontal and vertical components of the resultants are:

$$\sigma_{a,h} = \gamma z K_a \cos\beta\,,\ \sigma_{a,v} = \gamma z K_a \sin\beta \tag{4.20}$$

$$P_{a,h} = \frac{1}{2}\gamma H^2 K_a \cos\beta\,,\ P_{a,v} = \frac{1}{2}\gamma H^2 K_a \sin\beta \tag{4.21}$$

$$\sigma_{p,h} = \gamma z K_p \cos\beta\,,\ \sigma_{p,v} = \gamma z K_p \sin\beta \tag{4.22}$$

$$P_{p,h} = \frac{1}{2}\gamma H^2 K_p \cos\beta\,,\ P_{p,v} = \frac{1}{2}\gamma H^2 K_p \cos\beta \tag{4.23}$$

Example 4.3

Figure EX. 4.3a shows a 6 m high retaining wall in sand. The groundwater level is located 3 m below the surface. The unit weight of the soil above and below the groundwater level are 16 and 21 kN/m^3, respectively. The effective stress parameters (c', ϕ') are also shown in the figure. Compute the active earth pressure on the wall using Rankine's earth pressure theory.

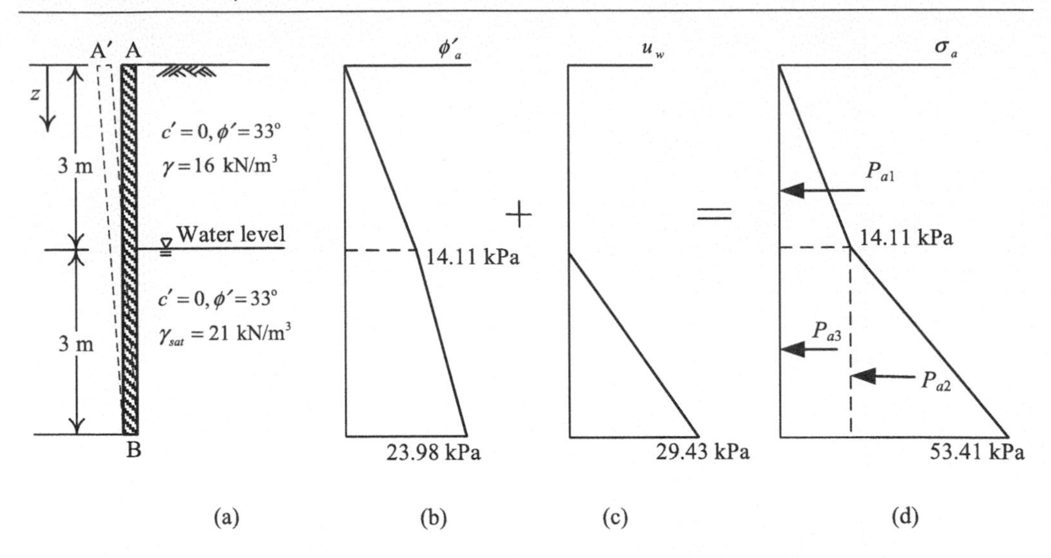

Figure EX 4.3 (a) A 6 m high retaining wall in sand, (b) variation in the effective active earth pressure with depth, (c) variation in the pore water pressure with depth, and (d) variation in the total active earth pressure with depth.

Solution

For $\phi' = 33°$, $K_a = \tan^2(45 - \frac{\phi'}{2}) = \tan^2(45 - \frac{33°}{2}) = 0.294$

At $z = 3$ m:

$\sigma'_a = 16 \times 3 \times 0.294 = 14.11 \text{ kN/m}^2$, $u_w = 0$

$\sigma_a = \sigma'_a + u_w = 14.11 \text{ kN/m}^2$

At $z = 6$ m:

$\sigma'_a = [16 \times 3 + (21 - 9.81) \times 3] \times 0.294 = 23.98 \text{ kN/m}^2$, $u_w = 9.81 \times 3 = 29.43 \text{ kN/m}^2$

$\sigma_a = \sigma'_a + u_w = 53.41 \text{ kN/m}^2$

Figures EX 4.3b, c, and d show the variations in the effective active earth pressure, pore water pressure, and total active earth pressure, respectively.

The total resultant force (Figure EX 4.3d) per unit length of the wall can be calculated as follows:

$$P_{a1} = 14.11 \times 3/2 = 21.17 \text{ kN/m}, \; P_{a2} = (53.41 - 14.11) \times 3/2 = 58.95 \text{ kN/m}$$

$$P_{a3} = 14.11 \times 3 = 42.33 \text{ kN/m}$$

$$P_a = P_{a1} + P_{a2} + P_{a3} = 21.17 + 58.95 + 42.33 = 122.45 \text{ kN/m}$$

Location of the line of action of the resultant force measured from the bottom of the wall:

$$\bar{z} = \frac{P_{a1} \times (3 + 3 \times 1/3) + P_{a2}(3 \times 1/3) + P_{a3}(3 \times 1/2)}{P_{a1} + P_{a2} + P_{a3}} = \frac{207.13}{122.45} = 1.69 \text{ m}$$

Example 4.4

Figure EX 4.4 shows a 6 m high retaining wall in clay where the soil condition is the same as in example 4.2. Compute the active earth pressure on the wall using Rankine's earth pressure theory.

Solution

The "$\phi = 0$ concept" is applied for clay, $K_a = \tan^2(45 - \frac{\phi}{2}) = \tan^2(45 - \frac{0°}{2}) = 1$.

1. Tension cracks are not formed, that is, the soil is able to bear tension stress.

At $z = 0$ m:

$\sigma_a = -2c\sqrt{K_a} = -40$ kN/m²

At $z = 2$ m⁻:

$\sigma_a = 15 \times 2 - 2 \times 20 = 30 - 40 = -10$ kN/m²

At $z = 2$ m⁺:

$\sigma_a = 15 \times 2 - 2 \times 30 = 30 - 60 = -30$ kN/m²

At $z = 6$ m:

$\sigma_a = 15 \times 2 + 18 \times 4 - 2 \times 30 = 42$ kN/m²

The variation in earth pressure with depth is shown in Figure EX 4.4a. From this figure, we have:

$P_{a1} = 30 \times 2 \times 0.5 = 30$ kN/m, $P_{a2} = 30 \times 4 = 120$ kN/m

$P_{a3} = 72 \times 4 \times 0.5 = 144$ kN/m

$P_{a4} = 2 \times 40 = 80$ kN/m

$P_{a5} = 4 \times 60 = 240$ kN/m

The resultant force per unit length of the wall is:

$$P_{a1} + P_{a2} + P_{a3} - P_{a4} - P_{a5} = 30 + 120 + 144 - 80 - 240 = -26 \text{ kN/m}$$

2. Tension cracks occur.

The depth of the tension crack can be calculated as:

$$0 = [2 \times 15 + (z_c - 2) \times 18] \times K_a - 2 \times 30, \; z_c = 3.67 \text{ m}$$

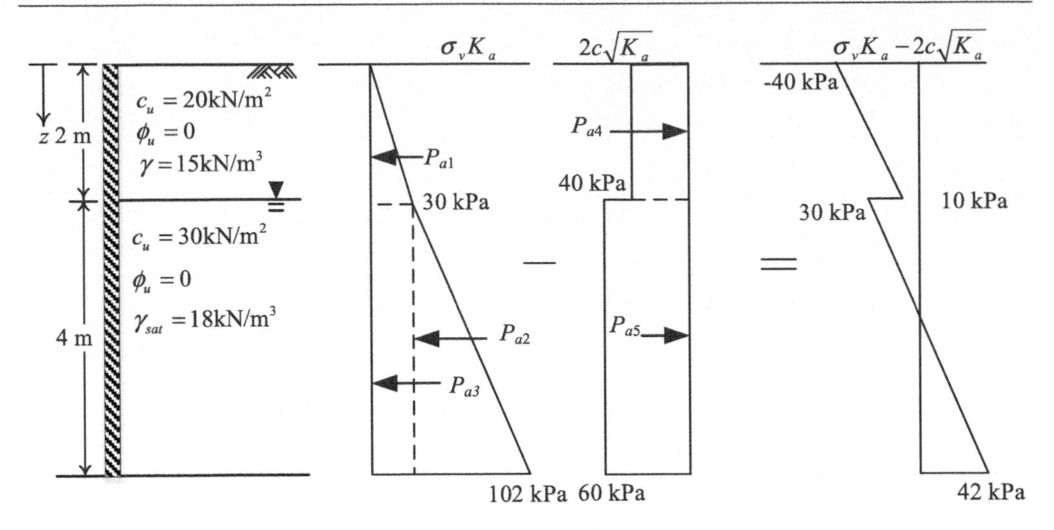

Figure EX 4.4 (a) Variation in the active earth pressure without the occurrence of tension cracks in a 6 m high retaining wall in clay.

The active earth pressure is:

At z = 3.67 m:

$$\sigma_a = 0$$

At z = 6 m:

$$\sigma_a = 2 \times 15 + 4 \times 18 - 2 \times 30 = 102 - 60 = 42 \text{ kN/m}^2$$

If the tension cracks are full of water due to the original groundwater, then the pore water pressure in the tension crack is:

At $z = 2$ m:

$$u_w = 0$$

At $z = 3.67$ m:

$$u_w = (3.67 - 2) \times 9.81 = 16.38 \text{ kN/m}^2$$

The variation in active earth pressure and pore water pressure with depth is shown in Figure EX 4.4b.

The resultant force per unit length of the wall is:

$$P_a = 42 \times (6 - 3.67) \times 0.5 = 48.93 \text{ kN/m}, \ P_w = 16.38 \times (3.67 - 2) / 2 = 13.68 \text{ kN/m}$$

$$P_a + P_w = 62.61 \text{ kN/m}$$

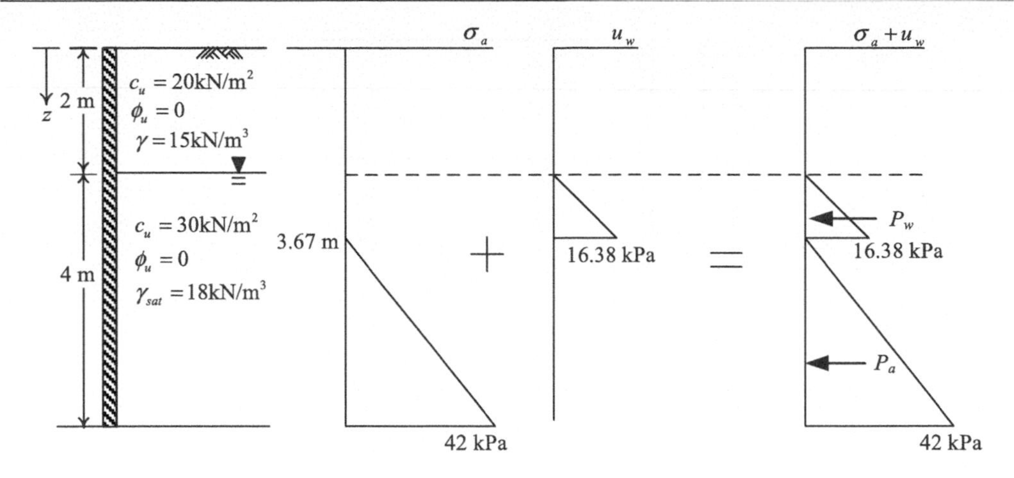

Figure EX 4.4 (b) Variation in the active earth pressure with tension cracks in a 6 m high retaining wall in clay.

Location of the resultant force from the bottom of the wall:

$$\bar{z} = \frac{P_a \times (2.33 \times 1/3) + P_w \times (2.33 + 1.67 \times 1/3)}{P_a + P_w} = \frac{77.44}{62.61} = 1.24 \text{ m}$$

Example 4.5

Figure EX. 4.5a shows a 9 m high retaining wall. The sand and clay deposits are behind the wall. The groundwater level is located on the surface. The saturated unit weight of the soils and strength parameters are indicated in the figure. Compute the active earth pressure on the wall using Rankine's earth pressure theory.

Solution

$$\text{Sand}(1): K_a = \tan^2(45 - \frac{\phi'}{2}) = \tan^2(45 - \frac{31°}{2}) = 0.32$$

$$\text{Sand}(2): K_a = \tan^2(45 - \frac{\phi'}{2}) = \tan^2(45 - \frac{33°}{2}) = 0.3$$

$$\text{Clay}: K_a = \tan^2(45 - \frac{\phi}{2}) = \tan^2(45 - \frac{0°}{2}) = 1$$

The earth pressure (σ'_a or σ_a) and pore water pressure (u_w) at each depth are calculated as follows.

The effective stress analysis with pore water pressure consideration is used for sand, while the total stress undrained analysis without pore water pressure consideration is used for clay. Therefore:

At $z = 3\text{ m}^-$:

$\sigma'_a = (21 - 9.81) \times 3 \times 0.32 = 10.74$ kN/m², $u_w = 9.81 \times 3 = 29.43$ kN/m²

$\sigma_a = \sigma'_a + u_w = 40.17$ kN/m²

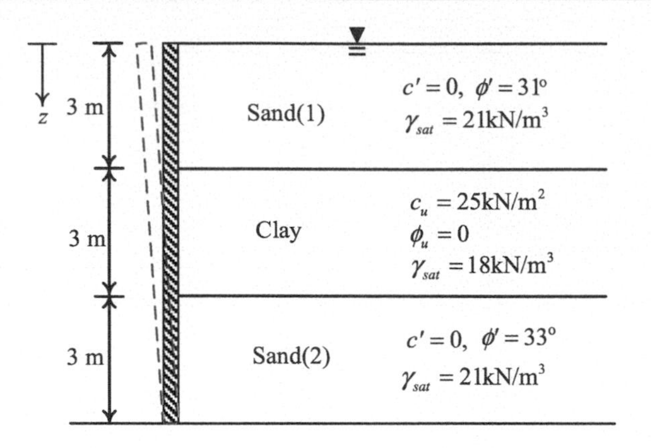

Figure EX 4.5 (a) A 9 m high retaining wall in a sand and clay deposit.

At $z = 3\,\text{m}^+$:

$\sigma_a = 21 \times 3 \times 1 - 2 \times 25 = 13\,\text{kN/m}^2$

At $z = 6\,\text{m}^-$:

$\sigma_a = 21 \times 3 + 18 \times 3 - 2 \times 25 = 67\,\text{kN/m}^2$

At $z = 6\,\text{m}^+$:

$\sigma'_a = (117\text{-}6 \times 9.81) \times 0.3 = 17.44\,\text{kN/m}^2,\ u_w = 9.81 \times 6 = 58.86\,\text{kN/m}^2$

$\sigma_a = \sigma'_a + u_w = 76.3\,\text{kN/m}^2$

At $z = 9\,\text{m}$:

$\sigma'_a = [117\text{+}3 \times 21\text{-}(9 \times 9.81)] \times 0.3 = 27.51\,\text{kN/m}^2,\ u_w = 9.81 \times 9 = 88.29\,\text{kN/m}^2$

$\sigma_a = \sigma'_a + u_w = 115.8\,\text{kN/m}^2$

The variations in earth pressure and pore water pressure are shown in Figure EX 4.5b. The resultant force per unit length of the wall is then calculated as:

$$
\begin{aligned}
P_a &= P_{a1} + P_{a2} + P_{a3} + P_{a4} + P_{a5} \\
&= 40.17 \times 3/2 + 13 \times 3 + (67 - 13) \times 3/2 + 76.3 \times 3 + (115.8 - 76.3) \times 3/2 \\
&= 60.26 + 39 + 81 + 228.9 + 59.25 = 468.41\ \text{kN/m}
\end{aligned}
$$

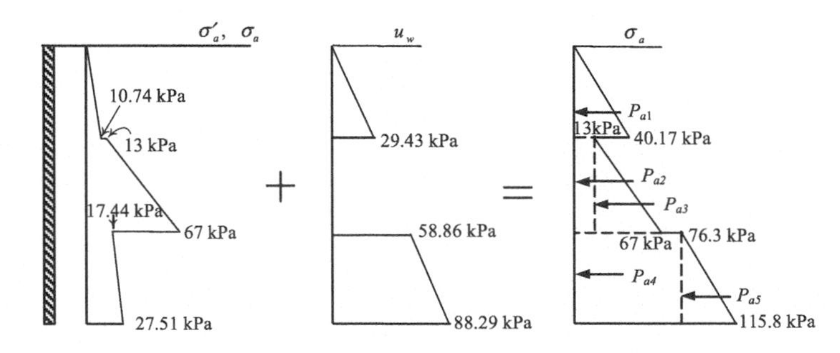

Figure EX 4.5 (b) Variation in the active earth pressure and pore water pressure with depth.

Location of the resultant force measured from the bottom of the wall:

$$\bar{z} = \frac{60.26\times(3/3+6)+39\times(3/2+3)+81\times(3/3+3)+228.9\times3/2+59.25\times3/3}{468.41}$$

$$= \frac{1323.92}{468.41} = 2.83 \text{ m}$$

Example 4.6

The wall and soil conditions are the same as those shown in Figure EX 4.3a. Compute the passive earth pressure on the wall using Rankine's earth pressure theory.

Solution

For $\phi' = 33^\circ$, $K_p = \tan^2(45+\frac{\phi'}{2}) = \tan^2(45+\frac{33^\circ}{2}) = 3.39$

The effective earth pressure (σ_a') and pore water pressure (u_w) at each depth are calculated as follows:

At $z = 3$ m:

$\sigma_p = \sigma_p' = 16\times3\times3.39 = 162.72$ kN/m^2, $u_w = 0$

At $z = 6$ m:

$\sigma_p' = [16\times3+(21\text{-}9.81)\times3]\times3.39 = 276.52$ kN/m^2, $u_w = 9.81\times3 = 29.43$ kN/m^2

$\sigma_p = \sigma_p' + u_w = 305.95$ kN/m^2

The variations in passive earth pressure and pore water pressure are shown in Figure EX 4.6. The resultant force per unit length of the wall can be calculated as:

$P_{p1} = 162.72\times3/2 = 244.08$ kN/m, $P_{p2} = (305.95-162.72)\times3/2 = 214.85$ kN/m

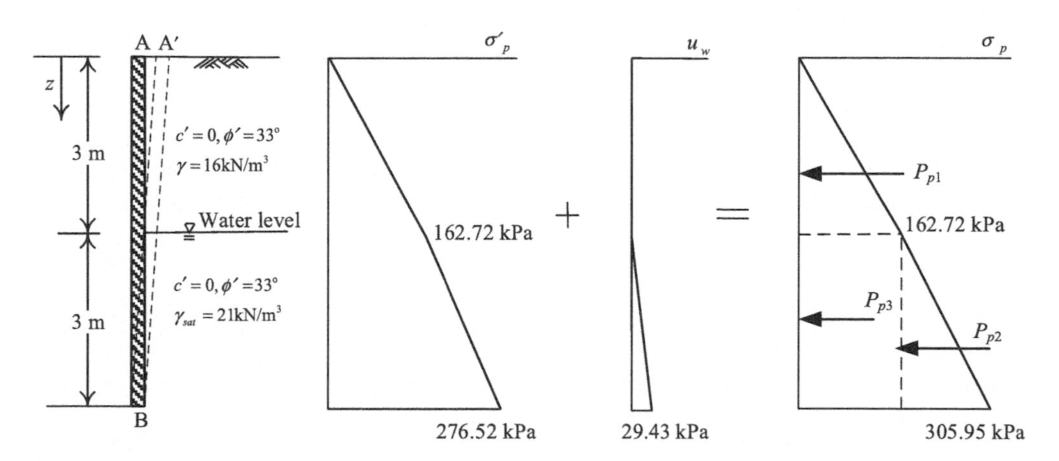

Figure EX 4.6 Variation in passive earth pressure and pore water pressure with depth.

$$P_{p3} = 162.72 \times 3 = 488.16 \text{ kN/m}$$

$$P_P = P_{P1} + P_{P2} + P_{P3} = 947.09 \text{ kN/m}$$

Location of the resultant force from the bottom of the wall:

$$\bar{z} = \frac{244.08 \times (3 + 3 \times 1/3) + 214.85 \times (3 \times 1/3) + 488.16 \times (3 \times 1/2)}{947.09} = 2.03 \text{ m}$$

4.4 Coulomb earth pressure theory

4.4.1 Active earth pressure

Coulomb's earth pressure theory (1776) assumes that the soil in the back of the retaining wall is homogeneous and cohesionless, the failure surface is a plane, the wedge between the wall and the failure surface is rigid material, and the weight of the wedge, the reaction of the soil, and the reaction of the wall are in equilibrium.

Figure 4.7a shows a retaining wall of height H, retaining a soil with an angle of friction ϕ and an angle of friction between the wall and soil δ. The BC_1 line is an assumed failure surface that intersects the horizontal direction at an angle of α_1. Figure 4.7b illustrates the force polygon formed by the reaction (P) of the wall against the wedge, the reaction (R) of the soil against the wedge, and the weight (W) of the soil wedge. The directions and magnitudes of P and R can then be determined from the force polygon.

By assuming different failure surfaces, for example, BC_2 and BC_3, and with a similar procedure, we can yield the corresponding reaction forces (P) of the wall. The largest reaction force, P_{max}, is then the active force on the wall, as shown in Figure 4.7a.

If the failure surface is treated as a variable, expressed by α_1, according to the force polygon in Figure 4.7b, we can derive the following equation:

$$\frac{W}{\sin(90^\circ + \theta + \delta - \alpha + \varphi)} = \frac{P}{\sin(\alpha - \phi)} \tag{4.24}$$

$$P = \frac{W \sin(\alpha - \phi)}{\sin(90^\circ + \theta + \delta - \alpha + \phi)} \tag{4.25}$$

$$= \frac{1}{2}\gamma H^2 \left[\frac{\cos(\theta - \alpha)\cos(\theta - \beta)\sin(\alpha - \phi)}{\cos^2\theta \sin(\alpha - \beta)\sin(90^\circ + \theta + \delta - \alpha + \phi)} \right] \tag{4.25a}$$

where γ is the unit weight of the soil; the parameters γ, ϕ, δ, θ, β, and H are constants; α is a variable because BC is an assumed failure surface; P varies with α; and the maximum P represents the active force P_a, that is:

$$\frac{dP}{d\alpha} = 0 \tag{4.26}$$

Substituting the critical α value derived from eq. 4.26 into eq. 4.25, we obtain the active force (P_a) as follows:

$$P_a = \frac{1}{2}\gamma H^2 K_a \tag{4.27}$$

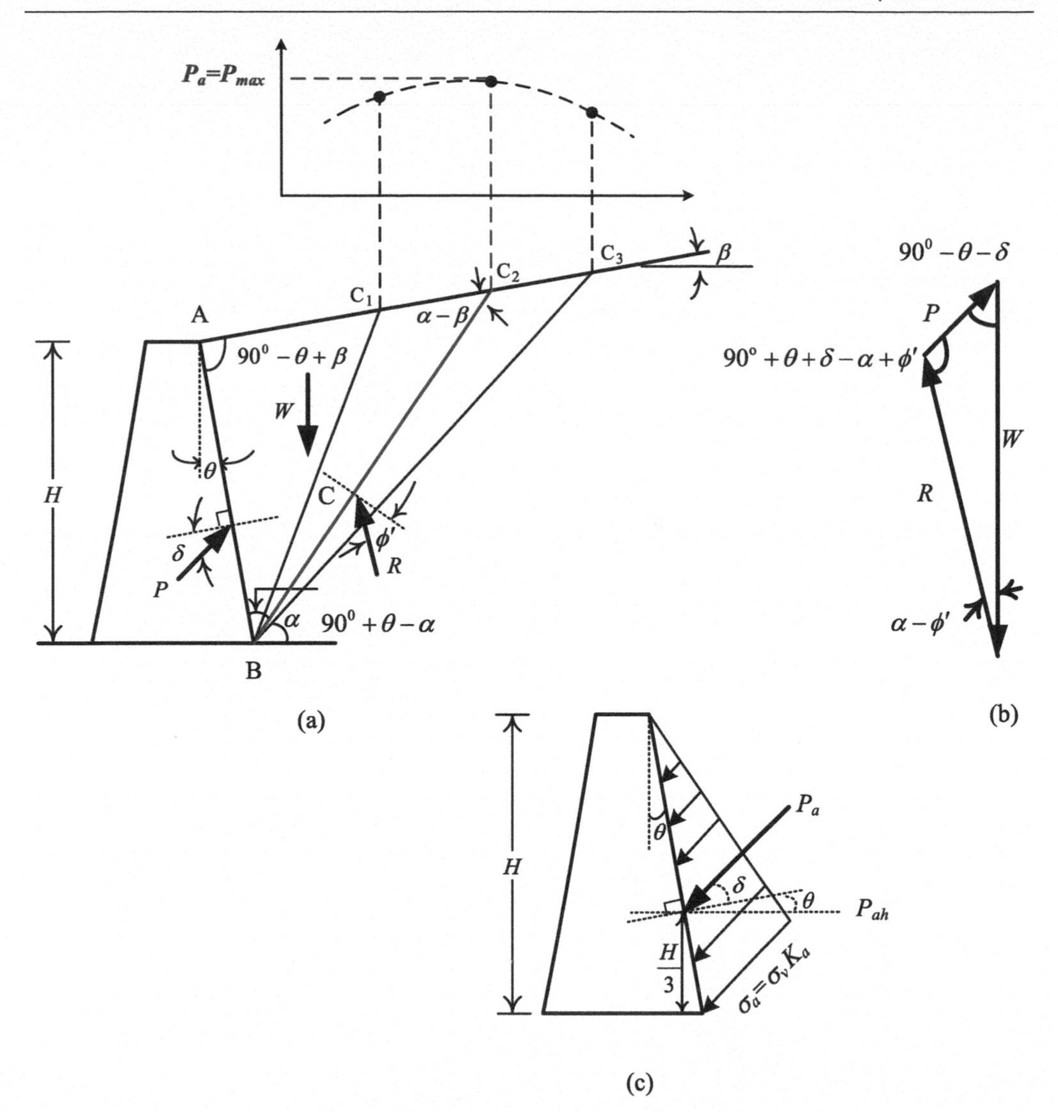

Figure 4.7 Coulomb's active earth pressure: (a) assumed failure surface and the corresponding reaction of the wall against the failure wedge, (b) force polygon, and (c) distribution and resultant of active earth pressure.

$$K_a = \frac{\cos^2(\phi - \theta)}{\cos^2\theta\cos(\delta + \theta)\left[1 + \sqrt{\frac{\sin(\delta + \phi)\sin(\phi - \beta)}{\cos(\delta + \theta)\cos(\theta - \beta)}}\right]^2} \tag{4.27a}$$

where K_a is the coefficient of Coulomb's active earth pressure or active force.

The direction of the active earth pressure and its distribution are shown in Figure 4.7c. The horizontal components of the active force and active earth pressure are $P_{ah} = P_a \cos(\delta + \theta)$ and $\sigma_{ah} = \sigma_a \cos(\delta + \theta)$, respectively. When $\theta = 0$, $\beta = 0$, and $\delta = 0$, eq. 4.27a can be rewritten as $K_a = \tan^2(45^\circ - \phi/2)$, which is identical to Rankine's.

4.4.2 *Passive earth pressure*

Figure 4.8 illustrates the passive soil failure in the back of the retaining wall, which is pushed backward by an external force. The BC line is an assumed failure surface. The directions of the soil reaction force (R) and the reaction (P) of the wall are determined on the basis that the wall is pushed backward and the wedge (ABC) moves upward. Following a similar method used to obtain the active force, the passive force (P_p) can be derived as shown in Figure 4.8b, which can also be expressed by the following equation:

$$P_p = \frac{1}{2}\gamma H^2 K_p \tag{4.28}$$

$$K_p = \frac{\cos^2(\phi + \theta)}{\cos^2\theta\cos(\delta - \theta)\left[1 - \sqrt{\dfrac{\sin(\phi + \delta)\sin(\phi + \beta)}{\cos(\delta - \theta)\cos(\beta - \theta)}}\right]^2} \tag{4.28a}$$

where K_p is the coefficient of Coulomb's passive earth pressure or force.

The direction of the passive force and its distribution are shown in Figure 4.8c. The horizontal components of the passive force and the passive earth pressure are $P_{ph} = P_p \cos(\delta - \theta)$ and $\sigma_{ph} = \sigma_p \cos(\delta - \theta)$, respectively. When $\theta = 0$, $\beta = 0$, and $\delta = 0$, eq. 4.28a can be rewritten as $K_p = \tan^2(45^\circ + \phi/2)$, which is identical to Rankine's.

Example 4.7

Similar to example 4.3. Compute the active earth pressure and its line of action of the resultant force using Coulomb's earth pressure theory with $\delta = 2\phi'/3$.

Solution

According to eq. 4.27a, $K_a = 0.26$.

The horizontal component of the coefficient of active earth pressure is:

$$K_{ah} = K_a \times \cos\delta = 0.26 \times \cos\left(\frac{2}{3} \times 33\right) = 0.24$$

At $z = 3$ m:

$$\sigma_{a,h} = \sigma'_{a,h} = 16 \times 3 \times 0.24 = 11.52\ \text{kN/m}^2$$

At $z = 6$ m:

$$\sigma'_{a,h} = [16 \times 3 + (21 - 9.81) \times 3] \times 0.24 = 19.58\ \text{kN/m}^2,\ u_w = 9.81 \times 3 = 29.43\ \text{kN/m}^2$$

$$\sigma_{a,h} = \sigma'_{a,h} + u_w = 49.01\ \text{kN/m}^2$$

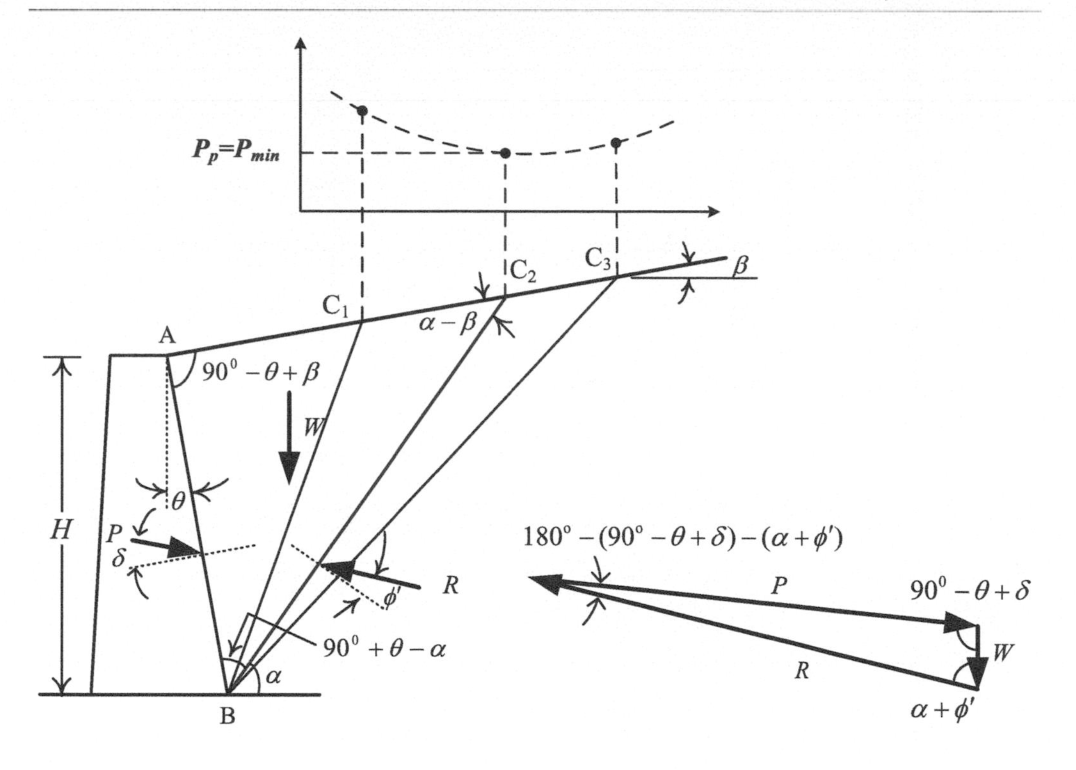

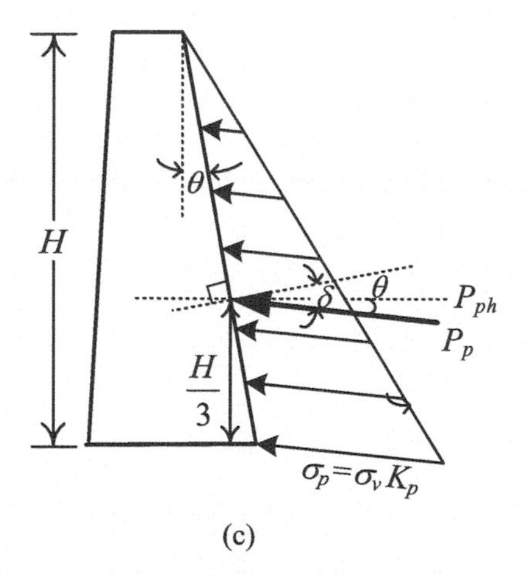

Figure 4.8 Coulomb's passive earth pressure: (a) assumed failure surface and the corresponding reaction of the wall against the failure wedge, (b) force polygon, and (c) distribution and resultant of passive earth pressure.

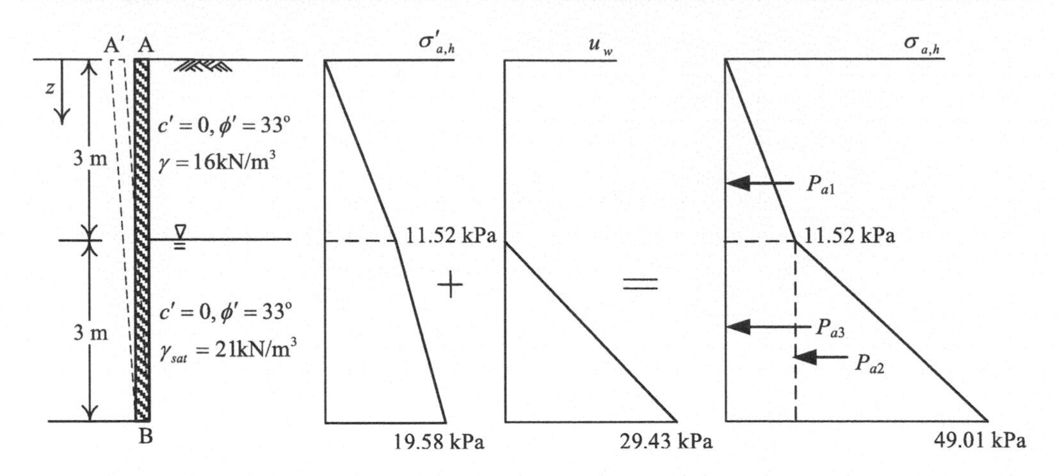

Figure EX 4.7 Variation in the active earth pressure and pore water pressure with depth.

The variations in acitve earth pressure and pore water pressure with depth are shown in Figure EX 4.7.
The resultant force per unit length of the wall can be calculated as:

$$P_{a1} = 11.52 \times 3/2 = 17.28 \text{ kN/m}$$
$$P_{a2} = (49.01 - 11.52) \times 3/2 = 56.24 \text{ kN/m}$$
$$P_{a3} = 11.52 \times 3 = 34.56 \text{ kN/m}$$
$$P_a = P_{a1} + P_{a2} + P_{a3} = 108.08 \text{ kN/m}$$

Location of the resultant force measured from the bottom of the wall:

$$\bar{z} = \frac{17.28 \times (3 + 3 \times 1/3) + 56.24 \times (3 \times 1/3) + 34.56 \times (3 \times 1/2)}{108.08} = 1.64 \text{ m}$$

Example 4.8

Similar to example 4.3. Compute the passive earth pressure and its line of action of resultant force using Coulomb's earth pressure theory with $\delta = 2\phi'/3$.

Solution

According to eq. 4.28a, $K_p = 8.08$.
The horizontal component of the coefficient of active earth pressure is:

$$K_{ph} = K_p \times \cos\delta = 8.08 \times \cos(\frac{2}{3} \times 33) = 7.49$$

At $z = 3$ m:

$$\sigma_{p,h} = \sigma'_{p,h} = 16 \times 3 \times 7.49 = 359.52 \text{ kN/m}^2$$

At $z = 6$ m:

$$\sigma'_{p,h} = [16 \times 3 + (21\text{-}9.81) \times 3] \times 7.49 = 610.96 \text{ kN/m}^2, \; u_w = 9.81 \times 3 = 29.43 \text{ kN/m}^2$$

$$\sigma_{p,h} = \sigma'_{p,h} + u_w = 640.39 \text{ kN/m}^2$$

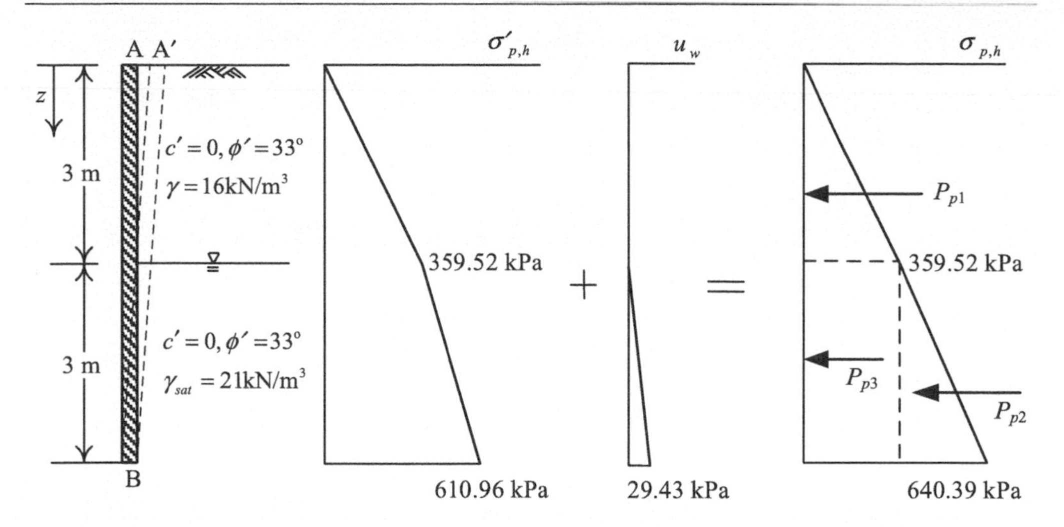

Figure EX 4.8 Variation in the passive earth pressure and pore water pressure with depth.

The variations in passive earth pressure and pore water pressure with depth are shown in Figure EX 4.8.

The resultant of passive earth pressure per unit length can be computed as:

$$P_{p1} = 359.52 \times 3/2 = 539.28\,\text{kN/m}$$

$$P_{p2} = (640.39 - 359.52) \times 3/2 = 421.31\,\text{kN/m}$$

$$P_{p3} = 359.52 \times 3 = 1078.56\,\text{kN/m}$$

$$P_p = P_{p1} + P_{p2} + P_{p3} = 2039.15\,\text{kN/m}$$

Location of the resultant force measured from the bottom of the wall:

$$\bar{z} = \frac{539.28 \times (3 + 3 \times 1/3) + 421.31 \times (3 \times 1/3) + 1078.56 \times (3 \times 1/2)}{2039.15} = 2.06\text{ m}$$

4.5 Displacement and earth pressure

According to the loading condition of the soil, when the strain state of the soil changes from the K_0 state to the active state, the direction of its principal stresses remains unchanged. However, if the direction is from the K_0 state to the passive state, the principal direction rotates by 90°. That is, σ_1 is originally vertical ($\sigma_1 = \sigma_v$) and then becomes horizontal ($\sigma_1 = \sigma_h$). Both the major and minor principal stresses rotate by 90°.

Thus, the strain of the soil required to reach passive failure should be greater than that required to reach active failure. Regarding the problem of retaining walls, many experiments have explored the relationships between wall displacement and earth pressure. Figure 4.9 shows the results from experiments on rigid gravity walls provided by NAVFAC DM7.2 (1982). As shown in Figure 4.9, the necessary wall displacement inducing the passive condition for loose and dense sands is much larger than that inducing active conditions. Similar behavior can be found for cohesive soil.

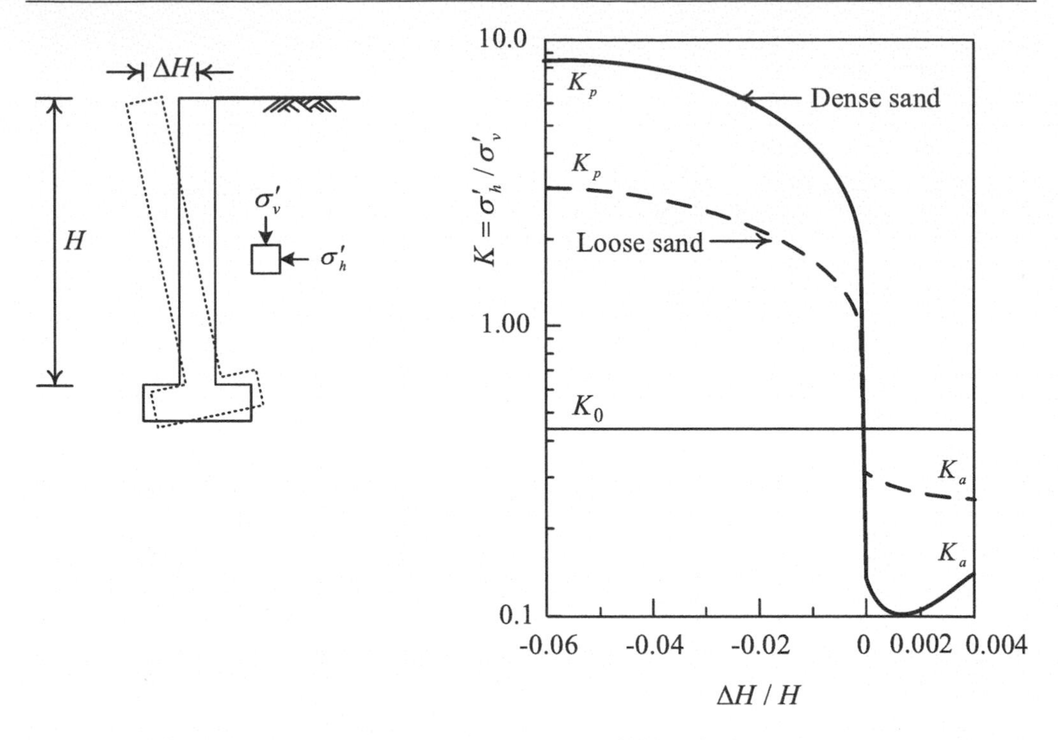

Figure 4.9 Effect of wall movement on earth pressure (ΔH is the lateral movement at the top of the wall, and H is the wall height).

Source: Redrawn from NAVFAC DM7.2 (1982).

4.6 Caquot and Kerisel's solution

Rankine's earth pressure theory assumes no friction between the retaining wall and soil. Coulomb's earth pressure theory takes into account the friction between the wall and soil but assumes that its failure surface is a plane. In fact, the wall surface may be rough, and friction between the wall and soil does exist, causing the failure surface to no longer be a plane. Whether the earth pressure theories are sufficiently accurate to represent the actual earth pressure depends on how close the assumed failure surfaces are to the actual condition, that is, the nonplane condition (Peck and Ireland, 1961; Rowe and Peaker, 1965; Mackey and Kirk, 1967; James and Bransby, 1970; Rehnman and Broms, 1972). As Figure 4.10a shows, Coulomb's active failure surface is close to the actual failure surface. Thus, Coulomb's active earth pressure is close to the actual earth pressure. Coulomb's earth pressure theory can then be used to estimate the active earth pressure.

However, as shown in Figure 4.10b, Coulomb's passive failure surface is rather different from the actual surface. The larger the friction between the retaining wall and the surrounding soil, the farther the failure surface to a plane and the greater the difference between Coulomb's passive earth pressure from the actual one. In addition to Rankine's and Coulomb's theories, several earth pressure theories are available in the literature. Figure 4.11 shows a commonly used model for calculating the passive earth pressure, which considers the friction between the wall and the soil. BC can be assumed to be an arc of a circle, an ellipse, or a logarithm spiral, while CD can be a plane. Then, following a method similar to Coulomb's earth pressure theory, the forces

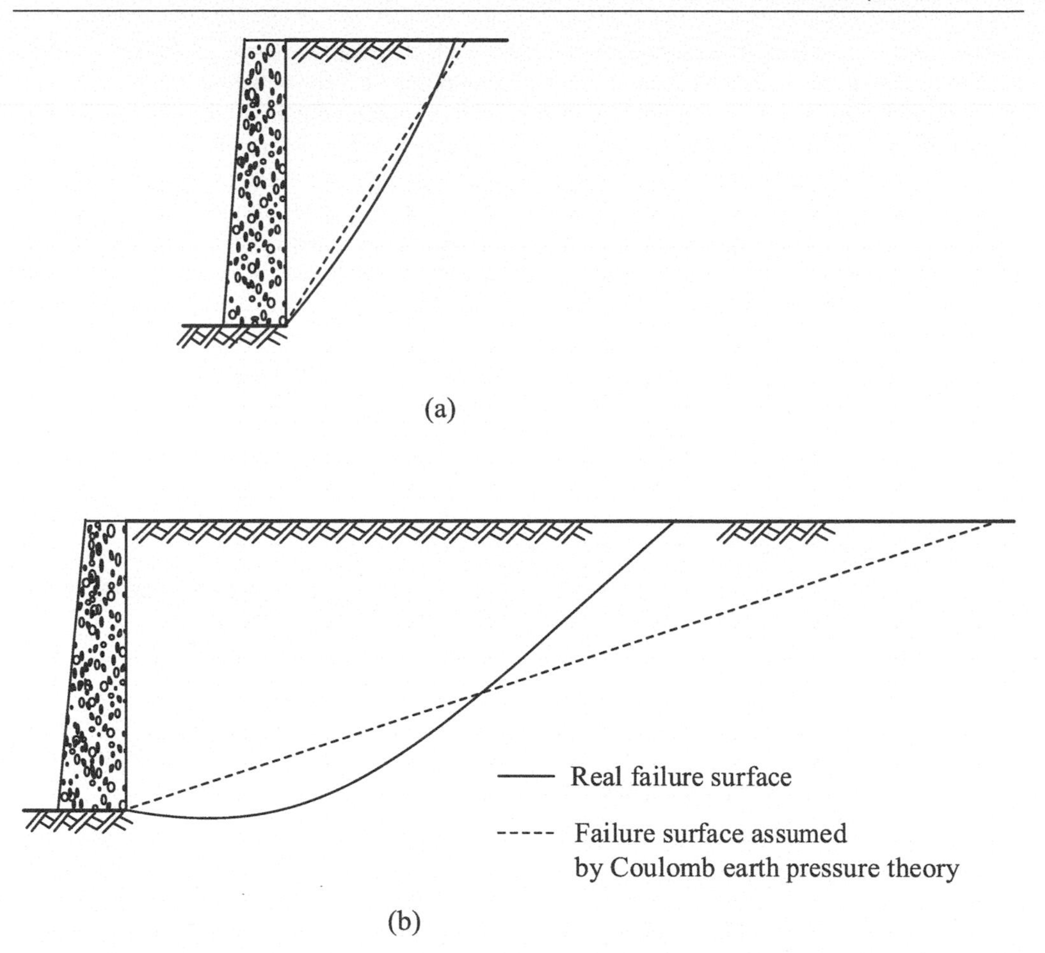

Figure 4.10 Real failure surface and assumed failure surface in Coulomb's earth pressure theory: (a) active condition and (b) passive condition.

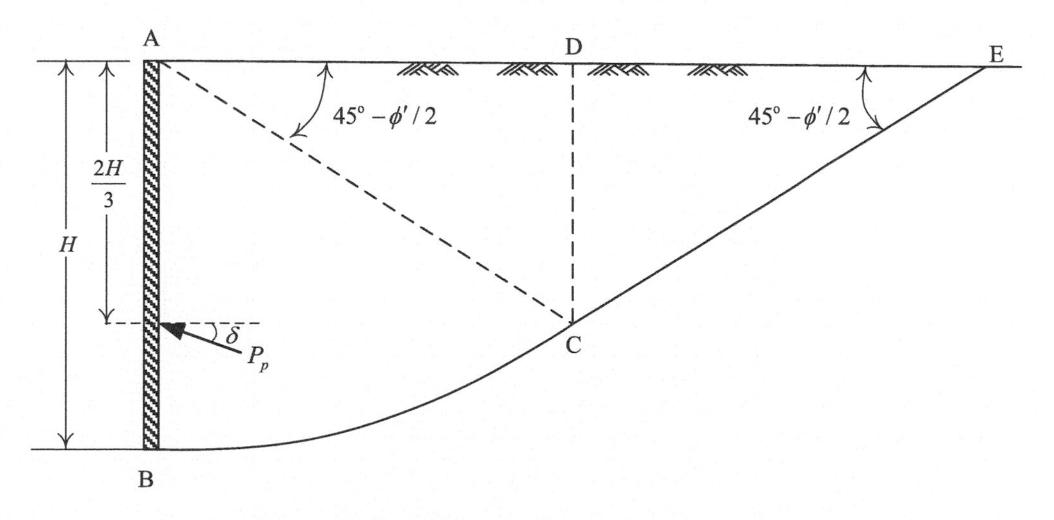

Figure 4.11 Passive earth pressure with curved failure surface.

in the wedge, including the reaction force of the wall, the reaction force of the soil, and the weight of the wedge, should be in equilibrium and form a close force polygon. The passive earth pressure can then be solved. Caquot and Kerisel (1948) assumed that BC is a function of the logarithm spiral and derived the passive earth pressure for various conditions. NAVFAC DM7.2 (1982) simplifies Caquot–Kerisel's solutions by figures and reduction factors. Figures 4.12 and 4.13 show Caquot–Kerisel's K_p for $\delta = \phi'$. For $\delta \neq \phi'$, a reduction factor is provided, as shown in Table 4.1. Figure 4.14 shows the values of K_p from Rankine's, Coulomb's, and Caquot–Kerisel's earth pressure theories for various ϕ' for $\delta = \phi'$. It can be found that the Rankine's K_p are the smallest, Coulomb's the largest, while Caquot–Kerisel's K_p are in between. Such a difference decreases with the reduction of the δ value.

The equation for Caquot–Kerisel's passive earth pressure has the same form as Coulomb's (i.e., eq. 4.28), but with different K_p values, as shown here:

$$P_p = \frac{1}{2}\gamma K_p H^2$$

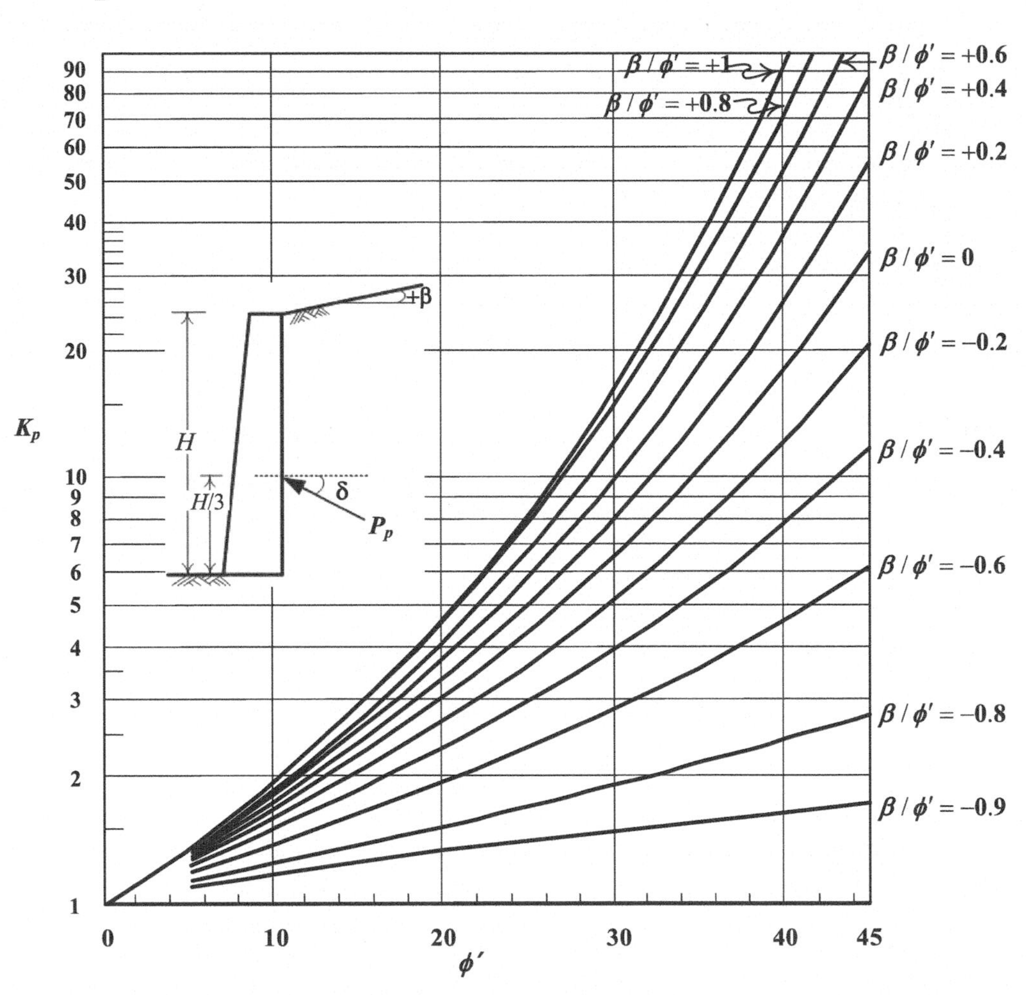

Figure 4.12 Caquot and Kerisel's passive earth pressure with vertical wall and sloping backfill for $\delta/\phi' = 1.0$.

Source: Redrawn from NAVFAC DM7.2 (1982).

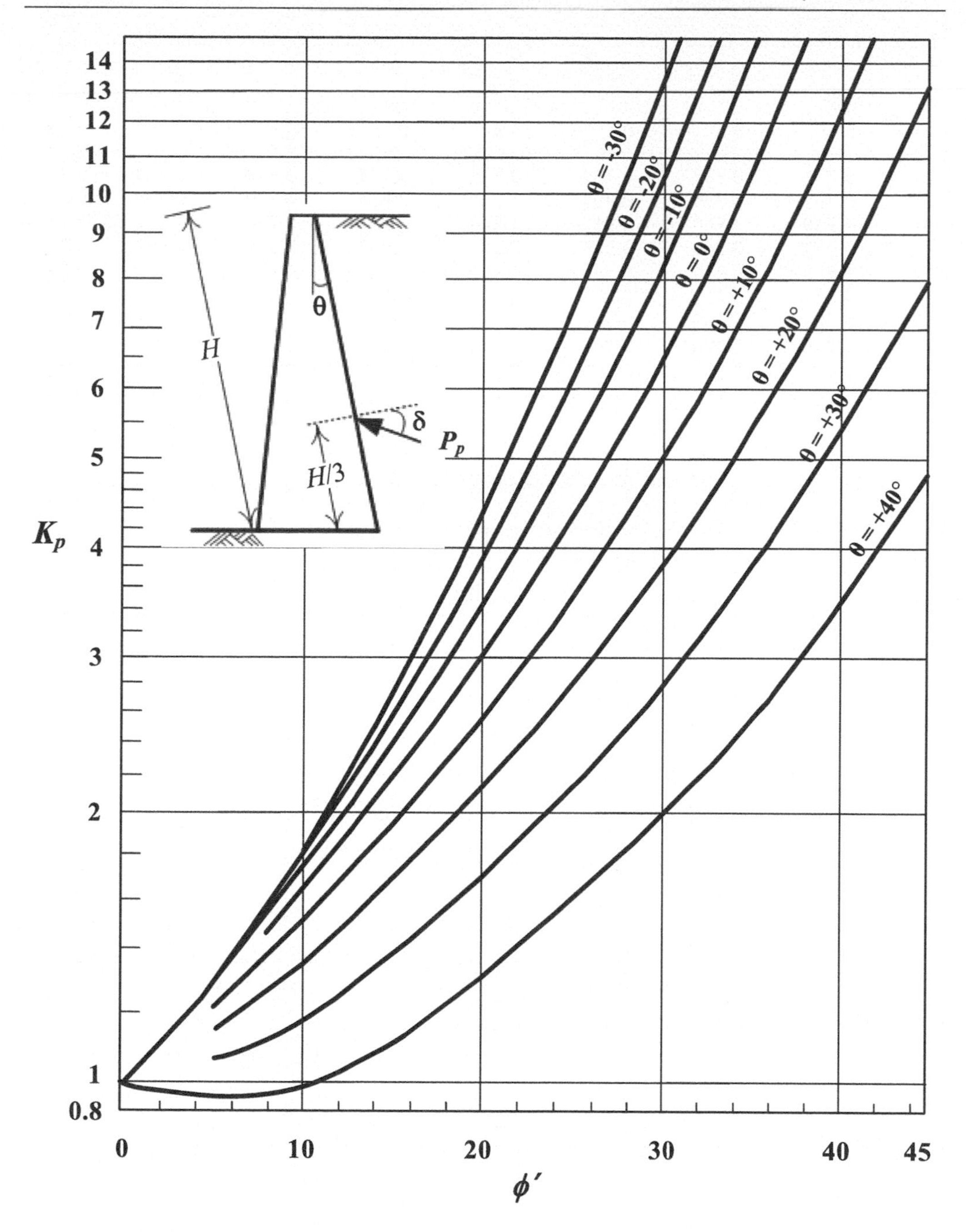

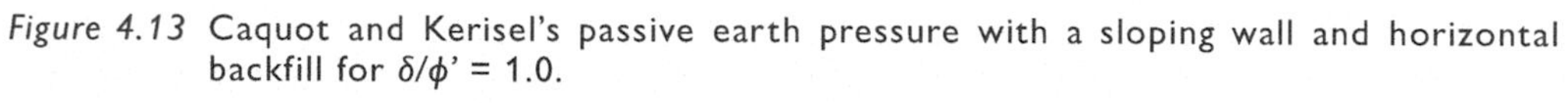

Figure 4.13 Caquot and Kerisel's passive earth pressure with a sloping wall and horizontal backfill for $\delta/\phi' = 1.0$.

Source: Redrawn from NAVFAC DM7.2 (1982).

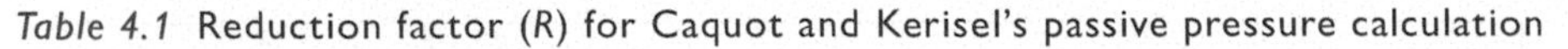
Table 4.1 Reduction factor (R) for Caquot and Kerisel's passive pressure calculation

ϕ'	δ/ϕ'							
	0.7	0.6	0.5	0.4	0.3	0.2	0.1	0.0
10	0.978	0.962	0.946	0.929	0.912	0.898	0.881	0.864
15	0.961	0.934	0.907	0.881	0.854	0.830	0.803	0.775
20	0.939	0.901	0.862	0.824	0.787	0.752	0.716	0.678
25	0.912	0.860	0.808	0.759	0.711	0.666	0.620	0.574
30	0.878	0.811	0.746	0.686	0.627	0.574	0.520	0.467
35	0.836	0.752	0.674	0.603	0.536	0.475	0.417	0.362
40	0.783	0.682	0.592	0.512	0.439	0.375	0.316	0.262
45	0.718	0.600	0.500	0.414	0.339	0.276	0.221	0.174

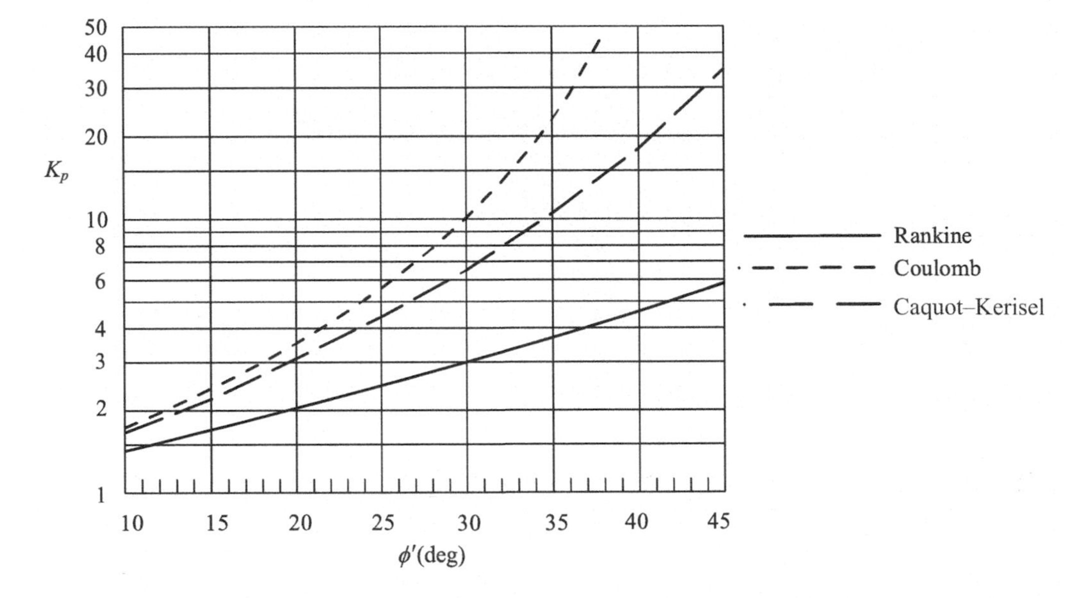

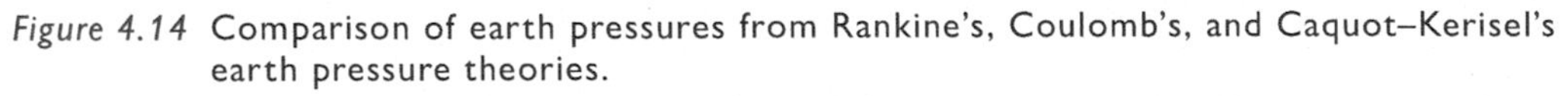
Figure 4.14 Comparison of earth pressures from Rankine's, Coulomb's, and Caquot–Kerisel's earth pressure theories.

Source: Ou (2022).

If $\delta = \phi'$, K_p can be obtained directly from Figures 4.13 or 4.14, while if $\delta \neq \phi'$, K_p can be estimated as:

$$K_p(\delta \neq \phi') = R \times K_p(\delta = \phi') \tag{4.29}$$

Example 4.9

Similar to example 4.3. Compute the passive earth pressure and its line of action of resultant force using Caquot–Kerisel's earth pressure theory with $\delta = 2\phi'/3$.

Solution

According to Figure 4.12 or 4.13, $K_p = 8.5$ for $\delta/\phi' = 1.0$ and $\phi' = 33^\circ$.

Table 4.1 shows that the reduction factor R = 0.83 for $\delta/\phi' = 2/3$. Therefore, $K_p = 8.5 \times 0.83 = 7.05$.

$$K_{ph} = K_p \times \cos\delta = 7.05 \times \cos(2/3 \times 33) = 6.54$$

At $z = 3$ m:

$$\sigma_{p,h} = \sigma'_{p,h} = 16 \times 3 \times 6.54 = 313.92 \text{ kN/m}^2$$

At $z = 6$ m:

$$\sigma'_{p,h} = [16 \times 3 + (21\text{-}9.81) \times 3] \times 6.54 = 533.47 \text{ kN/m}^2,\ u_w = 9.81 \times 3 = 29.43 \text{ kN/m}^2$$

$$\sigma_{p,h} = \sigma'_{p,h} + u_w = 562.9 \text{ kN/m}^2$$

The variations in passive earth pressure and pore water pressure with depth are shown in Figure EX 4.9.

The resultant force per unit length of the wall is:

$$P_{p1} = 313.92 \times 3/2 = 470.88 \text{ kN/m}$$

$$P_{p2} = (562.9 - 313.92) \times 3/2 = 373.47 \text{ kN/m}$$

$$P_{p3} = 313.92 \times 3 = 941.76 \text{ kN/m}$$

$$P_p = P_{p1} + P_{p2} + P_{p3} = 1786.11 \text{ kN/m}$$

Resultant force location from the bottom of the wall:

$$\bar{z} = \frac{470.88 \times (3 + 3 \times 1/3) + 373.47 \times (3 \times 1/3) + 941.76 \times (3 \times 1/2)}{1786.11} = 2.06 \text{ m}$$

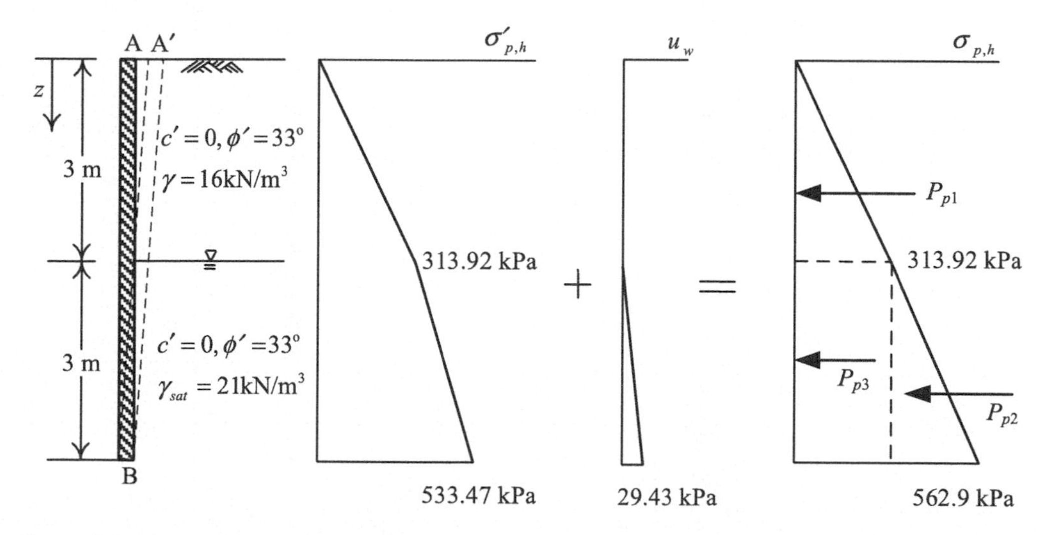

Figure EX 4.9 Variation in the passive earth pressure and pore water pressure with depth.

4.7 Earth pressure due to surcharge

Figure 4.15 shows a uniformly distributed load, q, acting on a very large area of the ground surface. If the wall is restrained from movement, the stress state should be in the at-rest condition. The lateral earth pressure is:

$$\sigma_h = qK_0 \tag{4.29}$$

where K_0 is the coefficient of earth pressure at rest.

If the wall moves forward or backward toward the soil, the lateral earth pressure induced by the uniformly distributed loads can be estimated as either active or passive earth pressure, respectively:

$$\sigma_a = qK_a \tag{4.30}$$

$$\sigma_p = qK_p \tag{4.31}$$

If a line load or strip load is applied to the surface, the wall will certainly move, but the movement may not be sufficiently large to make the soil achieve an active or passive state. Moreover, the solutions to lateral pressures induced by a line load or strip load in the active or passive state have not yet been derived. Under such conditions, the equations obtained from the theory of elasticity, with some modification, are adopted to estimate the lateral earth pressure induced by surface loads. Figure 4.16 illustrates the lateral earth pressure distribution caused by a line load of intensity Q_ℓ per unit length parallel to the retaining wall. The lateral earth

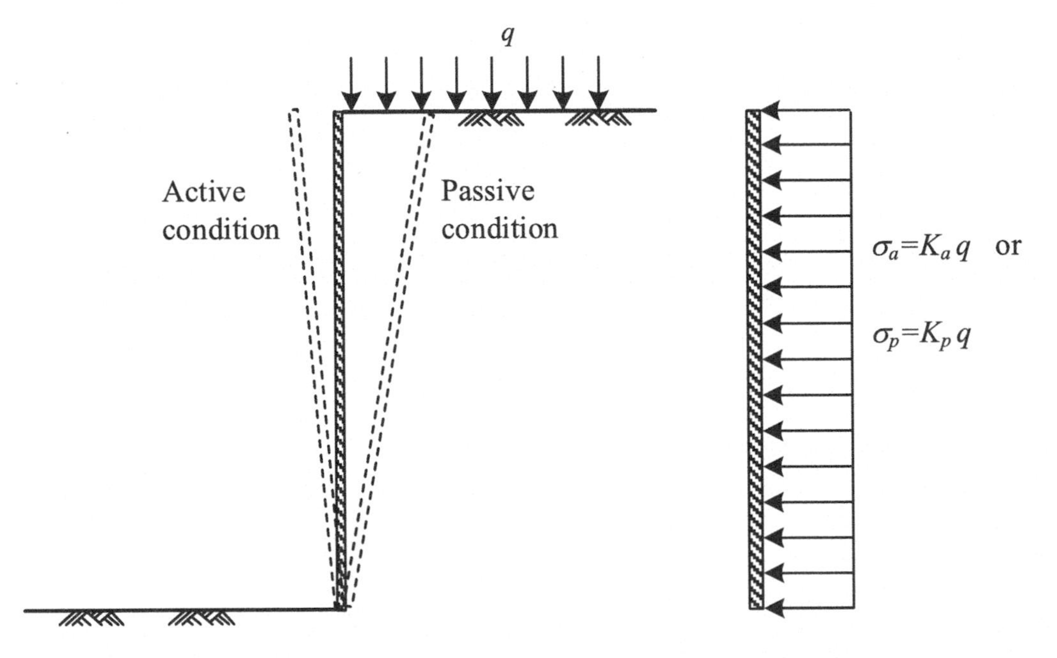

Figure 4.15 Lateral earth pressure produced by a uniformly distributed load.

pressure at any depth z and its resultant can be expressed as follows (NAVFAC DM7.2, 1982; USS Steel, 1975):

$m \le 0.4$

$$\sigma_h = \frac{0.203Q_\ell}{H}\frac{n}{(0.16+n^2)^2} \tag{4.32}$$

$m > 0.4$

$$\sigma_h = \frac{1.28Q_\ell}{H}\frac{m^2 n}{(m^2+n^2)^2} \tag{4.33}$$

And the resultant is:

$$P_h = \frac{0.64Q_\ell}{(m^2+1)} \tag{4.34}$$

Figure 4.16 illustrates the dimensionless diagram of the earth pressure distribution derived from the preceding equation, where $m = 0.1, 0.3, 0.5$, and 0.7. The corresponding point of action R is also marked in the figure.

Figure 4.17 illustrates a strip load with an intensity Q_s parallel to the retaining wall. The earth pressure acting on the wall caused by the strip load can be determined by the following equation (USS Steel, 1975):

$$\sigma_h = \frac{2Q_s}{H}(\beta - \sin\beta\cos 2\alpha) \tag{4.35}$$

where α and β are angles in radians.

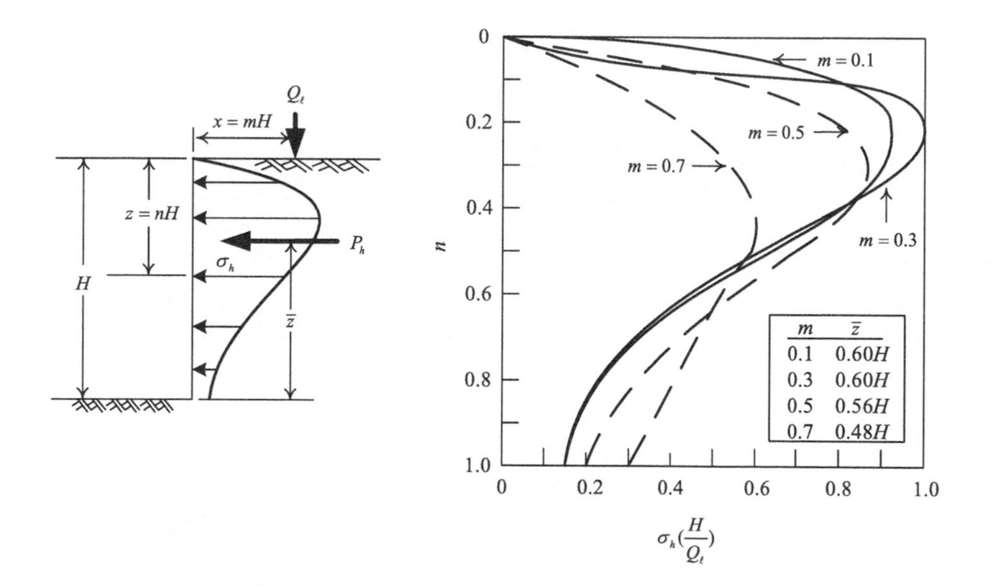

Figure 4.16 Lateral earth pressure produced by a line load.

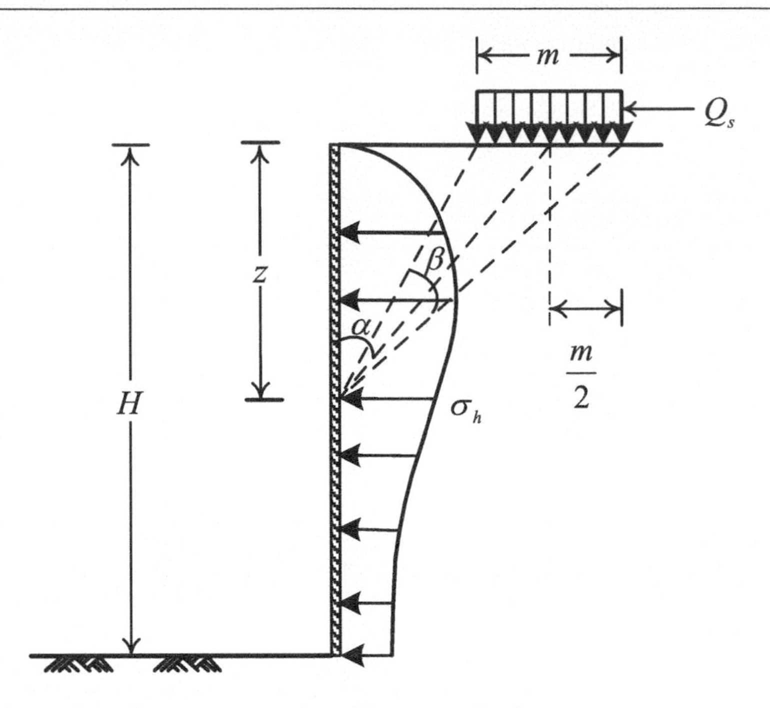

Figure 4.17 Lateral earth pressure produced by a strip load.

4.8 Earth pressure due to earthquakes

Earthquakes will generate lateral and vertical inertia forces on a soil mass. Let a represent the earthquake acceleration. Then, the inertia force (F) is:

$$F = ma = \frac{W}{g}(k_h \text{ or } k_v \times g) = Wk_h \text{ or } Wk_v \tag{4.36}$$

where m is the soil mass; W is the soil weight; g is the acceleration of gravity; a is the acceleration, which is a fraction of the acceleration of gravity, such as $a = k_v g$ or $k_h g$; k_h is the coefficient of horizontal acceleration; and k_v is the coefficient of vertical acceleration.

In the ultimate condition, earthquakes will increase the active earth pressure behind the retaining wall and decrease the passive earth pressure in front of the wall. The Mononobe–Okabe equation is generally adopted (Mononobe, 1929; Okabe, 1926) to compute the active and passive earth pressures under the influence of earthquakes. As shown in Figure 4.18, the total (static + dynamic) active force (P_{ae}) during an earthquake can be computed by the following equations:

$$P_{ae} = \frac{1}{2}\gamma H^2 (1 - k_v) K_{ae} \tag{4.37}$$

$$K_{ae} = \frac{\cos^2(\phi - \theta - \alpha)}{\cos\alpha \cos^2\theta \cos(\delta + \theta + \alpha)\left\{1 + \left[\frac{\sin(\phi + \delta)\sin(\phi - \beta - \alpha)}{\cos(\delta + \theta + \alpha)\cos(\beta - \theta)}\right]^{\frac{1}{2}}\right\}^2} \tag{4.37a}$$

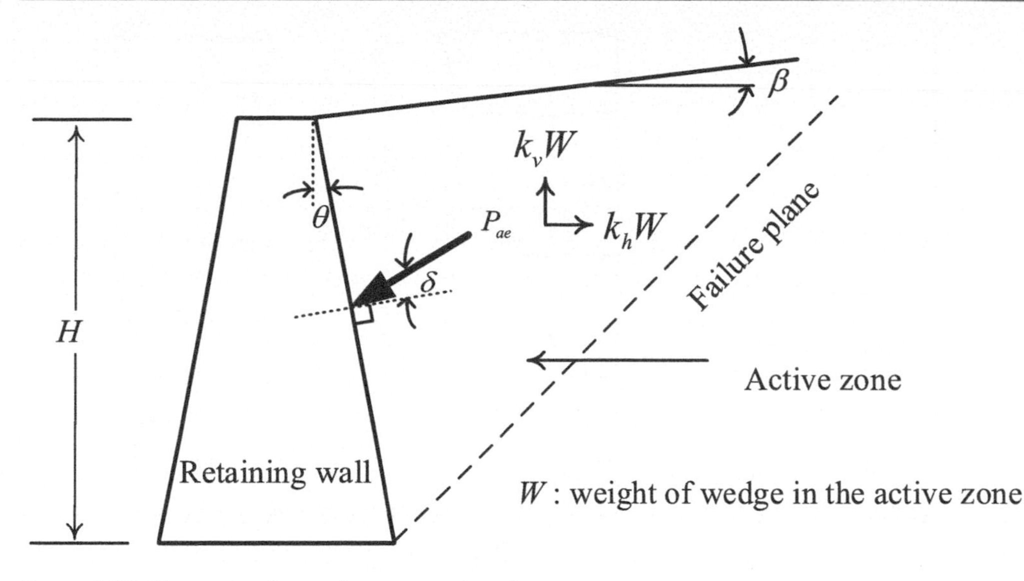

Figure 4.18 The active force due to earthquakes.

where $\alpha = \tan^{-1}\left[k_h/(1-k_v)\right]$.

In addition, based on many tests, Bowles (1988) suggested that δ be assumed to be 0 under dynamic conditions.

Seed and Whitman (1970) proposed that the total (static + dynamic) active force (P_{ae}) could be separated into two components: the static active force (P_a) and the dynamic increment (ΔP_{ae}) due to an earthquake, as illustrated in Figure 4.19, which can be expressed as:

$$P_{ae} = P_a + \Delta P_{ae} \tag{4.38}$$

The static active force as computed from Coulomb's theory is applied at $H/3$ from the base of the wall in a homogeneous cohesionless soil, resulting in a triangular distribution of earth pressure. Seed and Whitman (1970) also proposed that the dynamic increment was an inverted triangular pressure distribution with a force resultant acting at $0.6H$ or $2H/3$ from the base of the wall. The location of the total active force can be determined as follows:

1. Compute the dynamic increment of the active force ΔP_{ae}:

$$\Delta P_{ae} = P_{ae} - P_a \tag{4.39}$$

2. Locate P_a at $H/3$ from the base of the wall.
3. Locate ΔP_{ae} at $2H/3$ (or $0.6H$) from the base of the wall.
4. Compute the location ($\bar{z}$) of the total active force (P_{ae}):

$$\bar{z} = \frac{P_a(H/3) + \Delta P_{ae}(2H/3)}{P_{ae}} \tag{4.40}$$

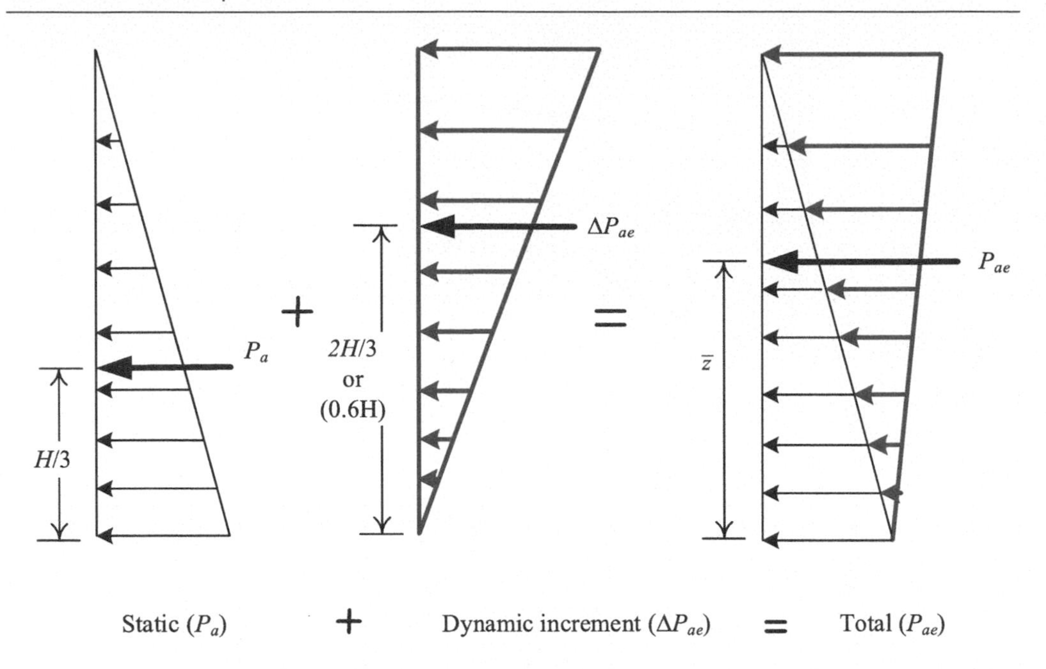

Figure 4.19 Total seismic earth pressure consisting of a static active earth force and a dynamic increment.

Moreover, the total (static + dynamic) passive force (P_{pe}) under the influence of earthquakes can be computed by the following equations (Figure 4.20):

$$P_{pe} = \frac{1}{2}\gamma H^2 (1 - k_v) K_{pe} \tag{4.41}$$

$$K_{pe} = \frac{\cos^2(\phi + \theta - \alpha)}{\cos\alpha \cos^2\theta \cos(\delta - \theta + \alpha)\left\{1 - \left[\dfrac{\sin(\phi+\delta)\sin(\phi+\beta-\alpha)}{\cos(\delta-\theta+\alpha)\cos(\beta-\theta)}\right]^{\frac{1}{2}}\right\}^2} \tag{4.41a}$$

The distribution and characteristics of the static passive force, dynamic increment, and total passive force are similar to those displayed in Figure 4.19. The locations of the static passive force and the dynamic increment are also at $H/3$ and $2H/3$ (or 0.6 H) from the base of the wall, respectively. Therefore, the location of the total passive force can be determined in a way similar to the total active force.

In design, the influence of earthquakes on the safety of a retaining wall as a permanent structure should be considered. In deep excavations, the outer walls of a basement are permanent structures. The design should thereby take into consideration the influence of earthquakes. If the diaphragm walls also serve as the outer walls of a basement, the influence of earthquakes should be considered accordingly. Otherwise, the influence of earthquakes can be ignored. For temporary retaining structures, such as soldier piles, steel sheet piles, and column piles, the effects of earthquake are ignorable.

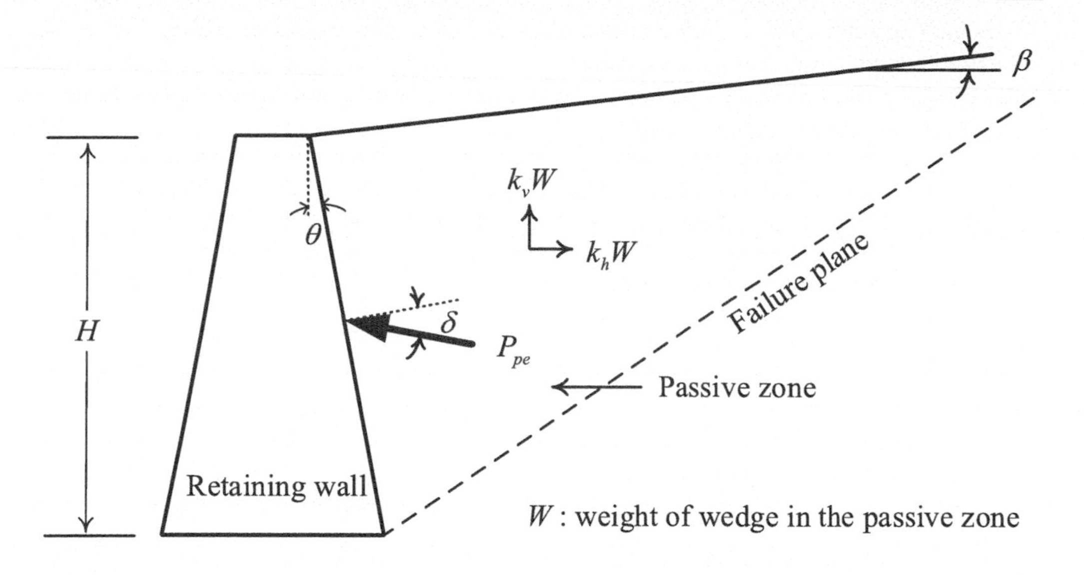

Figure 4.20 The passive force due to earthquakes.

4.9 Summary and general comments

This chapter can be summarized as follows:

1. For retaining walls where displacement does not occur, such as the outer wall of a basement, the lateral earth pressure at rest is adopted for design. The coefficients of the at-rest lateral earth pressure for normally consolidated cohesive soils and cohesionless soils can be estimated by Jaky's equation. For overconsolidated soils, this chapter provides an empirical equation estimation.
2. Rankine's earth pressure theory cannot consider the friction between the retaining wall and soil. Coulomb's earth pressure theory, on the other hand, can. Although the two theories are based on different assumptions, the obtained earth pressures are identical when a vertical and smooth wall surface is used. Both theories assume the failure surfaces as planes, not conforming to reality, and thus, their results cannot represent actual earth pressures.
3. As the friction between the retaining wall and soil is considered, the real failure surface is a curved surface. Thus, Rankine's active earth pressure would overestimate the real earth pressure very slightly, whereas its passive earth pressure would underestimate the real value. Coulomb's active earth pressure is usually treated as the real earth pressure because the assumed failure surface is close to the actual value. However, Coulomb's passive earth pressure significantly overestimates the real passive earth pressure for a large friction angle between the wall and soil. Caquot–Kerisel's earth pressures, both active and passive, are the closest to the real values and are thus regarded as the real earth pressures.
4. In retaining wall problems, the active earth pressure is the main force engendering failure. Caquot–Kerisel's active earth pressure, regarded as the real earth pressure, is adopted for

design. Coulomb's active earth pressure is close to Caquot–Kerisel's; therefore, Coulomb's active earth pressure can be used for design.

5. The passive earth pressure is often the force used to resist failure. Caqout–Kerisel's earth pressure is the first choice for analysis and design since it is regarded as the real earth pressure. Rankine's earth pressure is too small and differs significantly from the real value. Thus, Rankine's is not adopted. When $\delta \le 0.5\phi'$, Coulomb's coefficient of passive earth pressure is similar to that of Caqout–Kerisel's and can also be used for analysis and design. When $\delta > 0.5\phi'$, Coulomb's coefficient of passive earth pressure is obviously larger than Caqout–Kerisel's and will lead to an unsafe design.

Problems

4.1 A 9.5 m deep (measured from the ground surface to the bottom of the foundation) two-story basement is located in a normally consolidated clay. The depth of the groundwater level is $H = 2.5$ m, and the unit weights above and below the groundwater level are $\gamma = 16$ kN/m^3 and $\gamma_{sat} = 20$ kN/m^3 (saturated unit weight), respectively. The effective strength parameters are $c' = 0$ and $\phi' = 32°$. The undrained shear strength is $s_u = 55$ kN/m^2. Compute the total lateral force per unit length of the basement wall and the location of the resultant.

4.2 Similar to problem 4.1. The depth of the foundation is 9.5 m, and the groundwater level $H = 3$ m. The soil is silty sand. Assume the following parameters: $\gamma = 18$ kN/m^3, $\gamma_{sat} = 22$ kN/m^3 (saturated unit weight), $c' = 0$, $\phi' = 34°$. Compute the total lateral force per unit length of the outer wall of the basement and the location of the resultant.

4.3 Assume the ground shown in Figure P4.3 is clay. The groundwater level location $H_1 = 4$ m and $H_2 = 5$ m. Above the groundwater level, $s_{u1} = 20$ kN/m^2 and $\gamma_1 = 17$ kN/m^3, and below the groundwater level, $\gamma_{sat2} = 18$ kN/m^3 and $s_{u2} = 30$ kN/m^2. Determine the following:

a. The depth of the tension cracks
b. The variation in Rankine's active pressure with depth
c. The magnitude and location of the total active thrust per unit length of the wall, assuming tension cracks have occurred

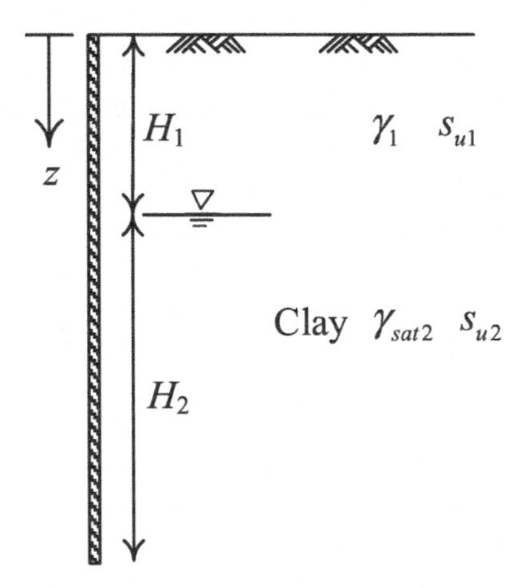

Figure P4.3

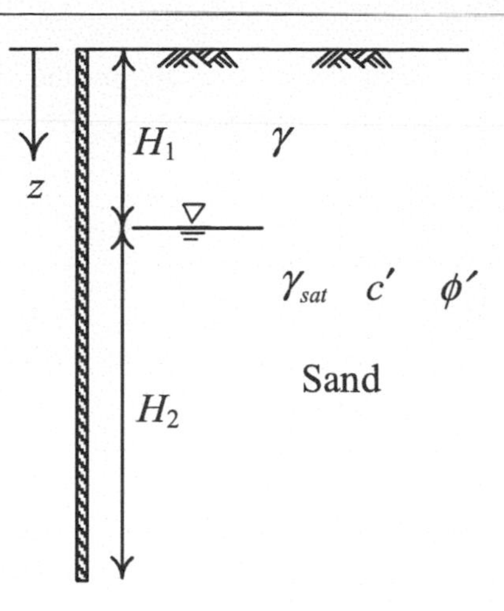

Figure P4.5

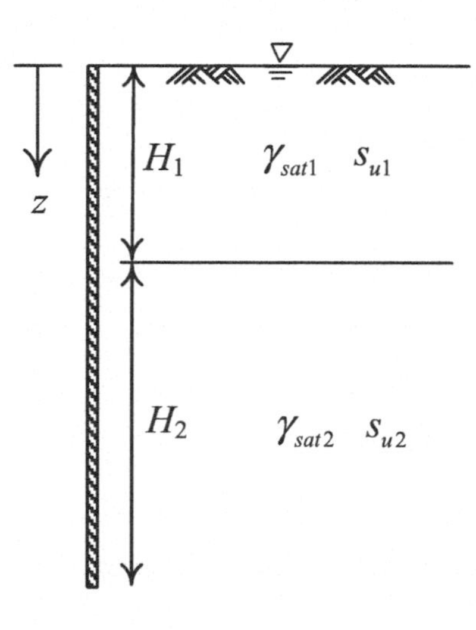

Figure P4.7

4.4 Redo problem 4.3 with the following parameters: $H_1 = 3.5$ m, $H_2 = 4.5$ m; $s_{u1} = 15\text{kN/m}^2$, $\gamma_1 = 16\text{ kN/m}^3$; $s_{u2} = 25\text{ kN/m}^2$, and $\gamma_{sat2} = 17\text{ kN/m}^3$.

4.5 Figure P4.5 shows a wall in sand where the groundwater level $H_1 = 2.0$ m and $H_2 = 4.0$ m. The effective strength parameters are $c' = 0$ and $\phi' = 34°$. The unit weight is $\gamma = 16\text{ kN/m}^3$, $\gamma_{sat} = 20\text{ kN/m}^3$. Use Rankine's earth pressure theory to compute the total active force per unit length of the wall and the location of the resultant.

4.6 Redo problem 4.5 with $H_1 = 3.0$ m and $H_2 = 4.0$ m. The effective strength parameters are $c' = 0$ and $\phi' = 35°$. The unit weight is $\gamma = 15\text{ kN/m}^3$, $\gamma_{sat} = 22\text{ kN/m}^3$.

4.7 Figure P4.7 shows a wall in two layers of clay with $H_1 = 4.0$ m and $H_2 = 4.0$ m. The groundwater level is on the surface. The undrained shear strengths and saturated unit weight are

$s_{u1} = 20\ \text{kN/m}^2$ and $\gamma_{sat1} = 16\ \text{kN/m}^3$, respectively, for the upper clay and $s_{u2} = 25\ \text{kN/m}^2$ and $\gamma_{sat2} = 18\ \text{kN/m}^3$, respectively, for the lower clay. Determine the following:

a. The depth of the tension cracks
b. The variation in Rankine's active pressure with depth
c. The magnitude and location of the total active thrust per unit length of the wall, assuming tension cracks do not occur

4.8 Redo problem 4.7 with the following parameters: $H_1 = 3.0$ m, $H_2 = 4.0$ m. $s_{u1} = 10\ \text{kN/m}^2$, $\gamma_{sat1} = 16\ \text{kN/m}^3$, $s_{u2} = 20\ \text{kN/m}^2$, and $\gamma_{sat2} = 17\ \text{kN/m}^3$.

4.9 Figure P4.9 shows a wall in two layers of sand with $H_1 = 2.0$ m and $H_2 = 3.0$ m. The groundwater level is on the surface. The effective strength parameters and saturated unit weight are $c_1' = 0$, $\phi_1' = 30°$, and $\gamma_{sat1} = 19\ \text{kN/m}^3$, respectively, for the upper sand and $c_2' = 0$, $\phi_2' = 32°$, and $\gamma_{sat2} = 21$, respectively, for the lower sand. Determine the following:

a. The variation in Rankine's active pressure and pore water pressure with depth
b. The magnitude and location of the total active thrust per unit length of the wall

4.10 Redo problem P4.9 with the following parameters: $H_1 = 3.0$ m, $H_2 = 3.0$ m, $c_1' = 0$, $\phi_1' = 32°$, $\gamma_{sat1} = 18\ \text{kN/m}^3$, $c_2' = 0$, $\phi_2' = 34°$, and $\gamma_{sat2} = 22\ \text{kN/m}^3$.

4.11 Figure P4.11 shows a wall in sand and clay with $H_1 = 3.0$ m and $H_2 = 4.0$ m. The groundwater level is on the surface. The effective strength parameters and saturated unit weight for the sand are $c' = 0$, $\phi' = 32°$, and $\gamma_{sat1} = 20\ \text{kN/m}^3$, respectively. The total stress undrained shear strength and saturated unit weight for the clay are $s_u = 20\ \text{kN/m}^2$ and $\gamma_{sat2} = 16\ \text{kN/m}^3$, respectively. Use Rankine's earth pressure theory to compute the following:

a. The variation in pore water pressure with depth
b. The variation in Rankine's active pressure with depth
c. The magnitude and location of the total active thrust per unit length of the wall

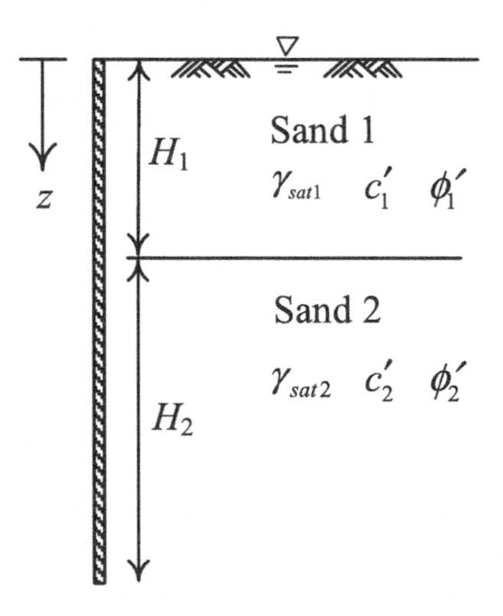

Figure P4.9

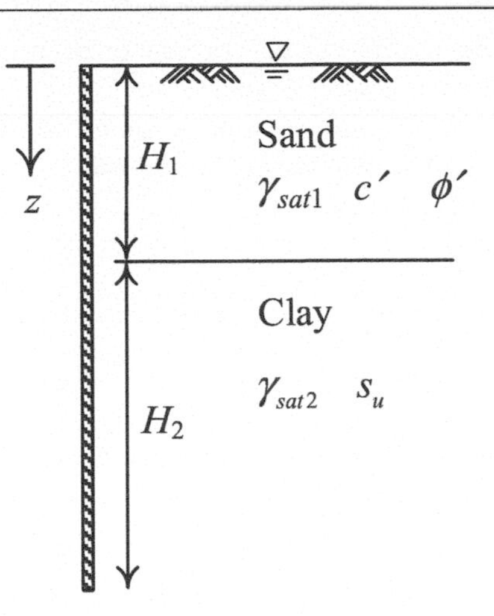

Figure P4.11

4.12 Redo problem 4.11 with the following parameters: $H_1 = 4.0$ m, $H_2 = 3.0$ m. $c' = 0$, $\phi' = 30^\circ$, $\gamma_{sat1} = 18\ \text{kN/m}^3$; $s_u = 15\ \text{kN/m}^2$, and $\gamma_{sat2} = 17\ \text{kN/m}^3$.

4.13 Same as problem 4.5. Assuming the friction angle between the wall and sand δ, use Coulomb's earth pressure theory to compute the total active force (horizontal component) per unit length of the wall and the location of the resultant for $\delta = \phi'/2$ and $\delta = 2\phi'/3$.

4.14 Calculate Coulomb's K_a for $\delta = 0$, $\delta = \phi'/3$, $\delta = \phi'/2$, $\delta = 2\phi'/3$, and $\delta = \phi'$, assuming $\phi' = 33^\circ$. For convenience, you can use software tools such as a spreadsheet to perform the calculation. Compare the results with Rankine's K_a and comment on their difference.

4.15 Same as problem 4.3. Assuming the retained clay is in the passive state, use Rankine's earth pressure theory to compute the magnitude and location of the total passive force per unit length of the wall against the wall.

4.16 Same as problem 4.4. Assuming the retained clay is in the passive state, use Rankine's earth pressure theory to compute the magnitude and location of the total passive force per unit length of the wall against the wall.

4.17 Same as problem 4.5. Assuming the retained sand is in the passive state, use Rankine's earth pressure theory to compute the magnitude and location of the total passive force (horizontal component) per unit length of the wall against the wall.

4.18 Same as problem 4.6. Assuming the retained sand is in the passive state, use Rankine's earth pressure theory to compute the magnitude and location of the total passive force (horizontal component) per unit length of the wall against the wall.

4.19 Redo problem 4.17. Use Coulomb's earth pressure theory for $\delta = \phi'/2$ and $\delta = 2\phi'/3$.

4.20 Calculate Coulomb's and Caquot–Kerisel's K_p for $\delta = 0$, $\delta = \phi'/3$ $\delta = \phi'/2$, $\delta = 2\phi'/3$, and $\delta = \phi'$, assuming $\phi' = 30^\circ$ and $\phi' = 35^\circ$. For convenience, you can use software tools such as a spreadsheet to perform the calculation. Compare the results with Rankine's K_p and comment on their difference.

4.21 In problem 4.1, if there is a surcharge of 30 kN/m^2 on the ground surface level, what would be the total lateral force per unit of the wall?

4.22 In problem 4.3, if there is a surcharge of 20 kN/m^2 on the ground surface level, what would be the total lateral force per unit of the wall?

4.23 In problem 4.15, if there is a surcharge of 40 kN/m^2 on the ground surface level, what would be the total horizontal force per unit of the wall?

References

Alpan, I. (1967), The empirical evaluation of the coefficient K_0 and K_{0R}. *Soils and Foundations*, Vol. VII, No. 1, pp. 31–40.

Bowles, J.E. (1988), *Foundation Analysis and Design*, 4th edition, New York: McGraw-Hill Book Company.

Caquot, A. and Kerisel, J. (1948), *Tables for the Calculation of Passive Pressure, Active Pressure, and Bearing Capacity of Foundations*, Paris: Gauthier-Villars.

Coulomb, C.A. (1776), Essai sur une application des regles de maximis et minimis a quelques problemes de statique, relatifs a l'architecture. *Memoires Royale des Sciences, Paris*, Vol. 3, p. 38.

Jaky, J. (1944), The coefficient of earth pressure at rest. *Journal of the Society of Hungarian Architects and Engineers (in Hungarian)*, Vol. 8, No. 22, pp. 355–358.

James, R.G. and Bransby, P.L. (1970), Experimental and theoretical investigation of a passive earth pressure problem. *Geotechnique*, Vol. 20, No. 1, pp. 17–37.

Ladd, C.C., Foote, R., Ishihara, K., Schlosser, F. and Poulous, H.G. (1977), Stress-deformation and strength characteristics, state-of-the-ART report. In *Proceedings of the Ninth International Conference on Soil Mechanics and Foundation Engineering*, Tokyo, Vol. 2, pp. 421–494.

Mackey, R.D. and Kirk, D.P. (1967), At rest, active and passive earth pressures. In *Proceedings of Southeast Asia Regional Conference on Soil Engineering*, Bangkok, pp. 187–199.

Mononobe, N. (1929), Earthquake-proof construction of masonary dams. In *Proceedings, World Engineering Conference*, Vol. 9, pp. 274–280.

NAVFAC DM7.2 (1982), *Foundations and Earth Structures*, Design Manual 7.2, USA: Department of the Navy.

Okabe, S. (1926), General theory of earth pressure. *Journal of the Japanese Society of Civil Engineers*, Vol. 12, No. 1.

Ou, C.Y. (2022), *Fundamentals of Deep Excavations*, London: CRC Press, Taylor and Francis Group.

Peck, R.B. and Ireland, H.O. (1961), Full-scale lateral load test of a retaining wall foundation. *Proceedings of 5th International Conference on Soil Mechanics and Foundation Engineering*, Vol. 2, pp. 453–458.

Rankine, W.M.J. (1857), *On Stability on Loose Earth*, London: Philosophic Transactions of Royal Society, Part I, pp. 9–27.

Rehnman, S.E. and Broms, B.B. (1972), Lateral pressures on basement wall: Results from full-scale tests. *Proceedings of 5th European Conference on Soil Mechanics and Foundation Engineering*, Vol. 1, pp. 189–197.

Rowe, P.W. and Peaker, K. (1965), Passive earth pressure measurements. *Geotechnique*, Vol. 15, No. 1, pp. 57–78.

Schmidt, B. (1967), *Lateral Stresses in Uniaxial Strains*, Bulletin, No. 23, Danish Geotechnical Institute, pp. 5–12.

Seed, H.B. and Whitman, R.V. (1970), Design of earth retaining structures for dynamic loads. In *ASCE Specialty Conference, Lateral Stresses in the Ground and Design of Earth Retaining Structures*, Cornell University, Ithaca, New York, pp. 103–147.

USS Steel (1975), *Sheet-Pile Design Manual*. July, ADUSS.

Chapter 5

Earth retaining structures

5.1 Introduction

Earth retaining structures are constructed to stabilize an unstable soil mass by providing external support or internal reinforcement to resist lateral earth pressure from the soil. Figure 5.1 illustrates the conditions in which the use of earth retaining structures should be considered for cohesionless slopes. When a slope is gentle, the slope inclination angle is less than the angle of repose of the soil (i.e., the angle of friction): $\beta < \phi'$. The mobilized soil shear strength is adequate to maintain slope stability, and the soil mass is stable (Figure 5.1a). When a slope is steep, $\beta \geq \phi'$, the soil mass is unstable because the mobilized soil shear strength is inadequate to maintain slope stability. Under this condition, the construction of earth retaining structures is necessary to maintain slope stability (Figure 5.1b). In general, constructing a retaining wall is more cost-intensive than forming a slope. Therefore, the need for a retaining wall should be assessed carefully during preliminary design, and efforts should focus on limiting the height of retaining structures to minimize construction costs and potential system instability.

Earth retaining structures are typically used for grade separation, bridge abutments, slope stabilization, and excavation support (Sabatini et al., 1997; Tanyu et al., 2008). Some applications include highway embankments, foundation walls, basement excavations, and cut slopes. Earth retaining structures are also widely used for slope stabilization or restoration to prevent and mitigate landslide hazards.

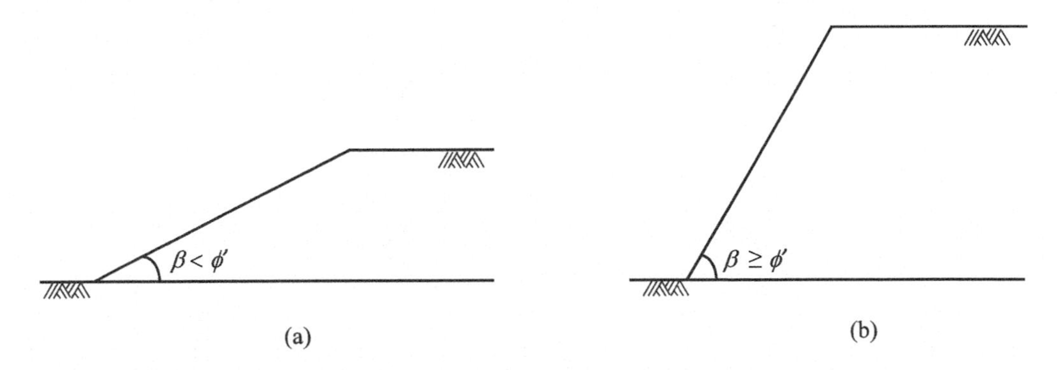

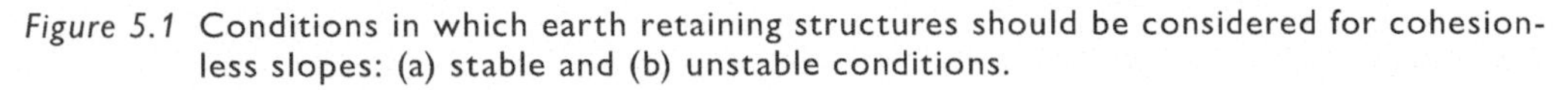

Figure 5.1 Conditions in which earth retaining structures should be considered for cohesionless slopes: (a) stable and (b) unstable conditions.

DOI: 10.1201/9781003350019-5

This chapter is organized as follows. Various earth retaining structures commonly used in engineering practice are introduced. These earth retaining structures are further classified based on their construction methods and stabilization mechanisms. The advantages and disadvantages of earth retaining structures are compared, and the wall selection flowchart and factors that govern the selection and use of various earth retaining structures are discussed. Subsequently, the design and analysis of three types of earth retaining structures, namely, gravity and semi-gravity walls, reinforced walls, and nongravity walls, are discussed. The typical dimensions, drainage systems, failure modes, stability analyses, and improvement measures for each wall type are discussed in detail.

5.2 Types of earth retaining structures

Numerous earth retaining structures have been developed and used in engineering practice. Despite the many options of earth retaining structures, often, only a few types of structures are the most efficient or economical for a certain project. Therefore, selecting the most suitable earth retaining structure for a specific case is important. Because retaining wall systems are diverse and have many benefits and limitations, this section provides information for readers to help them (1) identify the many retaining wall types on the basis of their construction methods and stabilization mechanisms; (2) understand the applications, limitations, advantages, and disadvantages of different earth retaining systems; and (3) select the most technically appropriate and cost-effective earth retaining structure for an application based on comprehensive knowledge of available wall systems.

5.2.1 *Classification*

A classification system for earth retaining structures is presented in Figure 5.2. According to Sabatini et al. (1997), earth retaining structure systems are classified by their construction

	Fill Wall		
Externally Stabilized Wall	Gravity, and Semi-Gravity Wall 1. Gravity wall 2. Semi-gravity cantilever wall Prefabricated Modular Gravity Wall 3. Crib/Bin wall 4. Gabion wall	Reinforced Wall 5. Mechanically stabilized earth (MSE) and geosynthetic-reinforced soil (GRS) walls	Internally Stabilized Wall
	Nongravity Wall 6. Sheet-pile wall 7. Soldier pile and lagging wall 8. Diaphragm (slurry) wall 9. Tangent/secant pile wall Anchored Wall 10. Soil anchor wall	In situ Reinforced Wall 11. Soil nail wall 12. Micropile wall	
	Cut Wall		

Figure 5.2 Classification of earth retaining structures.

methods (i.e., fill or cut construction) and stabilization mechanisms (i.e., external or internal stabilization). Accordingly, each earth retaining structure in Figure 5.2 is classified into four phases. For instance, a gravity wall is classified as an externally stabilized fill wall, whereas a soil nail wall is classified as an internally stabilized cut wall.

Figure 5.3 illustrates the construction of a fill wall, which involves a sequence of backfilling and compacting soil from the base to the top of the wall. Figure 5.4 presents the construction of a cut wall, which often involves excavating and stabilizing soil from the top to the base of the wall. Fill and cut walls typically cannot be used interchangeably because of differences in their construction methods and sequence. For example, a gravity wall cannot be used for a deep excavation project, and constructing a fill wall is infeasible for a cut-type project. However, the classification may have exceptions; for example, the sheet-pile wall for the waterfront structure can sometimes be constructed as a fill wall by backfilling the soil behind the wall.

Figure 5.5 presents the load transfer mechanism of an externally stabilized wall. The stability of externally stabilized walls is provided by the external resistance through (1) base shear friction induced by the weight of gravity walls or (2) passive earth pressure acting on the embedded wall length and the additional lateral load from supports for nongravity walls. Figure 5.6 presents the load transfer mechanism of an internally stabilized wall. The stability of internally stabilized walls is provided by internal resistance through the reinforcement tensile force mobilized by the soil–reinforcement interaction. Reinforcements should have sufficient embedment length, often extending beyond the potential failure surface, to prevent walls from

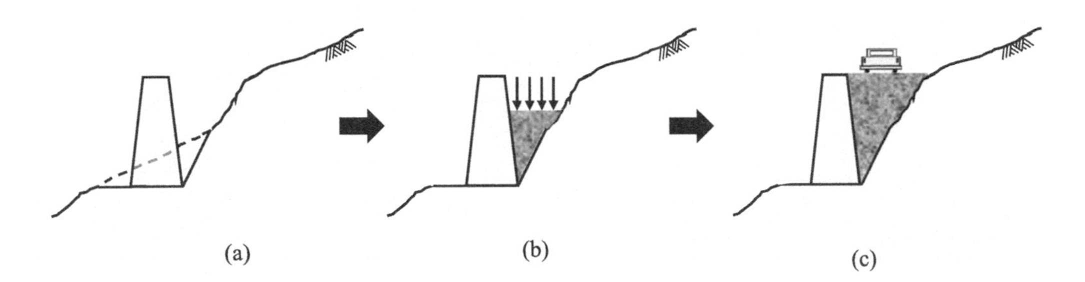

Figure 5.3 Fill wall construction: (a) retaining wall construction, (b) soil backfilling and compacting, and (c) repeated backfilling and compacting until the desired height is reached.

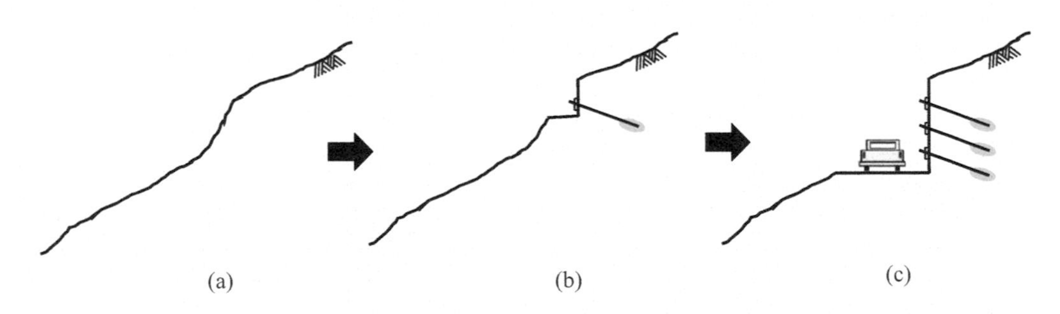

Figure 5.4 Cut wall construction: (a) original slope profile, (b) soil excavation and stabilization, and (c) repeated excavation and stabilization until the desired depth is reached.

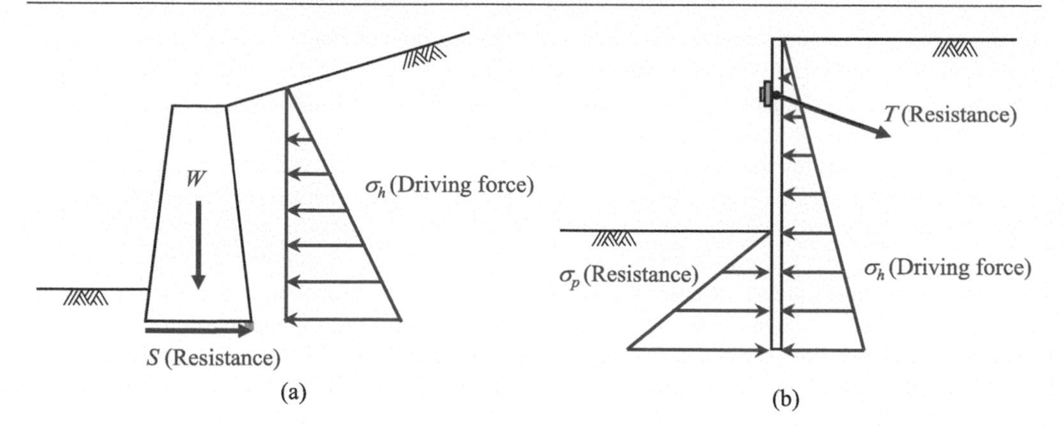

Figure 5.5 Externally stabilized mechanism: (a) gravity walls; (2) nongravity walls.

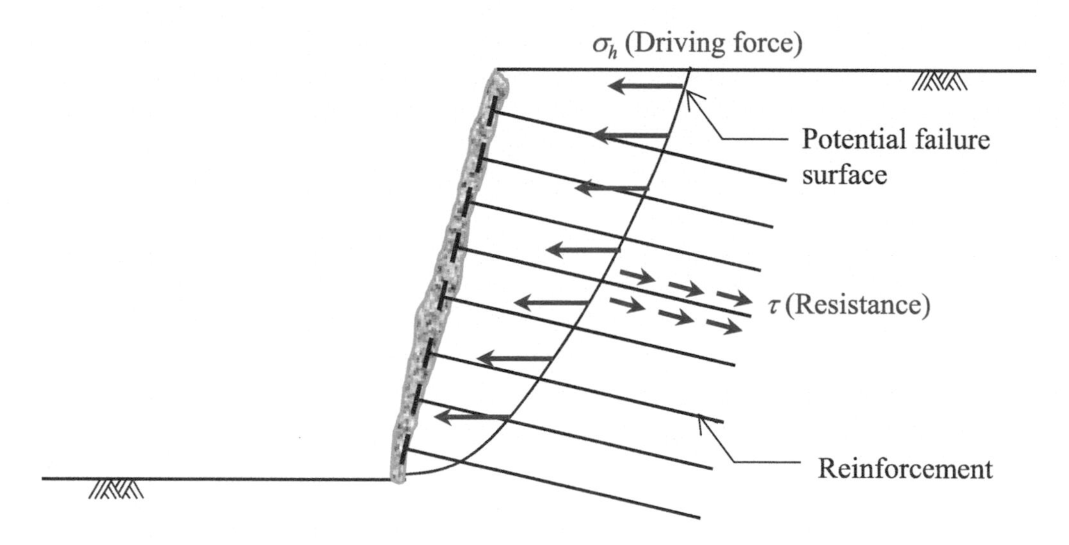

Figure 5.6 Internally stabilized mechanism.

losing anchorage efficiency and shifting with the soil failure mass (i.e., pullout failure). Because externally and internally stabilized walls have different load transfer mechanisms, their design methods are fundamentally different.

Figures 5.7–5.11 present illustrations and photographs of various types of fill walls. The typical dimensions and primary components of each fill wall are indicated in the figures. Gravity-type walls include gravity walls and semi-gravity cantilever walls. Prefabricated modular gravity walls include crib or bin walls and gabion walls. Gravity walls and prefabricated modular gravity walls rely on their own weight to resist overturning and sliding due to lateral pressure from retained soil behind the wall. Semi-gravity walls consist of steel-reinforced concrete wall stems and base slabs to form an inverted "T" shape. Triangular buttresses are sometimes installed at regular intervals along the length of a wall. The main purpose of both measures is to enhance wall stiffness and, hence, reduce wall deformation.

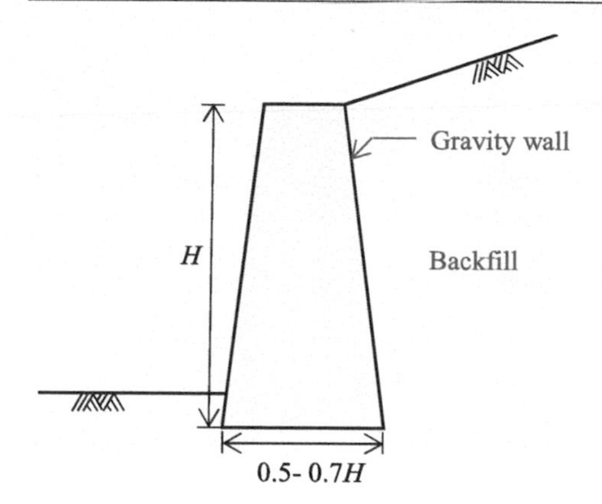

Figure 5.7 Gravity wall.

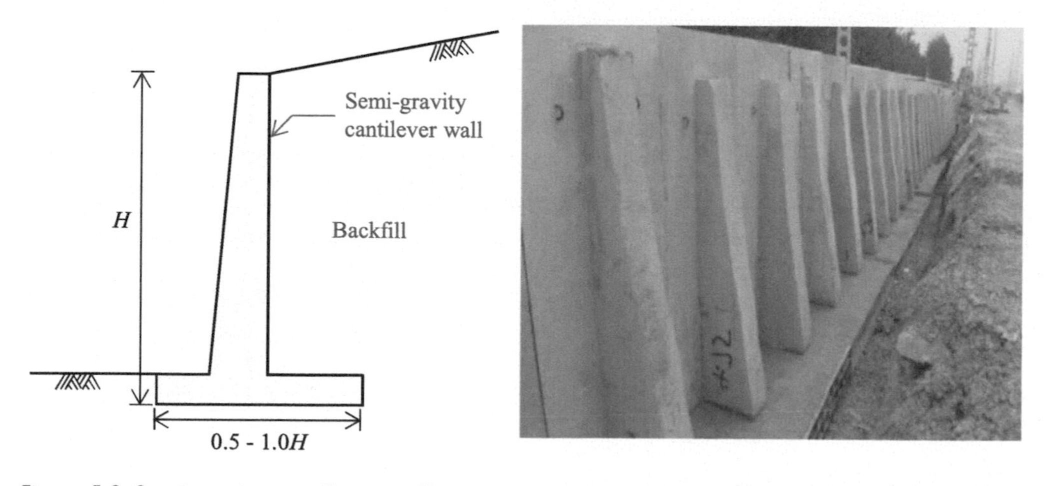

Figure 5.8 Semi-gravity cantilever wall.

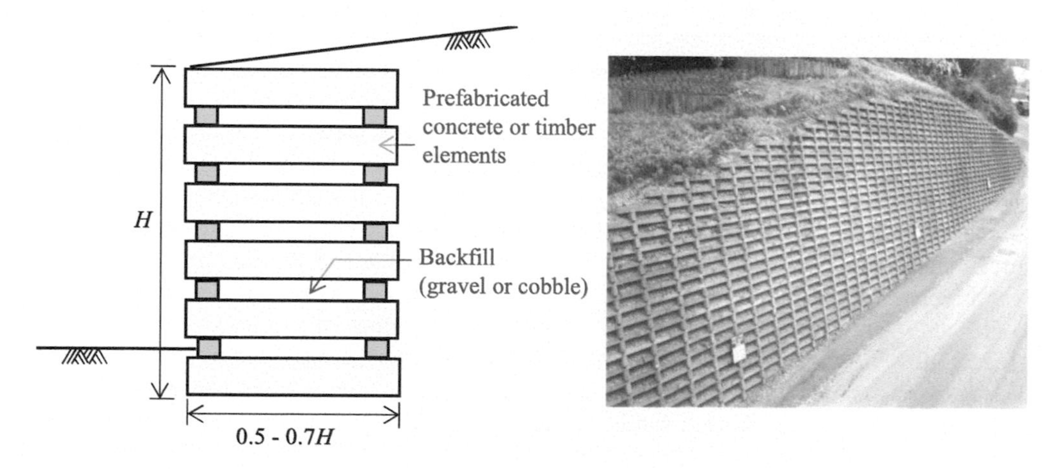

Figure 5.9 Crib/bin wall.

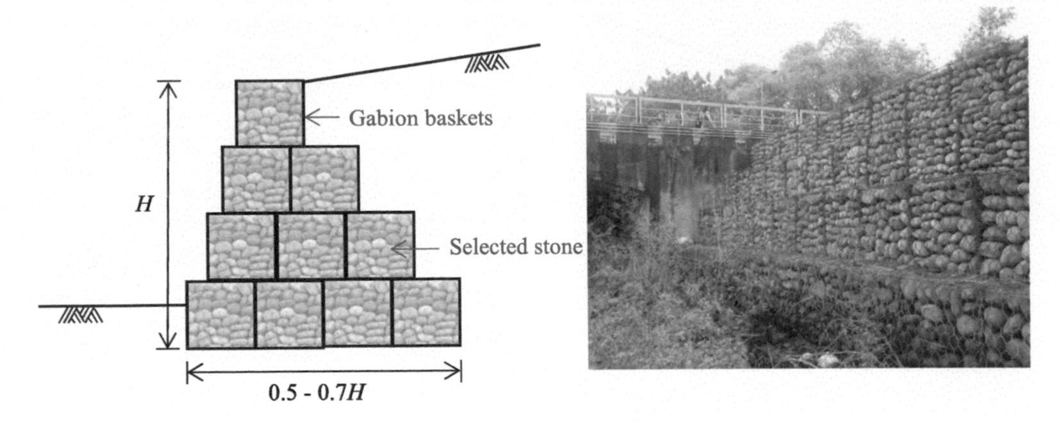

Figure 5.10 Gabion wall.

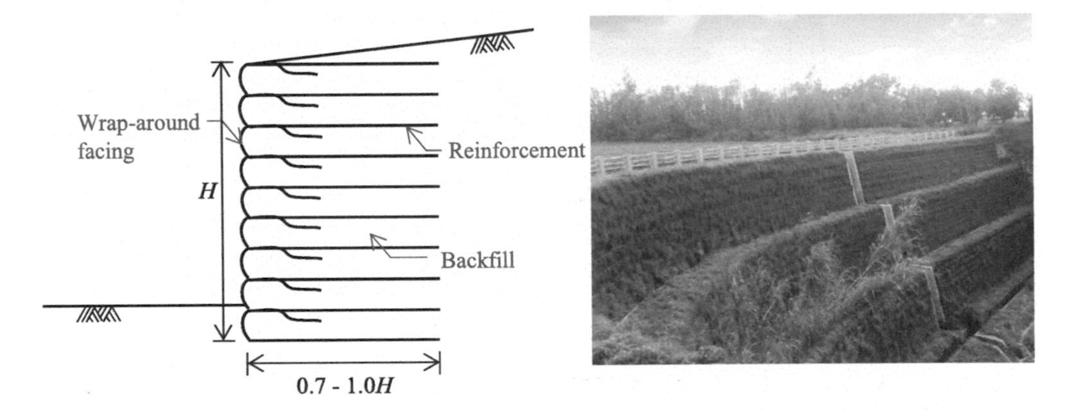

Figure 5.11 Reinforced wall.

Semi-gravity walls rely on their own weight plus the soil weight above the base slab to resist overturning and sliding caused by lateral earth pressure. Reinforced walls, generally known as mechanically stabilized earth (MSE) walls, employ either metallic reinforcements or geosynthetics (polymer reinforcements) in the backfill soil. Walls reinforced with geosynthetics are called geosynthetic-reinforced soil (GRS) walls. The lateral earth pressure within the reinforced wall is resisted internally by the reinforcement tensile force mobilized by the soil–reinforcement interaction. Reinforced walls are relatively flexible and sustain large deformations without considerable structural distress.

Fill walls typically require the use of granular, free-draining soil as backfill to prevent the accumulation of water pressure. Most fill walls are used for permanent applications because they are cost-intensive, and most of the wall components are nonreusable. Few MSE and GRS walls without permanent facings can be used for temporary applications. Compared with permanent walls, walls used for temporary applications typically have less-restrictive requirements on material durability, factors of safety (*Fs*), performance, and overall appearance. The service life of walls for temporary applications is based on how long they need to support the soil before permanent walls are installed. Temporary systems are commonly used for 18–36 months, depending on actual project conditions. Table 5.1 summarizes the cost-effective height ranges, right-of-way (ROW) requirements, advantages, and disadvantages of fill wall systems. The ROW is defined as the easement (or space) required to construct the retaining wall.

Figures 5.12–5.18 present illustrations and photographs of various types of cut walls. The primary components for each cut wall are indicated in these figures. Nongravity walls include sheet-pile walls, soldier pile and lagging walls, diaphragm walls, and tangent and secant pile walls. The resisting force of nongravity walls is derived mainly from passive earth pressure acting on the embedment length of the wall at the excavation side and the additional lateral resistance from several layers of support. Anchored walls derive resistance from ground anchors. One end of the ground, anchor is fixed and prestressed at the anchor head at the wall face, and the other end of the ground anchor is extended to a grouted zone beyond the potential failure surface. The prestressed load transfers from the ground anchor to the wall face to provide substantial lateral support. In situ reinforced walls include soil nail and micropile walls. For these walls, reinforcements, typically steel bars or beams, are drilled or driven to a depth beyond the failure

Table 5.1 Summary of fill wall types

Types	*Temporary/ permanent*	*Cost-effective height range*	*Required ROW*	*Advantages*	*Disadvantages*
Gravity wall	Permanent	1–4 m	0.5–0.7*H*	• Easy design • Design method is well-developed	• Differential settlement • Bearing capacity
Semi-gravity cantilever wall	Permanent	2–9 m	0.5–1.0*H*	• Easy design • Design method is well-developed	• Differential settlement • Bearing capacity
Crib/bin wall	Permanent	2–11 m	0.5–0.7*H*	• Easy design and fast construction • Design method is well-developed • High drainage capacity	• Difficult for compaction • Difficult to adjust the geometry of the wall • Bearing capacity
Gabion wall	Permanent	2–8 m	0.5–0.7*H*	• Erosion control • Ecology-friendly system • Flexible system • High drainage capacity	• Labor • Gabion wire backer corrosion • Source of rock and stone
Reinforced wall	Temporary/ permanent	2–20 m	0.7–1.0*H*	• Rapid construction • Different types of facing • Flexible system • High drainage capacity	• Granular soil for backfill • Metallic reinforcement corrosion • Reinforcement strength reduction due to creep, durability, and installation damage

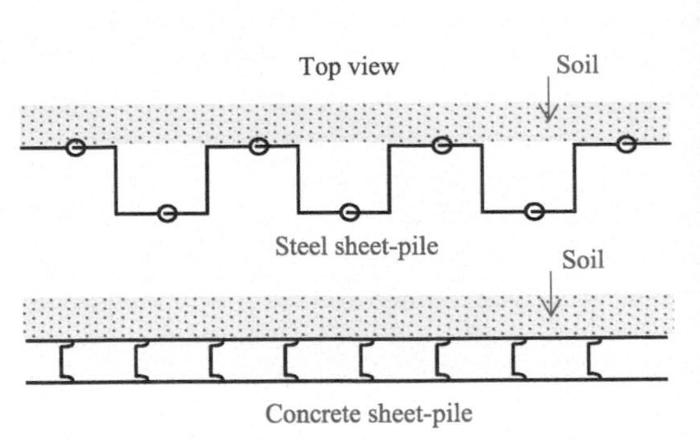

Figure 5.12 Sheet-pile wall.

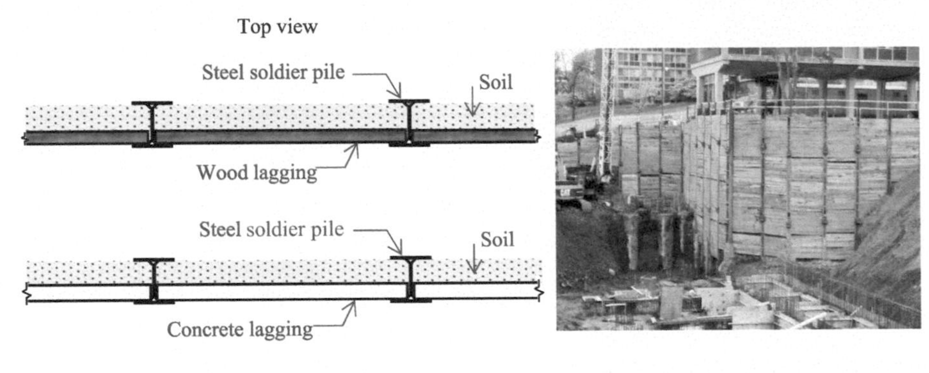

Figure 5.13 Soldier pile and lagging wall.

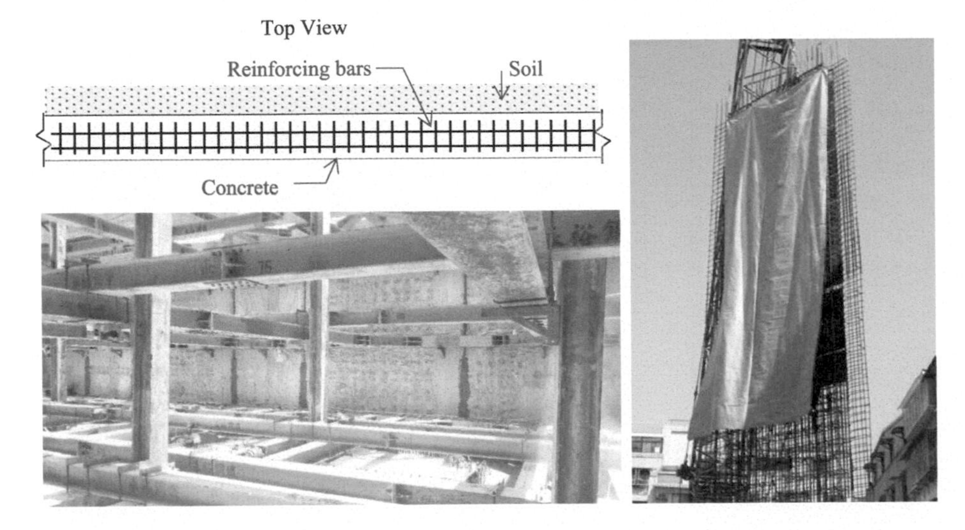

Figure 5.14 Diaphragm (slurry) wall.

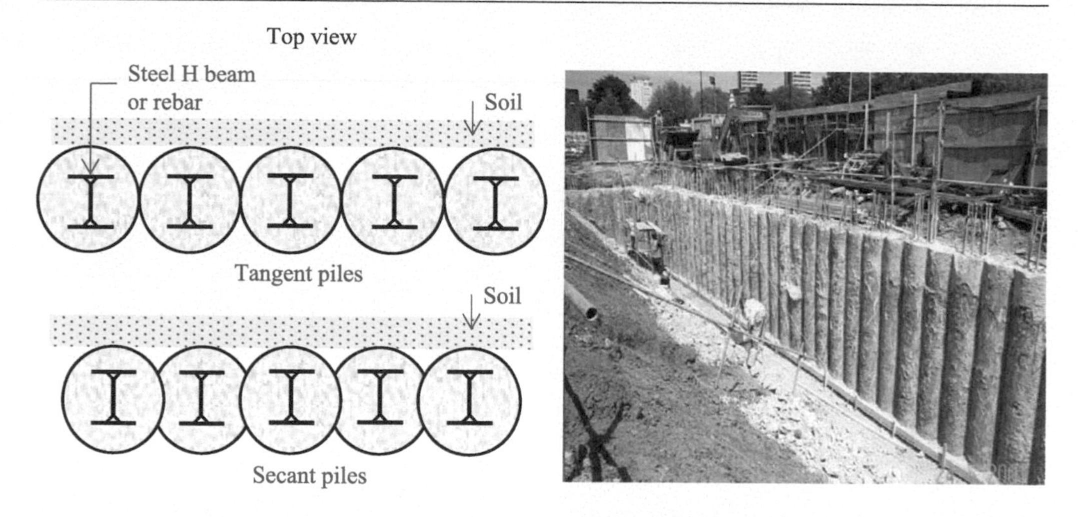

Figure 5.15 Tangent/secant pile wall.

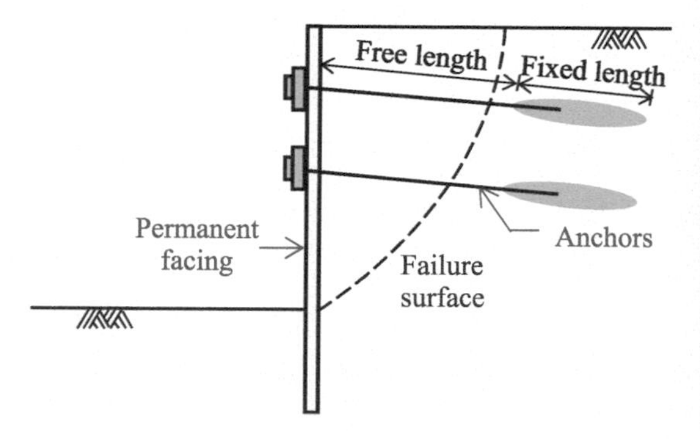

Figure 5.16 Soil anchor wall.

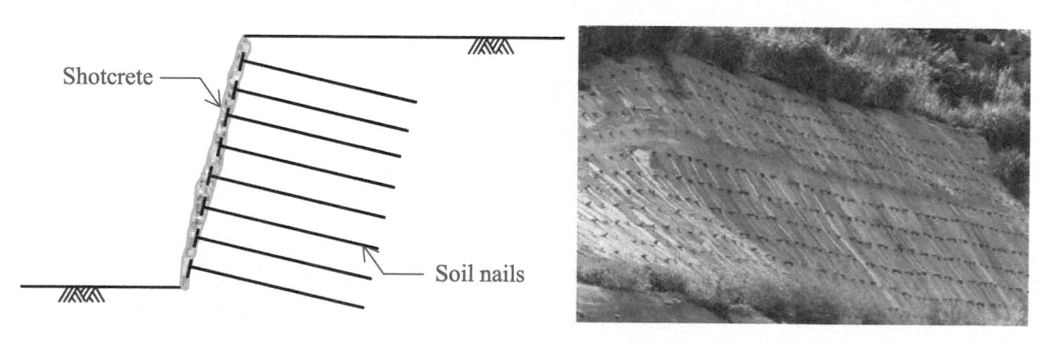

Figure 5.17 Soil nail wall.

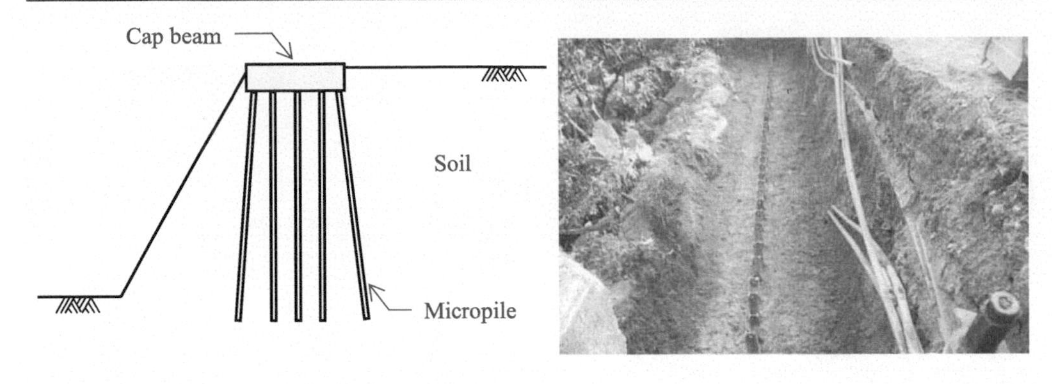

Figure 5.18 Micropile wall.

surface and grouted to prevent long-term corrosion of the metal. Steel reinforcements with high tensile strength prevent the development of a critical failure surface within the reinforced zone, thereby increasing the overall stability of the retaining wall system.

Nongravity walls typically must be watertight to prevent groundwater leakage through the wall. For some less-impermeable wall systems, such as sheet-pile walls, soldier pile and lagging walls, and tangent pile walls, a waterproofing layer must be installed behind the wall. In contrast to nongravity walls, soil anchor and soil nail walls must install drainage pipes to guide the groundwater out. If the metallic reinforcing elements of cut walls are submerged in groundwater, long-term corrosion protection measures are crucial to prevent the loss of tensile strength or even breakage. Some cut walls, such as sheet-pile walls, soldier pile and lagging walls, and soil nail walls, can be used for temporary support during construction. After excavation is complete, a permanent structure or facing is placed in front of these temporary walls. Soil anchor walls, soil nail walls, and micropile walls often require considerable underground space (or ROW) to enable the reinforcing elements to extend beyond the potential failure surface to achieve adequate anchorage. The feasibility of using these walls is influenced by the presence of buried utility lines and current foundations nearby and the cost of permanent underground easement for the placement of reinforcing elements. Table 5.2 summarizes the cost-effective height ranges, required ROWs, advantages, and disadvantages of cut wall systems.

5.2.2 Wall selection

The selection of a retaining wall for a certain construction project is complex because it involves matching several objectives of wall selection with the numerous retaining walls available to determine the most appropriate wall type. The objectives of wall selection include cost-effectiveness, practicality of construction, stability, watertightness, aesthetics and consistency with the surroundings, and environmental friendliness. First, the design engineer should be familiar with the applications, limitations, advantages, and disadvantages of various earth retaining systems, as discussed in the previous section. A rigorous and systematic evaluation and selection process should be conducted to determine the most appropriate wall type. Figure 5.19 presents a flowchart of the wall selection process suggested by Sabatini et al. (1997).

Table 5.2 Summary of cut wall types

Types	*Temporary/ permanent*	*Cost-effective height range*	*Required ROW*	*Advantages*	*Disadvantages*
Sheet-pile wall	Temporary/ permanent	<5 m	None	• Can be used below groundwater table • Design method is well-developed • Rapid construction	• Noisy and vibrates • Difficult to drive into hard, dense soil and gravels • Vibration-induced settlement in loose ground
Soldier pile and lagging wall	Temporary/ permanent	<5 m	None	• Rapid construction • Less piles are driven than a sheet-pile wall	• Difficult to maintain vertical tolerances in hard ground • Waterproof
Diaphragm wall	Permanent	6–25 m	None	• Rigid structure • Impermeable • Suitable for all soils	• Disposal of slurry may follow environmental restrictions
Tangent/ secant pile wall	Permanent	3–9 m/6 m above for wall with anchors or struts	None	• Rigid and impermeable • Suitable for all soils • Can control wall stiffness	• Requires specialized equipment • Significant spoil disposal
Soil anchor wall	Permanent	5–20 m	$0.6H$ + anchor bond length	• Can resist large horizontal pressures • Adaptable to varying site conditions	• Requires skilled labor and specialized equipment • Requires permanent easement • Anchors may require corrosion protection
Soil nail wall	Temporary/ permanent	3–20 m	$0.6–1.0H$	• Rapid construction	• Nails may require permanent easements • Difficult to construct design below water table
Micropile wall	Permanent	N/A	Varies	• Rapid construction • Suitable for slope stabilized and building protection	• Design method is not well-developed

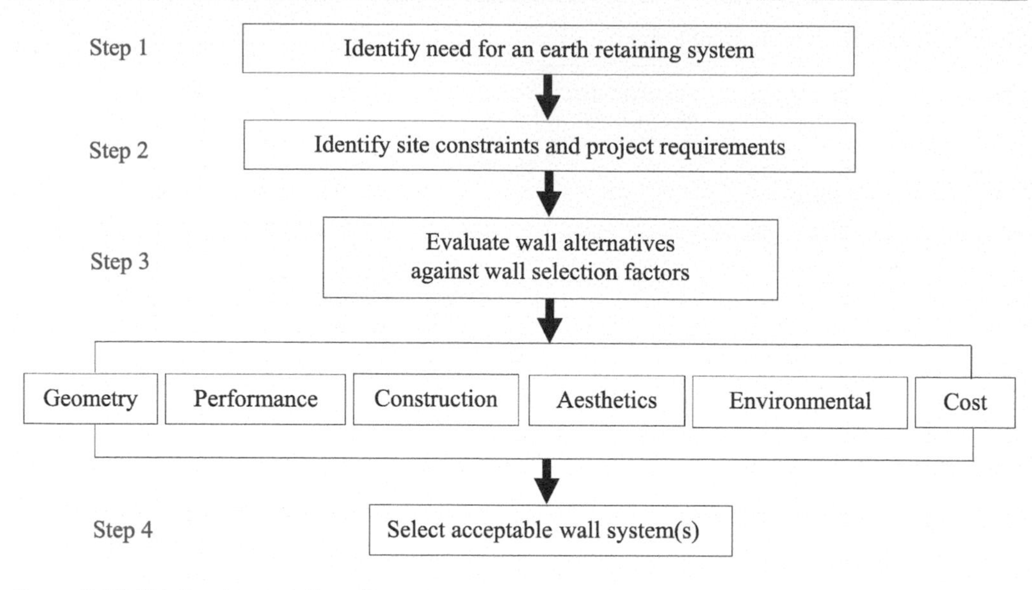

Figure 5.19 Wall selection flowchart.

Source: Redrawn from Sabatini et al. (1997).

The first step in wall selection is to identify the need for an earth retaining wall system. If the construction site has restricted or congested space (i.e., limited ROW), then an earth retaining system that uses abrupt changes in slope grades is warranted to stabilize the soil for projects. If space restrictions are not a problem at the construction site, simply grading a slope can achieve the same effect.

The second step is to identify project requirements and site constraints by performing a review of the specific site. Items that must be identified during the site review include (1) subsurface ground and groundwater conditions, (2) site accessibility and space restrictions, (3) locations of buried utilities and current foundations, (4) aesthetic requirements, and (5) environmental concerns. These items should be weighted by their priority during wall selection. On the basis of the site review results, several wall systems may be eliminated from consideration, and others may be recommended for further consideration.

In the third step, the remaining candidate wall systems are evaluated individually in detail against the priority items identified in the previous step. Specific factors for wall system evaluation include (1) wall geometry, (2) performance criteria, (3) construction consideration, (4), aesthetics, (5) environmental concerns, and (6) cost. Each factor is evaluated on the basis of its relevance and importance for each candidate wall system. The wall geometry factor is used to evaluate whether the cost-effective wall height and width of each wall candidate fulfill the project requirements and site ROW restrictions. The performance criteria factor ensures that the performance of the wall candidate (e.g., allowable differential settlement for a fill wall or allowable lateral movement for a cut wall) meets project requirements and legal regulations. The construction consideration factor is used to evaluate constructability, site accessibility, the availability of construction material and equipment, and on-site wall material storage, temporary dewatering requirements, construction speed, local contracting practices, and labor considerations. The aesthetic factor is used to ensure that the permanent retaining wall is aesthetically pleasing and consistent with its surroundings, particularly the natural environment. The environmental factor is used to assess the potential environmental impact of the candidate wall during and after construction, including the excavation and disposal of soil, use of construction techniques and materials that are environmentally detrimental or cause high CO_2 emissions, the discharge of large quantities of water or slurry fluids, and effects of construction noise, vibration, and dust. Regarding the cost factor, in addition to material and construction costs, long-term maintenance expenses should be considered in determining the total cost of a retaining wall. If a retaining wall is used for slope stabilization to mitigate landslide hazards, the effectiveness of the wall type on protecting different scales of landslides, as suggested by Liao (2008), should also be considered during wall system evaluation (Figure 5.20).

The final step is to select acceptable wall systems. This selection is typically achieved by analyzing the evaluation results in the third step in a wall selection matrix, and the wall (or walls) with the highest score is selected for the project (Tanyu et al., 2008). However, an introduction to the wall selection matrix is beyond the scope of this chapter. Based on the comparison results, the wall with the highest score is selected as the most appropriate wall system, and other high-scoring walls can be recommended as acceptable alternatives.

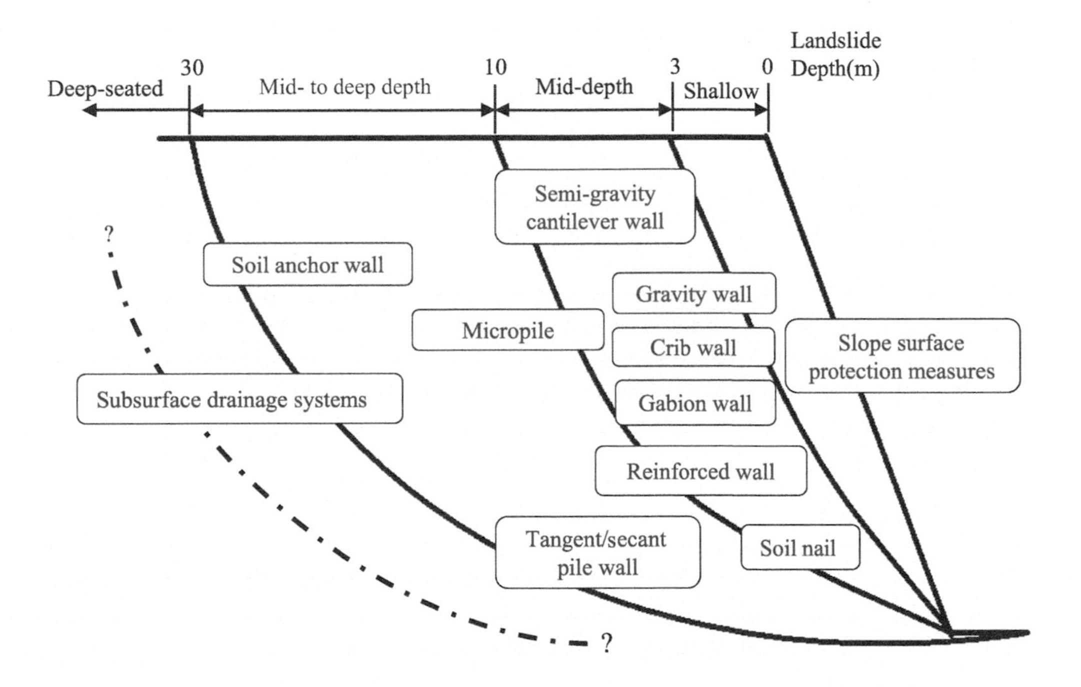

Figure 5.20 Effectiveness of different retaining walls for protecting against different scales of landslides.

Source: Modified from Liao (2008).

Example 5.1

Identify the type and classification of the retaining wall in the figure.

1 Which type of retaining wall is pictured?

(a) soil anchor wall, (b) gravity wall, (c) soil nail wall

2 How would the retaining wall be classified?

(a) internally stabilized cut wall, (b) externally stabilized cut wall, (c) internally stabilized fill wall

Figure EX 5.1 Photo of a retaining wall.

Solution

ANS: 1: (a); 2: (b)

The figure depicts a soil anchor wall because it features anchor heads and an RC grid. Compared with a soil nail wall, the anchor heads of a soil anchor wall are much larger. In addition, soil nail walls typically have shotcrete facings.

The soil anchor wall is classified as an externally stabilized cut wall. A slope is stabilized by the tensile force of the soil anchor. The tensile force is mobilized at the fixed end of the soil anchor and transferred to the anchor head through the anchor tendons. This force acting on the slope face suppresses slope displacement, therefore improving slope stability.

Example 5.2

A 13 m excavation is supported by a retaining wall at a construction site in an urban area. A site investigation reveals that the subsurface soil consists of mainly NC clay with an SPT-N value of 2–7. The water table is 2 m below the ground surface. According to the ground conditions in the question statement, which type of retaining wall is suitable for this construction project? (a) sheet-pile wall, (b) diaphragm wall, (c) tangent pile wall, (d) gravity wall

Solution

ANS: (b) and (c)

In this case, a cut wall should be selected because an excavation is needed. Thus, a gravity wall, which is classified as a fill wall, is unsuitable.

Because the construction of a sheet-pile wall could cause noise and vibration, it is also unsuitable, especially for construction in an urban area. Furthermore, in deep excavation, a sheet-pile wall, depending on its stiffness, may experience large displacement in the soft clay layer. Large ground deformation would affect adjacent buildings and might cause damage.

A diaphragm wall and tangent pile wall are most suitable for this construction project. Compared with the sheet-pile wall, these two retaining walls cause less noise and vibration during construction. In addition, these walls have high stiffness, producing relatively small lateral wall displacement when the subsurface soil is excavated.

Example 5.3

A typhoon and heavy rainfall caused many roads to collapse in a mountainous area. The roads were closed because of slope failures. A project was conducted to reconstruct retaining walls by using the nearby colluvium soil. The colluvium soil consists of mainly sand from weathered sandstone. According to the construction specifications, the relative compaction of backfill soil is regulated to be at least 90% to ensure the stability of the retaining wall. On the basis of these conditions, which type of retaining wall is suitable for this reconstruction project in the mountainous area? (a) tangent pile wall, (b) reinforced wall, (c) crib wall, (d) gravity wall

Solution

ANS: (b) and (d)

In this case, a fill wall is needed. Therefore, a tangent pile wall, which is classified as a cut wall, is unsuitable. A crib wall is constructed using precast components that must be transported by trucks from a factory. In this scenario, the road collapsed in many places; thus, transportation might be difficult for large trucks. This situation makes a crib wall unsuitable. Therefore, reinforced and gravity walls are the most suitable for the situation. In addition, the original slope was damaged by a typhoon and heavy rain. An improved drainage system should be planned and implemented in the reconstruction project to prevent the accumulation of water pressure behind the retaining walls, which could cause further retaining failure.

5.3 Gravity and semi-gravity walls

This section introduces the design of gravity and semi-gravity cantilever walls. The design methods and performance criteria of these walls are well-developed. This section also presents the procedures of stability analyses of wall stability against overturning, sliding, and bearing capacity failures.

5.3.1 Typical dimensions and drainage design

Gravity and semi-gravity cantilever walls are the most common earth retaining structures for various applications, including roadway embankments, bridge abutments, riverbank protection, and slope stabilization. The most desirable advantage of gravity and semi-gravity cantilever walls is their durability in many environments because they are made of mainly concrete.

Figures 5.21 and 5.22 illustrate the typical dimensions and drainage design of gravity and semi-gravity walls, respectively. Engineers initially use these typical dimensions as trial wall sections for stability analyses. If the stability analyses yield undesirable results, the wall sections can be adjusted until the calculated Fs satisfy the design requirements. The wall aspect

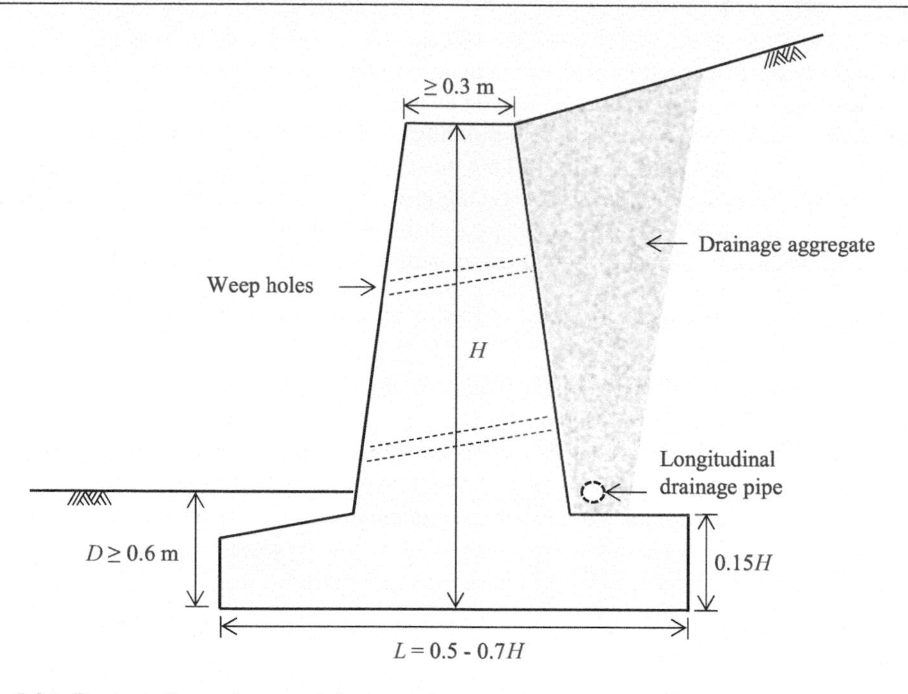

Figure 5.21 Typical dimensions and drainage layout of a gravity wall.

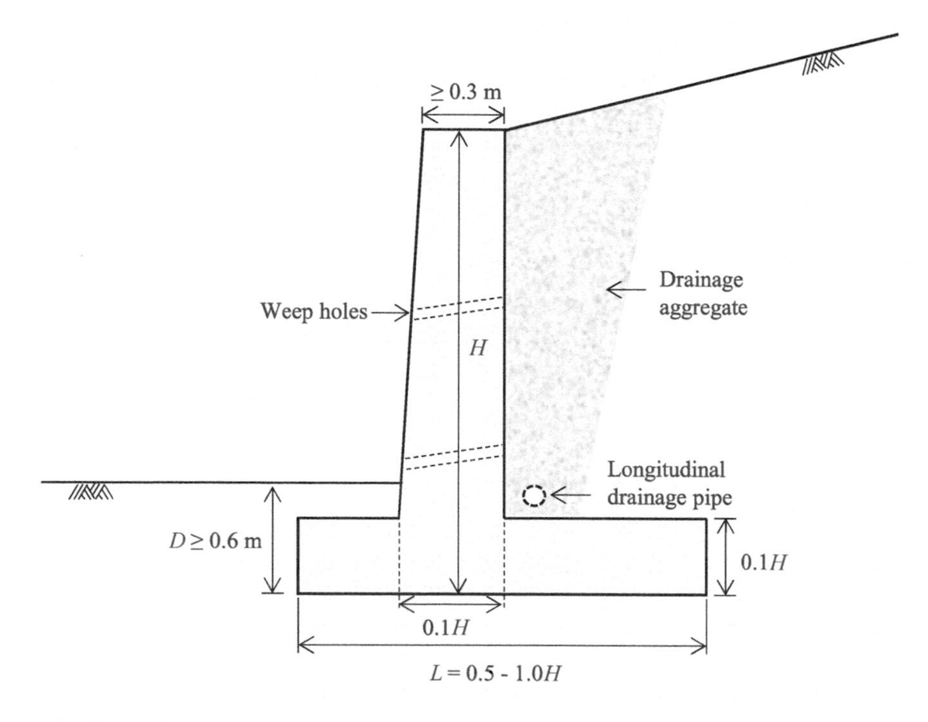

Figure 5.22 Typical dimensions and drainage layout of a semi-gravity cantilever wall.

ratios (L/H) are typically 0.5–0.7 for gravity walls and 0.5–1.0 for semi-gravity walls. The wall embedment depth D should be at least 0.6 m. The thickness at the top of the retaining wall should be >0.3 m to provide adequate space for proper concrete placement.

A drainage system is considered a primary component of gravity and semi-gravity walls. Because of rainfall infiltration and a high groundwater level, the backfill of a retaining wall might become saturated with water, and the water pressure can accumulate and act on the retaining wall as a driving force in addition to the lateral earth pressure. A properly constructed and maintained drainage system minimizes the possibility of poor wall performance resulting from water pressure accumulation. Typically, a layer of drainage aggregate with high permeability is placed behind the wall. If any water enters the drainage aggregate layer, it can easily drain out via the drainage pipe at the base. The outlet of the drainage pipe is connected to a stormwater drainage system outside the retaining wall. Weep holes are installed at the stem of the retaining wall to accelerate water drainage when the groundwater level is high. Weep holes should have a minimum diameter of approximately 0.1 m and be adequately spaced. Filters (i.e., geotextile) should be installed inside or behind the weep holes to prevent internal soil erosion by seepage, which could cause the loss of backfill soil or clog the weep holes in the long term.

5.3.2 Failure modes

A gravity or semi-gravity cantilever wall can fail if (a) it overturns about its toe (Figure 5.23a), (b) it slides along its base (Figure 5.23b), (c) the foundation soil has insufficient bearing capacity (Figure 5.23c), (d) global failure occurs (Figure 5.23d), or (e) excessive settlement or differential settlement

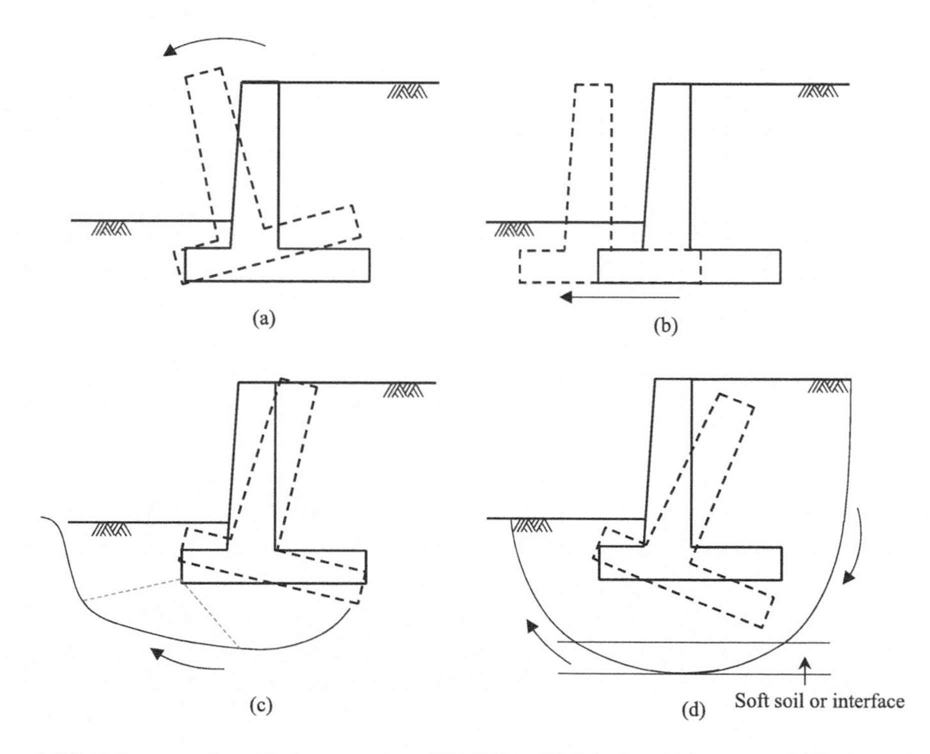

Figure 5.23 Failure modes: (a) overturning, (b) sliding, (c) bearing capacity, and (d) global failure.

occurs. Stability analyses to evaluate wall stability against overturning, sliding, and bearing capacity failure are described, and improvement measures are suggested in the subsequent sections.

Global failure, also known as deep shear failure, can occur when a soil layer with low shear strength is behind or under a wall (Figure 5.23d). A failure surface might develop along this weak soil layer, eventually causing global instability. Global failure is often analyzed with slope stability software by using limit equilibrium methods, which will be introduced in Chapter 8. Global failure can be avoided by conducting a detailed site investigation to identify weak soil layers and consider them in slope stability analysis. The principles of evaluating foundation settlement are discussed in Section 3.5. In addition, foundation settlement can be evaluated using advanced numerical approaches, such as the finite element method. If the calculated foundation settlement is excessively large, an improvement measure, such as ground improvement or the use of lightweight fill material (i.e., geofoam), should be implemented.

In seismic zones, walls can fail when subjected to seismic loadings during earthquakes. Consequently, the influence of earthquakes on the stability of permanent retaining walls should be considered. The dynamic earth pressure, as discussed in Section 4.8, should be used in stability analyses. When seismic loadings are included, the *Fs* under seismic conditions are typically required to be 0.75 times those under static conditions.

5.3.3 Force components for stability analyses

Figure 5.24 presents the force components of the stability analyses. The force components considered in the analyses include the weight of the soil above the heel (W_1 and W_2), the weight of the retaining wall (W_3 to W_6), the active lateral earth force behind the wall (P_a), the passive lateral earth force in front of the wall (P_p), and the interface shear resistance (S).

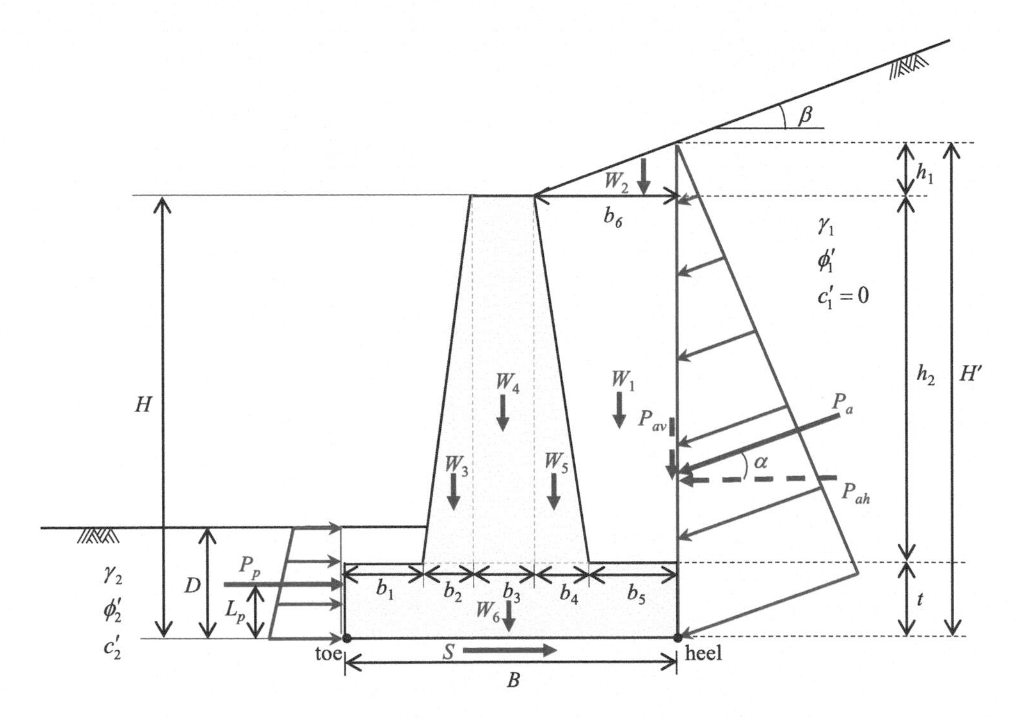

Figure 5.24 Force components of stability analyses.

To calculate the weight of the retaining wall, the polygon of the wall is divided into several triangles and rectangles to calculate their areas and centroids easily. P_a is calculated using the active lateral earth pressure, as expressed in eq. 5.1, and P_{av} and P_{ah} are the vertical and horizontal force components of P_a.

$$P_a = \frac{1}{2}\gamma_1 H'^2 K_a \tag{5.1}$$

where:

γ_1 = unit weight of backfill.
H' = height of retaining wall plus height of sloped backfill (as indicated in Figure 5.24).
K_a = active earth pressure coefficient.

The base shear, S, is estimated using the Mohr–Coulomb theory to consider the interface shear strength between the base of the wall and the foundation soil. S can be derived as follows:

On the basis of the Mohr–Coulomb failure criteria:

$$\tau_f = c'_a + \sigma' \tan\delta \tag{5.2}$$

where:

τ_f = interface shear strength.
σ' = normal stress.
δ and c_a = interface friction angle and cohesion, respectively, between the wall and the foundation soil.

The values of δ and c_a can be obtained experimentally from the direct shear test on the concrete–foundation soil interface. The values of δ and c_a typically range from 1/2 to 2/3 of ϕ' and c', respectively. Multiplying eq. 5.2 by the unit area of the wall base ($B \times 1$) converts stress to force as follows:

$$\tau_f (B \times 1) = Bc_a + B\sigma' \tan\delta \tag{5.3}$$

which yields:

$$S = Bc_a + V \tan\delta \tag{5.4}$$

where:

$\tau_f (B \times 1) = S$ = interface shear resistance.
$B\sigma' = V = \Sigma W + P_{av}$ = sum of vertical force acting at the base of the wall.

Tables 5.3 and 5.4 list the equations for horizontal and vertical force components and their perpendicular distances to the toe of the wall. The perpendicular distance to the toe of each force component listed in the tables represents the moment arm measured from the toe of the wall.

Table 5.3 Summary of horizontal force components

Horizontal force	Perpendicular distance to toe
$P_p = \frac{1}{2} K_p \gamma_2 D^2 + 2c_2' \sqrt{K_p} D$	L_P
$P_{ah} = P_a \cos\alpha = \frac{1}{2} \gamma_1 H'^2 K_a \cos\alpha$	$\frac{H'}{3} = \frac{h_1 + h_2 + t}{3}$
$S = Bc_a + V \tan\delta$	0

Table 5.4 Summary of vertical force components

Vertical force	Perpendicular distance to toe
$W_1 = \gamma_1 A_1 = \gamma_1 \frac{1}{2}(b_5 + b_6) h_2$	$x_1 = b_1 + b_2 + b_3 + b_4 + \frac{b_5^2 - (b_6 - b_5)^2/3}{b_6 + b_5}$
$W_2 = \gamma_1 A_2 = \gamma_1 \frac{1}{2} b_6 h_1$	$x_2 = B - \frac{b_6}{3}$
$W_3 = \gamma_c A_3 = \gamma_c \frac{1}{2} b_2 h_2$	$x_3 = b_1 + \frac{2b_2}{3}$
$W_4 = \gamma_c A_4 = \gamma_c b_3 h_2$	$x_4 = b_1 + b_2 + \frac{b_3}{2}$
$W_5 = \gamma_c A_5 = \gamma_c \frac{1}{2} b_4 h_2$	$x_5 = b_1 + b_2 + b_3 + \frac{b_4}{3}$
$W_6 = \gamma_c A_c = \gamma_c Bt$	$x_6 = \frac{B}{2}$
$P_{av} = P_a \sin\alpha = \frac{1}{2} \gamma_1 H'^2 K_a \sin\alpha$	B

Note: γ_1 = unit weight of backfill; γ_2 = unit weight of foundation soil; γ_c = unit weight of concrete.

These distance values are later used to calculate the resisting or driving moments in stability analyses.

5.3.4 Evaluation for overturning

The factor of safety against overturning (F_o) is defined as:

$$F_o = \frac{\sum M_R}{\sum M_D} = \frac{\sum W_i x_i + P_{av} B}{P_{ah} \times H'/3} \tag{5.5}$$

where:

ΣM_R = sum of the resisting moments with respect to the toe.
ΣM_D = sum of the driving moments with respect to the toe.

The moment caused by P_p is disregarded in the calculation. The wall embedment depth is typically shallow; therefore, the developed passive earth pressure resistance may be nonsignificant because little overburden pressure acts on the soil adjacent to the toe of the wall. In addition, the effect of P_p is typically ignored due to the potential for the soil to be removed through natural or manmade processes during the service life of the structure.

In general, $F_o \geq 2$ is required for design. If the required F_o value is not achieved, extending the wall base (i.e., increasing B) can increase it. Other improvement measures, as illustrated in Figure 5.25, can be implemented to increase system stability against overturning failure. A tier configuration (Figure 5.25a) can effectively reduce the lateral earth pressure acting on the walls as the offset distance between tier walls increases. When anchor or tieback systems (Figure 5.25b) are used, the prestressed load of the reinforcing element can increase the resisting moments of the wall system, thereby increasing the F_S against overturning failure.

5.3.5 Evaluation for sliding

The factor of safety against sliding failure along the base of the wall (F_s) is defined as:

$$F_s = \frac{\sum F_{H,R}}{\sum F_{H,D}} = \frac{S}{P_{ah}} \tag{5.6}$$

where:

$\Sigma F_{H,R}$ = sum of the horizontal resisting forces.
$\Sigma F_{H,D}$ = sum of the horizontal driving forces.

As presented in Figure 5.24, S and P_{ah} are the only horizontal resisting and driving forces, respectively. In the calculation, the resisting force from P_p is disregarded for the same reasons described in Section 5.3.4.

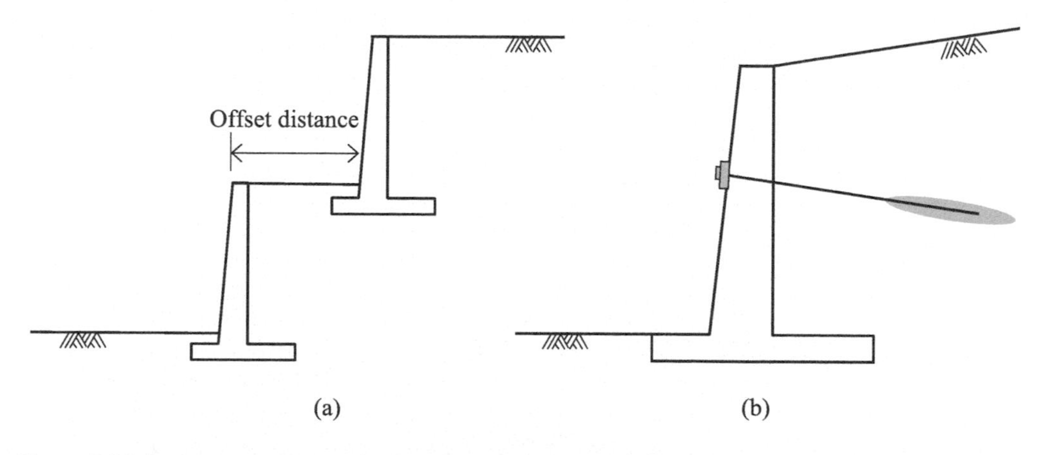

Figure 5.25 Improvement measures against overturning failure: (a) construction in a tier configuration and (b) use of anchor or tieback systems.

In general, $F_s \geq 1.5$ is required for design. If the required F_s value is not achieved, extending the wall base (i.e., increasing B) can increase S at the wall base, thus increasing the F_s value. Another effective improvement measure is implementing a shear key under the wall (Figure 5.26).

If a shear key is designed, the resistance from the passive earth pressure in front of the wall can be included in the calculation. The passive lateral force (P_p) is calculated as:

$$P_p = \frac{1}{2} K_p \gamma_2 D'^2 + 2c_2' D' \sqrt{K_p} \tag{5.7}$$

where:

K_p = passive earth pressure coefficient.
γ_2 = unit weight of foundation soil.
c_2' = cohesion of foundation soil.
D' = depth of the shear key.

The F_s for a wall with a shear key can be expressed as:

$$F_s = \frac{S + P_p}{P_{ah}} \tag{5.8}$$

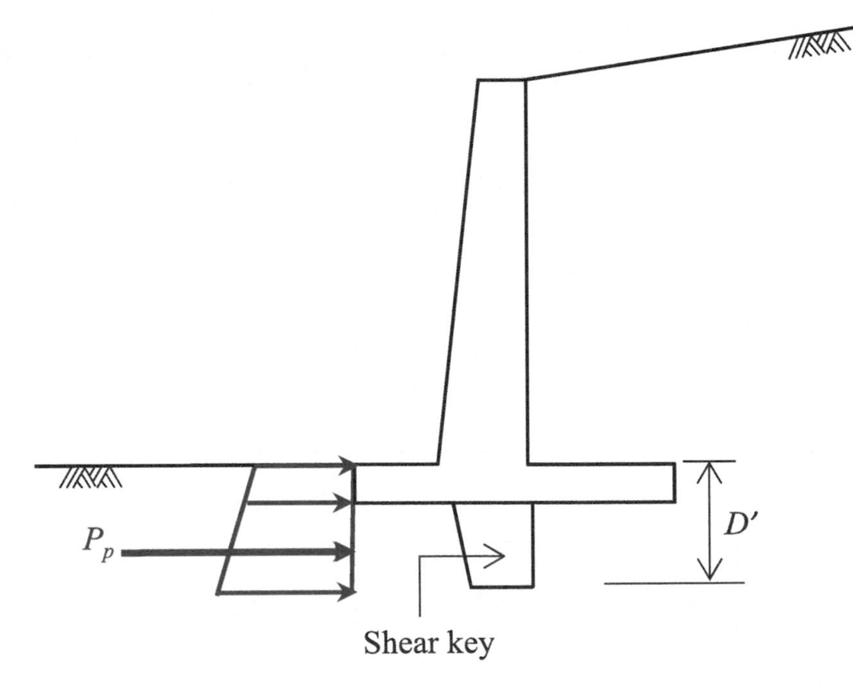

Figure 5.26 Shear key under the wall.

5.3.6 Evaluation for bearing capacity

The evaluation for bearing capacity failure uses the same calculation procedures for a shallow foundation subject to eccentric loading discussed in Section 3.5. In a retaining wall, eccentric loading at the wall base is caused by the lateral earth pressure, and the resultant force R acts on the wall base with a distance e from the centerline (Figure 5.27). This e, called eccentricity, can be evaluated using the following procedures:

The net moment M_{net} with respect to the toe is:

$$M_{net} = \sum M_R - \sum M_D \tag{5.9}$$

where ΣM_R and ΣM_o are the sums of the resisting and driving moments about the toe, respectively. The calculation of ΣM_R and ΣM_o is discussed in Section 5.3.4. The eccentricity e can be expressed as:

$$e = \frac{B}{2} - \frac{M_{net}}{V} \tag{5.10}$$

where M_{net}/V is the distance between the toe and the acting point of R at the wall base.

Because of the effect of eccentric loading, the vertical pressure distribution along the base of the wall is nonuniformly distributed (Figure 5.27). As the moment acts in a counterclockwise direction about the wall base, the maximum vertical pressure q_{max} occurs at the toe of the wall,

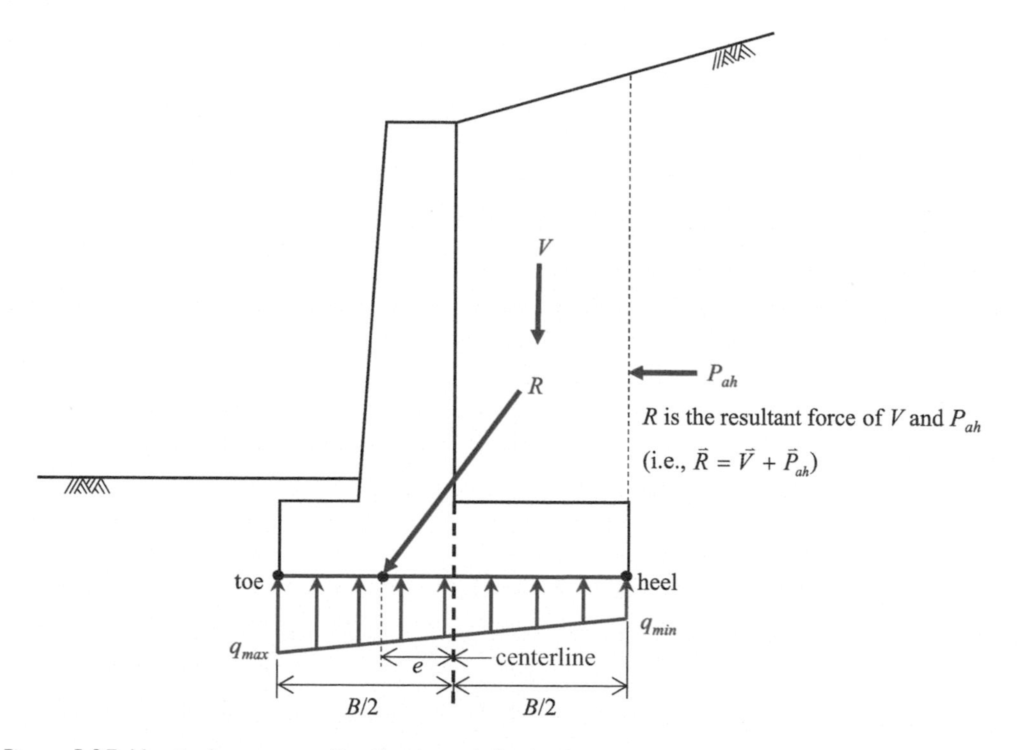

Figure 5.27 Vertical pressure distribution at the base.

and the minimum vertical pressure q_{min} occurs at the heel of the wall. The vertical stress distribution under the wall base can be determined using the following flexural formula for bending stress from the mechanics of materials:

$$q = \frac{V}{B \times 1} \pm \frac{M_{net} y}{I} \tag{5.11}$$

where:

y = perpendicular distance from any point to the neural plane of the wall base.
I = moment of inertia per unit area of the wall base.

For a rectangular wall base, I can be expressed as:

$$I = \frac{B^3 \times 1}{12} \tag{5.12}$$

To calculate q_{max} at the toe and q_{min} at the heel, eqs. 5.10, 5.11, and 5.12 are combined, and y is substituted with $B/2$.

$$q_{\max} = \frac{V}{B}\left(1 + \frac{6e}{B}\right) \tag{5.13a}$$

$$q_{\min} = \frac{V}{B}\left(1 - \frac{6e}{B}\right) \tag{5.13b}$$

When $e > B/6$, q_{min} changes from a positive to a negative value, meaning, that the vertical pressure changes from compression to tension. Thus, tensile stress occurs at the heel of the wall, meaning, that contact between the wall base and the underlying foundation soil is not tight. This situation is undesirable because the effective width of the wall base is reduced and vertical stress might become more concentrated. Accordingly, the following criterion for eccentricity should be fulfilled in the design of a retaining wall:

$$e < \frac{B}{6} \tag{5.14}$$

If the required e value is not achieved, the wall should be redesigned by extending its base.

On the basis of eq. 3.22, which expresses the ultimate bearing capacity of an eccentrically loaded foundation, the ultimate bearing capacity of a wall is:

$$q_u = c_2' N_c F_{cs} F_{cd} F_{ci} + q' N_q F_{qs} F_{qd} F_{qi} + \frac{1}{2}\gamma_2' B' N_\gamma F_{\gamma s} F_{\gamma d} F_{\gamma i} \tag{5.15}$$

where:

c_2' = cohesion of foundation soil.
q' = effective overburden pressure at the level of the wall base.

B' = effective width (= $B - 2e$).
N_c, N_q, N_γ = bearing capacity factors as defined in Chapter 3.
$F_{cs}, F_{qs}, F_{\gamma s}$ = shape factors.
$F_{cd}, F_{qd}, F_{\gamma d}$ = depth factors.
$F_{ci}, F_{qi}, F_{\gamma i}$ = load inclination factors.

The equations for the shape, depth, and load inclination factors are listed in Table 3.4. Notably, $F_{cs} = F_{qs} = F_{\gamma s} = 1$ because the retaining wall is considered a continuous foundation. To calculate F_{cd}, F_{qd}, and $F_{\gamma d}$, the original wall width, B is used instead of the effective width B'. B is not replaced with B' in the depth factor equations because B' is less than B and the use of B' in the denominator of the depth factor equations would cause large and unconservative values. Because the wall base is subject to an inclined loading R (the resultant force of V and P_{ah} in Figure 5.27), when F_{ci}, F_{qi}, and $F_{\gamma i}$ are calculated, the load inclination angle can be determined as $\beta_i = \tan^{-1}(P_{ah}/V)$ in degrees.

The factor of safety against bearing capacity failure (F_b) can then be evaluated as:

$$F_b = \frac{q_u}{q_{\max}} \tag{5.16}$$

or

$$F_b = \frac{q_u - q'}{q_{\max} - q'} \tag{5.17}$$

Eq. 5.16 is defined on the basis of the gross bearing capacity, and eq. 5.17 is defined on the basis of the net bearing capacity. In general, $F_b \geq 3$ is required for design. Because gravity and semi-gravity walls have low tolerance to total and differential settlement, a high F_b value is required to ensure that wall settlement presumably meets performance criteria. However, there are exceptions to this assumption. Wide walls (i.e., those with a large B) might fail because of substantial foundation settlement before bearing capacity failure occurs. Foundation settlement should always be evaluated using the methods discussed in Section 3.5. If the required F_b value is not achieved, improvement measures, such as increasing the wall width (B), improving the shear strength of the foundation soil (i.e., ground improvement), using lightweight fill material, or changing the foundation type (e.g., using a pile foundation), should be implemented to increase the bearing capacity of the foundation soil.

Example 5.4

A cross section of a semi-gravity cantilever retaining wall is presented in the figure. Evaluate the stability of the retaining wall by using Coulomb's lateral earth pressure method.

a Determine all the horizontal and vertical force components within the retaining wall and the moment arm of all the force components from the wall toe (assume γ_c = 24 kN/m³ and $\delta = 2\phi'/3$).

b Calculate the factors of safety against overturning, sliding, and bearing capacity failure (ignore the passive lateral earth pressure in front of the wall).

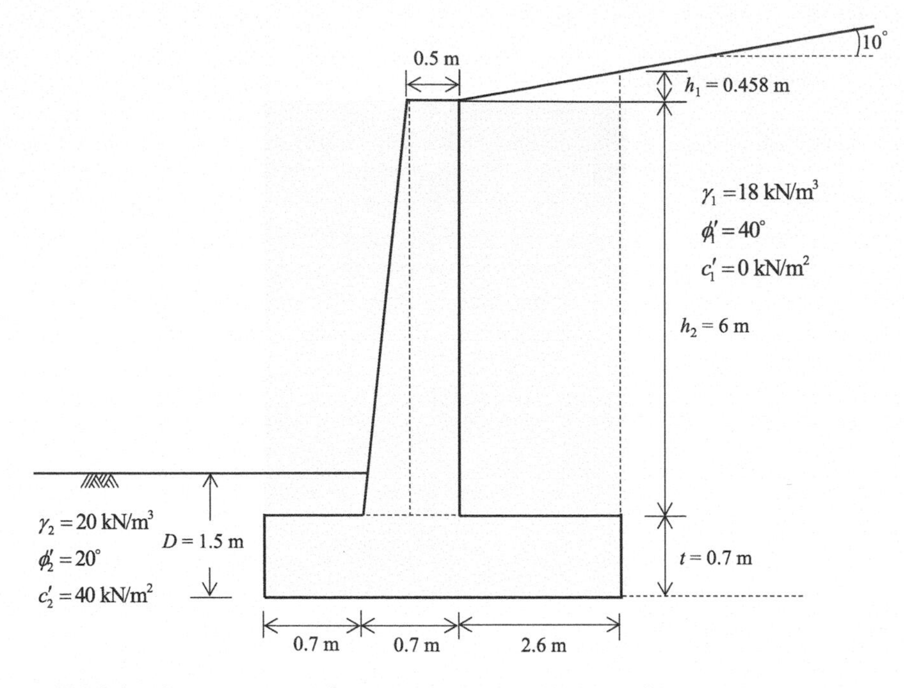

Figure EX 5.4a Cross section of a semi-gravity cantilever wall.

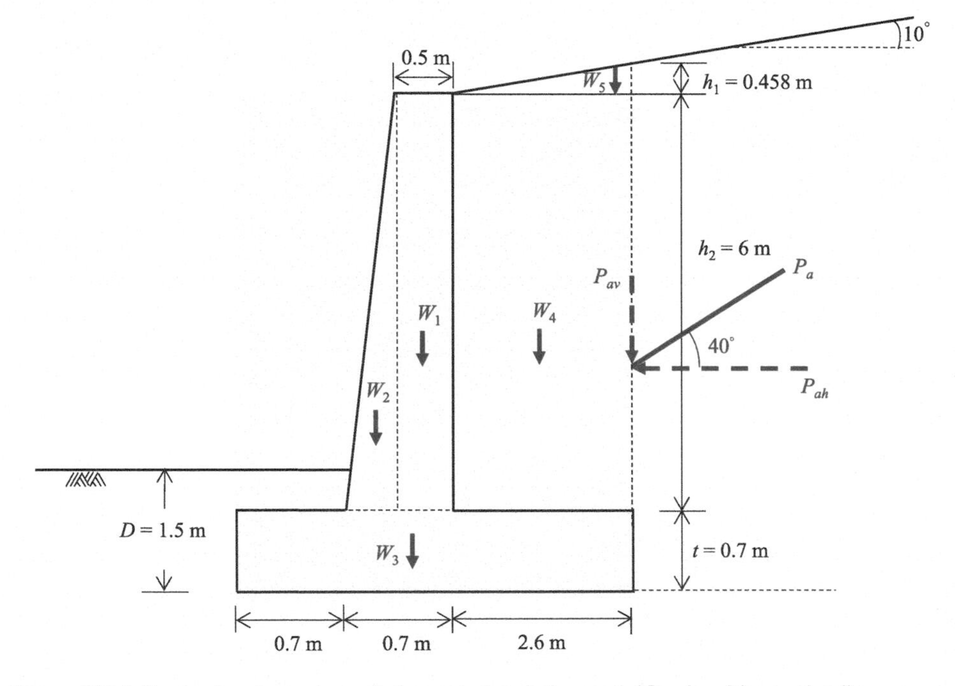

Figure EX 5.4b Active lateral earth force behind the wall (Coulomb's method).

Solution

Figure EX 5.4b indicates the active lateral earth force behind the wall. In Coulomb's method, the active lateral earth force P_a is typically assumed to directly act on the back face of the wall (gravity wall case) or on a vertical plane passing through the heel of the wall base (semi-gravity wall case). In this exercise, because P_a acts on the soil along the vertical plane above the wall heel, the soil–soil interface friction angle is considered ($\delta = \phi'$). The weight of concrete (W_1, W_2, and W_3) and the weight of the soil above the heel (W_4 and W_5) are also considered.

ANSWER TO PART A

Calculate the vertical force components:

$$W_1 = \gamma_c A_1 = 24 \times 3 = 72 \text{ kN/m}$$

$$W_2 = \gamma_c A_2 = 24 \times \left[\frac{1}{2} \times (0.7 - 0.5) \times 6\right] = 14.4 \text{ kN/m}$$

$$W_3 = \gamma_c A_3 = 24 \times \left[(0.7 + 0.7 + 2.6) \times 0.7\right] = 67.2 \text{ kN/m}$$

$$W_4 = \gamma_1 A_4 = 18 \times (2.6 \times 6) = 280.8 \text{ kN/m}$$

$$W_5 = \gamma_1 A_5 = 18 \times (\frac{1}{2} \times 2.6 \times 0.458) = 10.72 \text{ kN/m}$$

Calculate Coulomb's active earth pressure:

$$\beta = 10°,\ \theta = 0°,\ \phi' = 40°,\ \alpha = \phi' = 40°$$

$$K_a = \frac{\cos^2(\phi' - \theta)}{\cos^2\theta\cos(\delta + \theta)\left[1 + \sqrt{\dfrac{\sin(\delta + \phi')\sin(\phi' - \beta)}{\cos(\delta + \theta)\cos(\theta - \beta)}}\right]^2} = 0.221$$

$$P_{av} = P_a \sin\alpha = \frac{1}{2}\gamma_1 H'^2 K_a \sin\alpha$$

$$= \frac{1}{2} \times 18 \times (0.458{+}6{+}0.7)^2 \times 0.2214 \times \sin 40° = 65.63 \text{ kN/m}$$

Calculate the horizontal force components:

$$P_{ah} = P_a \cos\alpha = \frac{1}{2}\gamma_1 H^2 K_a \cos\alpha$$

$$= \frac{1}{2} \times 18 \times (0.458{+}6{+}0.7)^2 \times 0.2214 \times \cos 40° = 78.2 \text{ kN/m}$$

$$S = c_a B + V\tan\delta = Kc_2' B + V\tan K\phi_2'$$

$$= \frac{2}{3} \times 40 \times (0.7{+}0.7{+}2.6){+}510.75 \times \tan(\frac{2}{3} \times 20°) = 227.72 \text{ kN/m}$$

Table Ex 5.4a Vertical force components

Vertical force (kN/m)	*Perpendicular distance to toe (m)*	*Moment (kN · m/m)*
$W_1 = 72$	$0.7+0.7-0.5\times\frac{1}{2}=1.15$	82.8
$W_2 = 14.4$	$0.7+(0.7-0.5)\times\frac{2}{3}=0.833$	12
$W_3 = 67.2$	$(0.7+0.7+2.6)\times\frac{1}{2}=2$	134.4
$W_4 = 280.8$	$0.7+0.7+2.6\times\frac{1}{2}=2.7$	758.16
$W_5 = 10.72$	$0.7+0.7+2.6\times\frac{2}{3}=3.13$	33.55
$P_{av} = 65.63$	$0.7+0.7+2.6=4$	262.52
$V = 510.75$		$\sum M_R = 1283.43$

Table Ex 5.4b Horizontal force components

Horizontal force (kN/m)	*Perpendicular distance to toe (m)*	*Moment (kN · m/m)*
$P_{ah} = 78.2$	$\frac{(0.458+6+0.7)}{3}=2.38$	186.59
$S = 227.72$	0	0

ANSWER TO PART B

Calculate the factor of safety against overturning:

$$F_o = \frac{\sum M_R}{\sum M_D} = \frac{1283.43}{186.59} = 6.87 > 2 \quad \text{ok}$$

Calculate the factor of safety against sliding:

$$F_s = \frac{\sum F_{H,R}}{\sum F_{H,D}} = \frac{227.72}{78.2} = 2.91 > 1.5 \quad \text{ok}$$

Calculate the factor of safety against bearing capacity failure:

$$M_{net} = \sum M_R - \sum M_D = 1283.43 - 186.59 = 1096.84 \text{ kN}\cdot\text{m/m}$$

$$e = \frac{M_{net}}{V} - \frac{B}{2} = \frac{1096.84}{510.75} - \frac{4}{2} = 0.14 \text{ m} < \frac{B}{6} = 0.67 \text{ ok}$$

$$q_{max} = \frac{V}{B \times 1}\left(1 + \frac{6e}{B}\right) = 154.5 \text{ kN/m}^2$$

$$\phi' = 20°, N_c = 14.83, N_q = 6.4, N_r = 2.87$$

$$F_{cs} = F_{qs} = F_{rs} = 1$$

$$F_{cd} = 1.107, F_{qd} = 1.054$$

$$\beta_i = \tan^{-1}\left(\frac{P_{ah}}{V}\right) = \tan^{-1}\left(\frac{78.2}{510.75}\right) = 8.7°$$

$$F_{ci} = F_{qi} = 1 - \left(\frac{\beta_i}{90}\right)^2 = 1 - \left(\frac{8.7}{90}\right)^2 = 0.991, \; F_{ri} = 1 - \left(\frac{\beta_i}{\phi'}\right)^2 = 1 - \left(\frac{8.7}{20}\right)^2 = 0.81$$

$$q_u = 40 \times 14.83 \times 1 \times 1.107 \times 0.991 + 20 \times 1.5 \times 6.4 \times 1 \times 1.054 \times 0.99$$
$$+ \frac{1}{2} \times 20 \times (4 - 2 \times 0.14) \times 2.87 \times 1 \times 1.054 \times 0.81 = 942.45 \text{ kN/m}^2$$

$$F_b = \frac{q_u}{q_{max}} = \frac{942.45}{154.5} = 6.1 > 3 \text{ ok}$$

5.4 Reinforced walls

This section introduces the design of reinforced walls, generally known as mechanically stabilized earth (MSE) walls. Walls reinforced by geosynthetics are specifically called geosynthetic-reinforced soil (GRS) walls. Because reinforced walls are classified as internally stabilized fill walls, the internal failure modes and soil–reinforcement interaction mechanisms are discussed in this section. The steps of the stability analyses used to evaluate a wall's internal stability against reinforcement breakage and pullout and facing connection failure are presented.

5.4.1 *Typical dimensions and drainage design*

Reinforced walls are widely used in various projects, including highway embankment and bridge abutments, to improve load-bearing capacity and system stability. The use of reinforced walls is driven by numerous factors, including aesthetics, compatibility with the natural environment, cost-effectiveness, seismic performance, easy and rapid construction, and the ability to tolerate large deformations without structural distress. In addition to general applications of vertical loadings, reinforced walls have been used as barriers against lateral impact forces from natural disasters, such as floods, tsunamis, rockfalls, debris flows, and avalanches.

A reinforced wall consists of four elements: the reinforcement, the facing, the backfill, and the drainage system (Figure 5.28). A reinforced wall has typical dimensions of L/H = 0.7–1.0, where H is the wall height and L is the length of the reinforcement. The wall embedment depth D should be at least 0.6 m. Reinforcements have two types: metallic reinforcement (e.g., steel

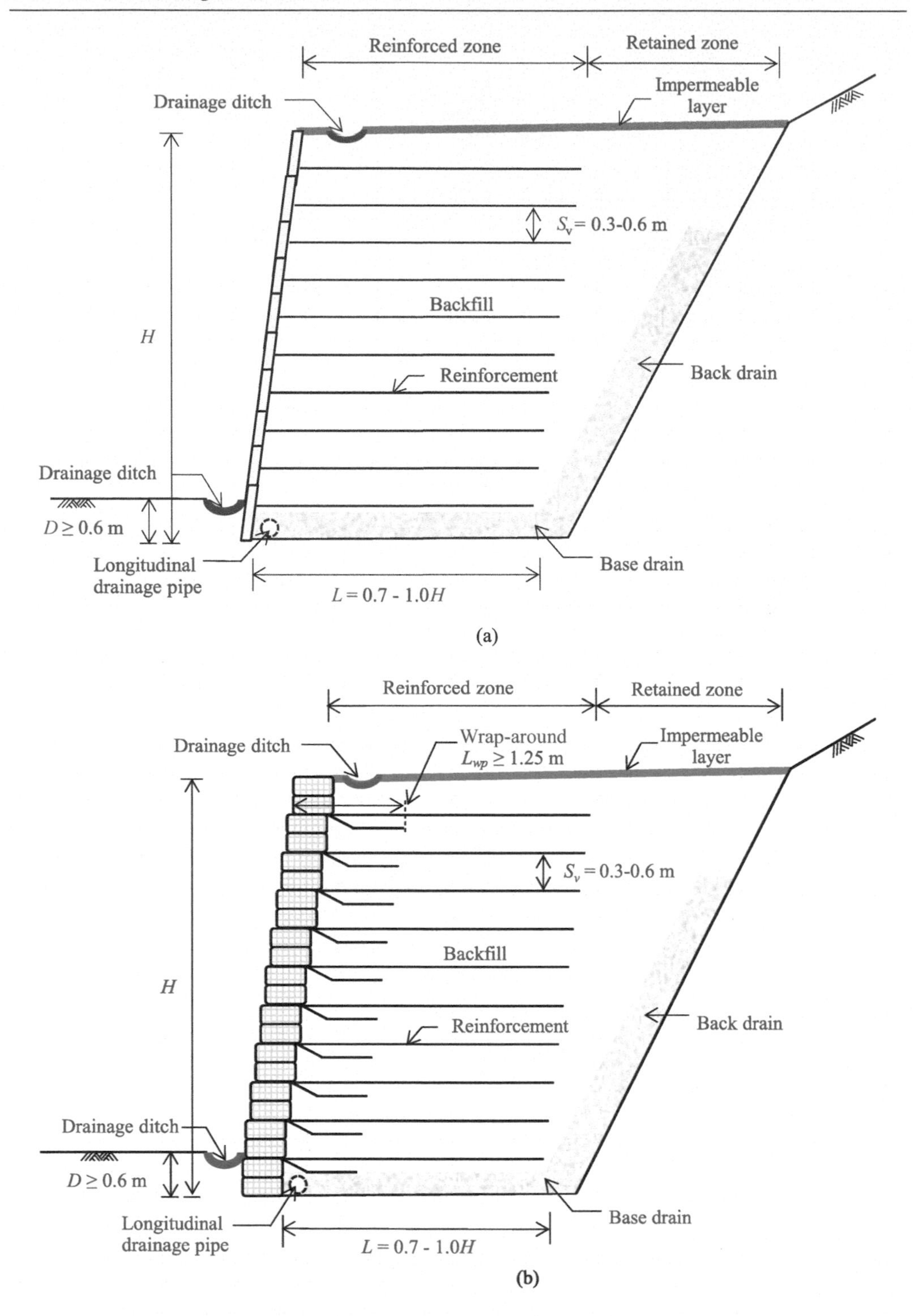

Figure 5.28 Typical dimensions and drainage layout of a reinforced wall: (a) concrete panel facing and (b) wraparound facing.

strips, steel grids, and welded wire mesh) and geosynthetic (e.g., geogrids and geotextiles) (Figure 5.29). Different reinforcement types have different extensibilities and cause different earth pressure distributions within the wall, which will be discussed later. Facing elements have various types, including segmental precast concrete panels, modular blocks, welded wire grids, gabion facings, and geosynthetic wraparound facings. The type of wall facing affects wall aesthetics and wall settlement tolerances and provides different levels of protection against backfill sloughing and erosion. Figure 5.30 shows the facing connection. The metallic reinforcement is connected to the facing element by means of mechanical or frictional connections (Figure 5.30a). The wraparound facing is formed by wrapping reinforcement around the soil bags at the wall face (Figure 5.30b). The wall facing can be vegetated for natural appearance.

A reinforced wall typically requires well-graded granular soil as backfill for constructability, durability, high drainage capacity, and favorable soil–reinforcement interaction. Table 5.5 lists suggested backfill properties in the design guidelines (Berg et al., 2009; Elias et al., 2001). In areas where granular backfill is not readily available, locally available soils containing certain

(a) (b)

Figure 5.29 Reinforcement types: (a) metallic reinforcement and (b) geosynthetics.

(a) (b)

Figure 5.30 Facing connection: (a) metallic reinforcement connected to concrete facing panel, (b) geosynthetic wraparound facing.

Table 5.5 Backfill properties for reinforced walls suggested for design

Item	*Symbol*	*Value*
Soil classification	USCS	GW, GP, GM SW, SP, SM
Maximum particle size	D_{max} (mm)	≤125 mm
Percent passing (%)	125 mm	100
	4.75 mm (no. 4)	20–100
	0.425 mm (no. 40)	0–60
	0.075 mm (no. 200)	0–15
Plasticity index	*PI* (%)	≤6
Relative compaction	R_c (%)	90–95

fines (typically referred to as marginal soils) have been used as alternative backfills to minimize the construction costs and environmental impact of transporting granular soil to a construction site. Fill materials outside the gradation and plasticity index requirements listed in Table 5.5 have been used successfully; however, problems, including significant distortion and structural failure, have been observed because the low permeability of marginal soils can cause pore water pressure to accumulate upon rainfall infiltration or seepage. Considerable care in drainage design is required when marginal soils are adopted as backfills to prevent undesirable accumulation of pore water pressure within the wall.

The drainage system is a crucial component of reinforced walls to dissipate the accumulation of pore water pressure. To ensure proper long-term functionality, drainage layers beneath and behind the reinforced and retained zone are strongly recommended. As illustrated in Figure 5.29, the back and base drains (also called chimney drains) are used to collect and guide rainfall or groundwater out into a longitudinal drain at the toe of the wall. The top of the back drain should be higher than the maximum groundwater level. An impermeable layer should cover the top of the wall to prevent rainfall infiltration, and a drainage ditch near the wall crest should be designed to collect surficial runoff.

5.4.2 Failure modes

A reinforced wall can fail because of (a) internal failure modes (Figure 5.31a–c), (b) external failure modes (Figure 5.31d–f), (c) global failure modes (Figure 5.31g), or (d) excessive wall deformation. A wall can fail internally when the reinforcement breaks, is pulled out, or is disconnected from the facing elements. Stability analyses for wall internal stability against reinforcement breakage, reinforcement pullout, and facing connection failure are described, and corresponding improvement measures are suggested in the subsequent sections.

External failure modes include overturning, sliding, and bearing capacity failures (Figure 5.31d–f). Stability analyses for the external stability of a wall consider the reinforced soil mass to be a rigid body, and active earth pressure from retained soils acts laterally on the reinforced wall. The calculation procedures are the same as those for gravity and semi-gravity walls in Section 5.3.4. Notably, the flexibility of reinforced walls makes overturning failure highly unlikely. However, overturning criteria aid in lateral deformation control by limiting tilting and should still be satisfied. Moreover, because reinforced walls can tolerate large deformations without structural distress, the F_S against bearing capacity failure is 2.5 for reinforced walls rather than the 3.0 used for gravity and semi-gravity walls.

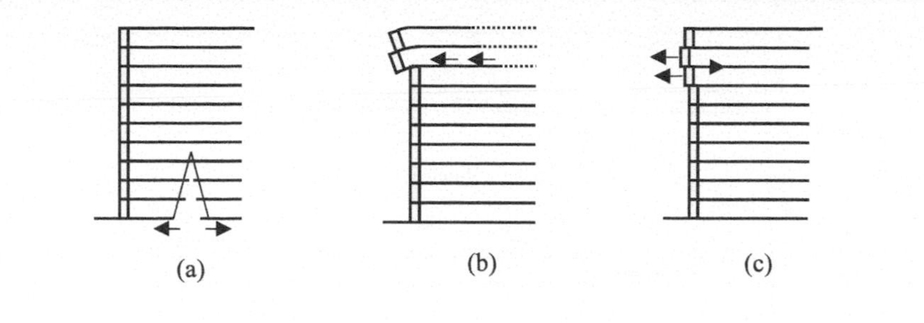

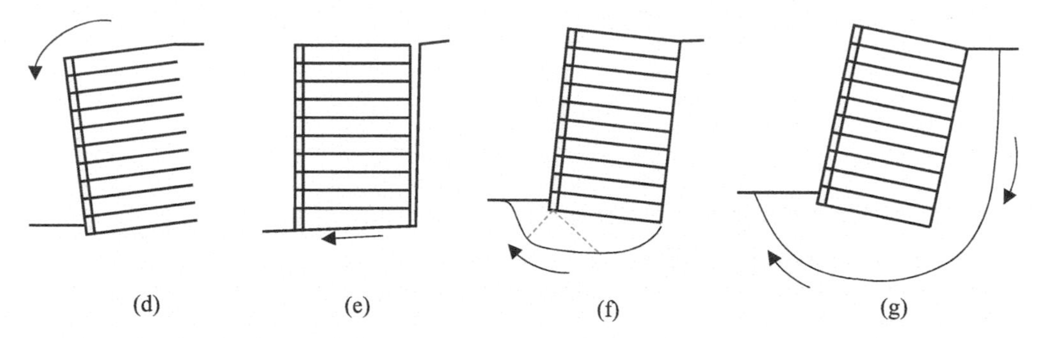

Figure 5.31 Failure modes: (a) reinforcement breakage, (b) reinforcement pullout, (c) facing disconnection, (d) overturning, (e) sliding, (f) bearing capacity failure, and (g) global failure.

Consistent with gravity and semi-gravity walls, global failure (Figure 5.31g) is often analyzed on the basis of limit equilibrium methods (introduced in Chapter 8) by using slope stability software. For excessive wall deformation, an evaluation of the anticipated wall deformation with respect to horizontal and vertical displacement should be performed. Vertical wall displacement is obtained through conventional settlement computations with particular emphasis on differential settlements longitudinally along the wall face and transversely from the face to the end of the reinforced soil volume. Horizontal wall displacement, especially wall-facing deflection, is obtained using the empirical correlation between the reinforcement type and the wall aspect ratio L/H. Advanced numerical approaches, such as the finite element method, can also be used. Finally, for walls located in seismic zones, the dynamic earth pressure, as discussed in Section 4.8, should be included in stability analyses.

5.4.3 Force components for stability analyses

Figure 5.32 illustrates the force components within a reinforced wall. The earth pressure σ_h pushes the soil mass within the potential failure surface forward and causes wall deformation, whereas the shear stress developed at the soil–reinforcement interface τ_{sr} beyond the potential failure surface pulls the soil mass backward to prevent failure. These two forces acting in opposite directions mobilize the reinforcement tensile force. The maximum reinforcement tensile force T_{max} is the peak of the reinforcement tensile force, where the reinforcement segment intercepts the potential failure surface.

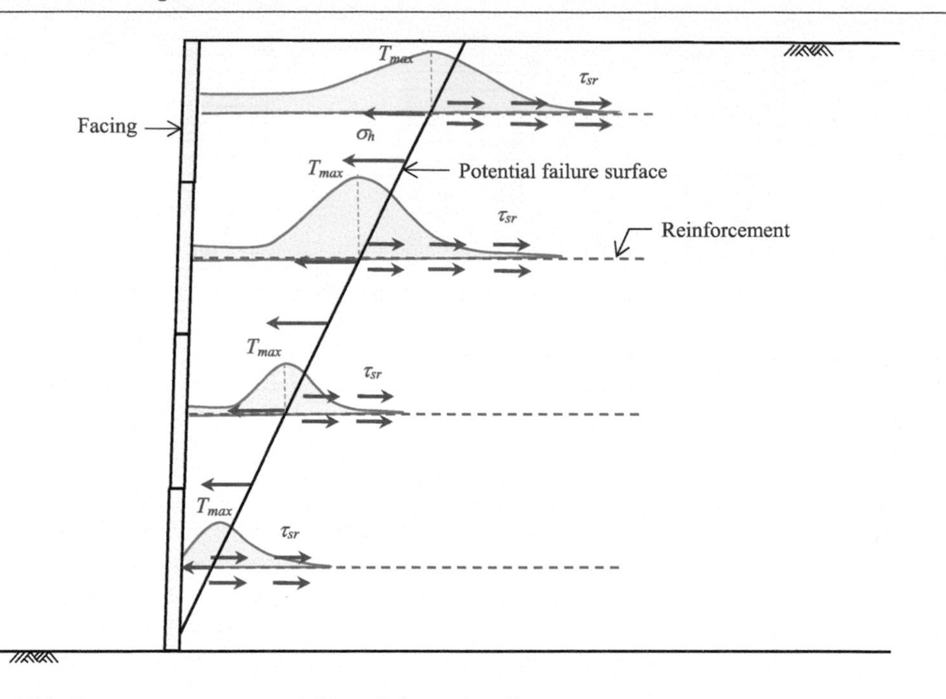

Figure 5.32 Force components within reinforced wall.

T_{max}, which is related to the lateral earth pressure, is a crucial factor in internal stability analyses of reinforced walls. For stability analyses of reinforcement breakage, engineers should select reinforcement with a tensile strength greater than T_{max}; otherwise, reinforcement rupture can occur. For stability analyses of reinforcement pullout, the reinforcement pullout resistance (i.e., anchorage strength) generated from τ_{sr} should be greater than T_{max} to prevent the reinforcement from being pulled out. For stability analyses of facing connection failure, the connection strength between the facing and reinforcement should be larger than the mobilized reinforcement force at the wall face. However, no simplified method has been developed to evaluate the mobilized reinforcement force at the wall face. In conservative designs, the mobilized reinforcement force at the wall face is often assumed to be T_{max}.

T_{max} at each reinforcement layer can be calculated on the basis of the horizontal force equilibrium at each reinforcement tributary area (Figure 5.33).

For planar reinforcement (geogrid, geotextile, and welded wire mesh):

$$T_{\max} = \sigma_h \times S_v \text{ (kN/m)} \tag{5.18a}$$

For strip reinforcement (steel strip and steel grid):

$$T_{\max} = \sigma_h \times S_v \times S_h \text{ (kN)} \tag{5.18b}$$

where:

σ_h = lateral earth pressure.
S_v = vertical spacing of reinforcement.
S_h = horizontal spacing of reinforcement.

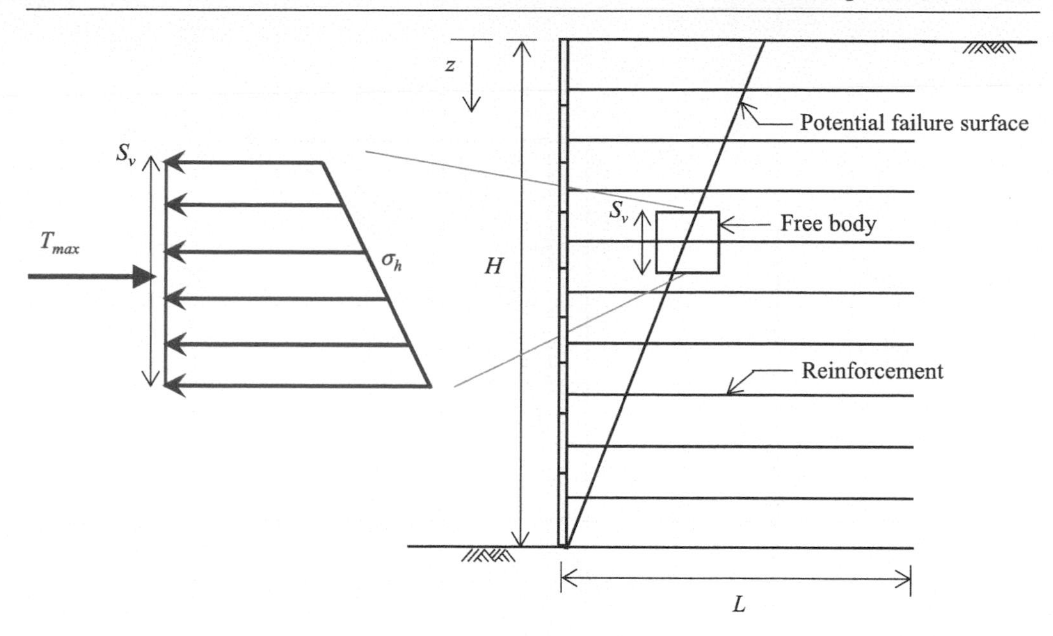

Figure 5.33 Horizontal force equilibrium at a reinforcement layer.

The value and distribution of σ_h depend on the extensibility of the reinforcement. For a wall reinforced with extensile reinforcements (i.e., geosynthetics), sufficient wall deformation can develop for the soil to reach active failure. Therefore, the active earth pressure is considered in the design. For a wall reinforced with inextensible reinforcements (i.e., metallic reinforcements), wall deformation is restrained by the reinforcements. Therefore, active soil failure cannot develop. In this case, the earth pressure between the at-rest and active conditions is assumed. Figure 5.34 presents the normalized earth pressure coefficient for a wall with different types of reinforcements (Berg et al., 2009; Elias et al., 2001). The figure was prepared through back analysis of the lateral stress ratio by using available field data. Stresses in the reinforcements were measured and normalized as a function of an active earth pressure coefficient K_a. On the basis of Figure 5.34, σ_h is calculated as:

$$\sigma_h = \left(\frac{K_r}{K_a}\right) K_a(\gamma z + q) \tag{5.19}$$

where:

K_r/K_a = normalized earth pressure coefficient.
K_r = earth pressure coefficient of the reinforced wall.
K_a = active earth pressure coefficient.
γ = backfill unit weight.
z = depth from the wall top.
q = surcharge.

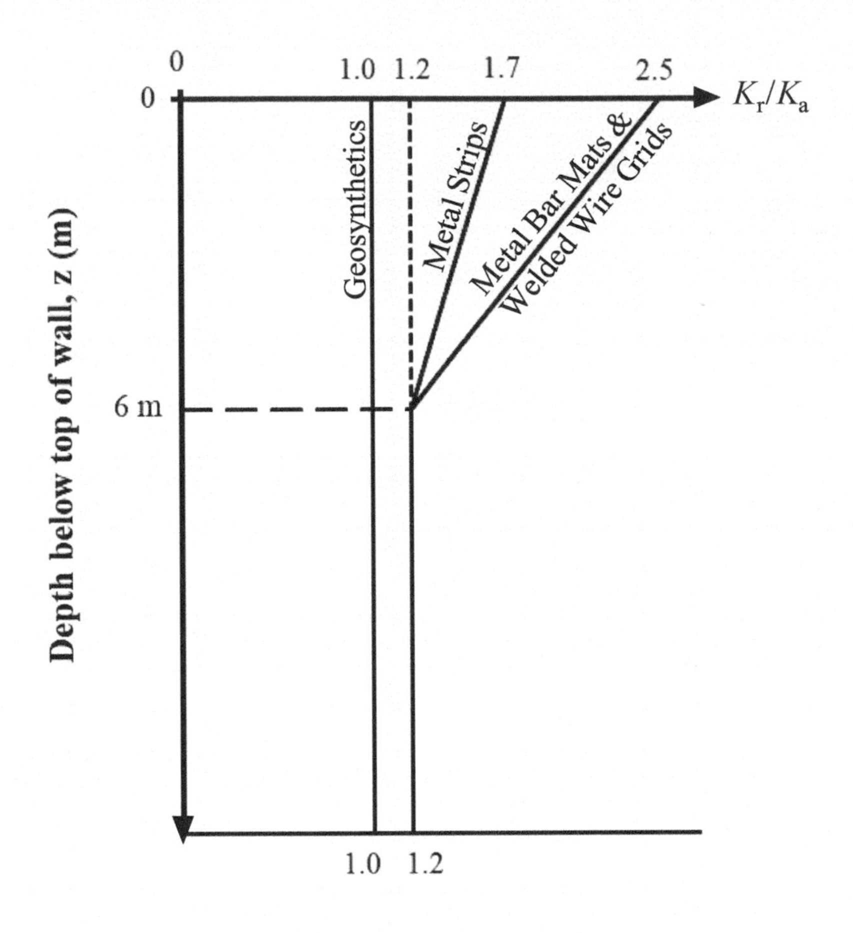

Figure 5.34 Normalized earth pressure coefficient.

5.4.4 Evaluation for reinforcement breakage

Because reinforcements are buried in the soil, the tensile properties of reinforcements, especially geosynthetics, can degrade because of factors such as creep, long-term aging, and mechanical damage. Given the aforementioned factors, the long-term reinforcement tensile strength is calculated as:

$$T_{al} = \frac{T_{ult}}{RF_{cr} \times RF_D \times RF_{ID}} \tag{5.20}$$

where:

T_{al} = long-term reinforcement tensile strength.
T_{ult} = ultimate reinforcement tensile strength (obtained experimentally from tensile test).

RF_{CR} = reduction factor for creep (obtain from creep test; typically ranges from 1.5 to 5, depending on polymer type).

RF_D = reduction factor for long-term durability (obtain from durability test; typically ranges from 1.1 to 2, depending on microorganism, chemical, thermal, oxidation, hydrolysis, and stress cracking).

RF_{ID} = reduction factor for installation damage (obtain from field installation damage test; typically ranges from 1.5 to 5, depending on the backfill gradation and product unit weight).

After the long-term reinforcement tensile strength is obtained, the factor of safety against reinforcement breakage (F_{br}) can be calculated as:

$$F_{br} = \frac{T_{al}}{T_{\max}} \tag{5.21}$$

In general, an $F_{br} \geq 1.5$ is required for design. If the required F_{br} value is not achieved, effective improvement measures include using a reinforcement with high tensile strength (increasing T_{ult}), reducing reinforcement spacing (reducing S_v or S_h), or using a backfill with high soil shear strength (increasing c' and ϕ').

5.4.5 Evaluation for reinforcement pullout

The reinforcement pullout resistance P_r is generated from the shear stress that develops at the soil–reinforcement interface τ_{sr}. As presented in Figure 5.35, P_r can be derived by integrating τ_{sr} with respect to the reinforcement embedment length L_e (the part of the reinforcement beyond the potential failure surface). In practice, P_r is calculated simply by using the laws of friction (i.e., normal stress × coefficient of friction × surface area), expressed as follows:

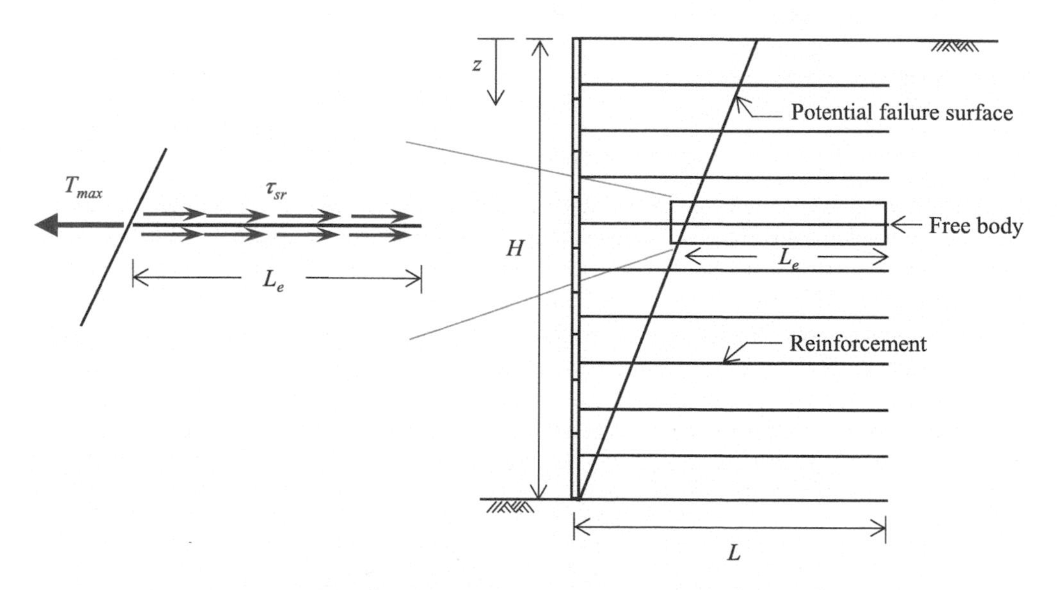

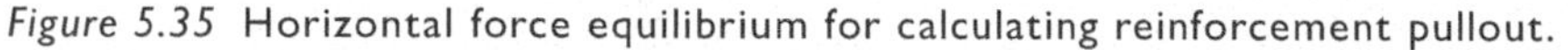

Figure 5.35 Horizontal force equilibrium for calculating reinforcement pullout.

For planar reinforcement (geogrid, geotextile, and welded wire mesh):

$$P_r = C \times F^* \times \alpha \times \sigma_v' \times L_e \text{ (kN/m)} \tag{5.22a}$$

For strip reinforcement (steel strip and steel grid):

$$P_r = C \times F^* \times \alpha \times \sigma_v' \times L_e \times b \text{ (kN)} \tag{5.22b}$$

where:

C = reinforcement effective unit perimeter ($C = 2$ for strips, grids, and sheets).
F^* = pullout resistance (or friction-bearing-interaction) factor (obtained experimentally from pullout test; $\tan 2\phi'/3$ if no pullout test is performed).
α = correction factor to account for reinforcement extensibility (1 for metallic reinforcement, 0.8 for geogrid, 0.6 for geotextile).
σ_v' = effective vertical stress on reinforcement (kPa) (live loads, such as traffic load, are excluded as resistance).
L_e = reinforcement embedment length (m) (calculated from failure surface, $L_e \geq 1.0$ m).
b = width of strip reinforcement (m).

After the reinforcement pullout resistance is obtained, the factor of safety against reinforcement pullout (F_{po}) can be calculated as:

$$F_{po} = \frac{P_r}{T_{max}} \tag{5.23}$$

In general, $F_{po} \geq 1.5$ is required for design. If the required F_{po} value is not achieved, effective improvement measures include increasing the reinforcement length (increasing L), reducing the reinforcement spacing (reducing S_v or S_h), or using reinforcement with high pullout resistance (increasing F^*).

5.4.6 Evaluation for facing connection

Reinforcements can be connected to the facing element by mechanical and frictional means. For example, geogrid reinforcements may be structurally connected to segmental precast panels by using a bodkin joint or frictionally connected to modular blocks by reinforcement and facing interface friction. As presented in Figure 5.36, facing connection failure occurs when the connection strength between the facing and reinforcement is insufficient for a reinforced wall with a concrete panel or modular block facing. Evaluations for facing connection failure are used only for reinforced walls with concrete panels or modular block facings. Facing connection failure is not a concern for reinforced walls with wraparound facing if the wraparound length is long enough (typically ≥ 1.25 m, as indicated in Figure 5.28b).

The long-term facing connection strength, considering creep and long-term aging, is calculated as:

$$T_{alc} = \frac{T_c}{RF_{cr} \times RF_D} \tag{5.24}$$

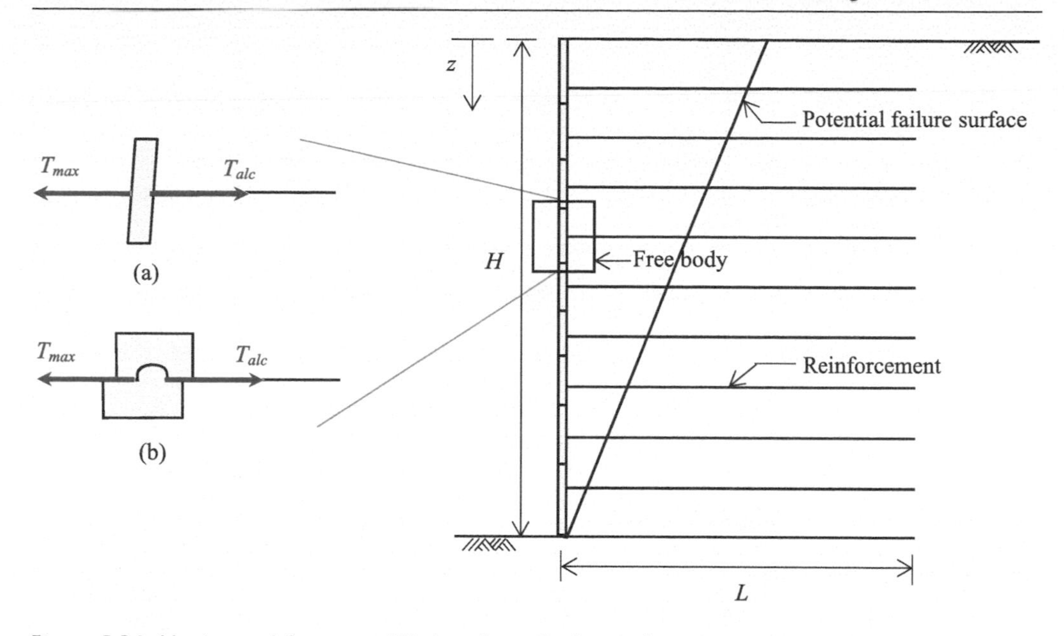

Figure 5.36 Horizontal force equilibrium for calculating facing connection: (a) concrete panel facing and (b) modular block facing.

where:

T_{alc} = long-term facing connection strength.
T_c = ultimate facing connection strength (obtained experimentally from connection test).
RF_{CR} = reduction factor for creep.
RF_D = reduction factor for durability.

Notably, the environment at the connection may not be the same as the environment in the backfill in the reinforced zone. Therefore, the reduction factor for the durability RF_D may be different than that used to calculate the long-term reinforcement tensile strength T_{al}. Moreover, installation damage is not considered (i.e., $RF_D = 1$) in the calculation of T_{alc} because heavy trucks and compaction rollers are not typically driven on the facing element during construction.

After the long-term facing connection strength is obtained, the factor of safety against facing connection failure (F_c) can be calculated as:

$$F_c = \frac{T_{alc}}{T_{\max}} \tag{5.25}$$

In general, an $F_c \geq 1.5$ is required for design. If the required F_c value is not achieved, effective improvement measures include increasing the facing connection strength (increasing T_{alc}), reducing the reinforcement spacing (reducing S_v or S_h), or using backfill with high soil shear strength (increasing c' and ϕ').

Example 5.5

A reinforced wall is planned to provide an access road in a mountainous area. A traffic surcharge q of 15 kPa is considered in the design. A cross section of the reinforced wall is illustrated in the figure. The wall dimensions, reinforcement layout, and material properties are also provided in the figure. Use the reduction factors of $RF_{cr} = 2$, $RF_D = 1.5$, and $RF_{ID} = 1.5$ to calculate the long-term reinforcement tensile strength.

a Calculate the factor of safety against reinforcement breakage and identify the most critical layer.
b Calculate the factor of safety against reinforcement pullout and identify the most critical layer.

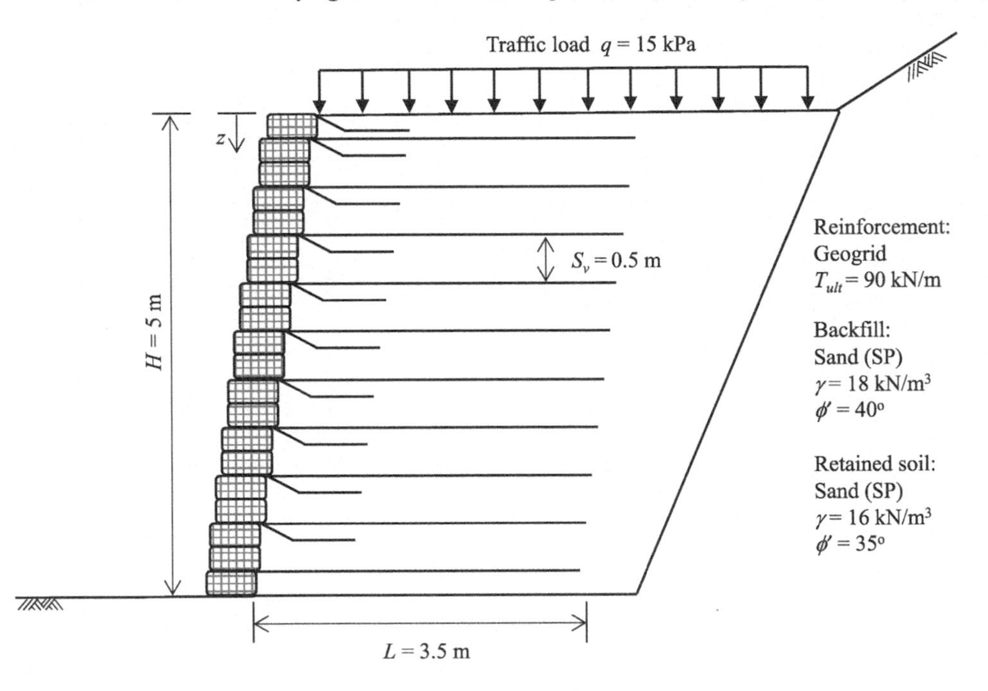

Figure EX 5.5 Reinforced wall cross section.

Solution

ANSWER TO PART A

Calculate the maximum reinforcement load at the top and bottom layers.
Top layer:

$$\frac{K_r}{K_a} = 1 \text{ for geosyntheitcs (Figure 5.34)}$$

$$\sigma_h = \left(\frac{K_r}{K_a}\right) \times K_a \times (\gamma z + q) = 1 \times \tan^2(45 - \frac{40}{2})(18 \times 0.25 + 15) = 4.24 \text{ kPa}$$

$$T_{\max} = \sigma_h \times S_v = 4.24 \times 0.5 = 2.12 \text{ kN/m}$$

Bottom layer:

$$\sigma_h = \left(\frac{K_r}{K_a}\right) \times K_a \times (\gamma z + q) = 1 \times \tan^2(45 - \frac{40}{2})(18 \times 4.75 + 15) = 21.85 \text{ kPa}$$

$$T_{\max} = \sigma_h \times S_v = 21.8 \times 0.5 = 10.93 \text{ kN/m}$$

Calculate the long-term reinforcement tensile strength.

$$T_{al} = \frac{T_{ult}}{RF_{CR} \times RF_D \times RF_{ID}} = \frac{90}{2 \times 1.5 \times 1.5} = 20 \text{ kN/m}$$

Calculate the factor of safety against breakage.

Top layer:

$$F_{br} = \frac{T_{al}}{T_{\max}} = \frac{20}{2.12} = 9.43 > 1.5 \text{ ok}$$

Bottom layer:

$$F_{br} = \frac{T_{al}}{T_{\max}} = \frac{20}{10.93} = 1.83 > 1.5 \text{ ok}$$

Table EX 5.5a Results of F_{br} for all reinforcement layers

No. of layer	*z (m)*	σ'_v *(kPa)*	σ_h *(kPa)*	T_{max} *(kN/m)*	F_{br}
10 (top layer)	0.25	4.5	4.24	2.12	9.43
9	0.75	13.5	6.20	3.10	6.45
8	1.25	22.5	8.15	4.08	4.90
7	1.75	31.5	10.11	5.06	3.95
6	2.25	40.5	12.07	6.03	3.32
5	2.75	49.5	14.03	7.01	2.85
4	3.25	58.5	15.98	7.99	2.50
3	3.75	67.5	17.94	8.97	2.23
2	4.25	76.5	19.90	9.95	2.01
1 (Bottom layer)	4.75	85.5	21.85	10.93	1.83

The bottom layer is the most critical for reinforcement breakage because the lateral earth pressure increases linearly with depth.

ANSWER TO PART B

Calculate the pullout resistance at the top and bottom layers.

Top layer:

$$P_r = C \times F^* \times \alpha \times \sigma'_v \times L_e = 2 \times 0.56 \times 0.8 \times 4.5 \times 1.29 = 5.2 \text{ kN/m}$$

where:

$$C = 2$$

$$F^* = \frac{2}{3}\tan\phi' = \frac{2}{3}\tan 40 = 0.56$$

$$\alpha = 0.8$$

$$\sigma_v' = \gamma z = 18 \times 0.25 = 4.5 \text{ kPa}$$

$$L_e = L - \frac{(H-z)}{\tan(45+\frac{\phi'}{2})} = 3.5 - \frac{(5-0.25)}{\tan(45+\frac{40}{2})} = 1.29 \text{ m}$$

Bottom layer:

$$P_r = C \times F^* \times \alpha \times \sigma_v' \times L_e = 2 \times 0.56 \times 0.8 \times 85.5 \times 3.38 = 258.9 \text{ kN/m}$$

where

$$\sigma_v' = \gamma z = 18 \times 4.75 = 85.5 \text{ kPa}$$

$$L_e = L - \frac{(H-z)}{\tan(45+\frac{\phi'}{2})} = 3.5 - \frac{(5-4.75)}{\tan(45+\frac{40}{2})} = 3.38 \text{ m}$$

Calculate the factor of safety against pullout.
Top layer:

$$F_{po} = \frac{P_r}{T_{max}} = \frac{5.2}{2.12} = 2.45 > 1.5 \text{ ok}$$

Bottom layer:

$$F_{po} = \frac{P_r}{T_{max}} = \frac{258.9}{10.93} = 23.69 > 1.5 \text{ ok}$$

Table EX 5.5b Results of F_{po} for all reinforcement layers

No. of layer	z (m)	L_e (m)	σ_v' (kPa)	P_r (kN/m)	T_{max} (kN/m)	F_{po}
10 (top layer)	0.25	1.29	4.5	5.2	2.12	2.45
9	0.75	1.52	13.5	18.3	3.10	5.90
8	1.25	1.75	22.5	35.3	4.08	8.65
7	1.75	1.98	31.5	56.0	5.06	11.07
6	2.25	2.22	40.5	80.4	6.03	13.33
5	2.75	2.45	49.5	108.6	7.01	15.49
4	3.25	2.68	58.5	140.5	7.99	17.58
3	3.75	2.92	67.5	176.2	8.97	19.64
2	4.25	3.15	76.5	215.7	9.95	21.68
1 (bottom layer)	4.75	3.38	85.5	258.9	10.93	23.69

The top layer is the most critical for reinforcement pullout because the reinforcement embedment length is the shortest and the overburden pressure is the lowest in this layer.

5.5 Nongravity walls

This section introduces the design of nongravity walls, which include sheet-pile walls, soldier pile and lagging walls, diaphragm walls, and tangent and secant pile walls. These walls are constructed to stabilize the soil for excavation support. Nongravity walls typically have two types: cantilevered and supported walls. Accordingly, these two types of nongravity walls have different earth pressure distributions and analytical methods. The stability analysis procedures for evaluating wall stability against rotational failure and determining whether a wall has sufficient embedment depth are presented in this section.

5.5.1 Types and applications

Nongravity walls are used for (a) cofferdams on land and in water for foundation construction; (b) waterfront structures for building wharfs, quays, and piers; (c) temporary excavation support; (d) excavation support for cut-and-cover roadway tunnels, culverts, and utility trenches; and (e) excavation support for building basements and foundations.

Figure 5.37 illustrates the types of nongravity walls. Depending on whether additional supports (e.g., strut, anchor, deadman anchor) are used, nongravity walls are typically categorized into cantilevered and supported walls. Cantilevered walls have no support during excavation (i.e., free unsupported conditions). The resistance is derived from the passive earth pressure on the wall embedment length on the excavation side. Wall deformation depends primarily on the stiffness of the structural elements. Because cantilevered walls have no additional support, they typically require longer embedment depths below the excavation surface. Supported walls have additional support from several layers of struts, anchors, or deadman anchors to minimize the

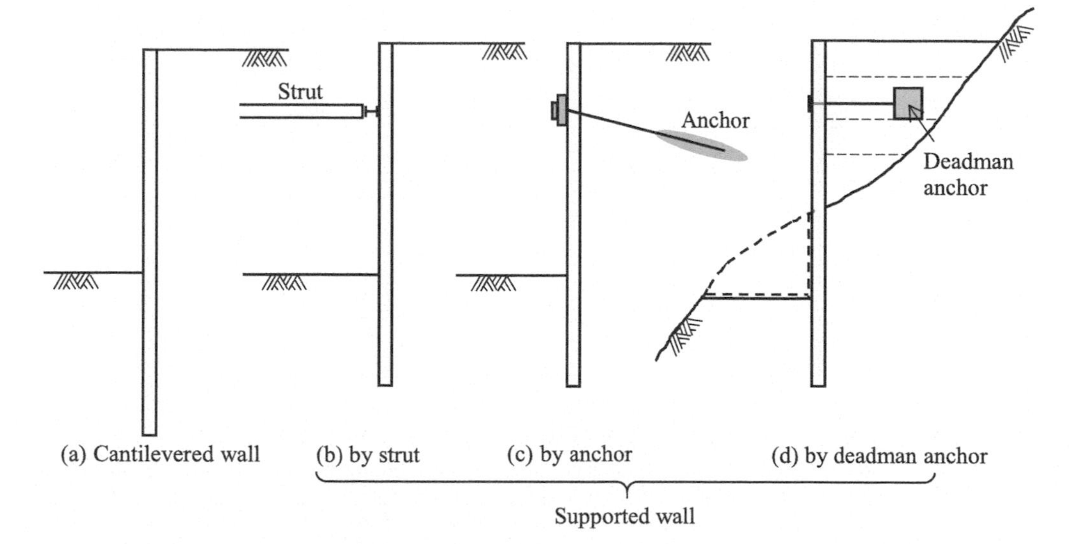

Figure 5.37 Nongravity wall types.

size (i.e., depth and thickness) of the wall and maintain wall deformation within certain required limits. The ratio of the embedment length H_p to the excavated depth (or unembedded length) H_e varies depending on subsurface soil conditions. In general, H_p/H_e should be equal to or greater than 0.8 for sandy soils, 1.0 for stiff clays, and 1.2–1.4 for soft clays.

5.5.2 Failure modes

Figure 5.38 illustrates the failure modes of a nongravity wall. The first failure mode is rotational failure, which occurs when a cantilevered wall has insufficient embedment length (Figure 5.38a). The wall can rotate forward considerably, causing the entire retaining wall system to fail. The second and third failure modes involve rotational failure in the supported walls (Figures 5.38b and c). The second failure mode occurs when the support resistance is insufficient, for example, insufficient anchor lengths or insufficient strut axial resistance. Supports can lose their anchorage efficiency and move together when soil mass failure occurs. Consequently, the wall rotates forward about the toe, and the wall system ultimately fails. Because

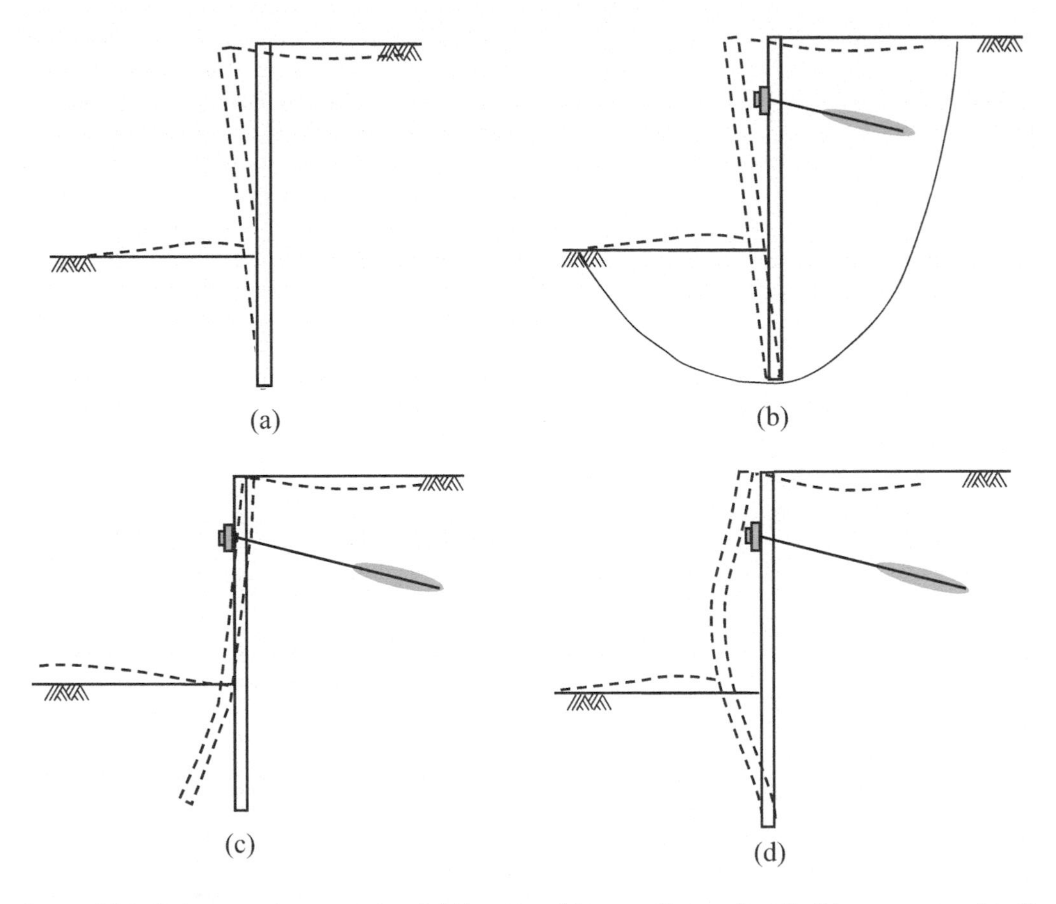

Figure 5.38 Failure modes: rotational failure for (a) a cantilevered wall, (b) a supported wall with insufficient support length, (c) a supported wall with insufficient wall embedment length, and (d) a supported wall with excessive bending deformation

the support is ineffective, wall deformation in the second failure mode is similar to that in the first failure mode. In design, support length should be evaluated to ensure that the support is sufficiently long, which often necessitates extending the support beyond the potential failure surface. The third failure mode occurs when the wall has insufficient embedment length. This failure occurs in the lower part of the wall. The wall rotates backward about the supported point, and the wall toe is kicked out.

In general, the first three failure modes are caused by inequilibrium in the earth pressure on both sides of the wall. The wall is then moved a large distance toward the excavation zone, causing the soil to move so much that the entire excavation site collapses. This rotational failure is also called base shear failure because it involves the generation of shear stress that reaches the soil shear strength in most of the soil below the excavation bottom, causing considerable soil displacement or heave and leading to the failure of the entire excavation system. A detailed discussion of base shear failure is presented in Chapter 6.

The fourth failure mode is excessive wall deformation by bending (Figure 5.38d). Excessive wall deformation occurs when a wall has low stiffness or when the space between support layers is large. Excessive wall deformation must be analyzed with the finite element method or beam–spring model (Ou, 2022). If the excavation site encounters groundwater, the water pressure may cause sand boiling failure in sandy soils and upheaval failure in clayey soils. Stability analyses for evaluating wall stability against rotational failure are described in the subsequent sections. Other stability analyses for sand boiling failure, upheaval failure, and excessive wall deformation are discussed in detail in Chapter 6.

For walls located in seismic zones, the influence of earthquakes on the stability of permanent retaining walls should be considered. The influence of earthquakes on temporary retaining structures can be ignored. For example, if diaphragm walls serve as the permanent outer walls of a basement, the influence of earthquakes should be considered accordingly. For sheet-pile walls and soldier pile and lagging walls that serve as temporary retaining structures, the effects of earthquakes can be disregarded.

5.5.3 Earth pressure distribution

Cantilevered and supported walls have different earth pressure distributions because the wall embedment length and additional supports influence the magnitude and shape of wall deformation. Accordingly, two analytical methods have been developed, fixed earth and free earth support methods (Tanyu et al., 2008; Ou, 2022; Padfield and Mair, 1984), as illustrated in Figures 5.39 and 5.40 using excavations in sandy soils as examples.

The design of cantilevered walls is based on the fixed earth support method. Figure 5.39 illustrates the deformation and earth pressure distribution of a cantilevered wall using the fixed earth support method, which assumes that the wall embedment length is sufficiently long and that a fixed point (no wall displacement) below the excavation surface can be identified. The embedded part may rotate about the fixed point (Figure 5.39a). Thus, at the soil limit state, the lateral earth pressure near the fixed point on both sides of the retaining wall varies gradually between active and passive conditions (Figure 5.39b). The simplified earth pressure distribution is used for stability analyses (Figure 5.39c).

The design of supported walls is based on the free earth support method. Figure 5.40 illustrates the deformation and earth pressure distribution of a supported wall, where the anchor behind the wall offers the anchorage force. The anchor force comes from the prestressed load transfers from the ground anchor to the wall face to provide substantial lateral support. The

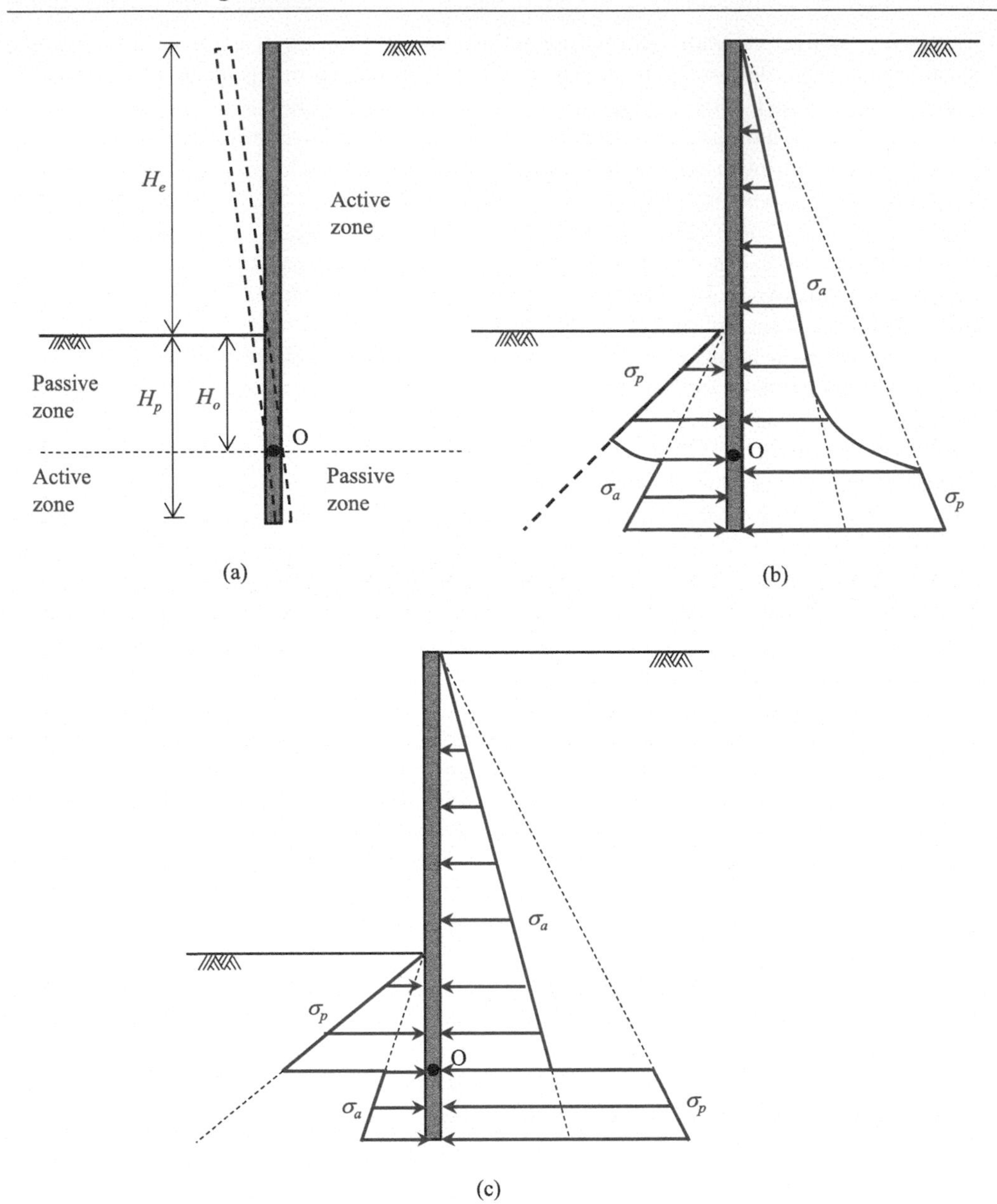

Figure 5.39 Cantilevered wall: (a) wall deformation, (b) real earth pressure distribution, and (c) simplified earth pressure distribution.

analysis of the supported wall is based on the free earth support method, which assumes that the wall embedment length is relatively short, resulting in insufficient embedment to prevent rotation of the toe of the wall. The embedded part of the wall can move a certain distance before the soil reaches the limit state (Figure 5.40a). Therefore, passive earth pressure on the excavation side and active earth pressure on the other side can be assumed to act on the wall (Figure 5.40b).

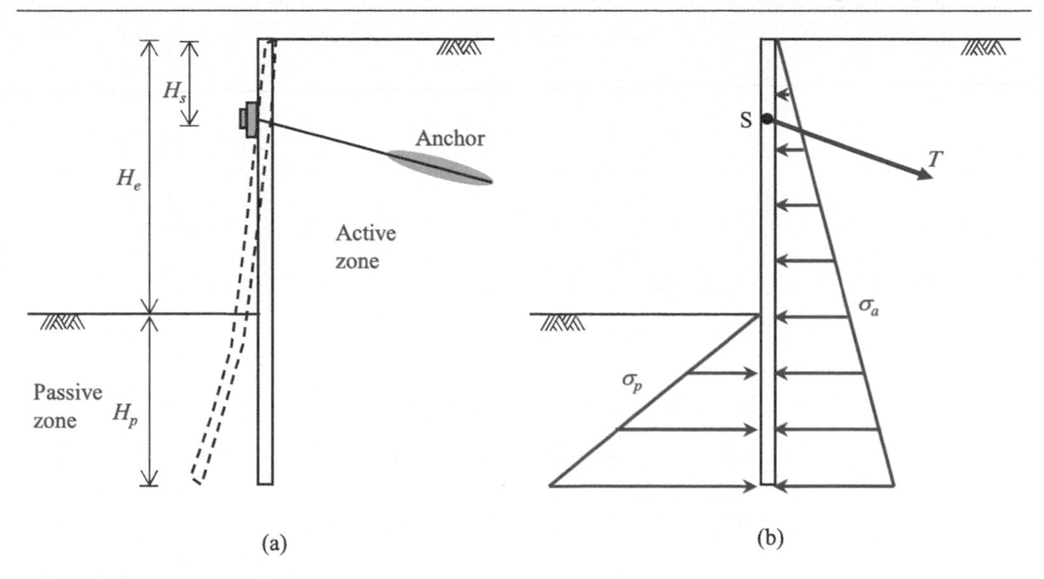

Figure 5.40 Supported wall: (a) wall deformation and (b) earth pressure distribution.

The applications of these two analytical methods are not interchangeable. If the free earth support method is used to design a cantilevered wall, external forces on the wall, which are only passive and active forces without the loads from the supports, may not reach equilibrium. Therefore, the free earth support method is not applicable to cantilevered walls. However, if the fixed earth support method is used to design a supported wall, the design might be too conservative, leading to a wall embedment depth that is too long to be economical.

In addition to earth pressure, nongravity walls generally must be watertight to prevent groundwater leakage through the wall into the construction site. Therefore, the water pressure might act on the wall if groundwater is encountered during excavation. This consideration should be included in stability analyses of excavations in sandy soil with groundwater on both sides of the wall. In clayey soils, even when groundwater is present, water pressure can be ignored because it is considered implicitly in the undrained analyses.

5.5.4 Stability analyses for cantilevered walls

This section introduces a simplified calculation procedure for a cantilevered wall commonly adopted in engineering designs to determine the H_p value to ensure that the wall has a sufficient embedment depth. The design of cantilevered walls is based on the fixed earth support method. The earth pressure distribution of the cantilevered wall is shown in Figure 5.39 and discussed in the previous section. Figure 5.41 presents the force components and free-body diagram for the stability analyses. The force components are assumed on the basis of the fixed earth support method.

In Figure 5.41, P_{au} and P_{pu} are the resultant forces of the active and passive earth pressures, respectively, above the fixed point O.

$$P_{au} = \frac{1}{2}\gamma\left(H_e + H_o\right)^2 K_{ah} \tag{5.26}$$

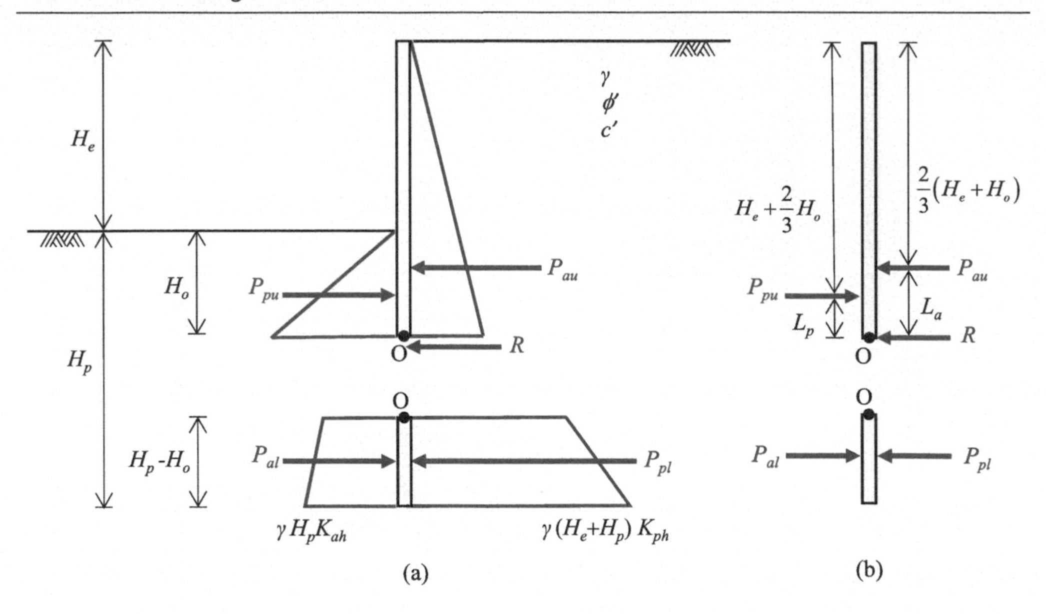

Figure 5.41 Stability analyses for a cantilevered wall: (a) force components and (b) free-body diagram.

$$P_{pu} = \frac{1}{2} \gamma H_o^2 K_{ph} \tag{5.27}$$

where:

γ = unit weight of soil.
H_e = excavated depth (or unembedded length).
H_o = distance from the excavation surface to the fixed point O.
K_{ah} = horizontal component of the active earth pressure coefficient.
K_{ph} = horizontal component of the passive earth pressure coefficient.
L_a and L_p represent the distance from P_{au} and P_{pu} to point O, respectively.

$$L_a = \frac{1}{3}(H_e + H_o) \tag{5.28}$$

$$L_p = \frac{1}{3} H_o \tag{5.29}$$

The moment with respect to point O is used to solve for H_o as follows:

$$F_S = \frac{M_r}{M_d} = \frac{P_{pu} \times L_p}{P_{au} \times L_a} = \frac{P_{pu} \times \frac{1}{3} H_o}{P_{au} \times \frac{1}{3}(H_e + H_o)} \tag{5.30}$$

In general, an $F_S \geq 1.5$ is required for design. Because the simplification of this analysis does not rigorously satisfy all equilibrium conditions, the required embedment depth for design H_p should be slightly (typically 20%) larger than H_o. Therefore:

$$H_p = 1.2H_o \tag{5.31}$$

The assumption in eq. 5.31 should be evaluated by comparing the forces above and below point O to ensure that the wall embedment length is sufficient. R in Figure 5.41 is the imbalance force between P_{pu} and P_{au} above point O.

$$R = P_{pu} - P_{au} \tag{5.32}$$

To maintain the equilibrium of the wall, the resistant force from P_{pl} and P_{al} below point O should be equal to or larger than R.

$$P_{pl} - P_{al} \geq R \tag{5.33}$$

where P_{pl} and P_{al} are the resultant forces of active and passive earth pressures, respectively, below point O.

$$P_{pl} = K_{ph}\gamma\left[\frac{(H_o + H_e) + (H_p + H_e)}{2}\right](H_p - H_o) \tag{5.34}$$

$$P_{al} = K_{ah}\gamma\frac{H_o + H_p}{2}(H_p - H_o) \tag{5.35}$$

When $P_{pl} - P_{al} \geq R$, the assumption in eq. 5.31 is valid, and the wall embedment length is sufficient to maintain the equilibrium of the wall.

5.5.5 Stability analyses for supported walls

This section introduces a calculation procedure for a supported wall. The design of supported walls is based on the free earth support method. The earth pressure distribution of the supported wall is shown in Figure 5.40 and discussed in the previous section. Figure 5.42 illustrates the force components for stability analyses based on the free earth support method. The method introduced here is applied only to walls with one level of support. The design of multilevel supported walls (more than two levels) involves different considerations and is detailed in Chapter 6.

In Figure 5.42, P_a and P_p are the resultant forces of the active and passive earth pressures on the wall, respectively.

$$P_a = \frac{1}{2}\gamma(H_e + H_p)^2 K_{ah} \tag{5.36}$$

$$P_p = \frac{1}{2}\gamma H_p^2 K_{ph} \tag{5.37}$$

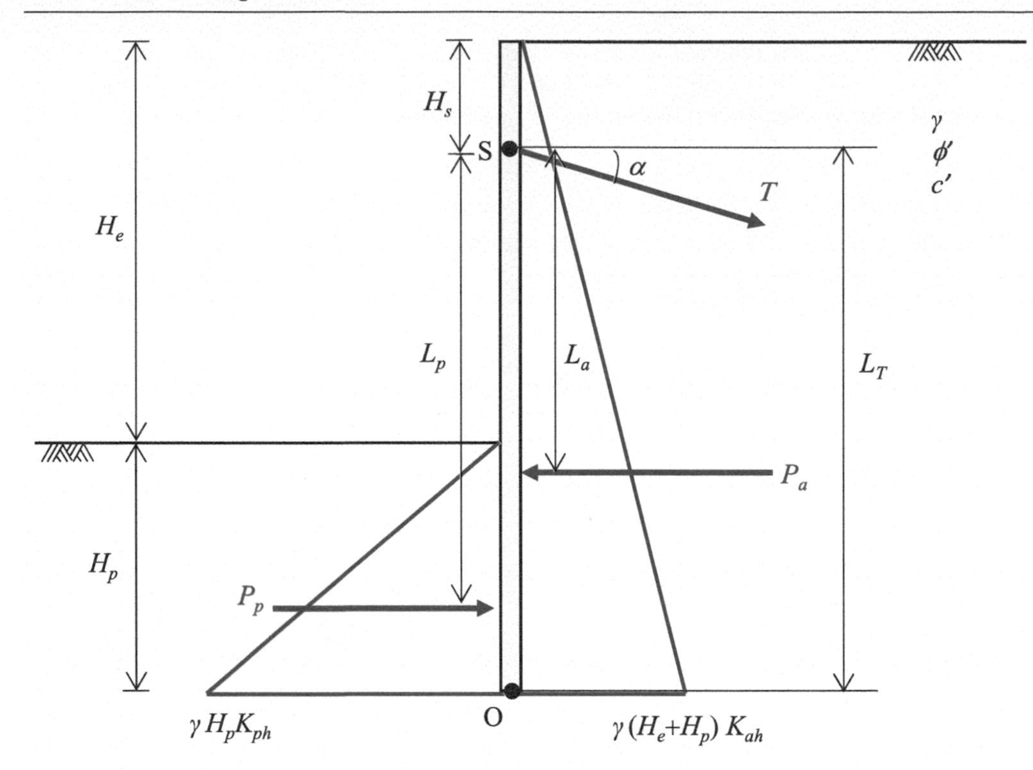

Figure 5.42 Force components of a supported wall for stability analyses.

where H_p is wall embedment (penetration) length below the excavation surface, and the other parameters were defined previously. L_a and L_p are the distances from P_a and P_p to point S, respectively.

$$L_a = \frac{2}{3}(H_e + H_p) - H_s \tag{5.38}$$

$$L_p = \frac{2}{3}H_p + (H_e - H_s) \tag{5.39}$$

where H_s is the distance from the top of the wall to the supported point S.

To determine the wall embedment length, take the moment with respect to point S and then solve for H_p. The moment induced by T, the load of the support per unit length, is not included in eq. 5.40 because T acts directly on point S. Similar to cantilevered wall designs, an $F_S \geq 1.5$ is typically required in the design guidelines for supported walls.

$$F_S = \frac{M_r}{M_d} = \frac{P_p \times L_p}{P_a \times L_a} = \frac{P_p \times \left[\frac{2}{3}H_p + (H_e - H_s)\right]}{P_a \times \left[\frac{2}{3}(H_e + H_p) - H_s\right]} \tag{5.40}$$

To determine the load of the support, take the moment with respect to point O at the end of the wall, and the same F_S value is applied. The moment induced by the horizontal component of T (i.e., $T\cos\alpha$) is included in eq. 5.41.

$$F_S = \frac{M_r}{M_d} = \frac{P_p(L_T - L_p)}{P_a(L_T - L_a) - TL_T} = \frac{P_p \times \frac{1}{3} H_p}{P_a \times \frac{1}{3}(H_e + H_p) - T\cos\alpha \times (H_e + H_p - H_s)} \tag{5.41}$$

where L_T is the distance from point S to point O.

Then, solve for T:

$$T = \frac{P_a(L_T - L_a) - \frac{P_p}{F_S}(L_T - L_p)}{L_T} = \frac{P_a \times \frac{1}{3}(H_e + H_p) - \frac{P_p}{F_S} \times \frac{1}{3} H_p}{\cos\alpha \times (H_e + H_p - H_s)} \text{ (kN/m)} \tag{5.42}$$

Moreover, eq. 5.41 can be rearranged as:

$$M_d = P_a(L_T - L_a) - TL_T = \frac{M_r}{F_S} = \frac{P_p(L_T - L_p)}{F_S} \tag{5.43}$$

In the preceding equation, the F_S is only applied to P_p (or K_p) to account for the uncertainty in the soil parameter and the associated passive earth pressure.

To rigorously satisfy all the equilibrium conditions, T can also be evaluated based on the horizontal force equilibrium. The F_S is also applied on P_p for the same reasons as discussed previously.

$$T\cos\alpha = P_a - \frac{P_p}{F_S} \tag{5.44}$$

And then:

$$T = \frac{P_a - \frac{P_p}{F_S}}{\cos\alpha} \text{ (kN/m)} \tag{5.45}$$

The load of the support can be obtained as follows:

$$F_T = T \times S_h \text{ (kN)} \tag{5.46}$$

where S_h is the horizontal spacing of supports.

The calculated F_T is the minimum required load of the support, that is, anchor force, for maintaining the equilibrium of the retaining wall system. For design, the allowable load of the support is determined by applying an F_S to the calculated F_T to account for uncertainties due to the influence of the construction method, soil properties, groundwater, and design life on the material of the support. This step is also taken to ensure that the designed supports perform under working stress conditions.

$$F_d = \frac{F_T}{F_{anchor}} \tag{5.47}$$

The F_S value for design depends on the materials of the support. For a ground anchor, an $F_{anchor} \geq 3$ for a permanent anchor and an $F_{anchor} \geq 2$ for a temporary anchor are generally required for design. Importantly, the F_{anchor} applied to the support in eq. 5.47 and the F_S applied to maintain the stability of the wall system in eqs. 5.40 and 5.41 have different values and meanings.

To determine the length of the support, the embedment length of the support should extend beyond the potential failure surface to achieve adequate anchorage, as shown in Figure 5.43.

$$L\cos\alpha \geq \left(H_e + H_p - H_s - L\sin\alpha\right)\tan(45 - \frac{\phi'}{2}) \tag{5.48}$$

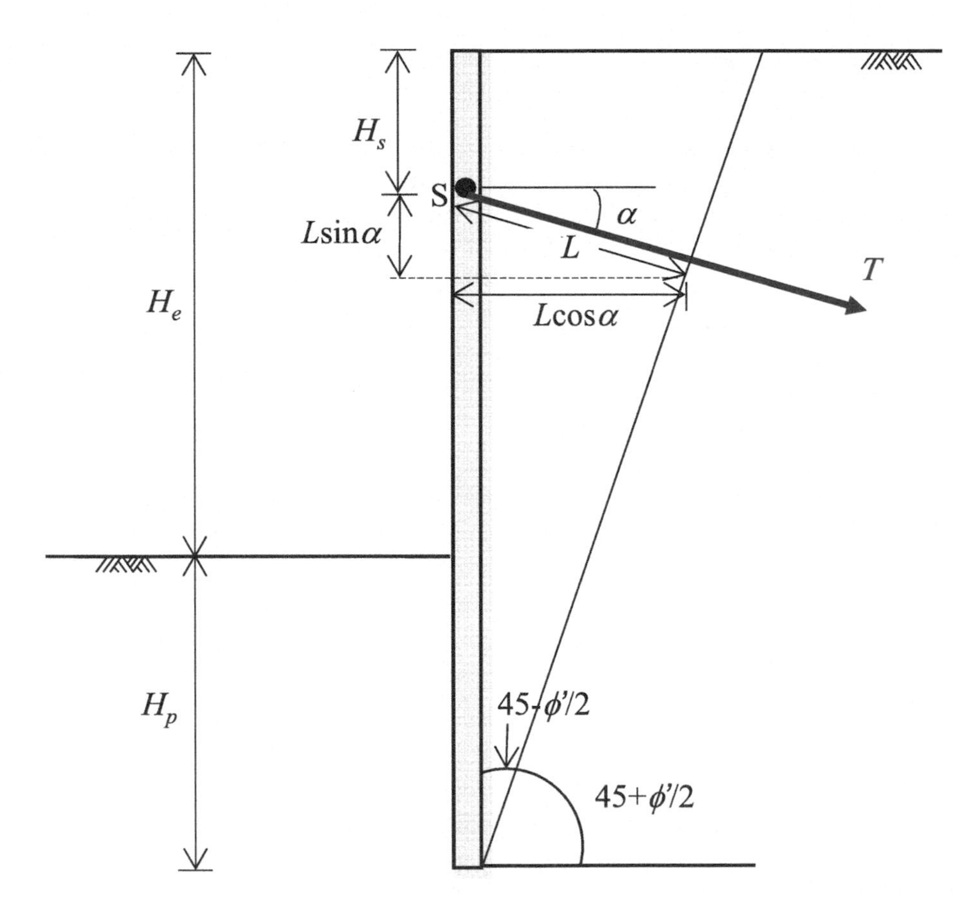

Figure 5.43 Length of the support related to the potential failure surface.

Because the inclination of the support is usually gentle ($10° \leq \alpha \leq 20°$ in general), for simplicity, $L\sin\alpha$ can be ignored in the calculation. The embedment length of the support can be calculated as:

$$L \geq \frac{\left(H_e + H_p - H_s\right)\tan(45 - \frac{\phi'}{2})}{\cos\alpha} \tag{5.49}$$

Finally, the total length of the support can then be determined as L plus the length of the fixed part of the support.

Example 5.6

A cross section of a cantilevered wall is presented in the figure. Assume that the groundwater level is very deep. Determine the embedment length of this wall by using the fixed earth support method with a required F_S of 1.5.

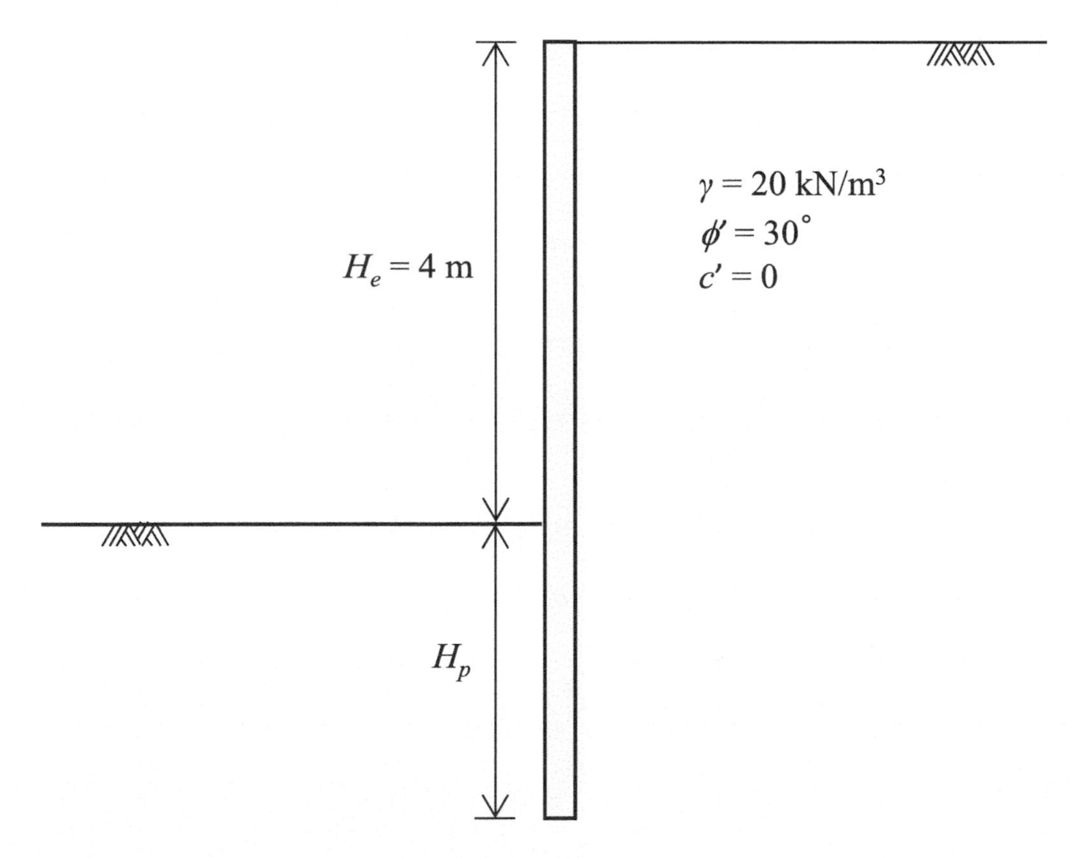

Figure EX 5.6 Cross section of a cantilevered wall.

Solution

Calculate Rankine's active and passive earth pressure coefficients:

$$K_{ah} = \tan^2(45 - \frac{\phi'}{2}) = \tan^2(45 - \frac{30}{2}) = 0.333$$

$$K_{ph} = \tan^2(45 + \frac{\phi'}{2}) = \tan^2(45 + \frac{30}{2}) = 3$$

Calculate the lateral earth forces:

$$P_{au} = \frac{1}{2}\gamma\left(H_e + H_o\right)^2 K_{ah} = \frac{1}{2} \times 20\left(4 + H_o\right)^2 \times 0.333$$
$$= 3.33H_o^{\ 2} + 26.64H_o + 53.28$$

$$P_{pu} = \frac{1}{2}\gamma H_o^{\ 2} K_{ph} = \frac{1}{2} \times 20 \times H_o^{\ 2} \times 3 = 30H_o^{\ 2}$$

Determine the wall embedment length to satisfy the required F_S:
Take the moment on point O.

$$F_S = \frac{M_r}{M_d} = \frac{P_{pu} \times \frac{1}{3}H_o}{P_{au} \times \frac{1}{3}\left(H_e + H_o\right)}$$

$$= \frac{30H_o^{\ 2} \times \frac{1}{3}H_o}{\left(3.33H_o^{\ 2} + 26.64H_o + 53.28\right) \times \frac{1}{3}\left(4 + H_o\right)} \geq 1.5$$

$$H_o \geq 4.9 \text{ m}$$
$$H_p = 1.2H_o = 5.88 \text{ m}$$

Compare the forces above and below point O to ensure that the embedment length of the wall is sufficient:

$$R = P_{pu} - P_{au} = \frac{1}{2}K_{ph}\gamma H_o^2 - \frac{1}{2}K_{ah}\gamma(H_e + H_o)^2$$
$$= \frac{1}{2} \times 3 \times 20 \times 4.9^2 - \frac{1}{2} \times 0.333 \times 20 \times (4 + 4.9)^2$$
$$= 456.5 \text{ kN/m}$$

$$P_{pl} - P_{al} = K_{ph}\gamma\left[\frac{\left(H_o + H_e\right) + \left(H_p + H_e\right)}{2}\right]\left(H_p - H_o\right) - K_{ah}\gamma\frac{H_o + H_p}{2}\left(H_p - H_o\right)$$

$$= 3 \times 20\left[\frac{(4.9 + 4) + (1.2 \times 4.9 + 4)}{2}\right](1.2 \times 4.9 - 4.9) - 0.333 \times 20 \times \frac{4.9 + 1.2 \times 4.9}{2} \times (1.2 \times 4.9 - 4.9)$$
$$= 516.9 \text{ kN/m}$$

$P_{pl} - P_{al} \geq R$ ok

Example 5.7

A supported wall is built to support the soil in an excavation project. A cross section of the supported wall is presented in the figure. The total excavated depth is 5 m below the ground surface. The subsurface in this construction site consists of two soil layers: the top fill layer that is 0–3 m deep, and the underlying original soil layer below 3 m deep. Determine the embedment length of this wall by using the free earth support method with a required F_S of 1.5. Additionally, find the anchor load for the anchor with a horizontal spacing of 3 m.

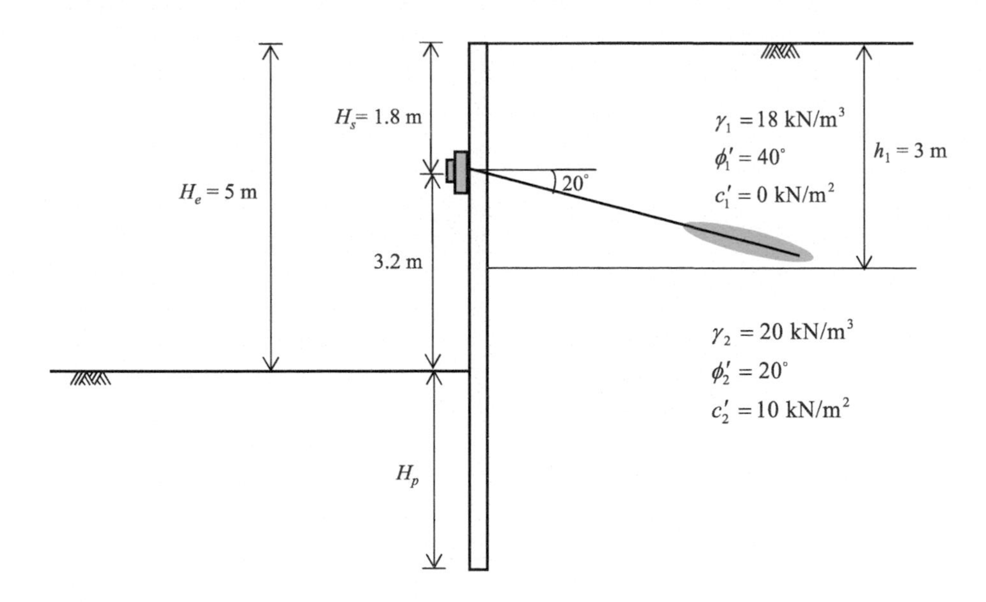

Figure EX 5.7a Cross section of a supported wall.

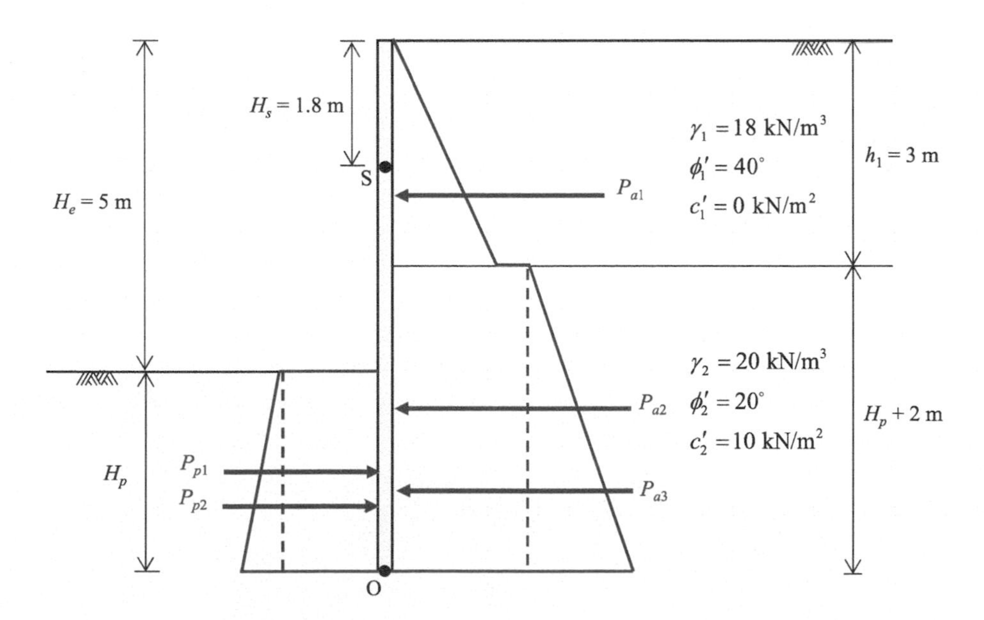

Figure EX 5.7b Earth pressure distribution on both sides of the wall.

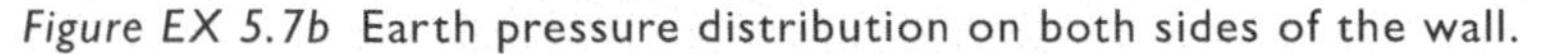

Solution

Figure EX 5.7b shows the lateral earth pressure distribution acting on both sides of the wall.
Calculate Rankine's active and passive earth pressure coefficients:

$$K_{ah1} = \tan^2(45 - \frac{\phi'}{2}) = \tan^2(45 - \frac{40}{2}) = 0.217$$

$$K_{ah2} = \tan^2(45 - \frac{\phi'}{2}) = \tan^2(45 - \frac{20}{2}) = 0.490$$

$$K_{ph2} = \tan^2(45 + \frac{\phi'}{2}) = \tan^2(45 + \frac{20}{2}) = 2.040$$

Calculate the lateral earth forces:

$$P_{a1} = \frac{1}{2}\gamma_1 h_1^2 K_{ah1} = \frac{1}{2} \times 18 \times 3^2 \times 0.217 = 17.57 \text{ kN/m}$$

$$\begin{aligned} P_{a2} &= \left(\gamma_1 h_1 K_{ah2} - 2c_2\sqrt{K_{ah2}}\right) \times \left(H_p + 2\right) \\ &= \left(18 \times 3 \times 0.49 - 2 \times 10 \times \sqrt{0.49}\right) \times \left(H_p + 2\right) \\ &= 12.46 H_p + 24.92 \end{aligned}$$

$$\begin{aligned} P_{a3} &= \frac{1}{2}\left((\gamma_1 h_1 + \gamma_2(H_p + 2))K_{ah2} - 2c_2\sqrt{K_{ah2}} - \left(\gamma_1 h_1 K_{ah2} - 2c_2\sqrt{K_{ah2}}\right)\right) \times \left(H_p + 2\right) \\ &= \frac{1}{2}\left((18 \times 3 + 20(H_p + 2)) \times 0.49 - 2 \times 10 \times \sqrt{0.49} - \left(18 \times 3 \times 0.49 - 2 \times 10 \times \sqrt{0.49}\right)\right) \times \left(H_p + 2\right) \\ &= 4.9 H_p^2 + 19.6 H_p + 19.6 \end{aligned}$$

$$\begin{aligned} P_{p1} &= 2c_2\sqrt{K_{ph2}} H_p \\ &= 2 \times 10 \times \sqrt{2.04} H_p \\ &= 28.57 H_p \end{aligned}$$

$$\begin{aligned} P_{p2} &= \frac{1}{2}\left(\gamma_2 H_p K_{ph2}\right) H_p \\ &= \frac{1}{2}\left(20 \times H_p \times 2.04\right) H_p \\ &= 20.4 H_p^2 \end{aligned}$$

Determine the wall embedment length to satisfy the required F_S:
Take the moment on point S and solve the equation to obtain H_p:

$$F_S = \frac{M_r}{M_d}$$

$$= \frac{P_{p1}\left[\frac{1}{2}H_p + \left(H_e - H_s\right)\right] + P_{p2}\left[\frac{2}{3}H_p + \left(H_e - H_s\right)\right]}{P_{a1}\left(\frac{2}{3}h_1 - H_s\right) + P_{a2}\left[\frac{1}{2}\left(H_p + 2\right) + \left(h_1 - H_s\right)\right] + P_{a3}\left[\frac{2}{3}\left(H_p + 2\right) + \left(h_1 - H_s\right)\right]}$$

$$= \frac{P_{p1}\left(\frac{1}{2}H_p + 3.2\right) + P_{p2}\left(\frac{2}{3}H_p + 3.2\right)}{P_{a1} \times 0.2 + P_{a2}\left(\frac{1}{2}H_p + 2.2\right) + P_{a3}\left(\frac{2}{3}H_p + 2.53\right)}$$

$$H_p \geq 2.43 \text{ m} \approx 2.5 \text{ m}$$

Summary of the resultant forces of the active and passive earth pressures:

$$P_{a1} = 17.57 \text{ kN/m}$$

$$P_{a2} = 12.46H_p + 24.92 = 56.07 \text{ kN/m}$$

$$P_{a3} = 4.9H_p^2 + 19.6H_p + 19.6 = 99.22 \text{ kN/m}$$

$$P_a = P_{a1} + P_{a2} + P_{a3} = 172.86 \text{ kN/m}$$

$$P_{p1} = 28.57H_p = 71.42 \text{ kN/m}$$

$$P_{p2} = 20.4H_p^2 = 127.5 \text{ kN/m}$$

$$P_p = P_{p1} + P_{p2} = 198.92 \text{ kN/m}$$

Determine the anchor load.

Take the moment on point O, and solve the equation to obtain T.

$$F_S = \frac{M_r}{M_d}$$

$$= \frac{P_{p1} \times \frac{1}{2}H_p + P_{p2} \times \frac{1}{3}H_p}{P_{a1} \times \left(\frac{1}{3}h_1 + H_p + 2\right) + P_{a2} \times \frac{1}{2}(H_p + 2) + P_{a3} \times \frac{1}{3}(H_p + 2) - T\cos\alpha \times (H_e + H_p - H_s)}$$

$$T = \frac{\left[P_{a1} \times \left(\frac{1}{3}h_1 + H_p + 2\right) + P_{a2} \times \frac{1}{2}(H_p + 2) + P_{a3} \times \frac{1}{3}(H_p + 2)\right] - \left(P_{p1} \times \frac{1}{2}H_p + P_{p2} \times \frac{1}{3}H_p\right)/F_S}{\cos\alpha \times (H_e + H_p - H_s)}$$

$$= \frac{[17.57 \times 5.5 + 56.07 \times 2.25 + 99.225 \times 1.5] - (71.425 \times 1.25 + 127.5 \times 0.83)/F_S}{\cos 20 \times (5 + 2.5 - 1.8)}$$

$$= 45.09 \text{ kN/m}$$

Solve T again based on the horizontal force equilibrium:

$$T\cos\alpha = P_a - \frac{P_p}{F_S}$$

and then

$$T = \frac{P_a - \frac{P_p}{F_S}}{\cos\alpha} = \frac{172.86 - \frac{198.92}{1.5}}{\cos 20} = 42.83 \text{ kN/m}$$

Notably, the T values calculated using both methods (moment and horizontal force equilibrium) are close.

For the anchor load with an anchor horizontal spacing of 3 m:

$$F_T = 45.09 \times 3 = 135.18 \text{ kN}$$

5.6 Summary and general comments

This chapter introduced various types of earth retaining structures commonly used in engineering. The designs of gravity, semi-gravity, reinforced, and nongravity walls were described. For each wall type, the typical dimensions, failure modes, and stability analyses were discussed in detail. A summary and general comments regarding this chapter are as follows:

1 Earth retaining systems are classified by construction method (i.e., fill or cut construction) and by stabilization mechanism (i.e., external or internal stabilization). In general, fill and cut walls cannot be used interchangeably because they involve different construction methods. Because externally and internally stabilized walls have different load transfer mechanisms, their design methods are fundamentally different. The applications, limitations (e.g., cost-effective height and required ROW), advantages, and disadvantages of various earth retaining systems are compared and summarized in Tables 5.1 and 5.2.
2 Selecting a retaining wall for a construction project can be complex because it involves matching several objectives of wall selection with the numerous retaining walls available in practice. The objectives of wall selection are cost-effectiveness, practicality of construction, stability, watertightness, aesthetic appropriateness and consistency with the surroundings, and environmental friendliness. To determine the most appropriate wall type, a rigorous and systematic wall system evaluation and selection process should be conducted.
3 A drainage system, typically comprising a layer of drainage aggregate with high permeability placed behind the wall and drainage pipes at the bottom, is considered a primary component of gravity and semi-gravity walls. A properly constructed and maintained drainage system minimizes the possibility of poor wall performance resulting from the accumulation of pore water pressure.
4 For the design of gravity and semi-gravity cantilever walls, stability analyses should be performed to evaluate wall stability against overturning, sliding, bearing capacity, global failure, and excessive settlement. In the calculation of the F_S against sliding failure, the resisting force from passive earth pressure P_p acting in front of the wall is disregarded if the shear key is not included. In the calculation of the F_S against bearing capacity failure, because of the influence of lateral earth pressure, the eccentric loading at the wall base should be considered, and eccentricity should be evaluated to ensure that it is less than one-sixth of the wall width ($e < B/6$).
5 For the design of reinforced walls, stability analyses should be performed to evaluate the internal, external, and global stabilities as well as wall deformation. Internal stability is used to assess reinforcement breakage, reinforcement pullout, and facing connection failure. The earth pressure distributions are dependent on the type and extensibility of the reinforcement. In the calculation of wall internal stability, active earth pressure is assumed for reinforced walls with extensile geosynthetics, whereas earth pressure between the at-rest and active conditions is assumed for reinforced walls with inextensible metallic reinforcements.
6 Nongravity walls are categorized into cantilevered and supported types, depending on whether they have additional supports. Accordingly, the earth pressure distributions and analytical methods for these two types of walls are different. The analysis of cantilevered walls should use the fixed earth support method, which assumes that a fixed point on the wall below the excavation surface can be identified. At the soil limit state, the lateral earth

pressure near the fixed point on both sides of the wall varies between active and passive conditions. Analysis of supported walls should use the free earth support method, which assumes that the wall rotates about its toe. Passive earth pressure on the excavation side and active earth pressure on the other side can be assumed to act on the wall.

Problems

5.1 An earth embankment is planned for highway transportation. The geometry of the earth embankment is shown in Figure P5.1. The height of the embankment is 5 m, and the width at the top of the embankment is 8 m. The ROW for the embankment construction is 12 m. The embankment is backfilled with sand, with $\phi' = 40°$ and $c' = 0$ kPa. Please determine whether a retaining wall is needed to support the soil of the embankment.

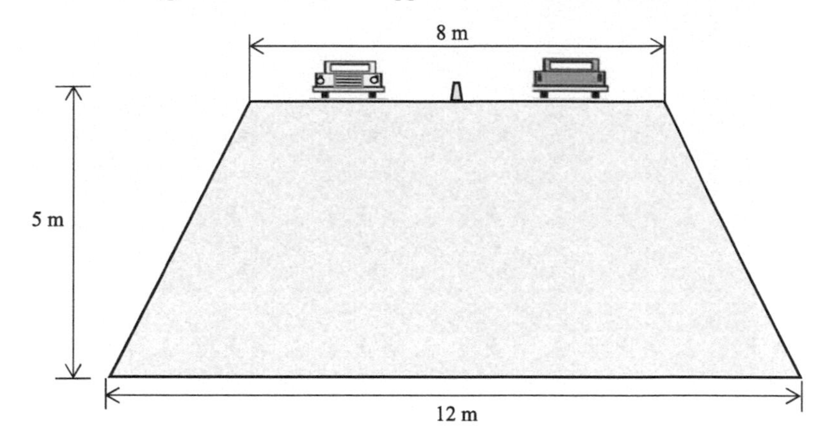

Figure P 5.1

5.2 Identify and classify the type of the retaining wall in the figure. Which type of retaining wall is pictured? How would the retaining wall be classified?

- **a** Which type of retaining wall is pictured? (1) reinforced wall, (2) diaphragm wall, (3) gabion wall
- **b** How would the retaining wall be classified? (1) internally stabilized fill wall, (2) externally stabilized cut wall, (3) externally stabilized fill wall

Figure P 5.2

5.3 Identify the type and classification of the retaining wall in the figure. Which type of retaining wall is pictured? How would the retaining wall be classified?

- **a** Which type of retaining wall is pictured? (1) reinforced wall, (2) gravity wall, (3) crib wall
- **b** How would the retaining wall be classified? (1) internally stabilized fill wall, (2) externally stabilized cut wall, (3) externally stabilized fill wall

Figure P 5.3

5.4 A cross section of a gravity wall is shown in Figure P5.4. The unit weight of concrete is $\gamma = 24$ kN/m^3. Assume δ is $2\phi'/3$. Use Rankine's earth pressure and consider the passive resistance.

- **a** Find the resultant of active earth pressure and passive earth pressure.
- **b** Determine the F_S against sliding and overturning.
- **c** Check if the bearing capacity of the retaining wall is sufficient.

5.5 A cross section of a cantilever wall is shown in Figure P5.5. The sand is used as a backfill. The unit weight of concrete is $\gamma = 24$ kN/m^3. Assume δ is $2\phi'/3$. Answer the following questions by using Rankine's lateral earth pressure method and ignore the passive lateral earth pressure in front of the wall.

- **a** Determine all the horizontal and vertical force components within the retaining wall and the moment arm of all the force components from the wall toe.
- **b** Calculate the factor of safety against sliding and check whether this design meets the specification requirements. If not, please propose an improvement measure.
- **c** Calculate the factor of safety against overturning and check whether this design meets the specification requirements. If not, please propose an improvement measure.

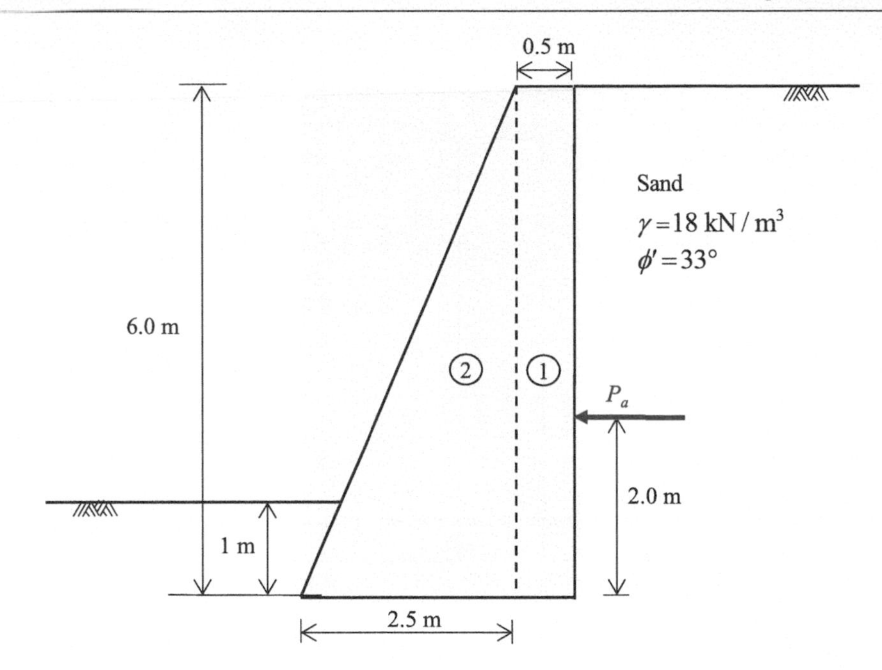

Figure P 5.4

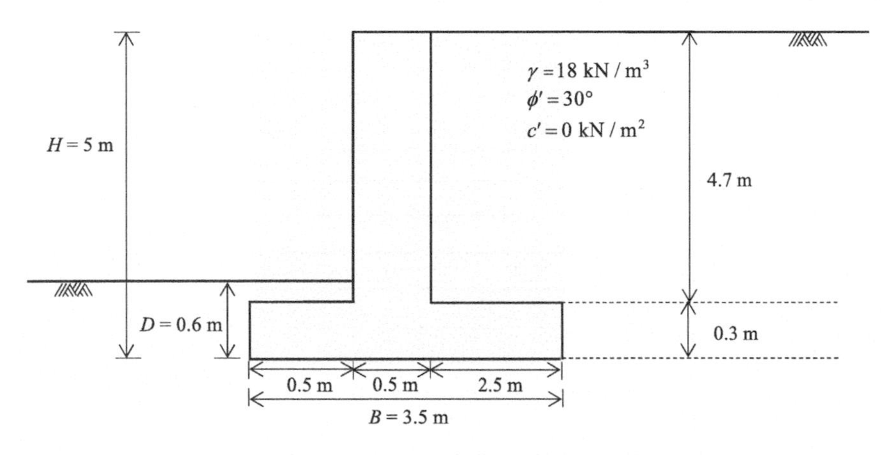

Figure P 5.5

5.6 Redo problem 5.5 using Coulomb's lateral earth pressure method.

5.7 Redo problem 5.5, considering a backslope behind the retaining wall with an inclination angle of $\beta = 10°$, as shown in Figure P5.7. Use Coulomb's method to calculate the lateral earth pressure.

5.8 Consider a reinforced wall, as shown in Figure P5.8, given $H = 8$ m, $\gamma = 16.8$ kN/m³, $\phi' = 34°$, $L = 4.5$ m, $S_v = 0.5$ m, and $T_{ult} = 150$ kN/m. The sand is used as backfill. The geogrid is

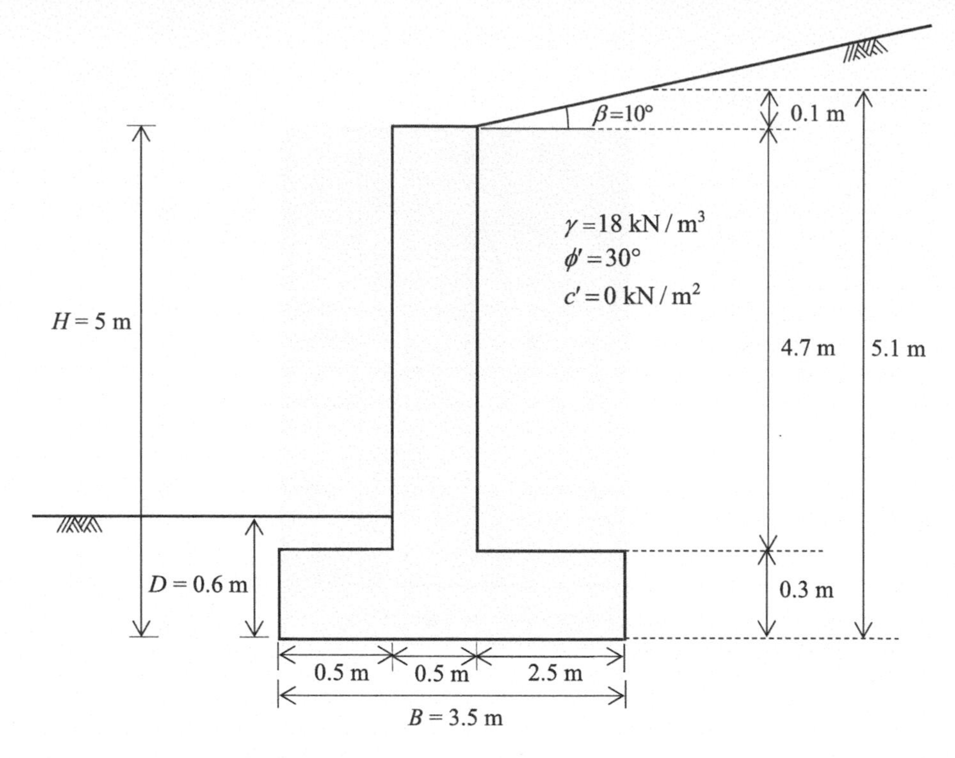

Figure P 5.7

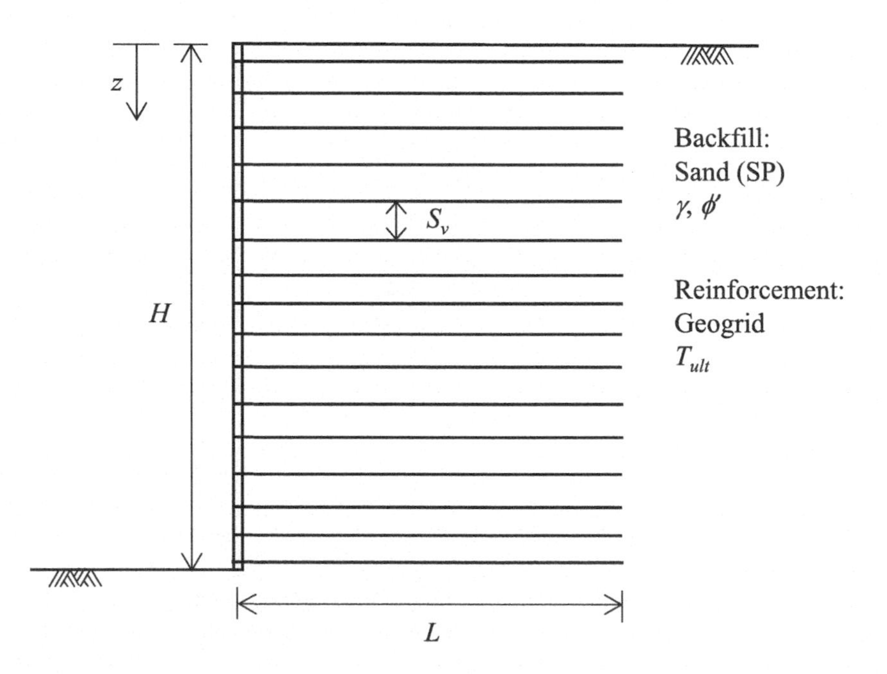

Figure P 5.8

used as reinforcement with reduction factors $RF_{CR} = 2$, $RF_D = 1.5$, and $RF_{ID} = 1.5$ obtained from laboratory tests. The top reinforcement layer is installed 0.25 m below the top of the wall.

a Calculate the factor of safety against reinforcement breakage and identify the most critical layer.
b Calculate the factor of safety against reinforcement pullout and identify the most critical layer.

5.9 Redo problem 5.8, considering a reinforced wall backfilled with a soil with soil strength properties of $\phi' = 31°$ and $c' = 5$ kPa.

a Calculate the factor of safety against reinforcement breakage, and identify the most critical layer.
b Calculate the factor of safety against reinforcement pullout and identify the most critical layer.

5.10 Consider the design of a reinforced wall, given $H = 6$ m, $\gamma = 15.7$ kN/m^3, $\phi' = 36°$, and $T_{ult} = 90$ kN/m. The geotextile is used as reinforcement with reduction factors $RF_{CR} = 2.5$, $RF_D = 1.5$, and $RF_{ID} = 1.2$ obtained from laboratory tests. The required Fs against reinforcement breakage and pullout are $F_{br} = 1.5$ and $F_{po} = 1.5$. Determine the optimal S_v and L that satisfy the required Fs.

5.11 Consider the design of a reinforced wall, given $H = 6$ m, $\gamma = 16.5$ kN/m^3, $\phi' = 31°$, $c' =$ 5 kPa, and $T_{ult} = 120$ kN/m. The geogrid is used as reinforcement with reduction factors $RF_{CR} = 2.5$, $RF_D = 1.5$, and $RF_{ID} = 1.2$ obtained from laboratory tests. The required Fs against reinforcement breakage and pullout are $F_{br} = 1.5$ and $F_{po} = 1.5$. Determine the optimal S_v and L that satisfy the required Fs.

5.12 Figure P5.12 shows a 4 m deep excavation with a cantilevered wall. The groundwater level is very deep. The unit weight of the soil is $\gamma = 20$ kN/m^3. The effective soil shear strength properties are $\phi' = 25°$ and $c' = 0$ kPa. Determine the embedment length of this wall by using the fixed earth support method with a required F_S of 1.5. Adopt Rankine's earth pressure method.

5.13 A cross section of a cantilevered wall is presented in the figure. Assume that the groundwater level is very deep. Determine the embedment length of this wall by using the fixed earth support method with a required F_S of 1.5.

5.14 A cross section of a cantilevered wall is presented in the figure. Determine the embedment length of this wall by using the fixed earth support method with a required F_S of 1.5.

5.15 The groundwater table lies at a depth of 10 m below the ground surface and consists entirely of sands. A 7 m deep excavation is to be constructed with anchored sheet piles to support the excavation. The tie rod of the anchor inclined at an angle of 15° is placed 1 m below the surface, with horizontal spacing of the tie rods being 3 m. The unit weights of sand above and below the water table are $\gamma_1 = 17.0$ kN/m^3 and $\gamma_2 = 20.0$ kN/m^3. The sand has $\phi' = 35°$.

a Determine the required depth of the sheet pile.
b Calculate the force in the tie rods if they are placed 3 m apart (inclined at an angle of 15°).
c Determine the length of the tie rods.

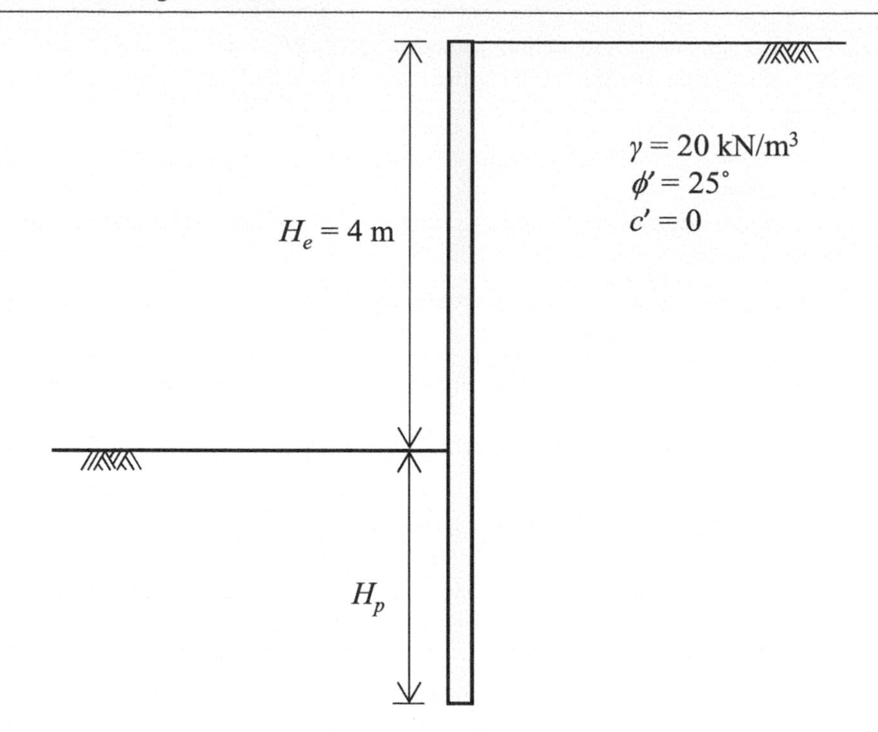

Figure P 5.12

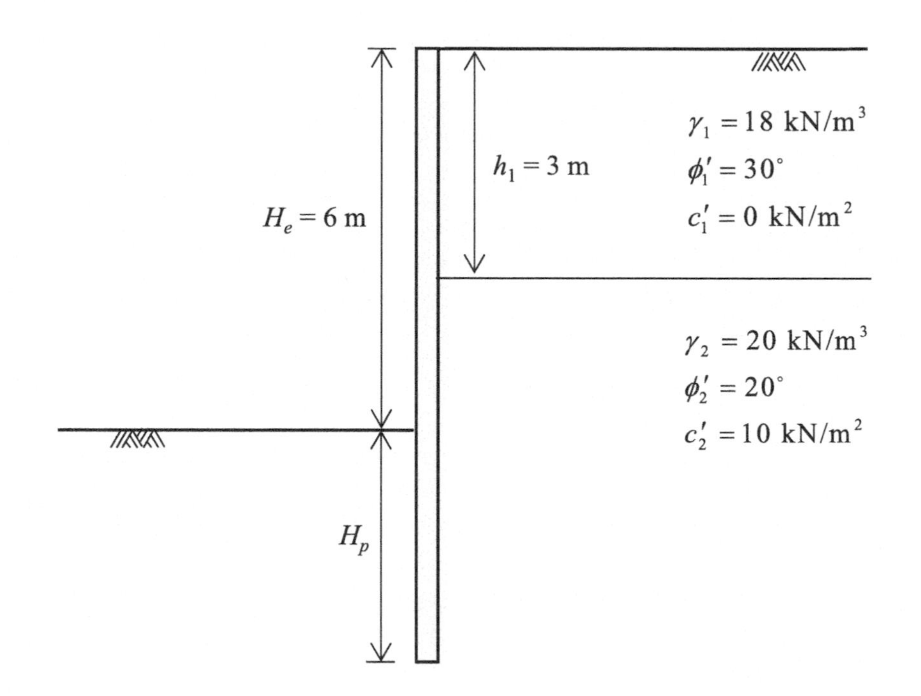

Figure P 5.13

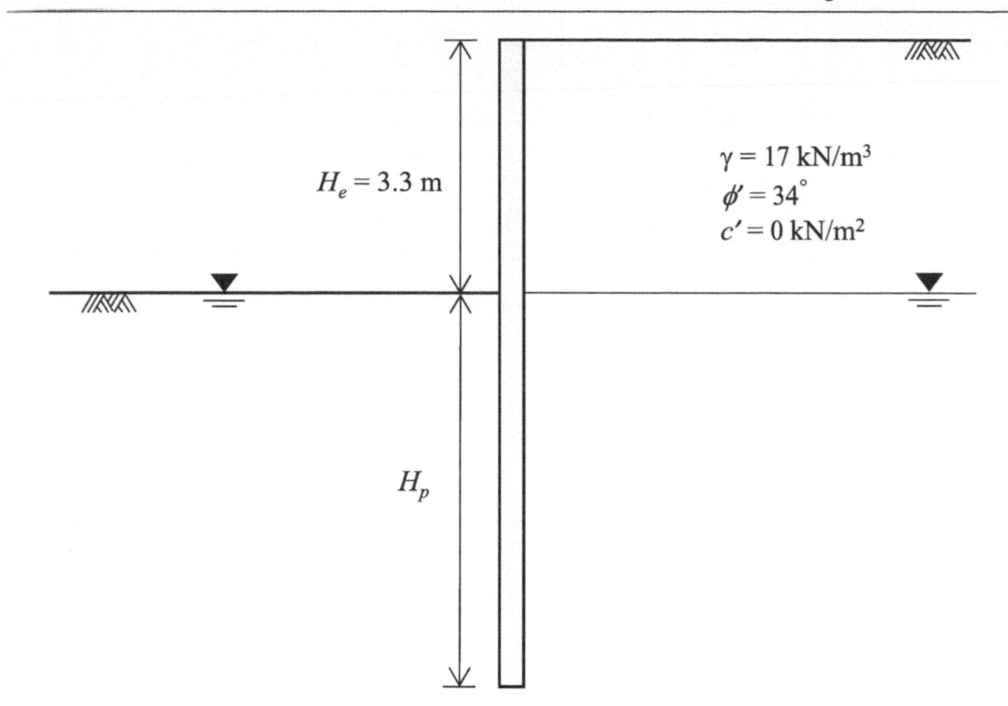

Figure P 5.14

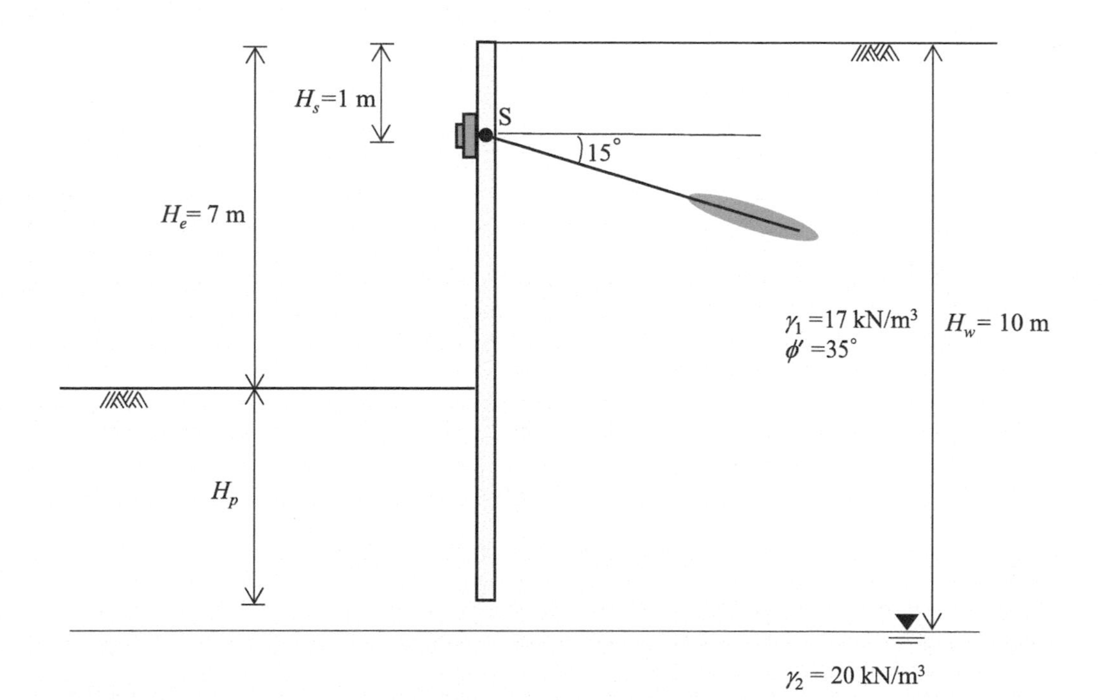

Figure P 5.15

5.16 A cross section of the sheet pile is presented in the figure. The subsurface soil consists of two soil layers: the depth of the top fill layer is 0–6 m, and the depth of the underlying original clay layer is below 6 m.

a Determine the embedment length of this wall by using the free earth support method with a required F_S of 1.5.

b Find the anchor load for the anchor with a horizontal spacing of 3 m.

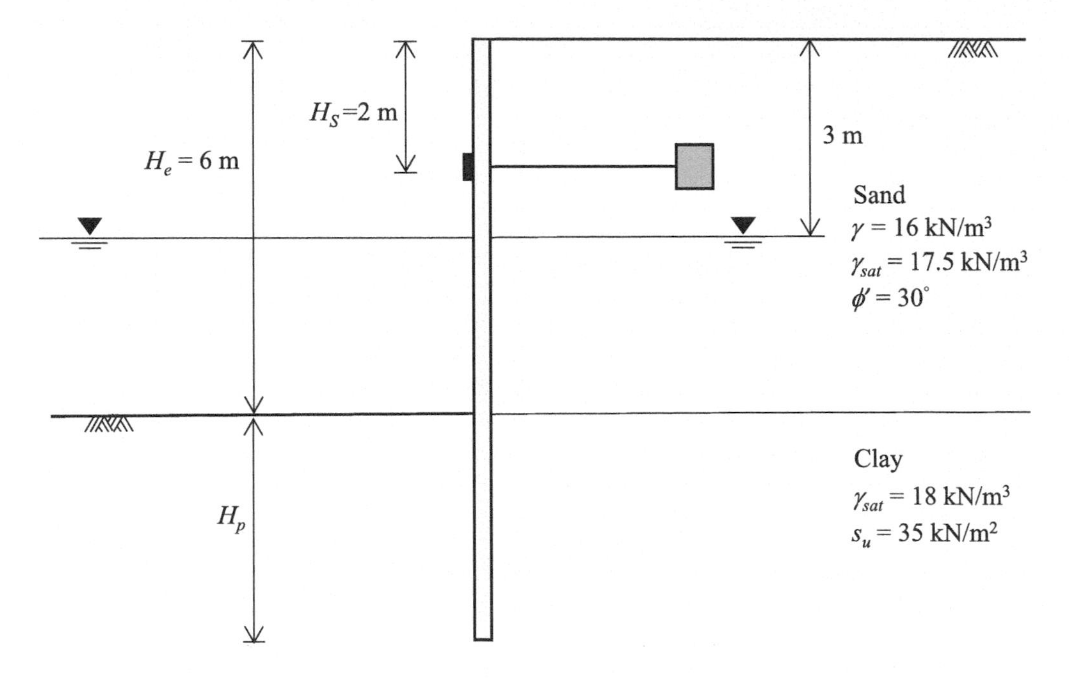

Figure P 5.16

References

Berg, R., Christopher, B.R. and Samtani, N. (2009), *Design and Construction of Mechanically Stabilized Earth Walls and Reinforced Soil Slopes*, Vol. 1. FHWA-NHI-10-024, Federal Highway Administration, Washington, DC, p. 306.

Elias, V., Christopher, B.R. and Berg, R. (2001), *Mechanically Stabilized Earth Walls and Reinforced Soil Slopes Design and Construction Guidelines*, FHWA-NHI-00-043, Federal Highway Administration, Washington, DC, p. 394.

Liao, R.T. (2008), *Countermeasures and Case Studies of Landslide Disaster Prevention and Mitigation*, Taipei: Techbook Ltd, p. 234.

Ou, C.Y. (2022), *Fundamentals of Deep Excavation*, London: CRC Press, p. 478.

Padfield, C.J. and Mair, R.J. (1984), *Design od Retaining Walls Embedded in Stiff Clay*. CIRIA Report 104, London: Construction Industry Research and Information Association, p. 146.

Sabatini, P.J., Elias, V.E., Schmertmann, G.R. and Bonaparte, R. (1997), *Geotechnical Engineering Circular No. 2, Earth Retaining Systems*, FHWA-SA-96-038, Federal Highway Administration, Washington, DC, p. 161.

Tanyu, B.F., Sabatini, P.J. and Berg, R. (2008), *Earth Retaining Structures*, FHWA-NHI-07-071, Federal Highway Administration, Washington, DC, p. 764.

Chapter 6

Excavation

6.1 Introduction

The construction of building basements or foundations or subways involves excavating soil down to several meters or even 40 m below the ground surface. To prevent the unexcavated soil from collapsing toward the excavation zone, a supported retaining system is needed. Figure 6.1 shows various structural components of a typical supported retaining system, which includes retaining walls, struts, wales, and center posts. As noted in Section 5.2, soldier piles with and without laggings, sheet piles, column piles, and diaphragm walls can be used as retaining walls.

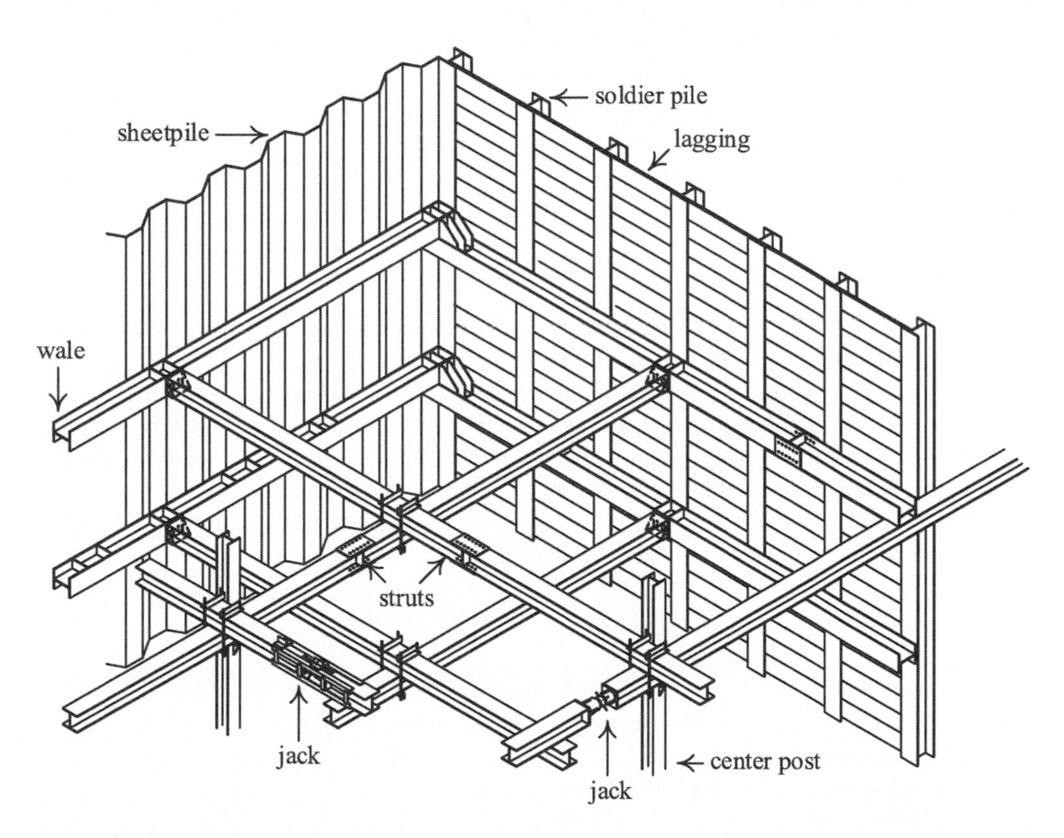

Figure 6.1 Components of a strutted retaining system.

DOI: 10.1201/9781003350019-6

When the soil is excavated, an unbalanced force acting on the excavation bottom and retaining wall is generated. The unbalanced force may cause an excavation collapse and sometimes a large amount of lateral deflection of retaining walls and ground settlement. An analysis of the excavation is necessary before the design. Excavation analysis is a typical soil–structure interaction problem. Soil is also a nonlinear, inelastic, and anisotropic material. Theoretically, the analysis of deep excavation involves simulations of the elastoplastic behavior of the soil, interface behavior between the soil and retaining walls, and the excavation process. However, for practical purposes, excavation analysis must be simplified. In this chapter, excavation analysis is simplified and divided into several individual subjects. The first important subject is stability analysis to prevent the excavation from collapse. Other subjects, such as deformation and stress analyses of retaining walls, wales, struts, and monitoring systems, are also introduced. In addition, the center post, dewatering system, excavation procedure, and deformation control should be analyzed or designed. However, considering the page limitation of this book, these subjects are not introduced here, and interested readers are advised to refer to Ou (2022).

6.2 Excavation methods

Excavation methods include the full open cut method, the braced excavation method, the island excavation method, the anchored excavation method, and the top-down construction method. The selection of an appropriate excavation method must consider many factors, such as the construction budget, construction period, existence of adjacent excavations, availability of construction equipment, area of construction site, conditions of adjacent buildings, foundation types of adjacent buildings, and so on. Experienced engineers are able to make good selections based on these factors. This section will introduce the two most commonly used methods: the braced excavation and top-down construction methods. For other methods, interested readers can refer to related references, such as Ou (2022).

6.2.1 Braced excavation

Installing struts in front of retaining walls to resist the earth pressure acting on the back of walls is called the braced excavation method. Figure 6.2 shows a typical arrangement of struts. Figure 6.3 shows a photo of a braced excavation. As shown in Figures 6.1 and 6.2, the braced excavation method includes struts, wales, and center posts. The function of wales is to transfer the earth pressure on the back of retaining walls to horizontal struts. Center posts support the strut, so the strut does not fall off due to its own weight.

In deep excavations, construction is often carried out in stages. The following is the construction procedure for the braced excavation method:

1. Place center posts in the construction area.
2. Proceed to the first stage of excavation.
3. Install wales above the excavation surface, then install struts and preload them.
4. Repeat procedures 2 to 3 until the designed depth is reached.
5. Build the foundation of the building.
6. Demolish the struts above the foundation.
7. Construct floor slabs.
8. Repeat procedures 6 to 7 until the construction of the ground floor is completed.

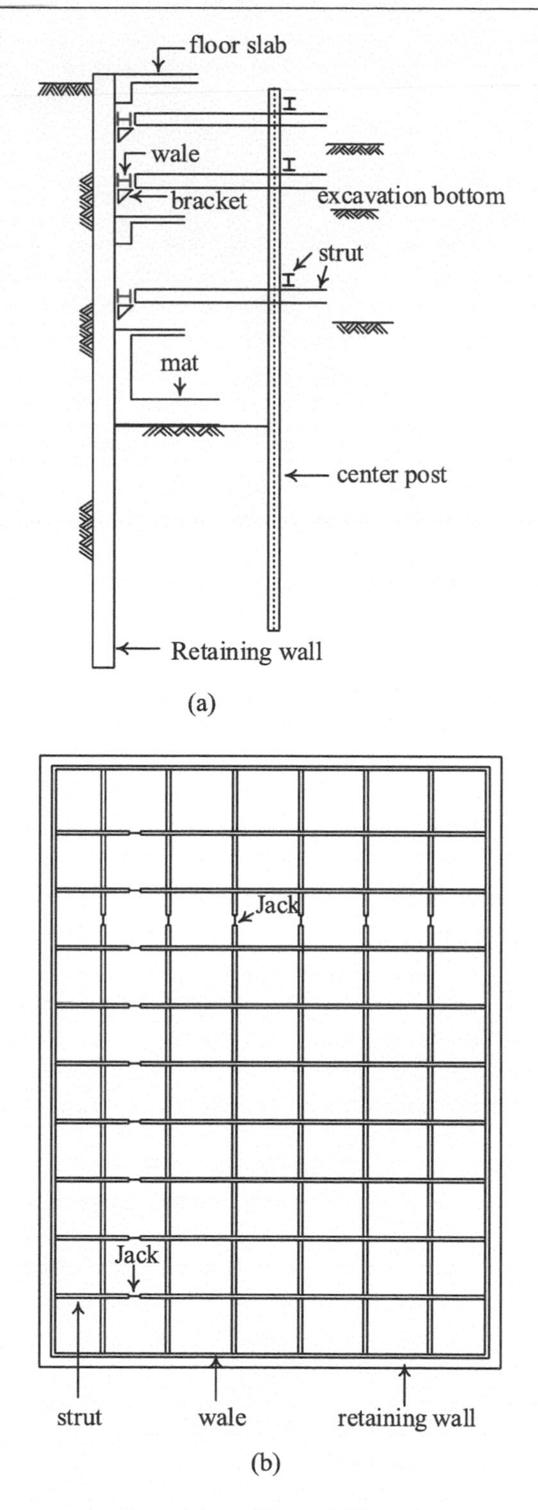

Figure 6.2 Braced excavation method: (a) profile and (b) plan.

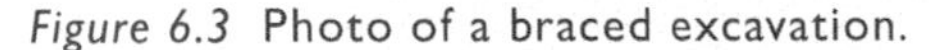

Figure 6.3 Photo of a braced excavation.

As excavation is carried out to the designed depth, after which raft foundations or foundation slabs are built, struts can then be removed level by level, and floor slabs are built accordingly. Thus, the underground construction is finished. As such, the underground structure is constructed from the bottom to the top, and the braced excavation method is thus also called the bottom-up construction method.

6.2.2 Top-down construction

The top-down construction method is to erect molds and construct floor slabs after each excavation. The floor slabs are permanent structures, which replace temporary steel struts in the braced excavation method to counteract the earth pressure acting on the back of the retaining wall. In this way, the underground structure construction is finished upon completion of the excavation process. The construction of the underground structure proceeds from top to bottom, the opposite of conventional foundation construction methods. The method is thus called the top-down construction method.

The floor slabs used in the top-down construction method are heavier than the steel struts used in conventional excavation methods. In addition, the superstructure, which is constructed simultaneously during excavation, places more weight on the column. Thus, the bearing capacity of the column must be examined. Consequently, pile foundations are often chosen as building foundations in the top-down construction method. The typical construction procedure of the top-down construction method is as follows (see Figure 6.4):

1 Construct the retaining wall.
2 Construct piles. Place the steel columns on top of the piles.
3 Proceed to the first stage of excavation.
4 Cast the floor slab of the first basement level (B1 slab).
5 Begin to construct the superstructure.

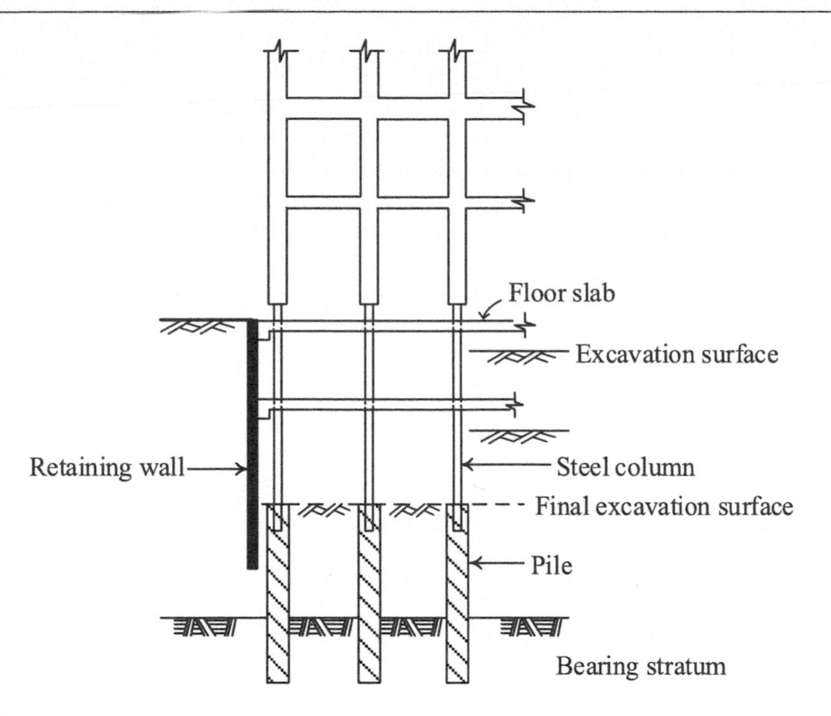

Figure 6.4 Top-down construction method.

6 Proceed to the second stage of excavation. Cast the floor slab of the second basement level (B2 slab).
7 Repeat the same procedures until the designed depth is reached.
8 Construct foundation slabs and ground beams. Complete the basement.
9 Keep constructing the superstructure until finished.

6.3 Stability analysis

An excavation may fail because the stress on the support system exceeds the strength of its materials, for example, when the strut load exceeds the buckling load of struts or the bending moment of the retaining wall exceeds the limiting bending moment. Failure can also arise from soil instability. The method for analyzing soil instability is called stability analysis. Stability analysis considers base shear failure, upheaval failure, and sand boiling.

6.3.1 Base shear failure

When the shear stress generated in most of the soil below the excavation bottom reaches the shear strength, the soil will be subject to a large amount of displacement or heave, leading to the failure of the entire strutted retaining system. This phenomenon is called base shear failure. Base shear failure may occur in either clayey or sandy soil. When the shear stress is equal to the shear strength, this phenomenon is called "failure" in conventional soil mechanics and the "plastic state" in the finite element method. Base shear failure in clayey soil will cause a large amount of soil to heave at the bottom of the excavation, and therefore, it is also called "plastic basal heave," or simply, "basal heave" or "plastic heave." Figure 1.1 shows an excavation case with base shear failure or basal heave.

As shown in Figure 6.5a, in the ultimate state, the earth pressures acting on the front and back of the retaining wall in excavations will reach the passive and the active earth pressures, respectively. Taking the retaining wall below the lowest level of the strut as a free body and conducting a force equilibrium analysis (Figure 6.5b), we can then calculate the factor of safety against base shear failure as follows:

$$F_b = \frac{M_r}{M_d} = \frac{P_p L_p}{P_a L_a - M_s} \tag{6.1}$$

where:

F_b = factor of safety against base shear failure.
M_r = resistant moment.
M_d = driving moment.
q_s = surcharge.
P_a = resultant of the active earth pressure on the back of the wall below the lowest level of the strut.
L_a = length from the lowest level of strut to the point of action, P_a.
M_s = allowable bending moment of the retaining wall.
P_p = resultant of the passive earth pressure on the front of the retaining wall below the excavation surface.
L_p = length from the lowest level of strut to the point of action, P_p.

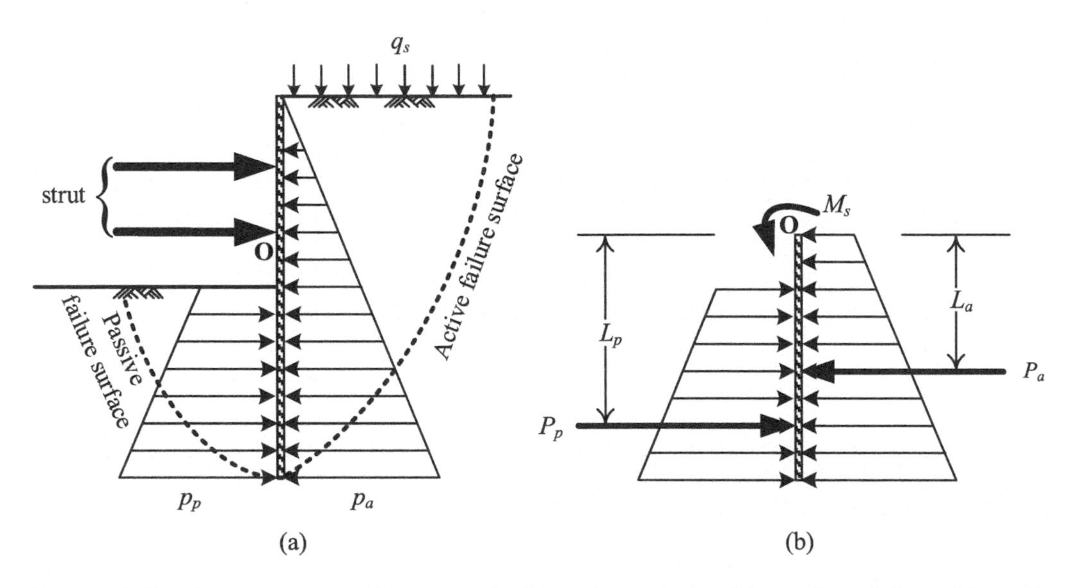

Figure 6.5 Earth pressure equilibrium method for the analysis of base shear failure: (a) earth pressure distribution and (b) forces on the free body.

Eq. 6.1 is computed based on the gross pressure distribution. The required F_b should be equal to or greater than 1.5. Nevertheless, when assuming $M_s = 0$, $F_b \geq 1.2$. Eq. 6.1 can be used to obtain either the factor of safety for a certain depth of wall or the required penetration depth of a retaining wall with a certain value of the safety factor. The method is often called the load factor earth pressure equilibrium method or the earth pressure equilibrium method (load factor). The earth pressure equilibrium method (load factor) can be applied to sandy soil, clayey soil, or layers of sandy and clayey soils. Since all the uncertainties are lumped into a single factor of safety, which is not necessarily reasonable, the results for some scenarios from the earth pressure equilibrium method (load factor), such as stiff clay with s_u = constant, are unreasonable (Ou, 2022).

If the factor of safety is located at the source where the largest uncertainty arises, that is, the shear strength, as shown in eq. 6.2 or 6.3, then ϕ'_m and c'_m are used to calculate the distributions of earth pressure on both sides of the wall.

For the effective stress analysis:

$$\tan\phi'_m = \frac{\tan\phi'}{F_b} \qquad c'_m = \frac{c'}{F_b} \tag{6.2}$$

For the undrained analysis:

$$s_{u,m} = \frac{s_u}{F_b} \tag{6.3}$$

Then, a factor of safety (F_b) is selected to ensure that the resistance moment is greater than or equal to the driving moment:

$$P_p L_p \geq P_a L_a - M_s \tag{6.4}$$

where the definitions of P_p, P_a, L_p, L_a, and M_s are the same as those in eq. 6.1.

Similar to eq. 6.1, M_s can be assumed to equal 0. The method is called the strength factor earth pressure equilibrium method, or the earth pressure equilibrium method (strength factor). The earth pressure equilibrium method (strength factor) can be applied to all kinds of soils, and normally, it can yield reasonable results for all scenarios.

Terzaghi (1943) assumes that the failure surface of base shear failure in clayey soil, that is, plastic heave, initiates with a circular arc failure surface below the excavation surface, then develops outward to the excavation surface level **ab**, extending upward to the ground surface. The soil weight within the width of $B/\sqrt{2}$ acting on plane **ab** is treated as a driving force that causes the excavation to fail.

According to Terzaghi's bearing capacity theory (Section 3.4.3), the bearing capacity of the clayey soil below plane **ab** can be denoted as $P_{\max} = 5.7s_u$. When the soil weight above plane ***ab*** is greater than the soil bearing capacity, the excavation will fail. In addition, the failure surface will be constrained by stiff soil. Let D represent the distance between the excavation surface and the stiff soil. Terzaghi's method can then be discussed in two parts.

When $D \geq B/\sqrt{2}$:

As shown in Figure 6.6a, the formation of the failure surface is not constrained by the stiff soil. Suppose the unit weight of the soil is γ. The soil weight (containing the surcharge q_s) range $B/\sqrt{2}$ on plane ***ab*** is:

$$W = (\gamma H_e + q_s)\frac{B}{\sqrt{2}} \tag{6.5}$$

The bearing capacity, Q_u, of the saturated clay below plane ***ab*** is:

$$Q_u = (5.7 s_{u2})\frac{B}{\sqrt{2}} \tag{6.6}$$

When a basal heave failure occurs, vertical failure plane ***bc*** can offer shear resistance ($s_{u1}H_e$), and the factor of safety against basal heave (F_b) is:

$$F_b = \frac{Q_u}{W - s_{u1}H_e} = \frac{5.7 s_{u2}\, B/\sqrt{2}}{(\gamma H_e + q_s)\, B/\sqrt{2} - s_{u1}H_e} \tag{6.7}$$

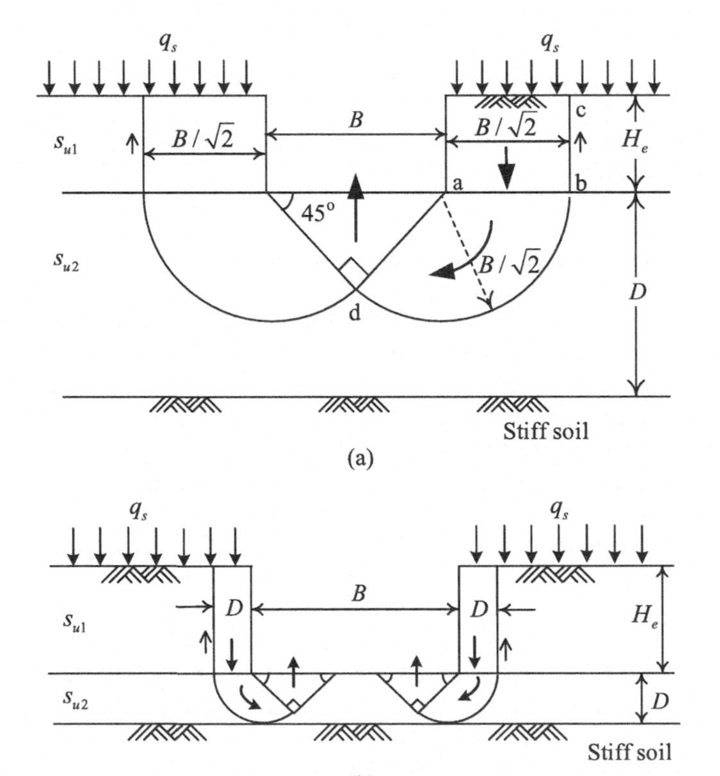

Figure 6.6 Terzaghi's method for the analysis of base shear failure in clayey soil: (a) $D \geq B/\sqrt{2}$, (b) $D < B/\sqrt{2}$.

where s_{u1} and s_{u2} represent the undrained shear strengths of the soils above and below the excavation surface, respectively, and q_s denotes the surcharge on the ground surface.

When $D < B/\sqrt{2}$, under such conditions, the failure surface is constrained by the stiff soil, as shown in Figure 6.6b, and its factor of safety (F_b) is:

$$F_b = \frac{Q_u}{W - s_{u1}H_e} = \frac{5.7 s_{u2} D}{(\gamma H_e + q_s)D - s_{u1}H_e} \tag{6.8}$$

Terzaghi's method can be applied only to excavations in clay. Moreover, Terzaghi's method does not take the influence of the penetration depth and stiffness of the retaining wall into account. That is, it assumes that the retaining wall does not exist. Terzaghi's method is applicable to wide excavations rather than narrow excavations because it overestimates the shear resistance along the vertical failure plane for narrow excavations. For most excavation cases, Terzaghi's factor of safety (F_b) should be greater than or equal to 1.5 (Mana and Clough, 1981; JSA, 1988).

Bjerrum and Eide (1956) assumed that the failure surface induced by excavation is similar to that of a pile foundation subjected to uplift loading, considering that the unloading behavior of the soil below the excavation surface caused by excavation can be analogous to that of a pile foundation subjected to uplift loading. Then, using the unloading bearing capacity equation for the pile foundation, we can obtain the unloading bearing capacity. The factor of safety is the ratio of the unloading bearing capacity to the unloading pressure. As shown in Figure 6.7, the failure surface initiates with a circular arc with a radius $B/\sqrt{2}$. The factor of safety against base shear failure can be calculated as follows:

$$F_b = \frac{N_c \cdot s_u}{\gamma \cdot H_e + q_s} \tag{6.9}$$

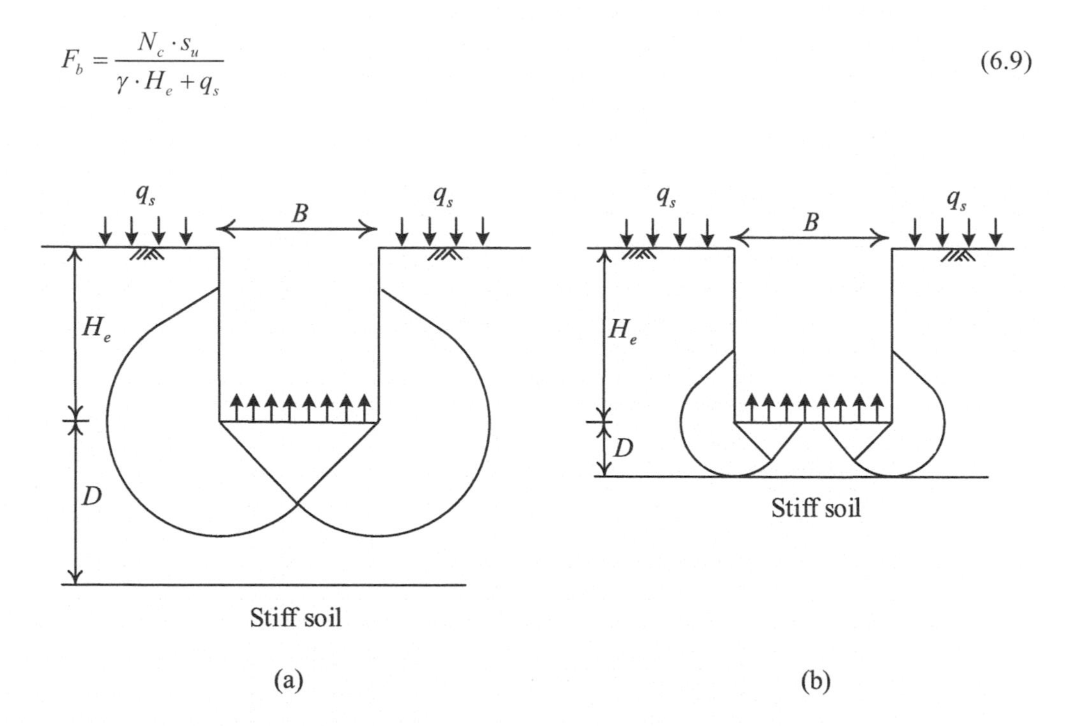

Figure 6.7 Bjerrum and Eide's method for the analysis of base shear failure in clayey soil: (a) $D \geq B/\sqrt{2}$, (b) $D < B/\sqrt{2}$.

where:

q_s = the surcharge on the ground surface.
γ = unit weight of the soil.
H_e = excavation depth.
s_u = undrained shear strength of the clayey soil.
N_c = Skempton's bearing capacity factor (1951), which can be determined from Figure 6.8 or calculated by the following equation:

$$N_{c(\text{rectangular})} = N_{c(\text{square})}(0.84 + 0.16\frac{B}{L}) \quad (6.10)$$

where:

B = the excavation width.
L = the excavation length.

Since N_c takes into account the effects of the excavation depth and excavation shape, eq. 6.9 is equally valid for shallow and deep excavations, as well as rectangular excavations.

Based on Reddy and Srinivasan's study (1967), NAVFAC DM 7.2 (1982) extended Bjerrum and Eide's method to excavations with stiff soils below the excavation surface (Figure 6.7b). The extension of Bjerrum and Eide's method can be expressed as follows:

$$F_b = \frac{s_u N_c f_d f_s}{\gamma H_e} \quad (6.11)$$

where:

N_c = bearing capacity factor that considers the stiff soil, which can be determined from Figure 6.9a.

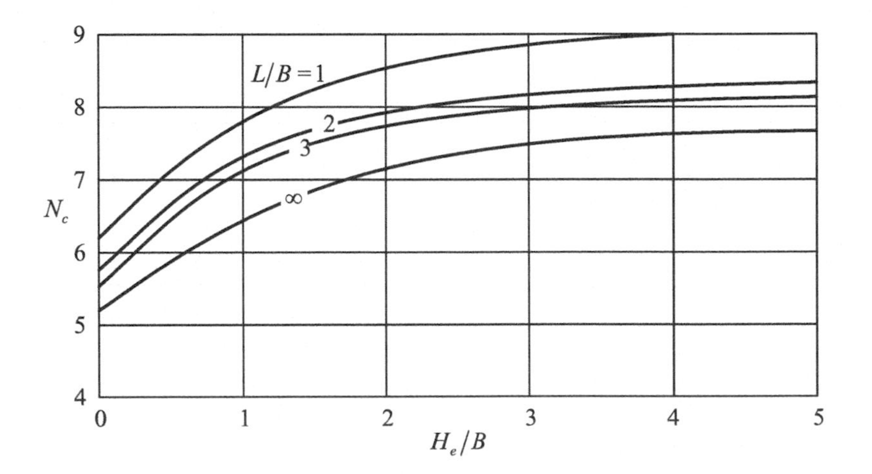

Figure 6.8 Skempton's bearing capacity factor.

Source: Skempton (1951).

f_d = depth modification factor, which can be found in Figure 6.9b.
f_s = shape modification factor, which can be estimated by the following equation:

$$f_s = 1 + 0.2\frac{B}{L} \tag{6.12}$$

where B refers to the excavation width and L is the excavation length.

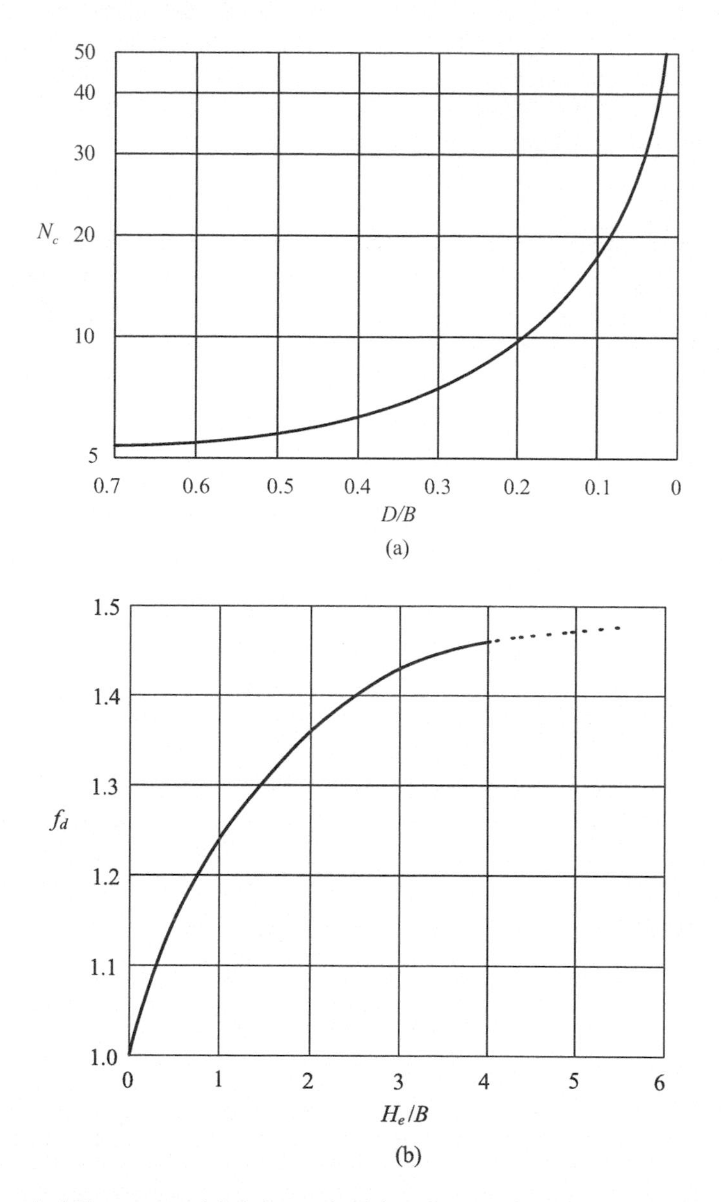

Figure 6.9 Extended Bjerrum and Eide's method for the analysis of base shear failure in clayey soil: (a) the bearing capacity factor considering stiff soil and (b) the depth modification factor.

Similar to Terzaghi's method, Bjerrum and Eide's method does not consider the influence of the penetration depth and stiffness of the retaining wall. That is, it assumes that the retaining wall does not exist. For most excavations, the factor of safety obtained with Bjerrum and Eide's method (F_b) should be larger than or equal to 1.2 (JSA, 1988).

The slip circle method assumes the main part of the trial failure surface below the excavation bottom is a circular arc, with its center at the lowest level of the strut. The failure surface then grows outward to the excavation surface level and extends vertically to the ground surface, as shown in Figure 6.10. The shear strength on the vertical failure plane (line ***bc*** in Figure 6.10a) is ignored. Take the retaining wall and soil below the lowest level of struts as well as above the circular arc as a free body (Figure 6.10b). The soil weight above the excavation surface in the back of the retaining wall can be treated as the driving force, and the shear strength along the failure surface generates the resistant force. The ratio of the resistance moment to the driving moment with respect to the lowest level of the strut is:

$$F_b = \frac{M_r}{M_d} = \frac{X \int_0^{\frac{\pi}{2}+\alpha} s_u (X d\theta)_s}{W \cdot \frac{X}{2} - M_s} \tag{6.13}$$

where:

M_r = resistance moment.
M_d = driving moment.
M_s = allowable bending moment of the retaining wall.

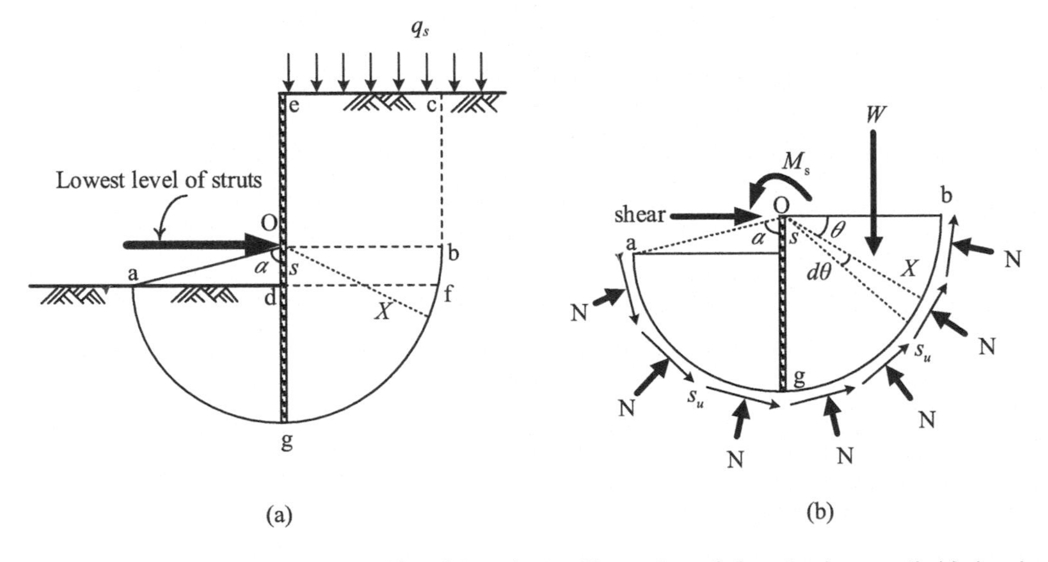

Figure 6.10 The slip circle method for the analysis of base shear failure in clayey soil: (a) development of the failure surface and (b) forces on the free body.

s_u = undrained shear strength of clay.
s = distance between the lowest level of the strut and the excavation surface.
X = radius of the failure circle, $X = H_p + s$.
W = total weight of the soil in front of the vertical failure plane and above the excavation surface, including the surcharge (q_s) on the ground surface.

The original source of the slip circle method is untraceable. Nevertheless, TGS (2022) and JSA (1988) adopt the method in their building codes. Both assume that $M_s = 0$ and recommend that the factor of safety (F_b) against base shear failure or basal heave should be greater than or equal to 1.2. According to design experience in some countries, the value is quite reasonable. In fact, the allowable bending moment value of the retaining wall M_s is far less than the resistance moment provided by shear strength. Thus, to simplify the computation, it is reasonable to assume $M_s = 0$.

When the safety factor computed as the soil shear strength increase with depth is insufficient, the penetration depth of the wall should be increased to ensure the computed safety factor meets the requirements. However, as the soil shear strength increases little with depth, an increase in the wall penetration depth contributes slightly to the safety factor. Under such conditions, soil improvement is an effective way to increase the safety factor.

Example 6.1

Figure EX. 6.1a shows an excavation in sandy soil with two levels of the strut. The groundwater levels behind the wall and in front of the wall are on the ground surface and on the excavation surface, respectively. The properties of sandy soil are as follows: the effective cohesion $c' = 0$, the effective angle of friction $\phi' = 33°$, and the saturated unit weight $\gamma_{sat} = 20\ \text{kN/m}^3$. The excavation width $B = 35$ m. The excavation depth $H_e = 10$ m, and the wall penetration depth $H_p = 8$ m. The strut locations are $h_1 = 2$ m, $h_2 = 4$ m, and $h_3 = 4$ m. The distance to the impermeable soil $D = 5$ m. Because of the difference between the levels of groundwater, seepage will occur. Assume that the friction angles (δ) between the retaining wall and soil on both the active and passive sides are equal to 0. Compute the factor of safety (F_b) against base shear failure using the earth pressure equilibrium method (load factor).

Solution

Since the groundwater level in front of the wall is different from the water pressure behind the wall, seepage will occur. The pore water pressure due to seepage can be analyzed with a flow net or numerical simulation. According to Figure EX 6.1a, the pore water pressure on the wall is evaluated, as shown in Figure EX 6.1b. The active and passive earth pressures can then be computed as follows.

According to Rankine's earth pressure theory, $K_a = 0.295$, $K_p = 3.392$.

Active earth pressure:

At GL-6 m:

$u_w = 36.2\ \text{kPa}$

$\sigma'_v = \sigma_v - u_w = 20 \times 6 - 36.2 = 83.8\ \text{kPa},\ \sigma'_a = \sigma'_v K_a = 24.7\ \text{kPa}$

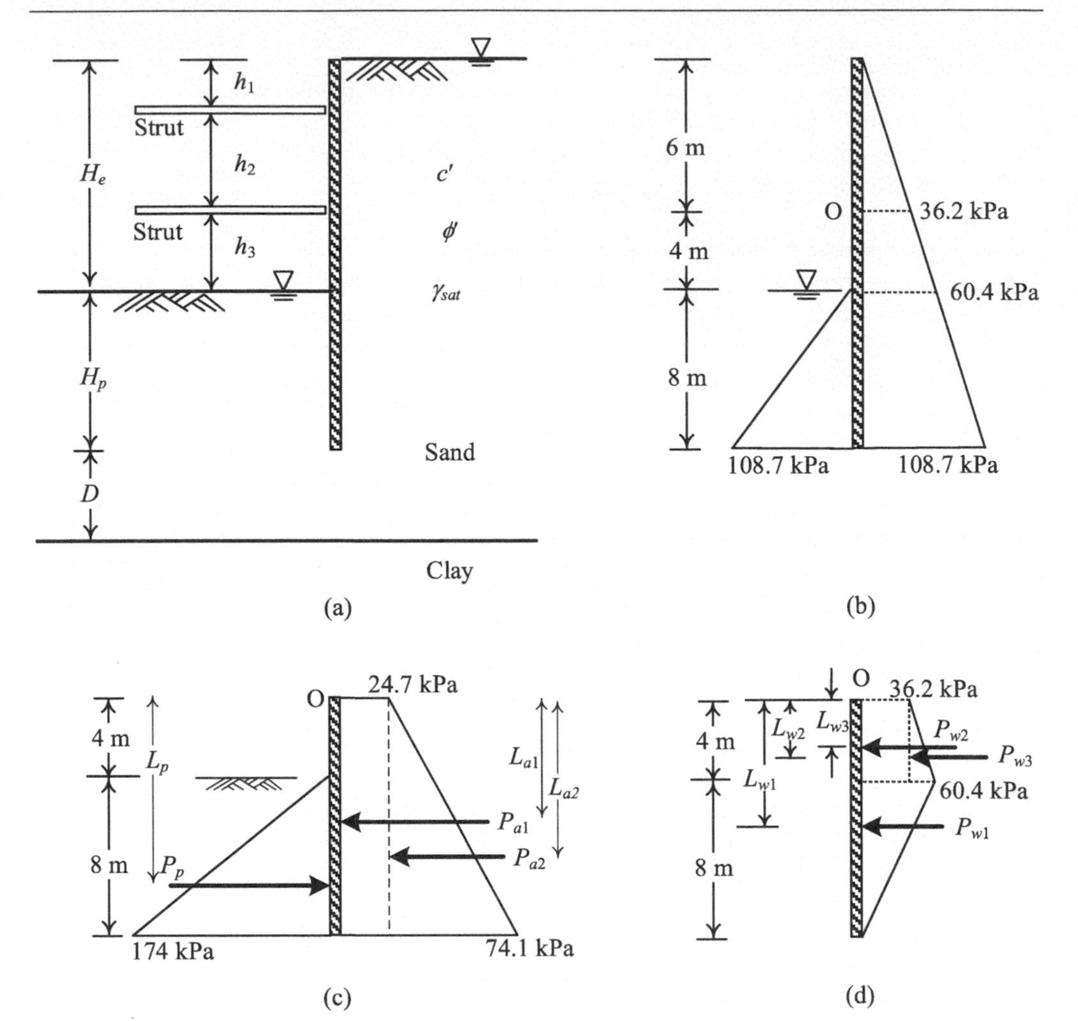

Figure EX 6.1 A 10 m excavation in sandy soil: (a) excavation profile, (b) pore water pressure distribution, and (c) lateral earth pressure distribution (including pore water pressure).

At GL-18 m:

$$u_w = 108.7 \text{ kPa}$$
$$\sigma'_v = \sigma_v - u_w = 20 \times 18 - 108.7 = 251.3 \text{ kPa}, \ \sigma'_a = \sigma'_v K_a = 74.1 \text{ kPa}$$

Passive earth pressure:

At GL-10 m:

$$u_w = 0, \ \sigma'_v = \sigma_v - u_w = 0, \ \sigma'_p = \sigma'_v K_p = 0$$
$$\sigma'_p + u_w = 0$$

At GL-18 m:

$u_w = 108.7$ kPa
$\sigma'_v = \sigma_v - u_w = 20 \times 8 - 108.7 = 51.3$ kPa, $\sigma'_p = \sigma'_v K_p = 174$ kPa

Figure EX 6.1c shows the distribution of effective earth pressure. A similar concept to eq. 5.43 in Chapter 5, the factor of safety, F_b, is only applied to P_p (or K_p) to account for the uncertainty in soil parameters related to passive earth pressure. Therefore, the net water pressure is treated as the driving force. Figure EX6.1d shows the net water pressure distribution. The resistance moment and driving moment are calculated as follows:

$$P_{a1} = 24.7 \times 12 = 296.4 \text{ kN/m}, \ L_{a1} = 6 \text{ m}$$

$$P_{a2} = (74.1\text{-}24.7) \times 12/2 = 296.4 \text{ kN/m}, \ L_{a2} = 8 \text{ m}$$

$$P_{w1} = 60.4 \times 8/2 = 241.6 \text{ kN/m}, \ L_{w1} = 6.67 \text{ m}$$

$$P_{w2} = 36.2 \times 4 = 144.8 \text{ kN/m}, \ L_{w2} = 2 \text{ m}$$

$$P_{w3} = (60.4 - 36.2) \times 4/2 = 48.4 \text{ kN/m}, \ L_{w3} = 2.67 \text{ m}$$

$$P_p = 174 \times 8/2 = 696 \text{ kN/m}, \ L_p = 9.3 \text{ m}$$

$$M_r = P_p L_p = 6473 \text{ kN-m/m}$$

$$M_d = P_{a1}L_{a1} + P_{a2}L_{a2} + P_{w1}L_{w1} + P_{w2}L_{w2} + P_{w3}L_{w3} = 6180 \text{ kN-m/m}$$

$$F_b = \frac{M_r}{M_d} = 1.05$$

Example 6.2

Figure EX 6.2a shows an excavation in clayey soil. The excavation depth $H_e = 15$ m, and the wall penetration depth $H_p = 15$ m. The excavation width and length are 30 m and 500 m, respectively. The groundwater levels behind the wall and in front of the wall are on the ground surface and on the excavation surface, respectively. The properties of clayey soils are:

Clay 1: $\gamma_{sat} = 15\text{kN/m}^3$, $s_u = 20\text{kN/m}^2$
Clay 2: $\gamma_{sat} = 20\text{kN/m}^3$, $s_u = 30\text{kN/m}^2$
Clay 3: $\gamma_{sat} = 18\text{kN/m}^3$, $s_u = 60\text{kN/m}^2$

Compute the factor of safety against base shear failure using the earth pressure equilibrium method (load factor).

Solution

Use the Rankine's earth pressure theory to calculate the earth pressure on the wall.

Active earth pressure, $\sigma_a = \sigma_v K_a \text{ - } 2c\sqrt{K_a}$

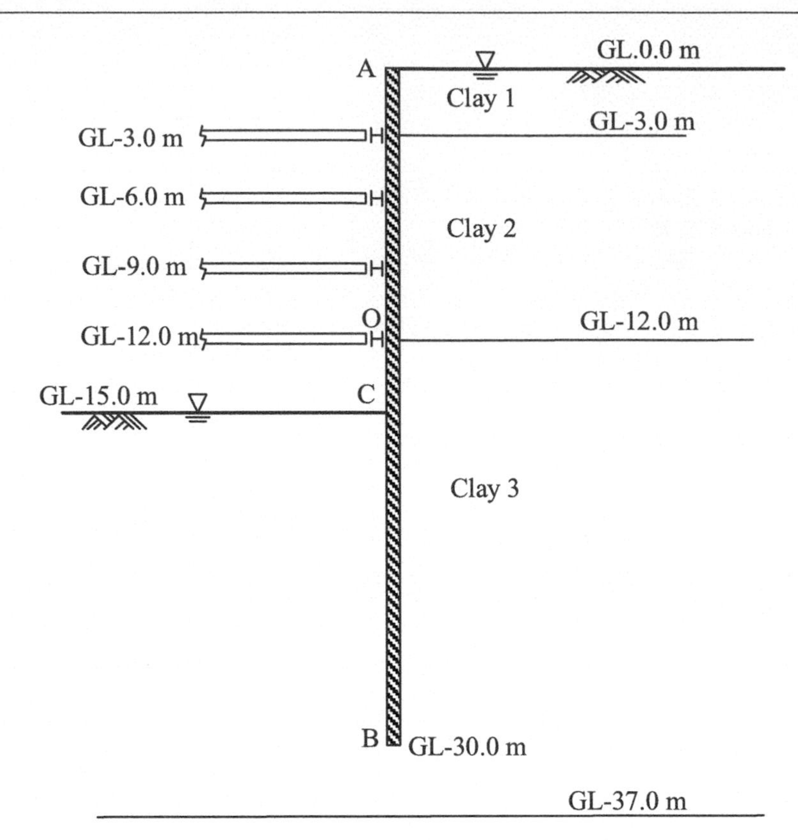

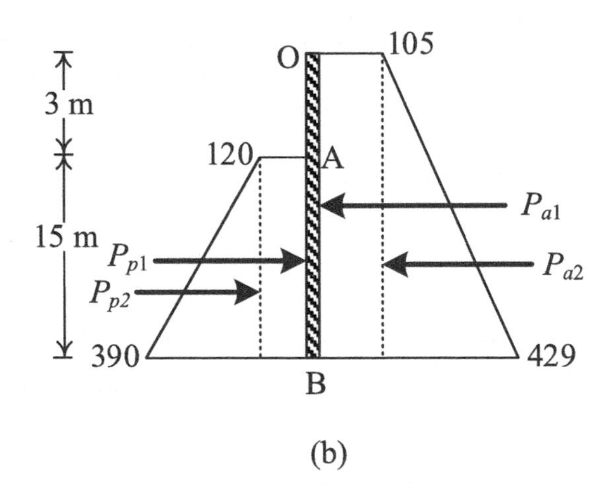

Figure EX 6.2 A 15 m excavation in clayey soil: (a) excavation profile and (b) lateral earth pressure distribution.

At GL-12 m (in clay 3):

$$\sigma_v = 20 \times 9 + 15 \times 3 = 225 \text{ kPa}$$
$$\sigma_a = 225 - 2 \times 60 = 105 \text{ kPa}$$

At GL-30 m:

$$\sigma_v = 225 + 18 \times 18 = 549 \text{ kPa}$$
$$\sigma_a = 549 - 2 \times 60 = 429 \text{ kPa}$$

Passive earth pressure: $\sigma_p = \sigma_v K_p + 2c\sqrt{K_p}$

At GL-15 m:

$$\sigma_v = 0$$

$$\sigma_p = 0 + 2 \times 60 = 120 \text{ kPa}$$

At GL-30 m:

$$\sigma_v = 15 \times 18 = 270 \text{ kPa}$$
$$\sigma_p = 270 + 2 \times 60 = 390 \text{ kPa}$$

Figure EX 6.2b shows the distribution of earth pressures on the front and back of the retaining wall.

$$M_d = 105 \times (3+15) \times \frac{3+15}{2} + (429 - 105) \times \frac{3+15}{2} \times \frac{2 \times (3+15)}{3} = 52{,}002 \text{ kN-m/m}$$

$$M_r = (120 \times 15) \times (\frac{15}{2} + 3) + (390 - 120) \times 15 \times \frac{1}{2} \times (15 \times \frac{2}{3} + 3) = 45{,}225 \text{ kN-m/m}$$

$$F_b = \frac{M_r}{M_d} = 0.87$$

Example 6.3

Same as example 6.2. Compute the factor of safety against base shear failure (basal heave) using the earth pressure equilibrium method (strength factor).

Solution

Assuming F is a strength reduction factor and following eq. 6.3, we have the reduced undrained shear strength $s_{u,m} = s_u/F$, for clay 3.

By trying various F values, the corresponding distribution of active and earth pressures, similar to Figure EX 6.2b, can be obtained with a similar computation procedure to that in example 6.2. The relationship between P_pL_p/P_aL_a and the F value can then be obtained as shown in Figure EX 6.3. As $P_pL_p/P_aL_a = 1.0$, then the F value is the factor of safety, $F_b = 0.85$.

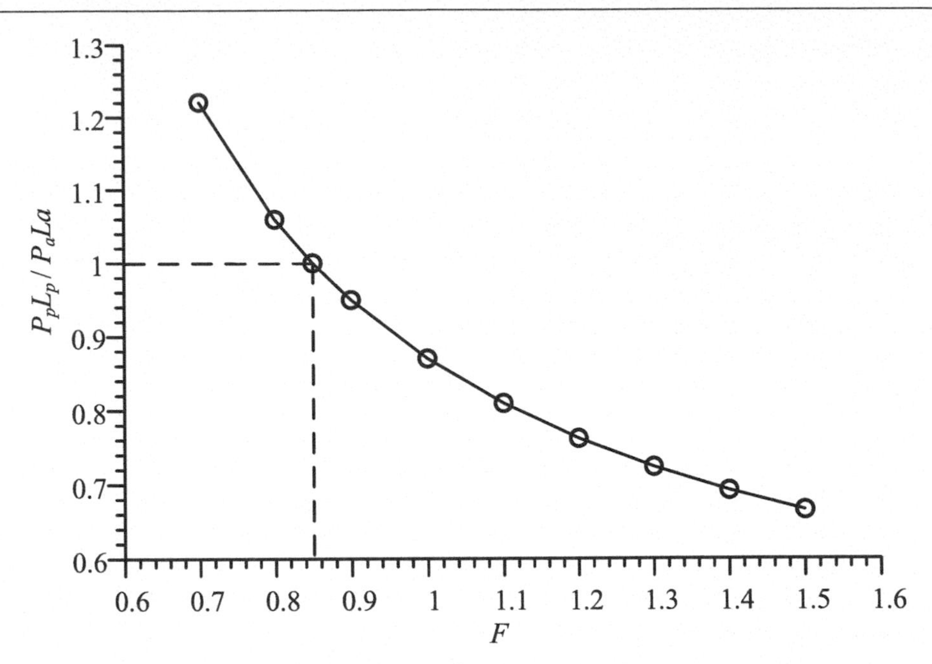

Figure EX 6.3 Variation in the ratio of the resistance moment to the driving moment with the strength reduction factor.

Example 6.4

Same as example 6.2. Compute the factor of safety against base shear failure (basal heave) using the slip circle method, Terzaghi's method, and Bjerrum and Eide's method.

Solution

1 SLIP CIRCLE METHOD

Assuming that the failure surface passes the wall toe, the radius of the failure surface is equal to 18 m.

$\alpha = \cos^{-1}(3/18) = 80.4^\circ = \pi/2.24$. The total vertical pressure at a depth of GL-15 m outside the excavation is:

$$\sigma_v = 15\times3 + 20\times9 + 18\times3 = 279 \text{ kPa}$$

The weight of the soil column is:

$$W = \sigma_{v,\text{GL}-15\text{m}} \times 18 = 279\times18 = 5{,}022 \text{ kN/m}$$

$$M_r = \int_0^{\pi/2+\pi/2.24} 18\times60\times(18\times d\theta) = 57{,}800 \text{ kN-m/m}$$

or

$$M_r = 18 \times (\frac{\pi}{2} + \frac{\pi}{2.24}) \times 60 \times 18 = 57,800 \text{ kN-m/m}$$

$$M_d = 5022 \times 18 / 2 = 45,198 \text{ kN-m/m}$$

$$F_b = \frac{M_r}{M_d} = \frac{57,800}{45,198} = 1.28$$

2 TERZAGHI'S METHOD

$D = 22$ m. According to Terzaghi's method, the radius of the failure surface is:

$$B/\sqrt{2} = 21.2 \text{ m} < D$$

$$F_b = \frac{Q_u}{W - s_{u1}H_e} = \frac{5.7 s_{u2} B/\sqrt{2}}{(\gamma H_e + q_s)B/\sqrt{2} - s_{u1}H_e}$$

$$= \frac{5.7 \times 60 \times 30/\sqrt{2}}{(15 \times 3 + 20 \times 9 + 18 \times 3) \times 30/\sqrt{2} - (20 \times 3 + 30 \times 9 + 60 \times 3)} = 1.34$$

3 BJERRUM AND EIDE'S METHOD

According to Figure 6.8, we can find $N_c = 5.8$. The radius of the failure surface = $B/\sqrt{2} = 21.2$. The failure surface will initiate from the intersection point of the wall and excavation bottom down to a depth of GL-(15 + 21.2) m or GL-36.2 m and then develop upward to the surface. The failure surface will pass through soil layers 1, 2, and 3. The average undrained surface along the failure surface is [60 × (36.2 – 12) + 30 × 9 + 20 × 3]/36.2 = 1782/36.2 = 49.2 kN/m².

$$F_b = \frac{Q_u}{Q} = \frac{(N_c s_u)B}{(\gamma H_e + q_s)B} = \frac{N_c s_u}{\gamma H_e + q_s} = \frac{5.8 \times 49.2}{15 \times 3 + 20 \times 9 + 18 \times 3} = 1.02$$

As shown in examples 6.2, 6.3, and 6.4, the F_b values obtained from the earth pressure equilibrium method (load factor), the earth pressure equilibrium method (strength factor), the slip circle method, Terzaghi's method, and Bjerrum and Eide's method are equal to 0.87, 0.85, 1.28, 1.34, and 1.02, respectively. Among those methods, the F_b value obtained from the earth pressure equilibrium methods is smaller than that obtained from Terzaghi's method and the slip circle method. This difference is because the shear strength or adhesion between the wall and clay in the earth pressure equilibrium methods is assumed to be 0 in examples 6.2 and 6.3. However, if it is included in the computation using the method as stated in Ou (2022), the F_b value obtained from the earth pressure equilibrium methods is close to that from the slip circle method and Terzaghi's method.

6.3.2 *Upheaval failure*

If there is a permeable layer (such as sandy or gravelly soils) underlying an impermeable layer below the excavation surface, the impermeable layer tends to be lifted up by the pore water pressure in the permeable layer. The safety against upheaval failure should be examined. As shown in Figure 6.11, the factor of safety against upheaval is:

$$F_{up} = \frac{\sum_{i} \gamma_{ti} \cdot h_i}{H_w \cdot \gamma_w} \tag{6.14}$$

where:

F_{up} = factor of safety against upheaval failure.
γ_{ti} = unit weight of soil in each layer above the bottom of the impermeable layer.
h_i = thickness of each soil layer above the bottom of the impermeable layer.
H_w = head of the pore water pressure in the permeable layer, which can be measured in piezometers.
γ_w = unit weight of the groundwater.

The factor of safety against upheaval failure, F_{up}, should be larger than or equal to 1.2.

When the computed safety factor does not satisfy the requirement, it is necessary to reduce the pore water pressure below the impermeable layer by dewatering. Improving the soil below the impermeable layer to increase the weight of the soils, that is, increase the numerator in eq. 6.14, is another alternative.

Example 6.5

Same as example 6.2. According to site investigation, the pore water pressure in the deposit is in the hydrostatic state. Compute the factor of safety against upheaval failure.

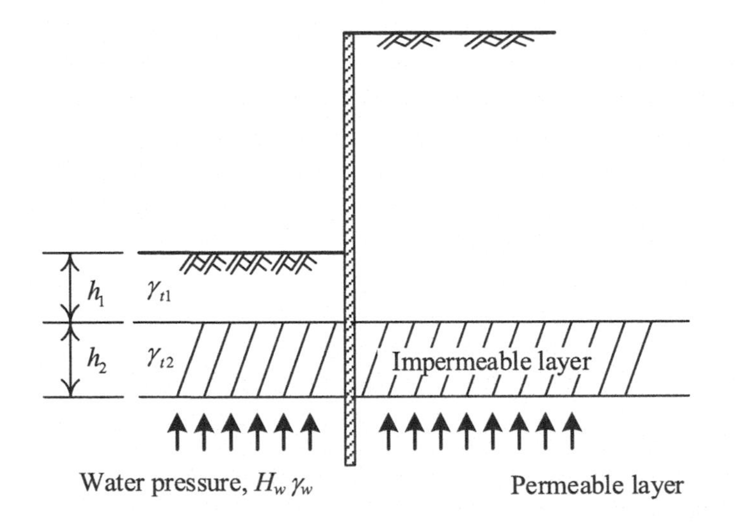

Figure 6.11 Stability analysis against upheaval failure.

Solution

Since the pore water pressure is in the hydrostatic state, the pore water pressure at a depth of GL-37.0 m is:

$$u_w = 9.81 \times 37 = 363 \text{ kPa}$$

The total vertical pressure above the permeable layer is:

$$\sigma_v = 18 \times 22 = 396 \text{ kPa}$$

The factor of safety against upheaval failure is:

$$F_{up} = \frac{396}{363} = 1.09$$

6.3.3 Sand boiling

If a retaining wall in sandy soil that has good permeability does not penetrate into the impermeable soil, dewatering in an excavation will cause a pressure difference behind the wall and in front of the wall, thereby inducing seepage. When the upward seepage water pressure is greater than or equal to the effective stress of the soil inside the excavation, the soil strength drops to 0 and cannot resist the earth pressure on the back of the wall, resulting in excavation failure. This phenomenon is called sand boiling. The safety of the excavation in sandy soil must be examined against sand boiling.

Commonly used methods for the analysis of sand boiling include Harza's method and Terzaghi's method. Figure 6.12 shows a watertight retaining wall, in which the maximum hydraulic

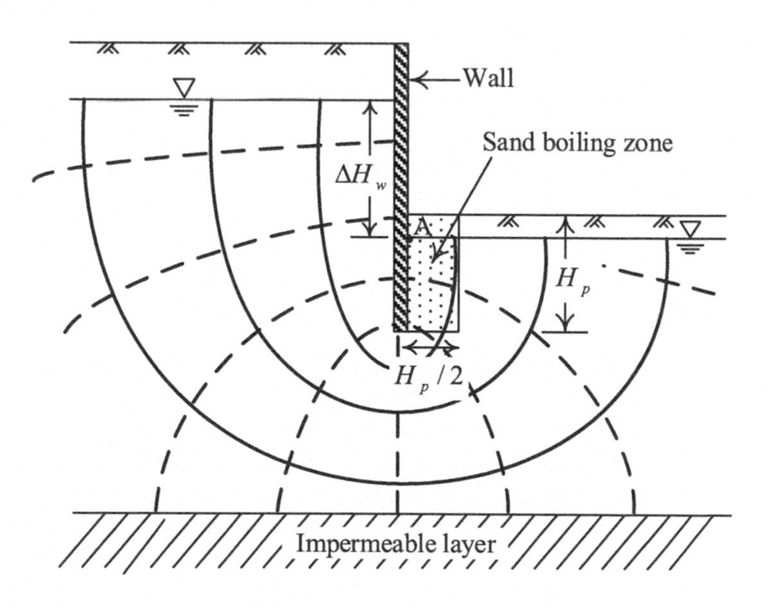

Figure 6.12 Stability analysis against sand boiling.

gradient at the exit of seepage normally occurs at the location near the wall, that is, point A. When the hydraulic gradient at point A is equal to or close to the critical value, sand boiling occurs. Harza (1935) calculates the safety factor against sand boiling as:

$$F_s = \frac{i_{cr}}{i_{\max(exit)}} \tag{6.15}$$

where i_{cr} is the critical hydraulic gradient and $i_{\max(exit)}$ is the maximum "exit" hydraulic gradient, which can be obtained with the flow net method or numerical analysis.

According to the phase relationship, the critical hydraulic gradient can be calculated as:

$$i_{cr} = \frac{\gamma'}{\gamma_w} = \frac{G_s - 1}{1+e} \tag{6.16}$$

Since the G_s value of sand is approximately 2.65 and its e value is between 0.57 and 0.95, the critical hydraulic gradient for most sands is close to 1 according to the preceding equation.

Terzaghi (1922) found from the results of many model tests with watertight sheet piles that sand boiling often occurs within a distance of approximately $H_p/2$ from the sheet piles (H_p refers to the penetration depth of the sheet piles). Thus, to analyze the stability of sheet piles against sand boiling, we can take the soil prism $H_p \times H_p/2$ in front of the sheet pile as the analytic object, as shown in Figure 6.12. The uplift force throughout the soil prism is:

$$U = \text{(the volume of the soil prism)} \times (i_{avg}\gamma_w) = \frac{1}{2}H_p^2 i_{avg}\gamma_w \tag{6.17}$$

where i_{avg} is the average hydraulic gradient throughout the soil prism. The downward force of the soil prism (i.e., the submerged weight) is:

$$W' = \frac{1}{2}H_p^2(\gamma_{sat} - \gamma_w) = \frac{1}{2}H_p^2\gamma' \tag{6.18}$$

Therefore, the factor of safety is:

$$F_s = \frac{W'}{U} = \frac{\gamma'}{i_{avg}\gamma_w} \tag{6.19}$$

According to Pratama et al. (2019), the required F_s for medium sand and very dense sand are 2.0 and 1.6, respectively, for Harza's method and 1.75 and 1.40, respectively, for Terzaghi's method. As the computed factor of safety is insufficient, sand boiling is likely to occur. Possible remedial measures are increasing the wall penetration depth, penetrating the wall into the impermeable layer, and lowering the water level behind the wall.

Example 6.6

Same as example 6.1. Compute the factor of safety against sand boiling using Harza's and Terzaghi's methods.

Solution

Figure EX 6.6a shows the flow net with 6 flow channels and 20 equipotential drops.
The critical hydraulic gradient is calculated as:

$$i_{cr} = \frac{\gamma'}{\gamma_w} = \frac{20-9.81}{9.81} = 1.039$$

The maximum hydraulic gradient at the exits is the one adjacent to the wall. The head loss at the exit is estimated as:

$$\Delta h = \frac{10}{20} = 0.5 \quad , \quad i_{\max(\text{exit})} = \frac{\Delta h}{L} = \frac{0.5}{1.54} = 0.325$$

With Harza's method:

$$F_s = \frac{i_{cr}}{i_{\max(\text{exit})}} = \frac{1.039}{0.325} = 3.20$$

With Terzaghi's method, the soil prism with the potential sand boiling zone has a cross section of 8 m × 4 m. Figure EX 6.6b shows the soil prism on an enlarged scale. By using the flow net, we can calculate the head loss across the prism as follows:

Along section *ba*, the head loss $= \frac{7}{20} \times 10\text{m} = 3.5$ m.

Along section *cd*, the head loss $= \frac{3.8}{20} \times 10\text{m} = 1.9$ m.

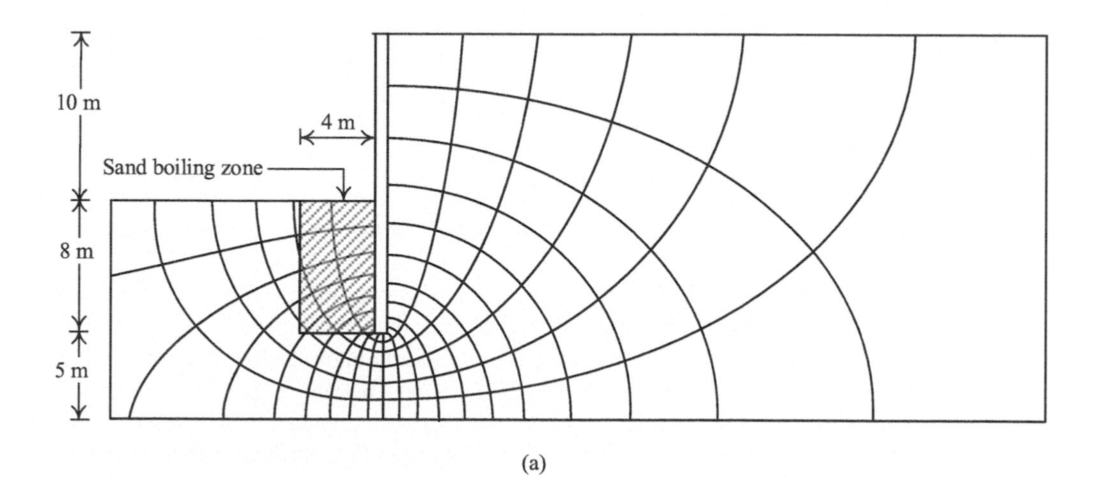

Figure EX 6.6a Analysis of sand boiling for Figure EX 6.1: (a) flow net and (b) soil prism.

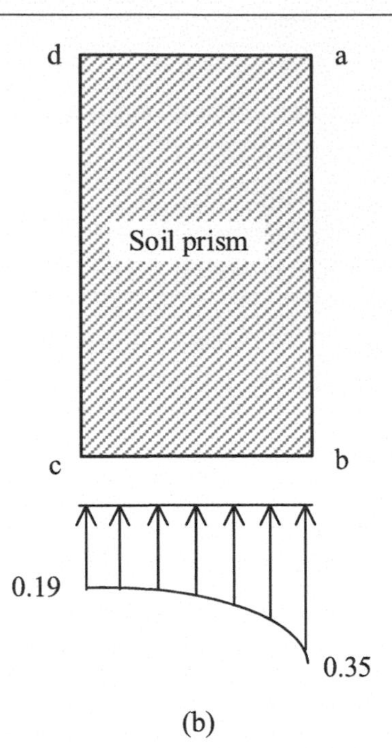

Figure EX 6.6b (Continued)

For other intermediate sections along *bc*, the approximate head loss can be similarly calculated. The average value of the head loss throughout the entire soil prism is 0.27 ′ 10 = 2.7 m, and the average hydraulic gradient is:

$$i_{avg} = \frac{2.7}{H_p} = 0.3375$$

$$F_s = \frac{\gamma'}{i_{avg}\gamma_w} = \frac{20 - 9.81}{0.3375 \times 9.81} = 3.08$$

6.4 Stress analysis

A typical supported retaining system is a highly indeterminate structure. The stress analysis can resort to finite element analysis or the beam–spring model. Basically, the finite element method with advanced soil models can obtain the necessary data for design, including the strut load, bending moment of the wall, wall deflection, ground settlement, and even the factor of safety against base shear failure. The beam–spring model assumes the retaining wall to be a beam supported by elastic soil. As shown in Figure 6.13, the interaction between the retaining wall and adjacent soils is simulated by a series of elastic soil springs. The lateral strut is also simulated by springs. Numerical analysis can then be conducted by simulating each construction activity. With the beam–spring model, one can obtain the strut load, bending moment, and shear force of the wall, but not the ground settlement. The detailed finite element method and the beam–spring

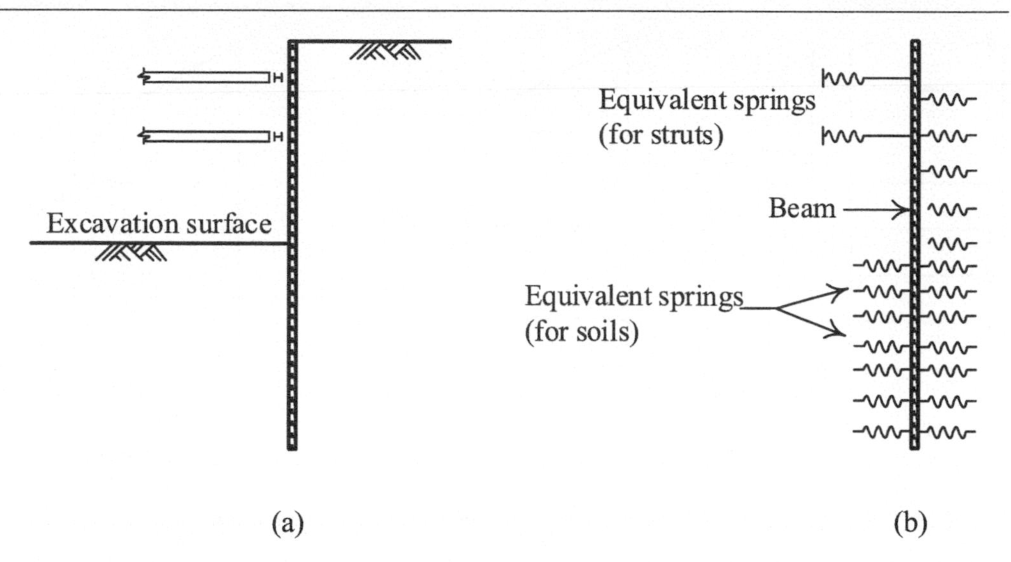

Figure 6.13 Beam–spring model: (a) profile of an excavation and (b) a beam–spring model.

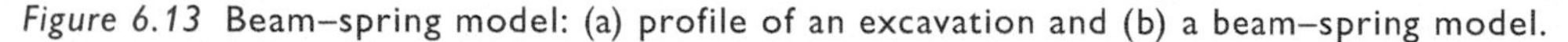

model can be found in Ou (2022). Considering the page limit, this section presents stress analysis using simple structural mechanics only.

6.4.1 Struts

Figure 6.14 shows diagrams of the apparent earth pressure as proposed by Peck (1969). As shown in the figure, when the soil in the back of the wall consists of mainly sandy soils, the apparent earth pressure, p_a, is:

$$p_a = 0.65\gamma H_e K_a \tag{6.20}$$

where:

γ = unit weight of sandy soils.
ϕ' = effective friction angle.
H_e = excavation depth.
K_a = coefficient of Rankine's earth pressure = $\tan^2(45° - \phi'/2)$.

The effective stress analysis method should be adopted for sand in eq. 6.20.

If the soil in the back of the wall is soft to medium clayey soil (i.e., $\gamma H_e / s_u > 4$), the apparent earth pressure, p_a, is the larger of

$$p_a = K_a \gamma H_e \text{ or } p_a = 0.3\gamma H_e \tag{6.21}$$

$$K_a = 1 - m\frac{4s_u}{\gamma H_e} \tag{6.21a}$$

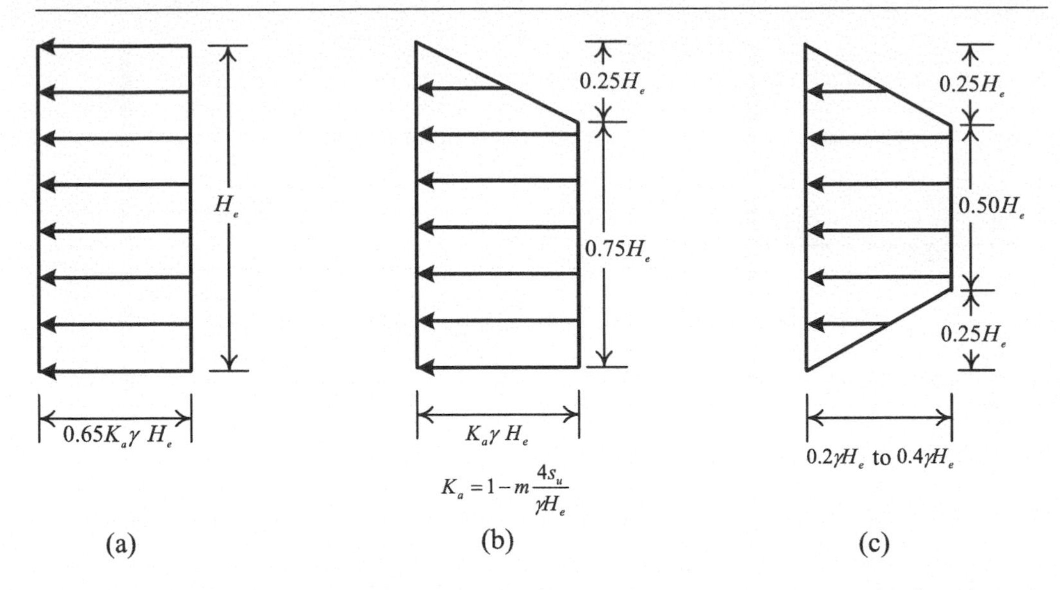

Figure 6.14 Apparent earth pressure diagram: (a) sand, (b) soft to medium clay $(\gamma H_e/s_u > 4)$, and (c) stiff clay $(\gamma H_e/s_u \leq 4)$.

where:

K_a = coefficient of earth pressure, $K_a = (1 - m\frac{4s_u}{\gamma H_e})$.
m = parameter considering a possible large deformation of the wall below the excavation bottom.
s_u = average undrained shear strength of clayey soil over the depth of excavation.

When $N_b = \gamma H_e / s_{ub} \leq 5.14$, m = 1.0 (Terzaghi et al., 1996), where s_{ub} is the undrained shear strength at the base of the excavation, that is, the undrained shear strength of the soil between the excavation bottom and the influence depth of the excavation. However, when $N_b > 5.14$, a large displacement at the bottom of the retaining wall is expected, m = 0.4 (Peck et al., 1977). Eq. 6.21 should be used based on the total stress analysis, that is, assume $\phi = 0$ without considering the pore water pressure.

If the soil in the back of the wall is stiff clayey soil ($\gamma H_e / s_u \leq 4$), the apparent earth pressure, p_a, is:

$$p_a = 0.2\gamma H_e \sim 0.4\gamma H_e \text{ (the average is } 0.3\,\gamma H_e) \quad (6.22)$$

Similarly, eq. 6.22 should be used based on total stress analysis. It does not consider the pore water pressure. In addition to Peck's diagrams of the apparent earth pressure, Peck et al. (1977), Terzaghi et al. (1996), and many other investigators have recommended similar types of diagrams.

For alternating layers of sandy and clayey soils, one can apply the concept of the equivalent cohesion and unit weight, respectively, to calculate the apparent earth pressure (Peck, 1943).

As shown in Figure 6.15a, where a sandy soil is above a clayey soil, the equivalent cohesion of alternating layers of sandy and clayey soils can be calculated as follows:

$$
\begin{aligned}
s_{u,eq} &= \frac{1}{H_e}\left[(K_s\gamma_s H_s \tan\delta)H_s/2 + H_c s_u\right] \\
&= \frac{1}{2H_e}\left[\gamma_s K_s H_s^2 \tan\delta + 2H_c s_u\right]
\end{aligned}
\tag{6.23}
$$

where:

H_e = excavation depth.
γ_s = unit weight of sandy soil.
H_s = height of the sandy soil.
H_c = height of the clayey soil.
K_s = coefficient of lateral earth pressure.
ϕ_s = friction angle of sandy soil.
δ = friction angle between the wall and the soil.
s_u = undrained shear strength of clayey soil.

The equivalent unit weight of the alternating layers is:

$$
\gamma_{eq} = \frac{1}{H_e}\left[\gamma_s H_s + H_c\gamma_c\right] \tag{6.24}
$$

where γ_c is the unit weight of clay.

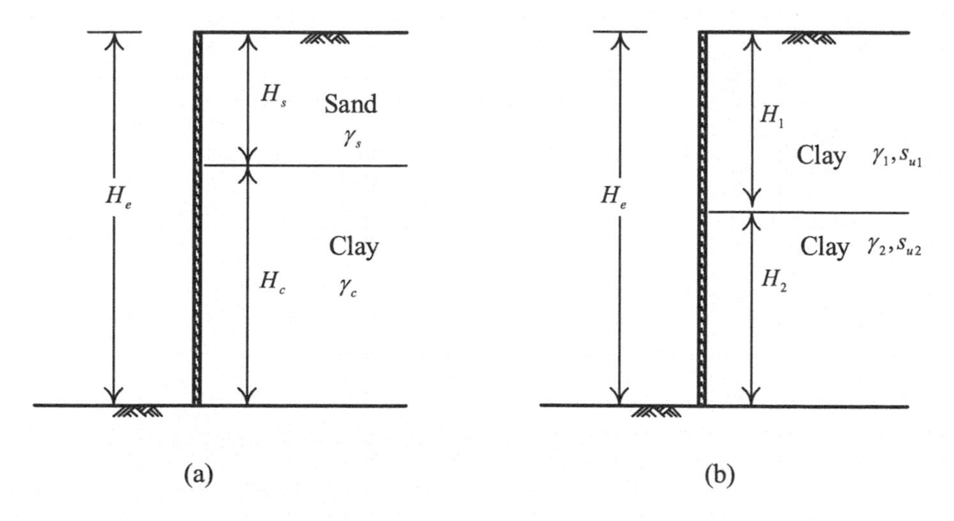

Figure 6.15 Multiple soil layers in excavations: (a) sand and clay and (b) clay.

Similarly, for layered clayey soils, the concept of equivalent values can also be used to calculate the strut load, as shown in Figure 6.15b. The equivalent cohesion of clayey soil is:

$$s_{u,eq} = \frac{1}{H_e}(s_{u1}H_1 + s_{u2}H_2) \tag{6.25}$$

where:

s_{u1} and s_{u2} = undrained shear strength of the first and second clayey layers, respectively.
H_1, H_2 = height of the first and second clayey layers, respectively.

The equivalent unit weight of clayey soil would be:

$$\gamma_{eq} = \frac{1}{H_e}(\gamma_1 H_1 + \gamma_2 H_2) \tag{6.26}$$

where γ_1, γ_2 is unit weight of the first and second clayey layers, respectively.

Providing the equivalent cohesion (or the undrained strength) and the equivalent unit weight are derived as earlier, Figure 6.14 can be used to choose a proper distribution of earth pressures. The strut load for each strut level can then be calculated using an appropriate simplified model. For example, the load on each level of struts can be the result of the earth pressure within the range covering half the vertical span between the upper and current struts and half the vertical span between the current and lower struts. This method can be referred to as the half method. The calculation procedure for the half method is shown in Figure 6.16.

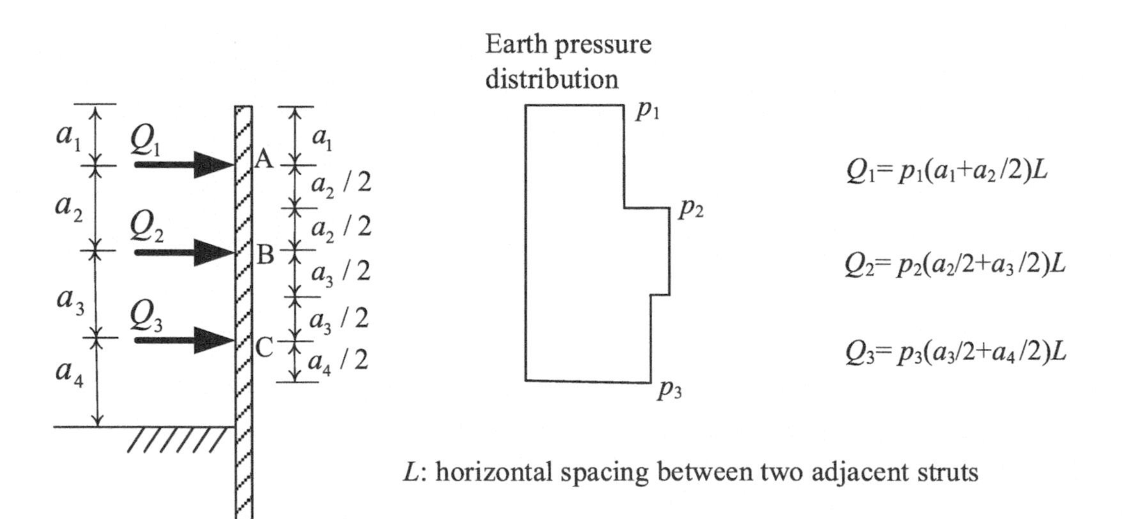

Figure 6.16 Computation of strut load.

6.4.2 Wales

The function of wales is to transfer the earth pressure acting on the retaining wall to the struts. For the purpose of analysis, the earth pressure can therefore be assumed to act on the wale directly. The earth pressure distribution can be obtained from the apparent earth pressure distribution. To compute the maximum bending moment and shear force of wales, the wales can be considered as simply supported beams with struts as supporting hinges or viewed as continuous beams, as shown in Figure 6.17. Details of the analysis can be found in Example 6.7.

If a wale is viewed as a simply supported beam, then:

$$M_{max} = \frac{1}{8} pL^2 \quad (6.27a)$$

$$Q_{max} = \frac{1}{2} pL \quad (6.27b)$$

If viewed as a continuous beam, then:

$$M_{max} = \frac{1}{12} pL^2 \quad (6.28a)$$

$$Q_{max} = \frac{1}{2} pL \quad (6.28b)$$

where:

M_{max} = maximum bending moment of the wale.
Q_{max} = maximum shear of the wale.
L = distance between struts.
p = earth pressure acting on the wall, which can be derived from the apparent earth pressure diagram (Figure 6.14) or can be transformed from the strut load (please see example 6.7).

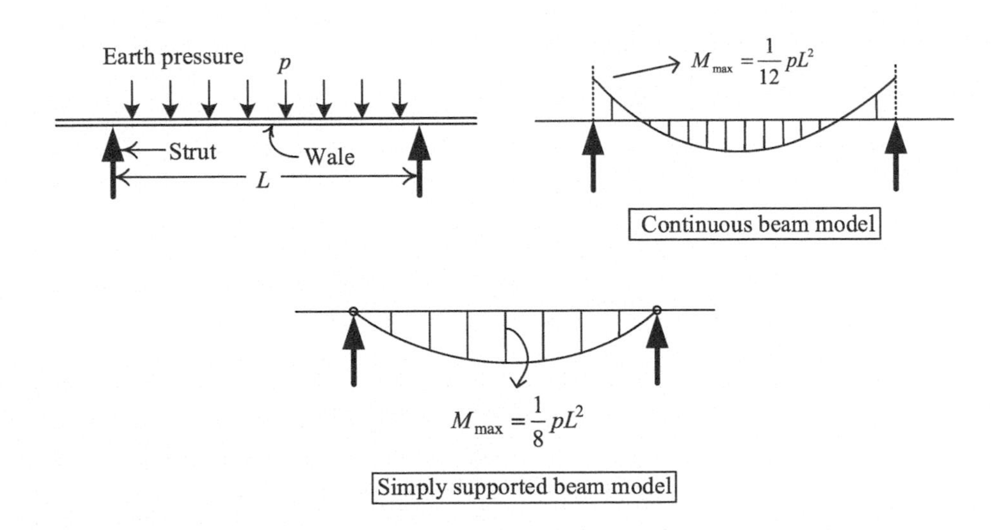

Figure 6.17 Computation of the bending moment of wales.

If a simply supported beam is assumed, it may not work out as economically. Nevertheless, if a continuous beam model is assumed, it tends to not be conservative. The real condition should be somewhere between a continuous beam and a simply supported beam. Further discussion can be found in Ou (2022).

6.4.3 Retaining walls

The assumed support method is a commonly used stress analysis method for supported retaining walls. Theoretically, the earth pressure on the back of the wall should be the active earth pressure rather than the apparent earth pressure. Therefore, in the computation of the wall bending moment and shear force, the active earth pressure is used. To simplify analysis, struts are assumed to be fulcrums, and the reaction force of the soil in front of the wall can be viewed as a concentrated force or roller, which is called the assumed support. When the moment formed by the active earth pressure and the moment formed by the passive earth pressure are balanced, the action point of the resultant passive earth pressure is the assumed support, as shown in Figure 6.18. The location of the assumed support (ℓ) can be determined as:

$$\ell = \frac{P_a \ell_a}{P_p} - s \tag{6.29}$$

After determining the distribution of earth pressures and location of the assumed support, the retaining wall can be seen as a simply supported beam. Figure 6.19 shows the computed wall bending moment and shear force using the assumed support method. The wall bending moment and shear force can also be computed using another beam model (Figure 6.20), but the active earth pressure should be used rather than the apparent earth pressure. Details of the analysis can be found in example 6.7.

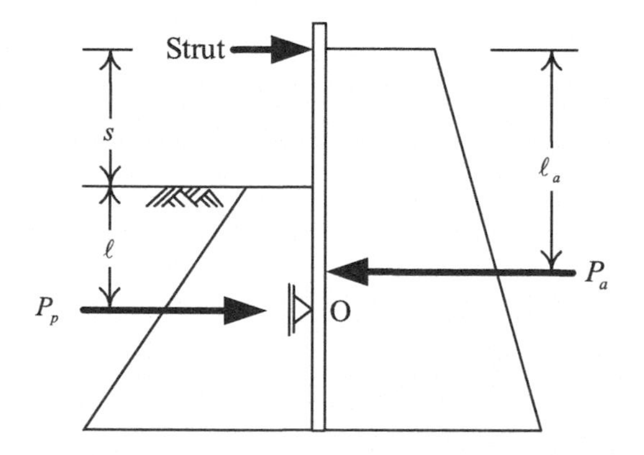

Figure 6.18 Estimation of the location of the assumed support.

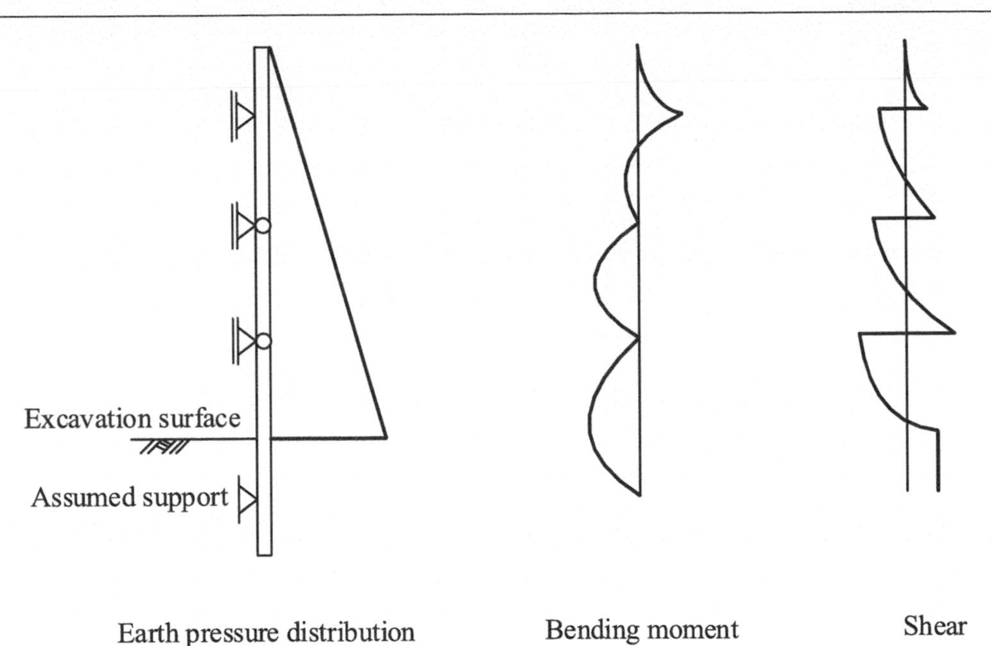

Figure 6.19 Computation of the wall bending moment and shear force.

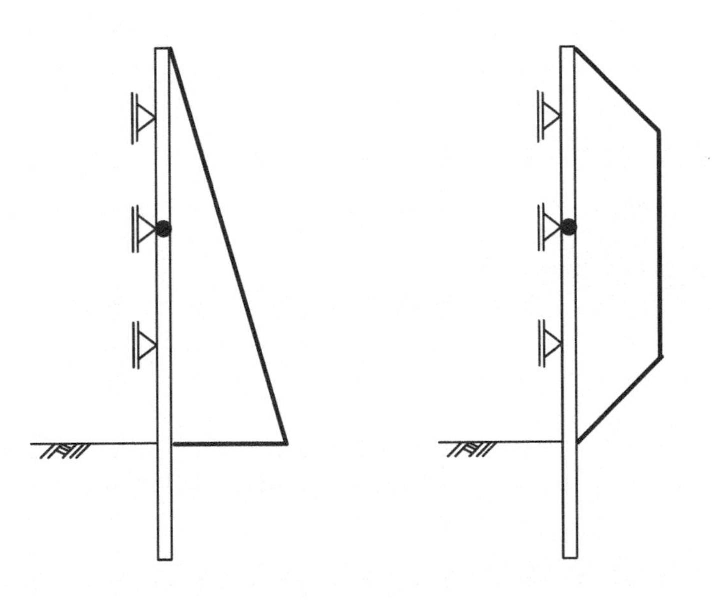

Figure 6.20 Alternative computation of the wall bending moment.

Example 6.7

Same as example 6.2. Assume that the sheet pile is used as the retaining wall. The average horizontal spacing between struts is 5 m. The allowable stress of the steel is $\sigma_{all} = 170 \times 10^3$ kN/m².

1 Draw the earth pressure envelope.
2 Compute the strut loads at the first, second, and third levels.
3 Determine the section modulus required for the wale at the second level.
4 Determine the section modulus required for the sheet pile.

Solution

1 Draw the earth pressure envelopes.

According eqs. 6.21 and 6.22, the average values (0–15 m) are:

$s_{u(avg)} = 34$ kPa, $\gamma_{sat\,(avg)} = 18.6$ kN/m³

$$N_b = \frac{\gamma H_e}{s_{ub}} = \frac{15\times3+20\times9+18\times3}{60} = 4.65$$

Since $N_b > 5.14$, $m = 1.0$.

$$K_a = 1 - m\frac{4s_u}{\gamma H_e} = 1 - \frac{4\times34}{18.6\times15} = 0.51$$

$$p_a = K_a\gamma H_e = 0.51\times18.6\times15 = 142 \text{ kN/m}^2$$

Alternatively, $p_a = 0.3\gamma H_e = 84$ kN/m².

Consider $p_a = 142$ kN/m².

The apparent earth pressure is shown in Figure EX 6.7a.

2 The strut load.

With the half method:

$$Q_1 = (3+1.5+0.75)\times0.5\times142\times5 = 1864 \text{ kN}$$
$$Q_2 = (1.5+1.5)\times142\times5 = 2130 \text{ kN}$$
$$Q_3 = (1.5+1.5)\times142\times5 = 2130 \text{ kN}$$
$$Q_4 = (1.5+1.5)\times142\times5 = 2130 \text{ kN}$$

3 Section modulus for the wale.

The strut load at the second level of the strut $Q_2 = 2130$ kN. The strut spacing $L = 5$ m. The uniform pressure acting on the wale $p = 2{,}130/L = 426$ kN/m.

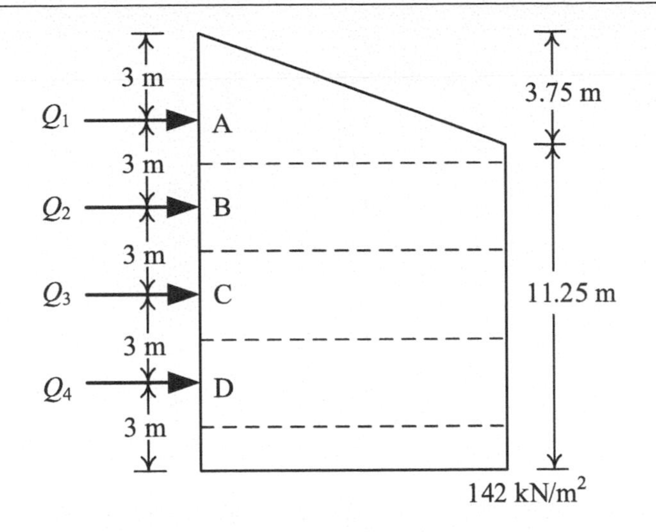

(a)

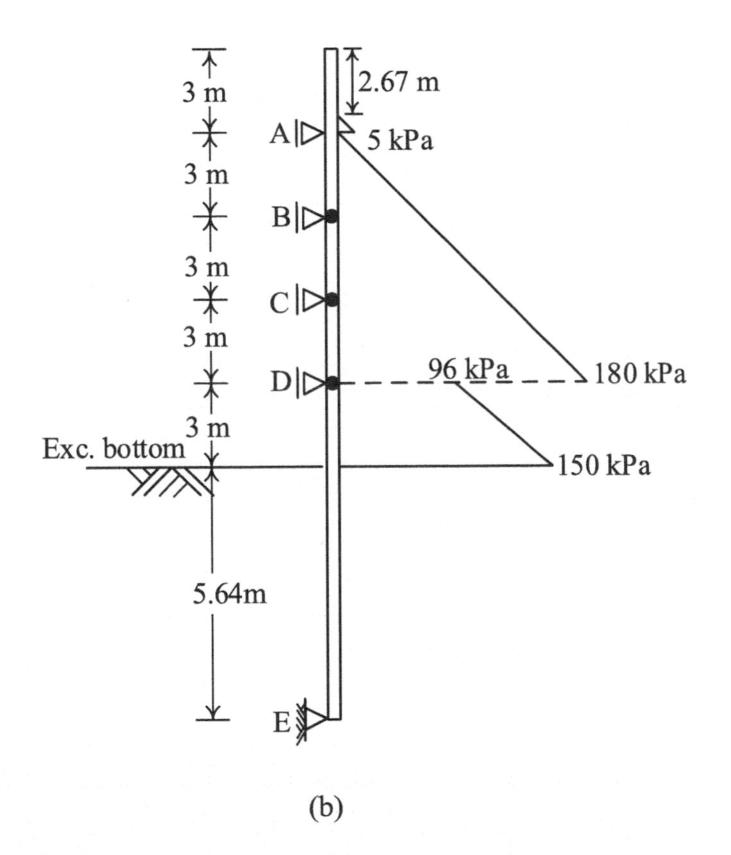

(b)

Figure EX 6.7 Design of various components of the retaining support system for Figure EX 6.2: (a) apparent earth pressure diagram, (b) analysis of bending moment and shear force using the assumed support method, and (c) calculation of the location of zero shear force.

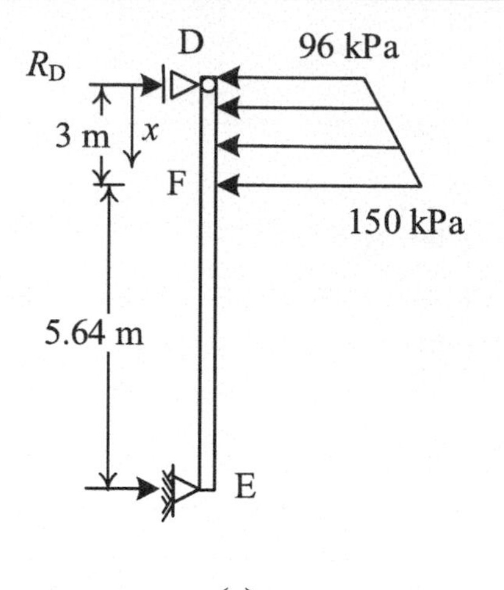

Figure EX 6.7 (Continued)

If the simply supported model is adopted:

$$M_{max} = \frac{1}{8}pL^2 = \frac{1}{8} \times 426 \times 5^2 = 1331 \text{ kN-m}$$

$$Q_{max} = \frac{1}{2}pL = \frac{1}{2} \times 426 \times 5 = 1065 \text{ kN}$$

The section modulus $S = \dfrac{M_{max}}{\sigma_{all}} = \dfrac{1331}{170 \times 1000} = 7.83 \times 10^{-3} \text{ m}^3$.

If the continuous end beam model is employed:

$$M_{max} = \frac{1}{12}pL^2 = \frac{1}{12} \times 426 \times 5^2 = 888 \text{ kN-m}$$

The section modulus $S = \dfrac{M_{max}}{\sigma_{all}} = \dfrac{888}{170 \times 1000} = 5.22 \times 10^{-3} \text{ m}^3$.

4 Section modulus for the sheet pile.

Use the Rankine's earth pressure to calculate the bending moment and shear force of the wall $K_a = K_p = 1$.

Depth of the tension crack, $z_c = \dfrac{2c_u}{\gamma_{sat}} = \dfrac{2 \times 20}{15} = 2.67 \text{ m}$.

Active side:

At GL-2.67 m, $\sigma_a = 0$.

At GL-3.0 m$^-$, $\sigma_a = \sigma_v K_a - 2c_u\sqrt{K_a} = (3\times15) - 2(20) = 5$ kPa.

At GL-3.0 m$^+$, $\sigma_a = \sigma_v K_a - 2c_u\sqrt{K_a} = (3\times20) - 2(30) = 0$.

At GL-12 m$^-$, $\sigma_a = \sigma_v K_a - 2c_u\sqrt{K_a} = (12\times20) - 2(30) = 180$ kPa.

At GL-12 m$^+$, $\sigma_a = \sigma_v K_a - 2c_u\sqrt{K_a} = (12\times18) - 2\times60 = 96$ kPa.

At GL-15 m, $\sigma_a = \sigma_v K_a - 2c_u\sqrt{K_a} = (15\times18) - 2\times60 = 150$ kPa.

At GL-30 m $\sigma_a = \sigma_v K_a - 2c_u\sqrt{K_a} = (30\times18) - 2\times60 = 420$ kPa.

Passive side:

At GL-15 m, $\sigma_p = \sigma_v K_p + 2c_u\sqrt{K_p} = (0\times18) + 2\times60 = 120$ kPa.

At GL-30 m, $\sigma_p = \sigma_v K_p + 2c_u\sqrt{K_p} = (15\times18) + 2\times60 = 390$ kPa.

According to the method introduced in Figure 6.18, the location of the assumed support can be estimated as:

$$\ell = \frac{P_a \ell_a}{P_p} - s = 5.64 \text{ m}$$

Figure EX 6.7b shows the earth pressure distribution that will be used for the analysis of the wall bending moment with the assumed support method. The maximum bending moment will occur at the location of the shear force equal to zero. Figure EX 6.7b shows that zero shear force should occur within span D–E. As shown in Figure EX 6.7c, the reaction force at point D is $R_D = 300$ kN/m.

The shear force at distance x from point D within span D–F is calculated as:

$$V_x = 300 - 96x - \frac{1}{2}\left[\frac{(150-96)x}{3}\right]x$$

When $V_x = 0, x = 2.53$ m.

The maximum bending moment at $x = 2.53$ m is calculated to be 404 kN-m.

The section modulus $S = \frac{M_{max}}{\sigma_{all}} = \frac{404}{170\times1000} = 2.38\times10^{-3}$ m^3.

The size and type of the sheet pile can then be selected according to the product manual.

6.5 Deformation analysis

The objective of deformation analysis is to analyze the deflection of the retaining wall and the ground settlement caused by excavation to ensure the safety of the retaining wall and adjacent building around the excavation. The items of deformation analysis include the lateral deflection of the retaining wall and ground settlement.

6.5.1 Wall deflection

Two types of deflection of the retaining wall usually occur in excavations: cantilever deflection and deep inward deflection, as shown in Figure 6.21. The cantilever deflection usually occurs in stiff soil or excavations with weak strut stiffness, and the maximum deflection occurs at the top of the retaining wall. The deep inward deflection normally occurs in soft clay or excavation with high strut stiffness, and the maximum deflection occurs at a certain depth of the retaining wall, usually near the excavation surface. To accurately predict the deflection pattern of the retaining wall, it must be analyzed by the finite element method or a beam–spring model (Ou, 2022). In this section, only the maximum wall deflection is discussed.

For soft to medium stiff clay, the maximum deflection of the retaining wall can be obtained by Clough and O'Rourke's chart (1990), as shown in Figure 6.22. The parameters used to derive the maximum wall deflection are the safety factor against base shear failure or basal heave and the system stiffness of the retaining wall. The safety factor must be calculated by Terzaghi's method (eq. 6.4 or 6.5). The system stiffness is a dimensionless parameter that is defined as:

$$S_w = \frac{EI}{\gamma_w h_{avg}^4} \tag{6.30}$$

here γ_w is the unit weight of water, EI is the stiffness of the retaining wall, and h_{avg} is the average vertical spacing of the struts.

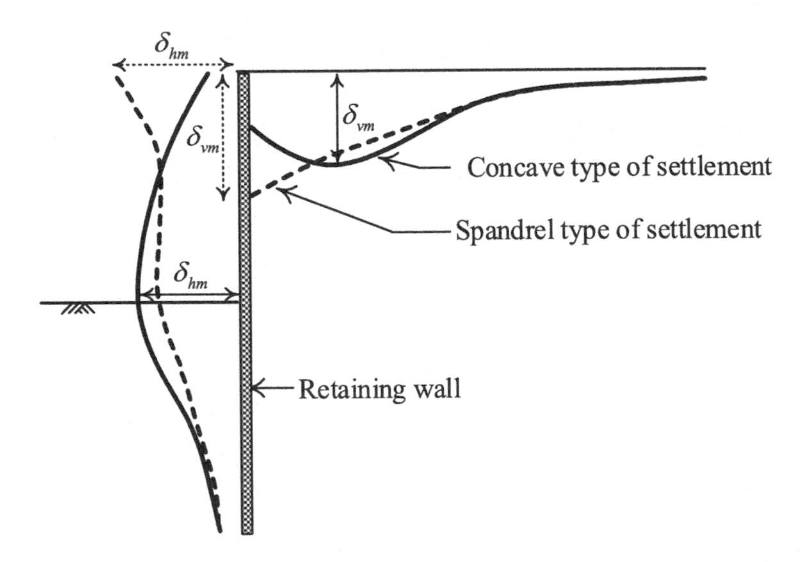

Figure 6.21 Types of wall deflection and ground surface settlement.

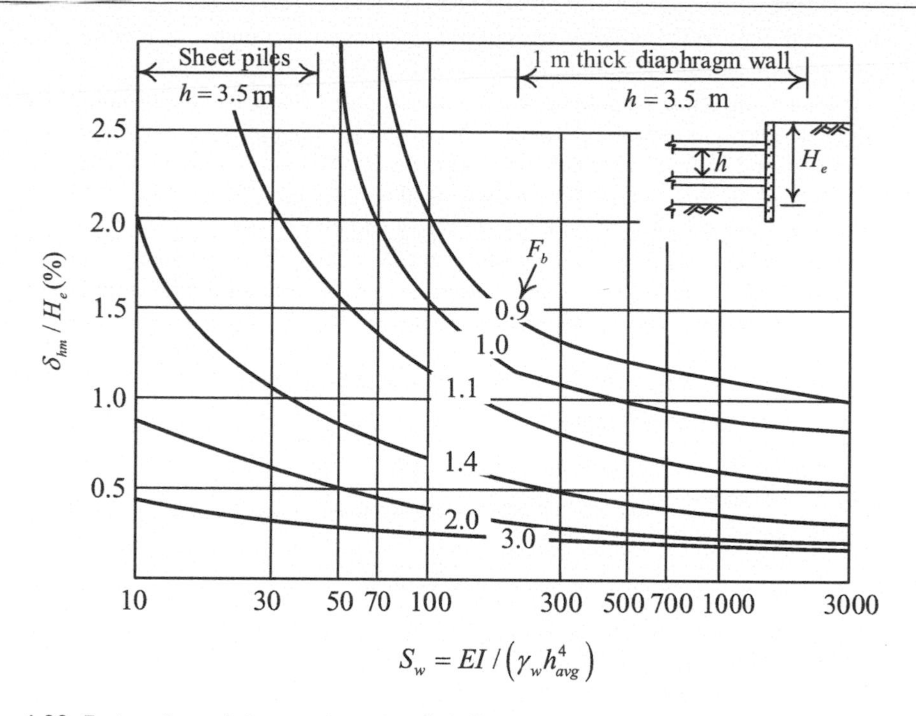

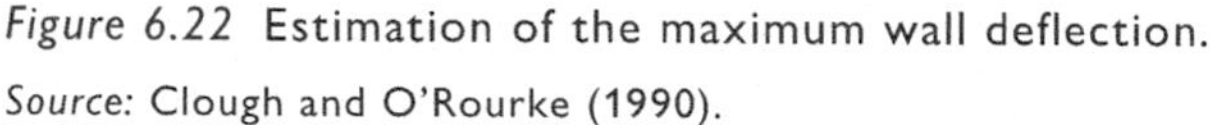

Figure 6.22 Estimation of the maximum wall deflection.
Source: Clough and O'Rourke (1990).

For sandy soil, the maximum deflection of the retaining wall can be estimated by an empirical equation $\delta_{hm} = (0.2-0.3)H_e$, where H_e is the excavation depth.

Notably, the wall deflection is also affected by the excavation width. The greater the excavation width, the greater the unloading force, which in turn causes greater wall deflection. However, Clough and O'Rourke's chart does not take the excavation width into account. Based on the authors' experience, Clough and O'Rourke's chart is suitable for excavation widths of approximately 40 m. If the excavation width is different from $B = 40$ m, the maximum wall deflection obtained from Figure 6.22 can be modified based on the fact that the maximum wall deflection is inversely proportional to the excavation width because the soil inside the excavation has unloading behavior, that is, elastic behavior.

6.5.2 Ground settlement

The shapes or types of ground surface settlement engendered by excavation can be categorized into the spandrel type and the concave type, as shown in Figure 6.21 (Hsieh and Ou, 1998). The main factors responsible for these two types of ground surface settlements are the magnitude and shape of the deflection of a retaining wall.

The spandrel type of settlement is more likely to relate to the cantilever type of wall deflection, while the concave type is more related to the deep inward type of wall deflection. Under normal construction conditions, excavation in soft clay will produce more wall deflection, which tends to bring about deep inward movement, leading to the concave type of settlement. Excavation in sandy soil or stiff clay, on the other hand, will produce less deflection of the retaining wall, and the spandrel type of settlement is more likely to occur (Ou and Hsieh, 2011).

Since the factors affecting the deflection of a retaining wall also affect the ground surface settlement, there should be a certain relationship between the maximum wall deflection (δ_{hm}) and the maximum ground surface settlement (δ_{vm}). Ou et al. (1993) found that $\delta_{vm} = (0.5 \sim 0.75)\delta_{hm}$ for most cases, with the lower limit for sandy soils, the upper limit for clays, and a limit somewhere between the two for alternating layers of sandy and clayey soils.

The ground surface settlement at the various distances behind the wall can be estimated using Peck's method (1969), which was derived from excavations with soldier piles and sheet piles. Peck's method classifies soil into three types according to the soil characteristics (Figure 6.23):

Type I: Sandy and soft to stiff clayey soil, with average workmanship
Type II (a): Very soft to soft clayey soil

1. Limited depth of clayey soil below the excavation bottom
2. Significant depth of clayey soil below the excavation bottom but with an adequate factor of safety against base shear failure, or $N_b < N_{cb}$

(b): settlements affected by construction difficulties

Type III: Very soft to soft clayey soil to a significant depth below the excavation bottom and with a low factor of safety against base shear failure, or $N_b \geq N_{cb}$.

where N_b, the stability number of the soil, is defined as $\gamma H_e / s_u$, where γ is the unit weight of the soil, H_e is the excavation depth, s_u is the undrained shear strength of the soil below the base of the excavation; N_{cb} is the critical stability number against base shear failure, corresponding to the factor of safety against base shear failure equal to 1. Under such conditions, $N_{cb} = 5.7$ (Terzaghi et al., 1996).

Clough and O'Rourke (1990) proposed various envelopes of settlements for different soils. Basically, excavations in sandy or stiff clayey soil tend to produce triangular (spandrel)

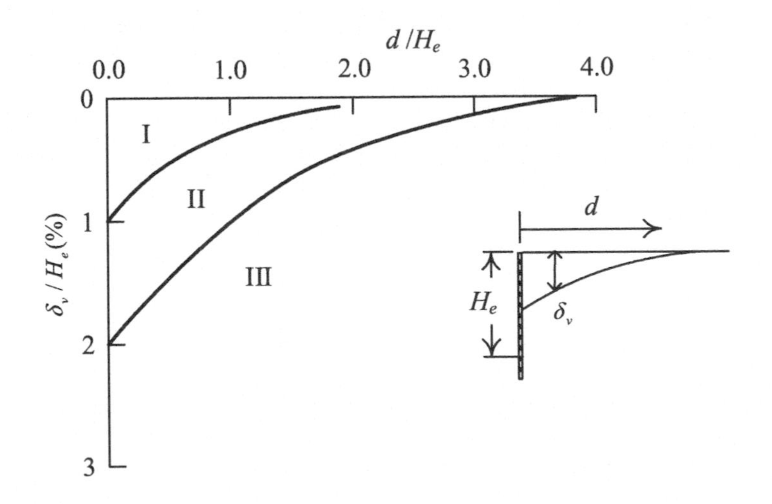

Figure 6.23 Peck's method for the estimation of ground surface settlement.
Source: Peck (1969).

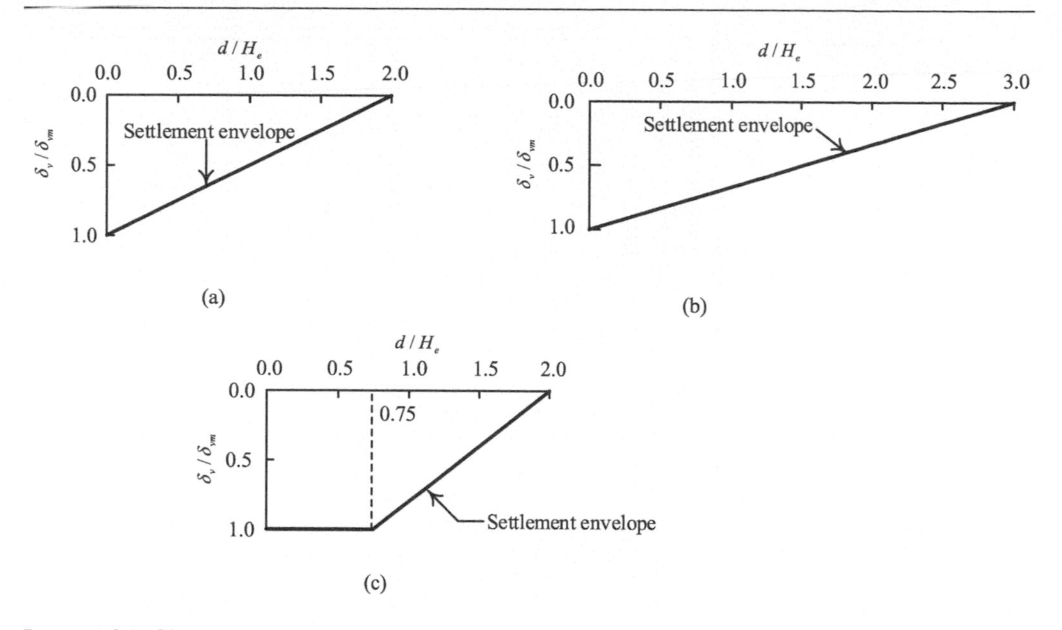

Figure 6.24 Clough and O'Rourke's method for the estimation of ground surface settlement: (a) sand, (b) stiff to very stiff clay, and (c) soft clay.

Source: Clough and O'Rourke (1990).

ground surface settlement. The maximum settlement will be found near the retaining wall. The envelopes of the ground surface settlement are shown in Figures 6.25a and 6.25b. Excavation in soft to medium clayey soil will generate a trapezoidal envelope of ground surface settlement, as shown in Figure 6.24c. However, as studied by Ou and Hsieh (2011), excavations in soft clayey soil normally produce the concave type of ground surface settlement, and the settlement predicted from Figure 6.24c is not very consistent with field observation, as shown in Figure 6.21. Regarding the prediction of spandrel and concave types of settlement profiles, please refer to Ou and Hsieh (2011).

6.6 Control of excavation-induced movement

Excessive wall deflections may damage the retaining wall itself, which in turn causes excessive ground settlements, leading to damage to adjacent buildings or public facilities. Wall deflections and/or ground settlement in excavations must be reduced to a tolerable value by improvement measures. Strengthening the bracing system and/or increasing the passive resistance are two commonly adopted improvement measures in engineering practice, and many successful case histories have been reported.

Soil improvement has been proven to be an effective way to increase the passive resistance of soil in front of the wall and therefore reduce the wall deflection and ground settlement. Figure 6.25 shows that many soil improvement piles are implemented in front of the wall to increase the passive resistance of the soil in front of the wall. Figure 6.26 shows a photo of an excavation with ground improvement piles. During excavation, both the improved soil and in

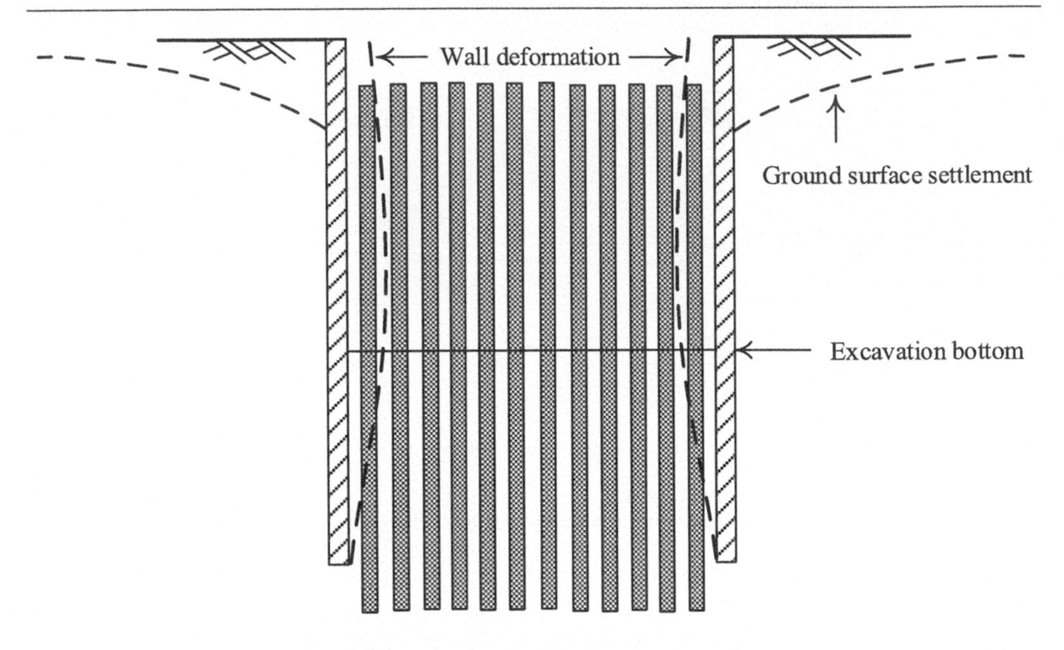

Figure 6.25 Nonoverlapping soil improvement piles in an excavation.

Figure 6.26 Photo of an excavation with ground improvement piles.

situ soil inside the excavation are excavated. Soil improvement can also be performed with overlapping improvement piles, which is certainly more effective to reduce movements but more expensive than nonoverlapping piles.

In addition, cross walls and buttress walls have been used extensively to reduce the wall deflection and ground settlement in excavations. Figure 6.27 shows a schematic arrangement

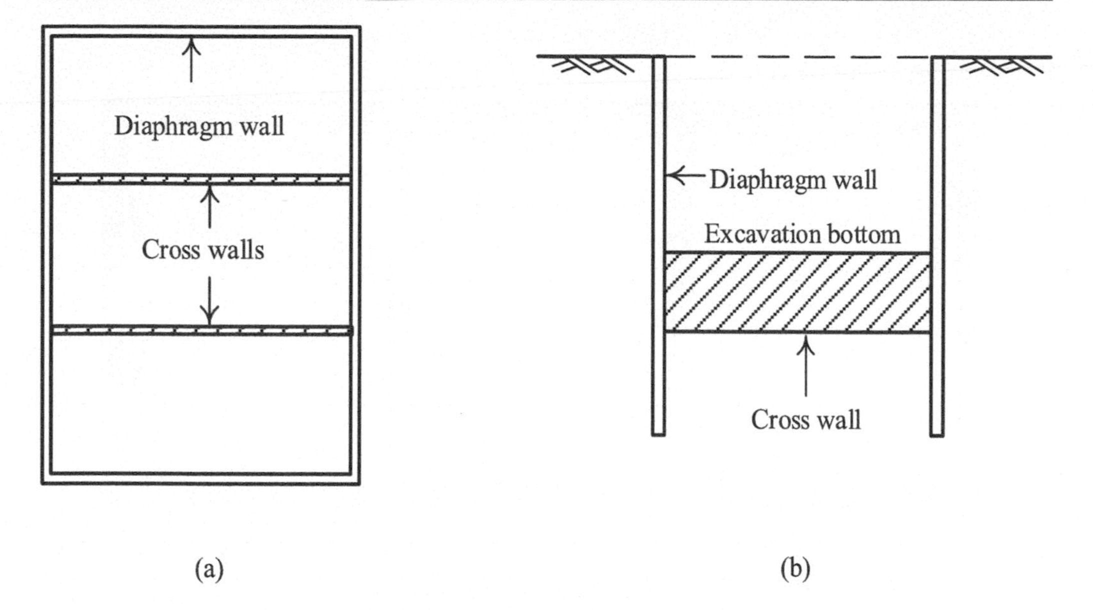

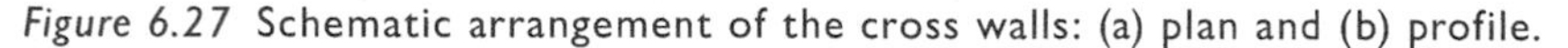

Figure 6.27 Schematic arrangement of the cross walls: (a) plan and (b) profile.

of the cross walls in excavations. As shown in this figure, prior to excavation, a concrete wall is constructed to connect two opposite retaining walls. The cross wall can be viewed as a lateral support or strut, which exists before excavation and bears a great deal of compressive strength. Therefore, the cross wall can reduce the lateral wall deformation and ground settlement to a very small value as long as suitable spacing and depth of the cross walls are designed. The case histories, performance, design, and analysis for cross walls have been reported in the literature (e.g., Ou et al., 2011; Hsieh et al., 2012; Hsieh et al., 2013; Ou et al., 2013). Figure 6.28 shows a possible wall deformation mode for an excavation with and without cross walls.

Moreover, the concrete wall is constructed with limited length, that is, not to the opposite side of the retaining wall. This concrete wall is called the buttress wall. Figure 6.29 shows a schematic arrangement of buttress walls in excavations. Buttress walls can be demolished step-by-step as excavation progresses; however, for safety reasons, they may be maintained until excavation is completed and then demolished. They may be maintained even after excavation and can be used as permanent structures or partition walls. Figure 6.30 shows a photo of an excavation with buttress walls where the buttress walls are maintained during excavation. Different treatments of buttress walls result in different mechanisms in reducing the wall deflection. In general, buttress walls are less expensive and less effective in reducing movements than cross walls. Figure 6.31 shows a possible deformation mode for an excavation with and without buttress walls. It is beyond the scope of this chapter to introduce those measures in detail. Interested readers are advised to refer to related references (e.g., Ou et al., 2011; Hsieh and Ou, 2018; Lim et al., 2019; Ou, 2022).

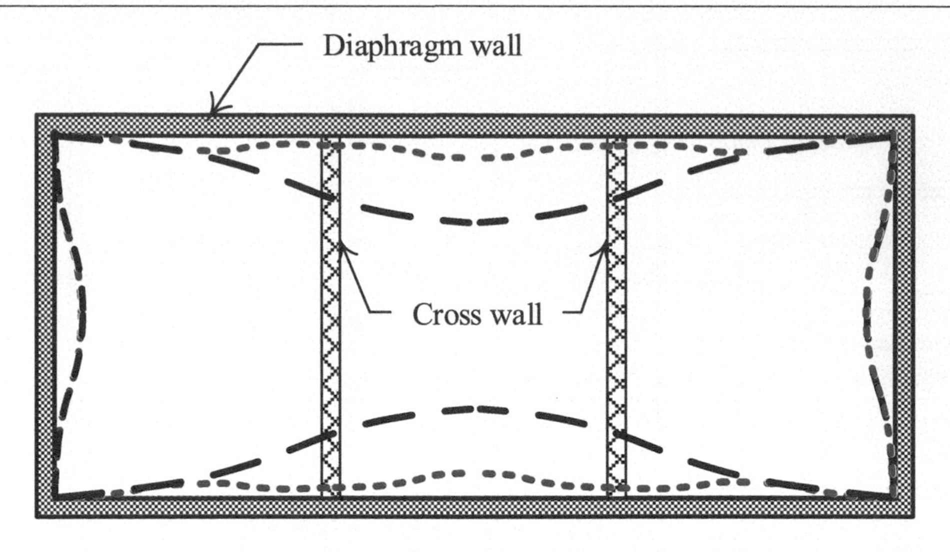

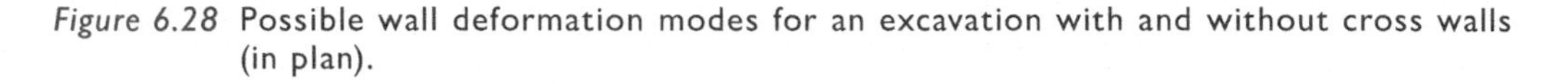

Figure 6.28 Possible wall deformation modes for an excavation with and without cross walls (in plan).

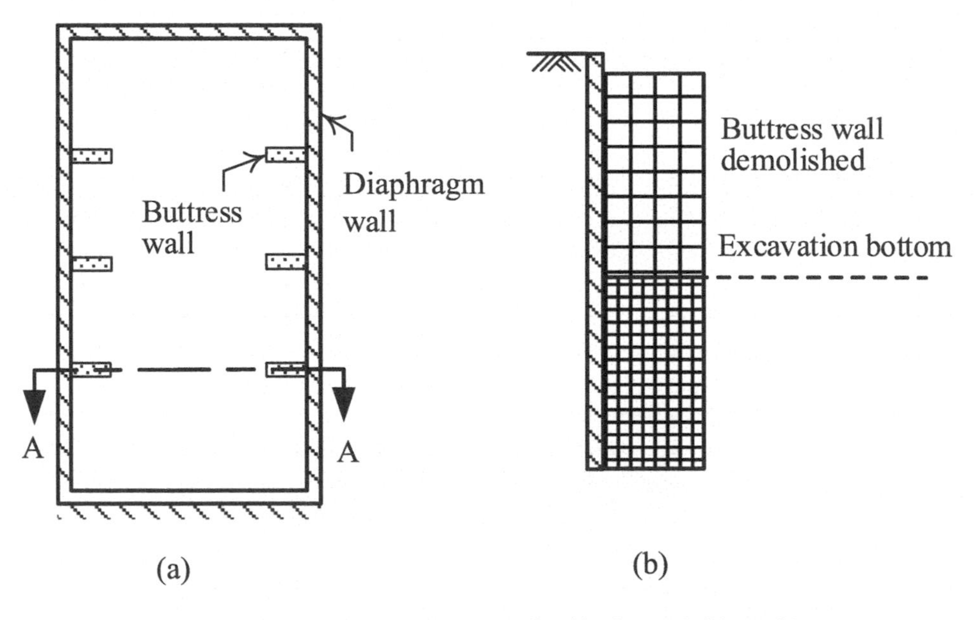

Figure 6.29 Schematic arrangement of buttress walls: (a) plan and (b) profile.

Figure 6.30 Photo of an excavation with buttress walls.

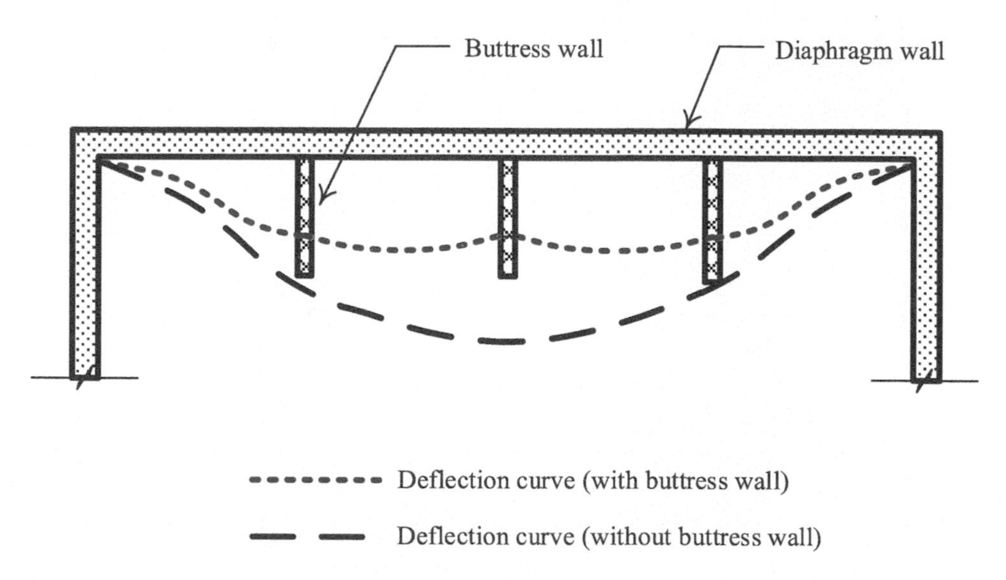

Figure 6.31 A possible wall deformation mode for an excavation with and without buttress walls.

6.7 Monitoring system

Although the technologies of analysis, design, and construction for excavations have advanced considerably, they are not capable of contending with all the changes or uncertainty in an excavation. A well-arranged monitoring system can be helpful to monitor the safety of excavations. Before a failure or damage occurs, there are signs, such as an extraordinary increase in the wall or soil displacement or stress. A monitoring system can issue an immediate warning to help engineers adopt effective measures to forestall a failure or damage when these signals appear.

6.7.1 Monitoring items

Monitoring field performance is basically the measurement of physical quantities of some objects, such as displacement, stress, and strain. The common monitoring items in excavations are (1) displacement of a structure or soil, (2) stress or strain of a structure or soil, and (3) pore water pressure and groundwater level.

The displacement of a retaining wall relates closely to the ground settlement or the building settlement. The inclinometer is a commonly used device for measuring the lateral displacement of retaining walls. An inclinometer is a device for measuring the displacement across a reference line in a ground. As shown in Figure 6.32, an inclinometer contains a tilt sensor inside a rigid tube. The tilt sensor measures the angle that the tube makes with the reference line in the vertical direction. As measurement is conducted, the inclinometer is placed inside a guide tube (known as an inclinometer casing) that is normally installed inside a borehole. An inclinometer

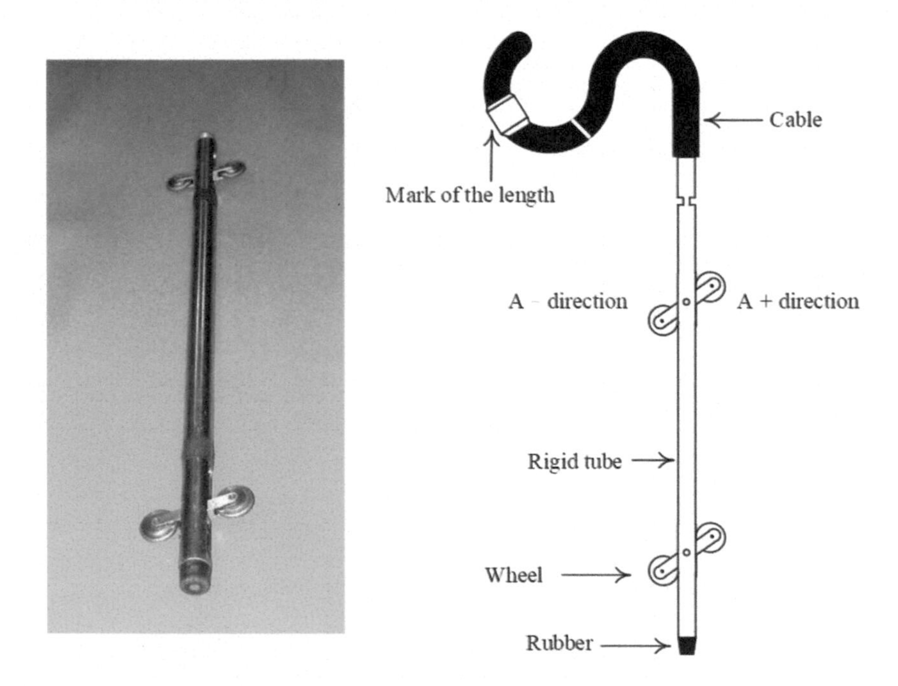

Figure 6.32 An inclinometer.

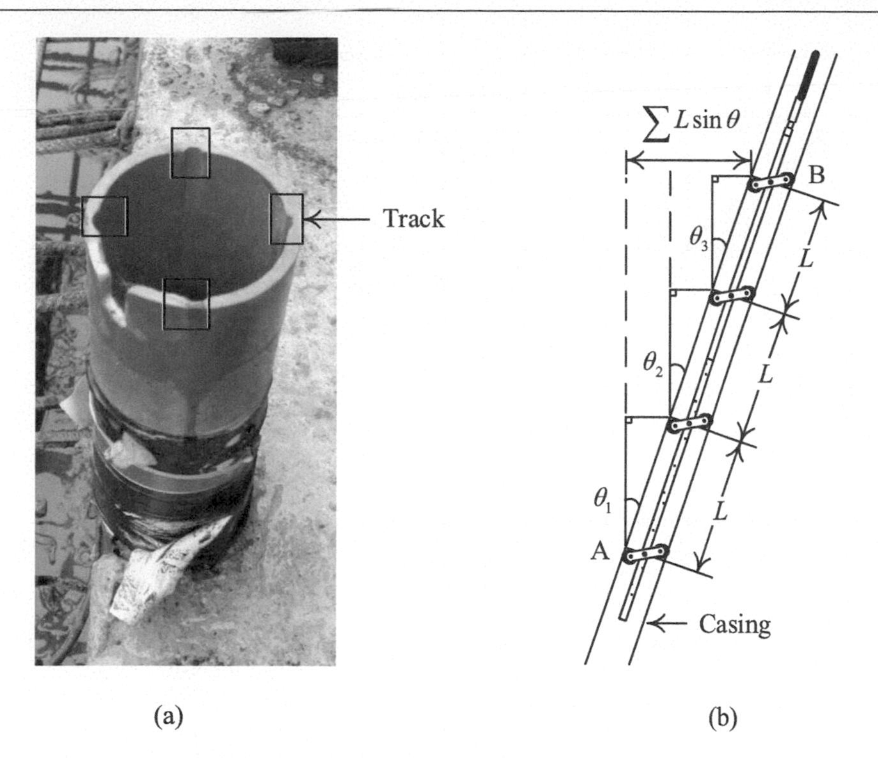

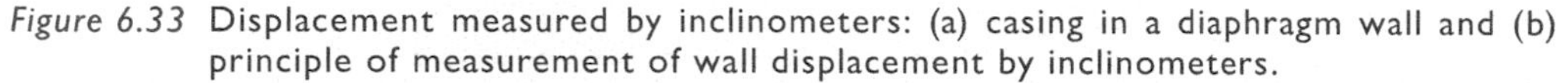
Figure 6.33 Displacement measured by inclinometers: (a) casing in a diaphragm wall and (b) principle of measurement of wall displacement by inclinometers.

casing has two pairs of internal tracks. Figure 6.33a shows a photo of an inclinometer casing installed in a diaphragm wall. To ensure accurate measurement, the space between the inclinometer casing and soil/concrete should be filled with a cement/bentonite grout. Measurement can be performed by running an inclinometer along a pair of tracks to measure the deviation at intervals equal to the length of the inclinometer. Figure 6.33b shows how the relative displacement between the top and bottom of an inclinometer casing is measured. Figure 6.34a shows the typical displacements for the diaphragm wall at various stages of excavation that were measured using an inclinometer in an excavation, namely, the TNEC excavation project (Ou et al., 1998).

For the ground settlement, the simplest method is the installation of steel nails or settlement nails and measuring the settlement with a level. Buildings will tilt as a result of ground settlement and thereby be damaged. A typical ground surface settlement that was measured with level and settlement nails is shown in Figure 6.34b.

The tilt of a building can be estimated by the relative settlement between two reference points with plane surveying, such as a level or theodolite. It can also be monitored using a tiltmeter, which provides the tilt angle. Figure 6.35 shows the photo of a datum plate that is fixed on a building or structure. Similar to an inclinometer, a tiltmeter contains a sensor that can measure the tilt of the datum plate or the building/structure.

The load on the struts must be monitored constantly during excavation so as not to exceed the allowable value and endanger the safety of the excavation. Strain gauges are widely used

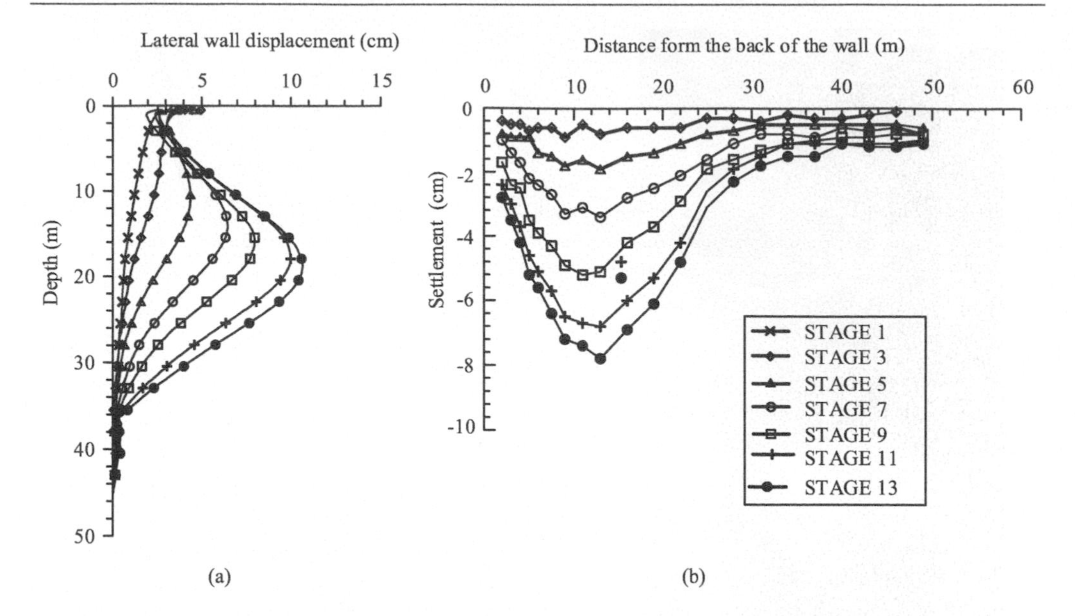

Figure 6.34 A typical measurement of movements in an excavation project, TNEC project: (a) lateral wall displacement and (b) ground surface settlement.

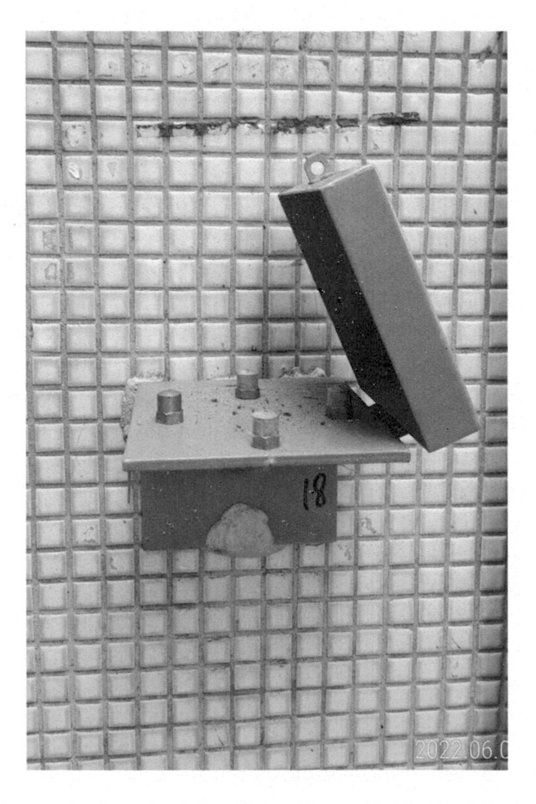

Figure 6.35 A datum plate of tiltmeters fixed on the wall of a building.

Figure 6.36 Photo of a strain gauge on a strut.

devices to measure strut loads. Figure 6.36 shows a photo of a strain gauge installed on the web of a strut.

Piezometers are usually installed in sand or gravel in excavations to measure the pore water pressure that may cause upheaval failure below the impermeable layer. The piezometer can also be used to monitor changes in water seepage during dewatering in an excavation. The instrument for measuring groundwater level is the water observation well, where a perforated standpipe is often used. A typical configuration for a piezometer and water observation well can be found in Figures 1.22 and 1.23, respectively, in Section 1.8.1.

In addition to the monitoring instruments mentioned previously, other types of monitoring instruments can also be employed in excavations. For example, crackmeters can be used to monitor the cracks on the walls or columns of buildings. The concrete stressmeter can be used to monitor the stress of concrete, earth pressure cells for the earth pressure on the retaining wall, and so on. Interested readers can refer to related literature for details (e.g., Ou, 2022).

6.7.2 Arrangement of the monitoring system

Monitoring items or devices should be set according to excavation characteristics. The frequency of measurements or readings should be determined based on the size of the excavation, soil conditions, uncertainty, and risk. It is beyond the scope to introduce these devices and measurements in detail. Interested readers are advised to refer to related references (e.g., Ou, 2022). An example of a monitoring system in an excavation project, namely, the TNEC excavation

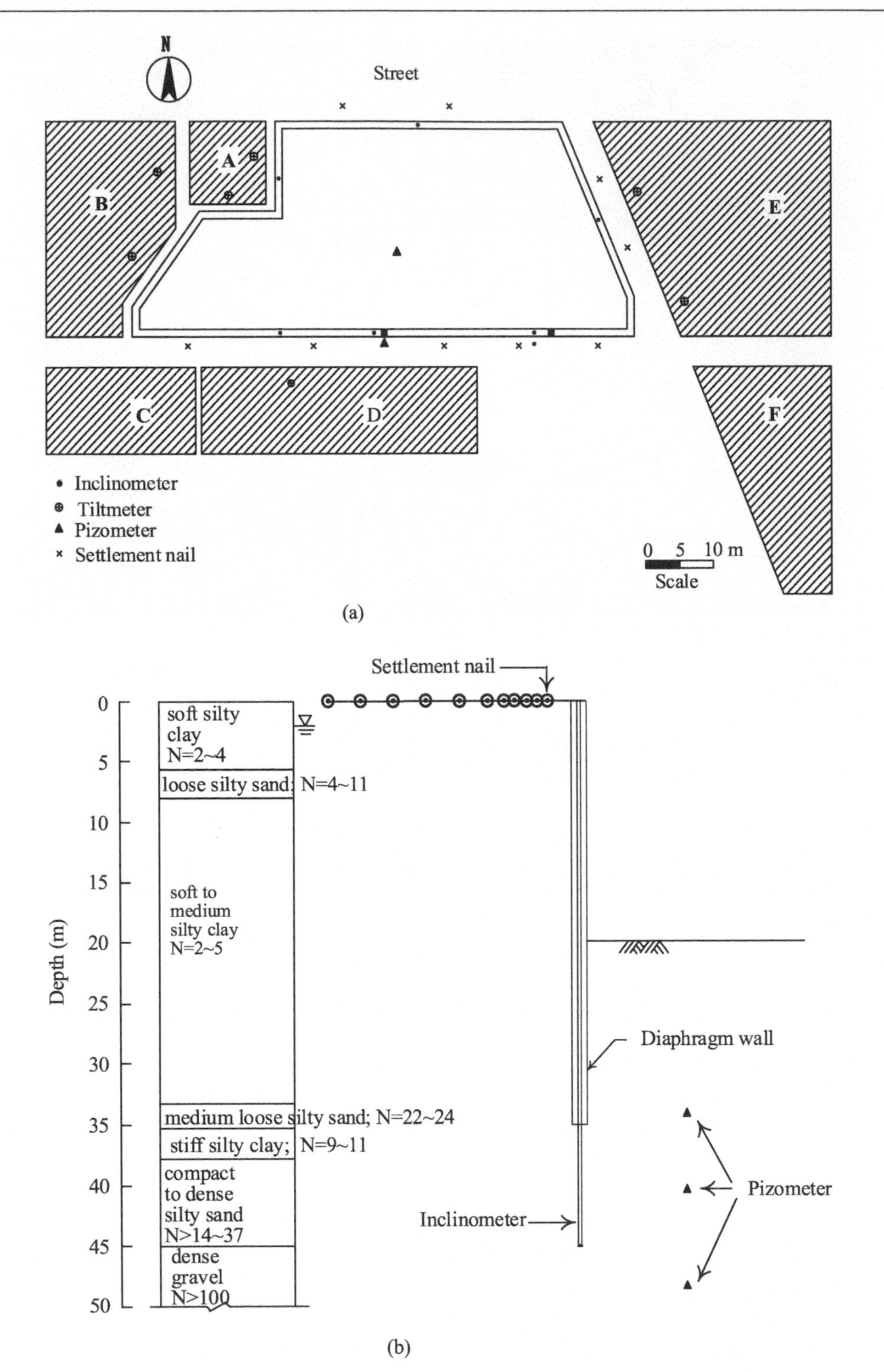

Figure 6.37 An example of the arrangement of monitoring devices in the TNEC excavation project (a) plan and (b) profile.

project, is illustrated in Figure 6.37 (Ou et al., 1998). The excavation depth was 19.7 m, and a diaphragm wall was used as an earth retaining wall. The top-down construction method was adopted. As shown in this figure, there were six buildings, A, B, C, D, E, and F, adjacent to the site. Buildings A, B, D, and E were installed with tiltmeters. Five inclinometers were installed in the diaphragm wall. Piezometers were installed in sandy and gravel soils. Several settlement nails were set around the excavation site.

6.8 Summary and general comments

This chapter introduces two commonly used excavation methods, various stability analysis methods, deformation analysis, stress analysis, control of excavation-induced movements, and monitoring systems. The summary and general comments of this chapter are as follows.

1 Stability analysis of excavations against base shear failure can be conducted with the earth pressure equilibrium method. In addition, Terzaghi's method, Bjerrum and Eide's method, and the slip circle method are commonly used for clayey soil.
2 For excavations in sandy soil, stability analysis of base shear failure and sand boiling should be carried out. In clayey soil, stability analysis includes base shear failure or basal heave. In alternating sandy and clayey soil, analysis of base shear failure, upheaval failure, and sometimes, sand boiling should be conducted.
3 The strut load can be calculated by the apparent earth pressure method. The apparent earth pressure is the earth pressure obtained by back-calculating the strut load obtained from the field measurement. Generally, the apparent earth pressure method should be applied only to a small scale or shallow depth of excavation.
4 A typical supported retaining system is a highly indeterminate structure. The stress analysis can resort to a sophisticated finite element analysis or beam–spring model. With the beam–spring model, one can obtain the strut load, bending moment diagram, and shear force diagram. However, for simplicity, each structural component can be analyzed with simple structural mechanics and based on experience. It is generally suggested that these simplified methods be applied only to small-scale or shallow depths of excavations.
5 There are two types of deflection for a retaining wall: cantilever type and deep inward type. The cantilever deflection usually occurs in stiff soil or excavations with weak strut stiffness, and the maximum deflection occurs at the top of the retaining wall. Deep inward deflection normally occurs in soft clayey soil or excavations with high strut stiffness, and the maximum deflection normally occurs a certain depth below the ground surface. The maximum deflection can be estimated using Clough and O'Rourke's chart (1990).
6 There are two types of ground surface settlement caused by excavation: spandrel type and concave type. The former is related to cantilever wall deflection, while the latter is associated mostly with deep inward wall deflection. The maximum ground surface settlement can be estimated using the methods recommended by Peck (1969) and Clough and O'Rourke (1990).
7 Soil improvement in an excavation is a common measure to increase the factor of safety against base shear failure, upheaval failure, or sand boiling. In addition, soil improvement on the passive side can reduce the lateral wall deflection and ground settlement. To reduce excavation-induced movement, cross walls and buttress walls have recently been adopted.

Problems

6.1 Figure P6.1 shows an excavation in layers of sandy and clayey soils. The groundwater level behind the wall is the same as that in front of the wall. Assume the following parameters: $H_e = 9.0$ m, $H_p = 8.0$ m; $h_1 = 2.0$ m, $h_2 = 3.5$ m, $h_3 = 3.5$ m; $\gamma = 16$ kN/m^3, $\gamma_{sat} = 20$ kN/m^3; $c' = 0$, $\phi' = 30°$; and $D = 5$ m. Compute the factor of safety (F_b) against base shear failure using both the load factor and strength factor earth pressure equilibrium methods with Coulomb's earth pressure theory, assuming $\delta / \phi' = 0.67$.

6.2 Redo problem 6.1 with the following parameters: $H_e = 10.0$ m, $H_p = 11.0$ m; $h_1 = 2.0$ m, $h_2 = 4.0$ m, $h_3 = 4.0$ m; $\gamma = 16$ kN/m^3, $\gamma_{sat} = 18$ kN/m^3; $c' = 0$, $\phi' = 33°$; and $D = 5$ m.

6.3 Same as problem 6.1, but the penetration depth H_p is unknown. Assuming the factor of safety (F_b) against base shear failure is equal to 1.2, compute the required H_p using both the load factor and strength factor earth pressure equilibrium methods.

6.4 Same as problem 6.2, but the penetration depth H_p is unknown. Assuming the factor of safety (F_b) against base shear failure is equal to 1.2, compute the required H_p using both the load factor and strength factor earth pressure equilibrium methods.

6.5 Figure P6.5 shows an excavation in three layers of clay. The groundwater levels behind the wall and in front of the wall are on the ground surface and on the excavation surface, respectively. The excavation width is 20 m. Assume the following parameters: $H_e = 9.0$ m, $H_p = 11.0$ m; $h_1 = 2.0$ m, $h_2 = 3.0$ m, $h_3 = 4.0$ m; $s_{u1} = 10$ kN/m^2, $\gamma_{sat1} = 16$ kN/m^3; $s_{u2} = 20$ kN/m^2, $\gamma_{sat2} = 17$ kN/m^3; $s_{u3} = 30$ kN/m^2, $\gamma_{sat3} = 18$ kN/m^3; and $D = 5$ m. Compute the factor of safety (F_b) against base shear failure using both the load factor and strength factor earth pressure equilibrium methods.

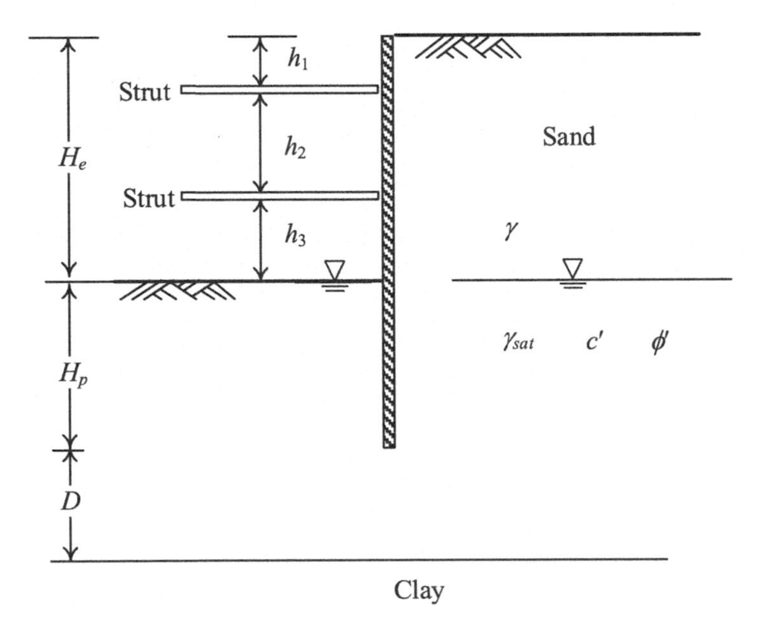

Figure P6.1

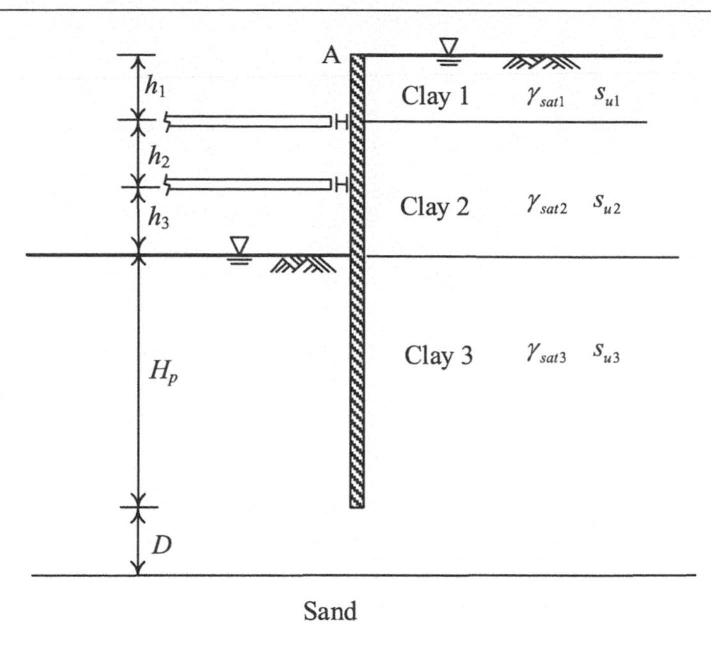

Figure P6.5

6.6 Redo problem 6.5 using the slip circle method, Terzaghi's method, and Bjerrum and Eide's method.

6.7 Refer to Figure P6.5. The excavation width is 30 m. Assume the following parameters: $H_e = 10.0$ m, $H_p = 11.0$ m; $h_1 = 2.0$ m, $h_2 = 4.0$ m, $h_3 = 4.0$ m; $s_{u1} = 12$ kN/m^2, $\gamma_{sat1} = 18$ kN/m^3; $s_{u2} = 25$ kN/m^2, $\gamma_{sat2} = 19$ kN/m^3; $s_{u3} = 30$ kN/m^2, $\gamma_{sat3} = 18$ kN/m^3; and $D = 8$ m. Compute the factor of safety (F_b) against base shear failure using both the load factor and strength factor earth pressure equilibrium methods.

6.8 Redo problem 6.7 using the slip circle method, Terzaghi's method, and Bjerrum and Eide's method.

6.9 Refer to Figure P6.9. The excavation width is 20 m. Assume the following parameters: $H_e = 9.0$ m, $H_p = 11.0$ m; $h_1 = 2.0$ m, $h_2 = 3.0$ m, $h_3 = 4.0$ m, $h_4 = 4.0$ m; $s_{u1} = 10$ kN/m^2, $\gamma_{sat1} = 16$ kN/m^3; $s_{u2} = 20$ kN/m^2, $\gamma_{sat2} = 17$ kN/m^3; $s_{u3} = 30$ kN/m^2, $\gamma_{sat3} = 18$ kN/m^3; and $D = 5$ m. Compute the factor of safety (F_b) against base shear failure using both the load factor and strength earth pressure equilibrium methods, slip circle method, Terzaghi's method, and Bjerrum and Eide's method.

6.10 Refer to Figure P6.9. The excavation width is 30 m. Assume the following parameters: $H_e = 10.0$ m, $H_p = 11.0$ m; $h_1 = 2.0$ m, $h_2 = 4.0$ m, $h_3 = 4.0$ m, $h_4 = 4.0$ m; $s_{u1} = 12$ kN/m^2, $\gamma_{sat1} = 18$ kN/m^3; $s_{u2} = 25$ kN/m^2, $\gamma_{sat2} = 19$ kN/m^3; $s_{u3} = 30$ kN/m^2, $\gamma_{sat3} = 18$ kN/m^3; and $D = 8$ m. Compute the factor of safety (F_b) against base shear failure using both the load factor and strength factor earth pressure equilibrium methods, slip circle method and Terzaghi's method, and Bjerrum and Eide's method.

6.11 Refer to problem 6.1. If the groundwater level behind the wall and in front of the wall are on the ground surface and on the excavation surface, respectively, seepage will occur.

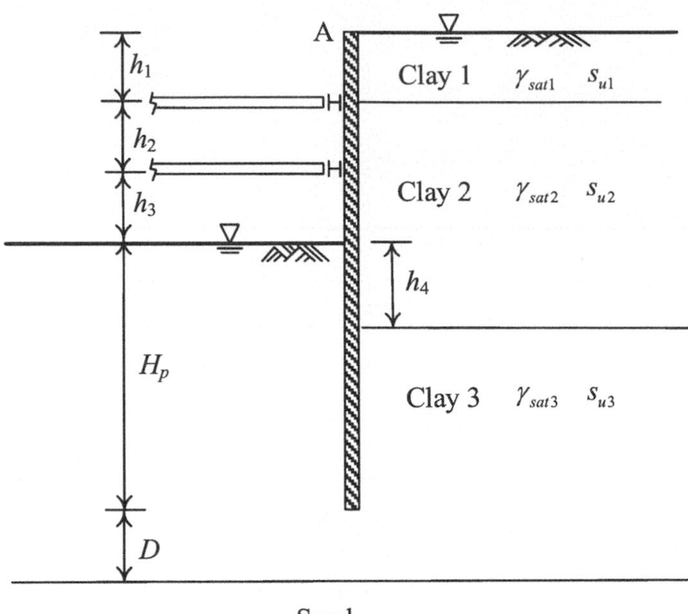

Figure P6.9

Compute the factor of safety (F_s) against sand boiling using Terzaghi's method and Harza's method, assuming excavation width equal to 60 m.

6.12 Refer to problem 6.2. If the groundwater level behind the wall and in front of the wall are on the ground surface and on the excavation surface, respectively, seepage will occur. Compute the factor of safety against sand boiling (F_s) using Terzaghi's method and Harza's method, assuming excavation width equal to 52 m.

6.13 Refer to problem 6.5. If the pore water pressure head in sand is also on the ground surface, compute the factor of safety (F_{up}) against upheaval failure.

6.14 Refer to problem 6.7. If the pore water pressure head in sand is also on the ground surface, compute the factor of safety (F_{up}) against upheaval failure.

6.15 Refer to problem 6.5. If the retaining wall is the sheet pile and the spacing between two adjacent struts is equal to 5 m, determine the following:

a Draw the apparent earth pressure.
b Compute the strut loads at the struts.
c Determine the section modulus required for the sheet pile.
d Determine the section modulus required for the wale at the second level.

6.16 Refer to problem 6.7. If the retaining wall is the sheet pile and the spacing between two adjacent struts is equal to 5 m, determine the following:

a Draw the earth pressure envelope.
b Compute the strut loads at the struts.
c Determine the section modulus required for the sheet pile.
d Determine the section modulus required for the wale at the second level.

6.17 Refer to problem 6.5. Assuming the stiffness of the retaining wall per unit meter is 65,000 kN-m^2. Estimate the maximum wall deflection induced by excavation, and plot the possible settlement profile using Clough and O'Rourke's method.

6.18 Refer to problem 6.7. Assuming the stiffness of the retaining wall per unit meter is 175,500 kN-m^2. Estimate the maximum wall deflection induced by excavation, and plot the possible settlement profile using Clough and O'Rourke's method.

References

Bjerrum, L. and Eide, O. (1956), Stability of strutted excavation in clay. *Geotechnique*, Vol. 6, pp. 32–47.

Clough, G.W. and O'Rourke, T.D. (1990), Construction-induced movements of insitu walls, *Design and Performance of Earth Retaining Structures*, ASCE Special Publication, No. 25, pp. 439–470.

Harza, L.F. (1935), *Uplift and Seepage Under Dams in Sand*, Transaction, ASCE, No. 100, pp. 1352–1406.

Hsieh, P.G. and Ou, C.Y. (1998), Shape of ground surface settlement profiles caused by excavation. *Canadian Geotechnical Journal*, Vol. 35, No. 6, pp. 1004–1017.

Hsieh, P.G. and Ou, C.Y. (2018), Mechanism of buttress walls in restraining the wall deflection caused by deep excavation. *Tunneling and Underground Space Technology*, Vol. 82, pp. 542–553.

Hsieh, P.G., Ou, C.Y. and Lin, Y.L. (2013), Three-dimensional numerical analysis of deep excavations with cross walls. *Acta Geotechnica*, Vol. 8, No. 1, pp. 33–48.

Hsieh, P.G., Ou, C.Y. and Shih, C. (2012), A simplified plane strain analysis of the lateral wall deflection for excavations with cross walls. *Canadian Geotechnical Journal*, Vo. 49, pp. 1134–1146.

JSA (1988), *Guidelines of Design and Construction of Deep Excavations*, Japanese Society of Architecture.

Lim, A., Ou, C.Y. and Hsieh, P.G. (2019), An innovative earth retaining supported system for deep excavation. *Computers and Geotechnics*, Vol. 114, p. 103135.

Mana, A.I. and Clough, G.W. (1981), Prediction of movements for braced cut in clay. *Journal of Geotechnical Engineering Division, ASCE*, Vol. 107, No. 6, pp. 759–777.

NAVFAC DM7.2 (1982), *Foundations and Earth Structures*, USA: Department of the Navy, Alexandria, Virginia.

Ou, C.Y. (2022), *Fundamentals of Deep Excavations*, London: CRC Press, Taylor and Francis Group.

Ou, C.Y. and Hsieh, P.G. (2011, December), A simplified method for predicting ground settlement profiles induced by excavation in soft clay. *Computers and Geotechnics*, Vol. 38, pp. 987–997.

Ou, C.Y., Hsieh, P.G. and Chiou, D.C. (1993), Characteristics of ground surface settlement during excavation. *Canadian Geotechnical Journal*, Vol. 30, pp. 758–767.

Ou, C.Y., Hsieh, P.G. and Lin, Y.L. (2011), Performance of excavations with cross walls. *Journal of Geotechnical and Geoenvironmental Engineering, ASCE*, Vol. 137, No. 1, pp. 94–104.

Ou, C.Y., Hsieh, P.G. and Lin, Y.L. (2013, May), A parametric study of wall deflections in deep excavations with the installation of cross walls. *Computers and Geotechnics*, Vol. 50, pp. 55–65.

Ou, C.Y., Liao, J.T. and Lin, H.D. (1998), Performance of diaphragm wall constructed using top-down method. *Journal of Geotechnical and Geoenvironmental Engineering, ASCE*, Vol. 124, No. 9, pp. 798–808.

Peck, R.B. (1943), Earth pressure measurements in open cuts Chicago (III) subway. *Transactions, ASCE*, Vol. 108, p. 223.

Peck, R.B. (1969), Advantages and limitations of the observational method in applied soil mechanics. *Geotechnique*, Vol. 19, No. 2, pp. 171–187.

Peck, R.B., Hanson, W.E. and Thornburn, T.H. (1977), *Foundation Engineering*, New York: John Wiley and Sons.

Pratama, I.T., Ou, C.Y. and Ching, J. (2019), Calibration of reliability-based safety factors for sand boiling in excavations. *Canadian Geotechnical Journal*, Vol. 57, No. 5, pp. 742–753.

Reddy, A.S. and Srinivasan, R.J. (1967), Bearing capacity of footing on layered clay. *Journal of the Soil Mechanics and Foundations Division, ASCE*, Vol. 93, No. 2, pp. 83–99.

Skempton, A.W. (1951), The bearing capacity of clays. *Proceeding of Building Research Congress*, Vol. 1, pp. 180–189.

Terzaghi, K. (1922), Der Grundbrunch on Stauwerken und Seine Verhutung. *Die Wasserkraft*, Vol. 17, pp. 445–449, Reprinted in From *Theory to Practice in Soil Mechanics*, New York: John Wiley and Sons, pp. 146–148, 1961.

Terzaghi, K. (1943), *Theoretical Soil Mechanics*, New York: John Wiley & Sons, Inc.

Terzaghi, K., Peck, R.B. and Mesri, G. (1996), *Soil Mechanics in Engineering Practice*, New York: John Wiley and Sons.

TGS (2022), *Design Specifications for the Foundation of the Building*, Taiwan Geotechnical Society, Taipei.

Chapter 7

Pile foundations

7.1 Introduction

Pile foundations, as deep foundations, are commonly used for high-rise buildings, bridges, and structures under complex loading conditions (such as retaining walls and offshore wind turbine foundations), as shown in Figure 7.1. Piles can be driven or bored into the ground. Notably, bored piles are usually called drilled shafts in some areas, such as North America. The functions of piles are as follows:

a. Transferring loads from the superstructure through weak compressible strata or through water onto stiffer or more compact and less compressible soils or onto rocks.
b. Carrying uplift loads when used to support tall structures subjected to overturning forces (from winds or waves).
c. Carrying combinations of vertical and horizontal loads to support retaining walls, bridge piers and abutments, and machinery foundations.

The capacity of piles is highly affected by the installation methods, ground conditions, and loading types. In this chapter, the pile installation, estimation of pile load capacity and settlement, and in situ pile load test will be introduced. Importantly, the estimation methods for pile load capacity introduced in the following sections are based on theories, field tests, and model tests, and in situ pile load tests should be carried out if possible. The pile load test result provides a check of the estimation.

7.2 Classification of piles

Piles can be generally classified in a number of ways, such as by function, material, and installation method. Each classification will be introduced in detail in the following sections.

7.2.1 *Functions of piles*

Piles provide resistance to loadings from the superstructures, which include both vertical (compression and tension) and lateral loadings. Thus, piles can be classified based on their functions as follows:

a. *Compression pile*. This category represents the majority of pile foundations. Piles resist the vertical downward loadings from the superstructures, as shown in Figure 7.2. The resistances

DOI: 10.1201/9781003350019-7

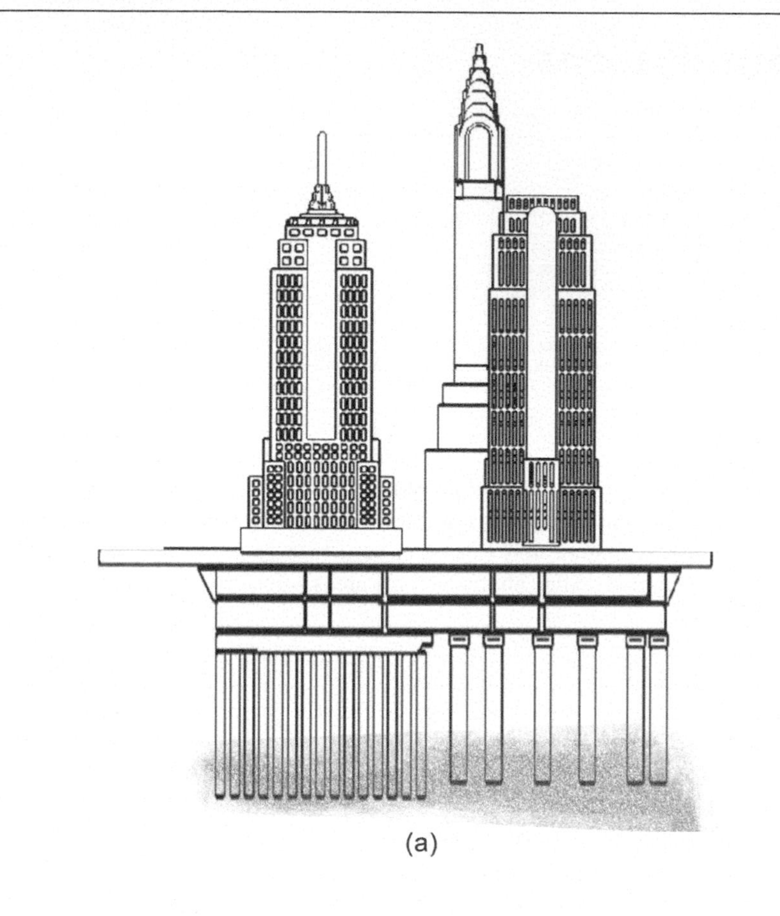

(a)

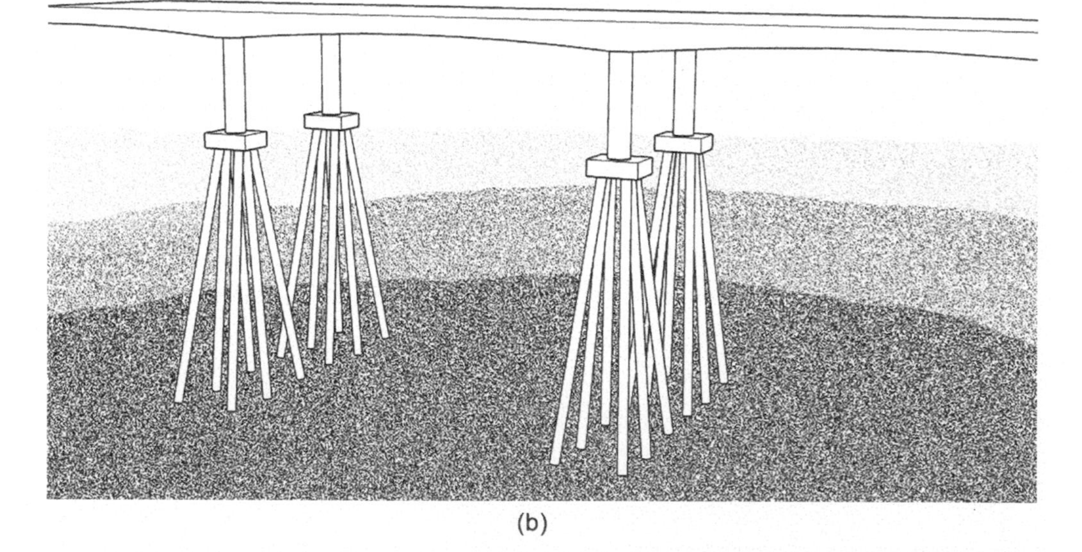

(b)

Figure 7.1 Functions of piles: (a) foundations for high-rise buildings, (b) foundations for bridges, and (c) in retaining walls.

(c)

Figure 7.1 (Continued)

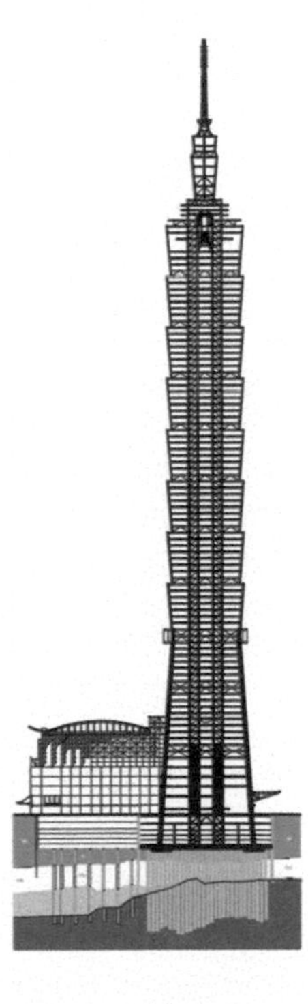

(a)

Figure 7.2 Pile in compression: (a) foundations for high-rise buildings and (b) foundations for bridges.

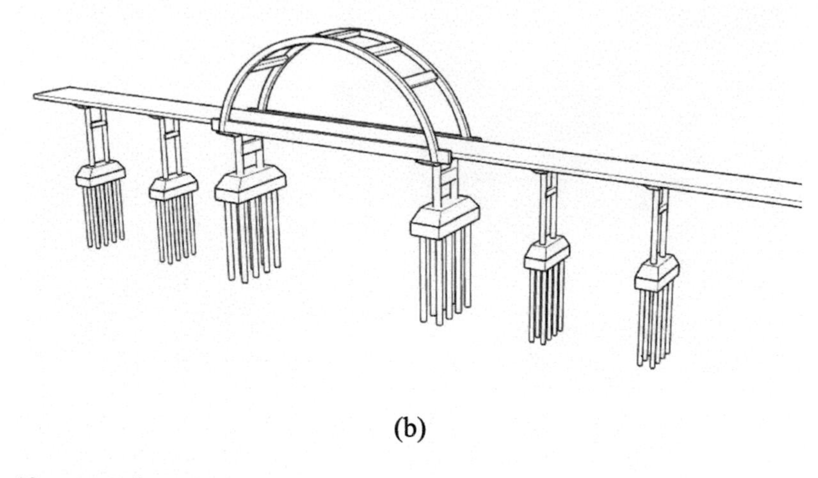

(b)

Figure 7.2 (Continued)

of the piles are provided by the mobilized side frictional force and end-bearing capacity. Settlements occur when loading is applied; thus, the allowable settlement is an important design factor for the compression pile.

b. *Tension pile*. This pile is also called an uplift pile and provides resistance against vertical upward forces in transmission line towers and foundations subjected to floating forces, as shown in Figure 7.3. The resistance is mainly from the tensile strength of the surrounding soils, which is generally lower than the compressive strength. Thus, the design load capacity of tension piles is generally lower than that of compression piles.
c. *Laterally loaded pile*. The laterally loaded pile can be classified into two subcategories, that is, pile-to-soil and soil-to-pile. As shown in Figure 7.4a, piles can be used as the foundation of wind turbines and to transmit the lateral loads from the wind turbine to soils. On the other hand, in Figure 7.4b, the piles are used as the retaining structures, which resist the lateral forces from the soil. The design considerations of lateral piles are the bending moment and deflections of piles and the lateral resistance of soils, which determine the ultimate horizontal force of piles.

7.2.2 Materials

In addition to their function, piles can also be classified by their materials, such as timber, steel, and concrete. The selection of pile material depends mainly on the required function, durability, and local supplier capacity. Common materials for piles are described here:

a. Timber piles are light and easy to handle. Timber comprises excellent natural materials in regions with rich timber resources; thus, they can be quite competitive in terms of cost. Even though timbers are convenient to collect and manufacture into piles, their dimension and loading capacity are limited due to their natural texture and defects. A timber pile head is damaged easily during the pile-driving process. Timber is effective when situated wholly below the groundwater level.

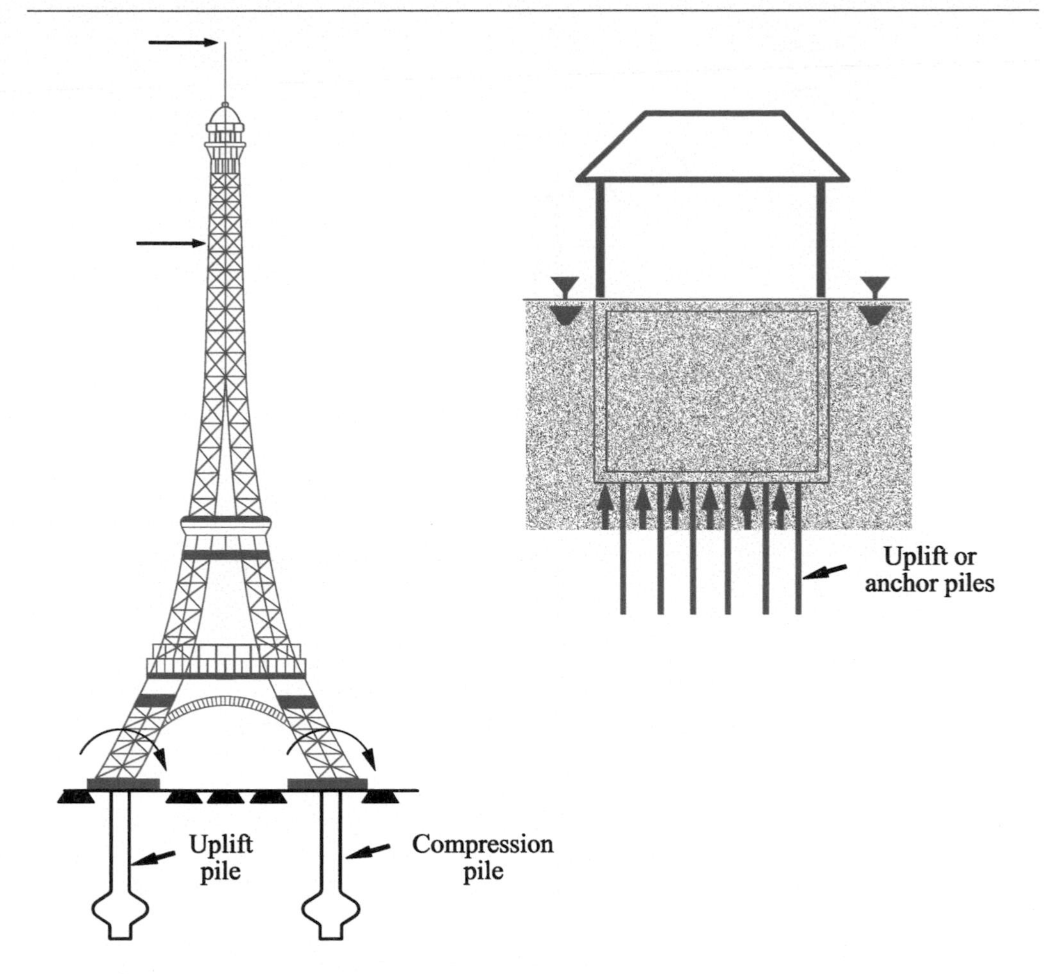

Figure 7.3 Pile in tension.

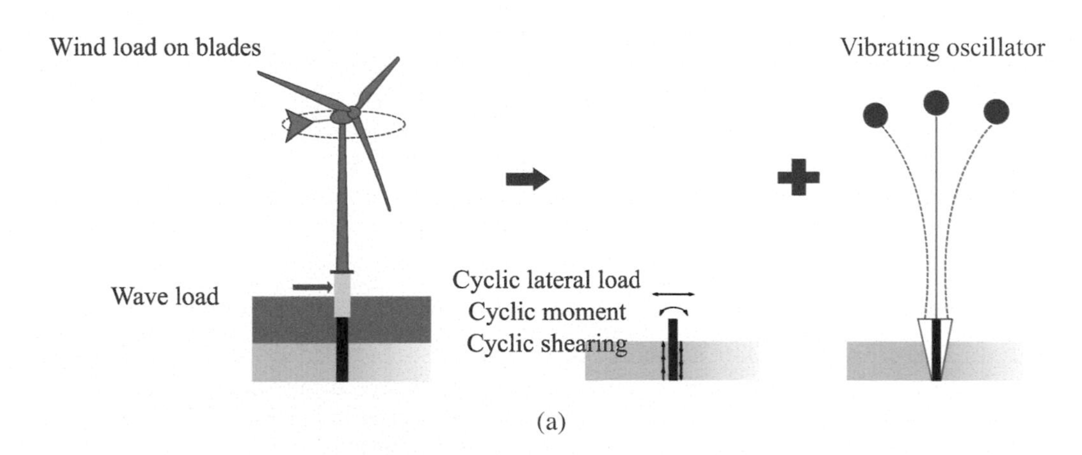

Figure 7.4 Laterally loaded pile: (a) pile-to-soil and (b) soil-to-pile.

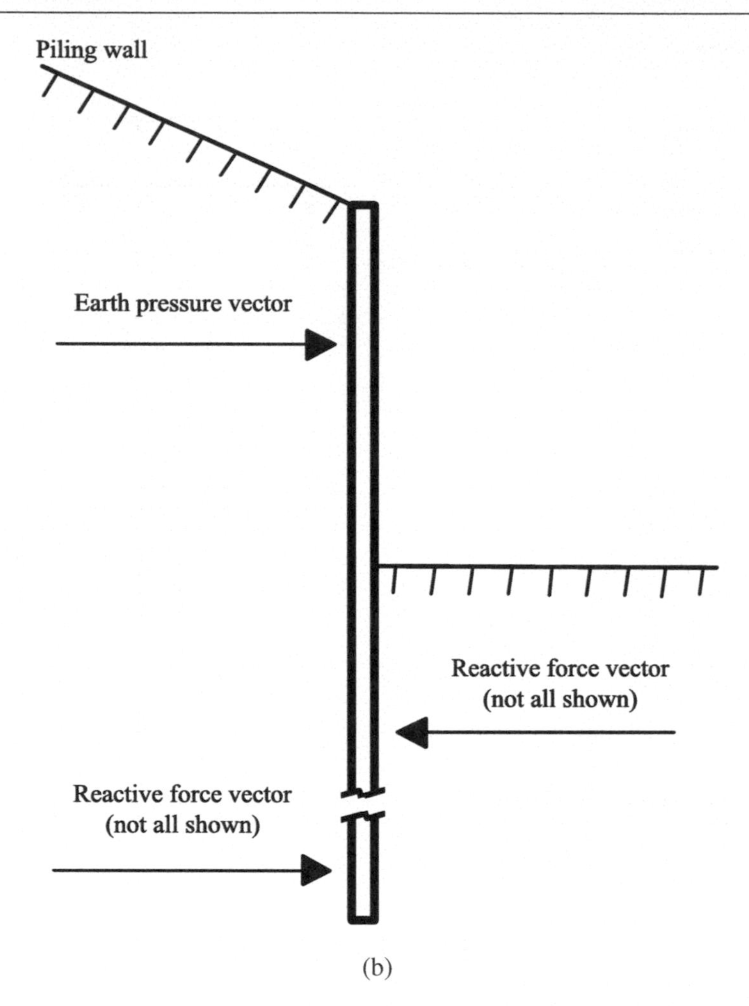

Figure 7.4 (Continued)

b. Steel piles are robust, easy to handle, and capable of carrying high compressive loads when driven onto a hard stratum. Steel piles have many shapes and cross sections, such as plain tubes, box sections, box piles built up from sheet piles, H-sections, and tapered and fluted tubes. However, compared to concrete piles, the cost is high.
c. Concrete piles are commonly used in practice due to their cost and performance. The installation of concrete piles is very flexible. Both precast and cast-in-place are available.

7.2.3 Method of installation

The pile load capacity is highly dependent on its installation. Installation with a driving device usually causes larger displacement to surrounding soils, namely, displacement piles, and thus compacts or squeezes the soils. Installation with an opening prior to inserting the pile is usually called a replacement pile, where the soils are removed first and then the pile is installed. The installation of piles will have different effects on the surrounding soils, which depend on the soil

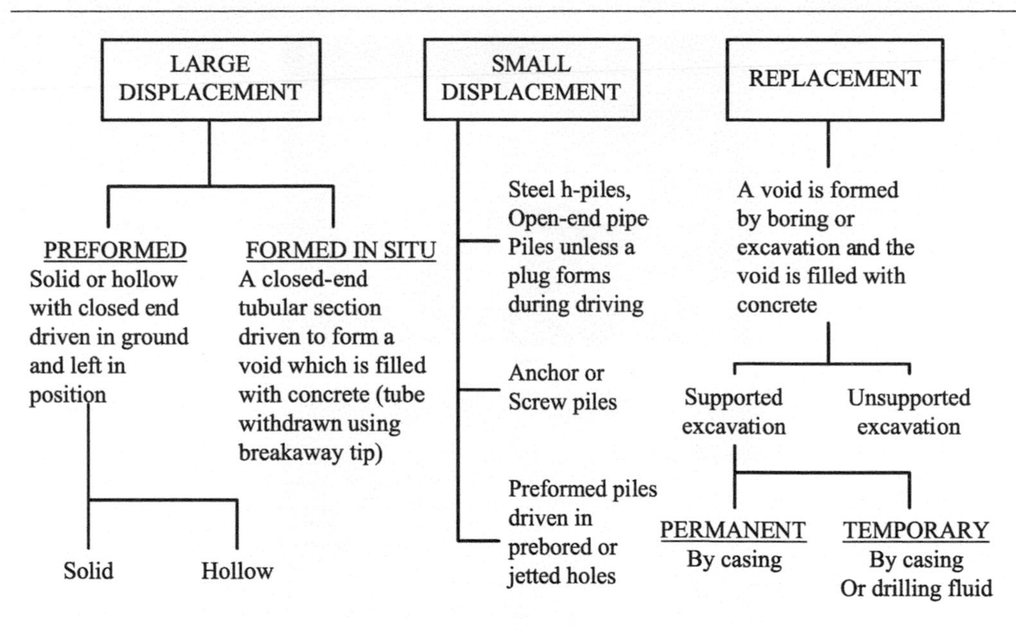

Figure 7.5 Method of pile installation.

Source: Redrawn from Poulos and Davis (1980).

types and installation methods. The effects will be introduced in the next section. The classification of pile installation is shown in Figure 7.5 and explained in what follows.

a. Displacement pile:
    i. Large displacement
       All types of driven piles, such as solid section piles or hollow section piles with a closed end, are driven or jacked into the ground and thus displace the soil.
    ii. Small displacement
       Also driven or jacked into the ground but have a relatively small cross section, such as a hollow section with an open end, H- or I-section, pipe, or box steel pile.
b. Replacement pile (non-displacement pile): Remove soils by boring first and then fill the borehole (either with casing or without casing) with concrete.

7.3 Installation of piles and its impact on soils

The ground condition and installation of piles predominantly determine the load capacity of piles. The installation of piles affects the status of the surrounding soils and thus results in different load capacities. In this section, installation methods and their effects on both cohesive and cohesionless soils will be introduced.

7.3.1 Driven pile

Driven installation is commonly used for steel piles and precast concrete piles. Figure 7.6 shows the precast concrete pile with a sharp tip and corresponding driven equipment. The main process

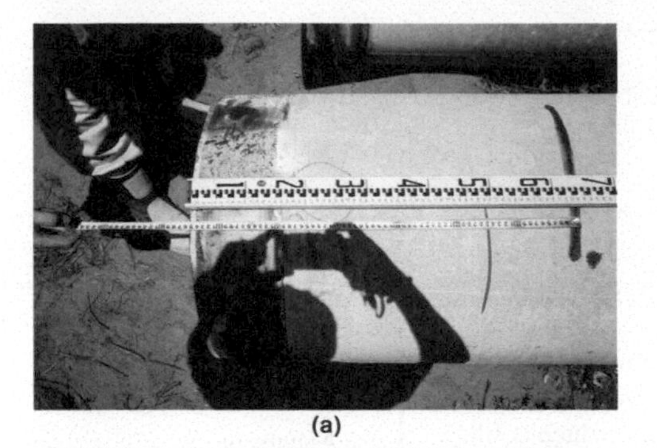

(a)

(b)

(c)

Figure 7.6 Driven piles and equipment: (a) pile head, (b) pile tip, and (c) pile driving equipment.

Source: Provided by CTCI Resources Engineering Inc.

for a driven pile includes transportation, unloading, placement, lifting, positioning, driving, and welding if necessary. Handling stress should be considered in driven piles due to the mentioned handle process. Piles can be driven by a drop hammer or vibrating devices. For prefabricated driven piles, the advantages are as follows:

a. Good quality on pile material with inspections before driving
b. Not liable to squeezing or necking, especially in soft ground
c. Construction operation not affected by groundwater

Potential problems include the following:

a. Noise and vibration due to driving are high.
b. Need to be jointed for the desired length and may break during driving.
c. May suffer unseen damage.
d. Displacement of soil during driving may lift adjacent piles/structures.
e. Jointed precast piles may not be suitable for tension and lateral loads.

For driven and cast-in-place displacement piles, the advantages and disadvantages are as follows.

Advantages:

a. Easily adjust length.
b. Driving tube driven with a closed end to exclude groundwater.
c. Enlarging the base is possible.

Disadvantages:

a. Concrete liable to be defective in soft squeezing soils where withdrawable tube types are used.
b. Noise and vibration due to driving are high.
c. Displacement of soil during driving may lift adjacent piles/structures.

7.3.2 *Effect of pile driving in clays*

When driving piles into clayey soils, the undrained shear strength of clays, excess pore water pressures, and soil structures will be affected and thus impact the performance of piles, especially in the short term. The effects of pile driving in clays can be classified into four major categories (de Mello, 1969):

a. Remolding or partial structural alteration of the soil surrounding the pile
b. Alteration of the stress state in the soil near the pile
c. Dissipation of the excess pore water pressure around the pile
d. Long-term strength regain in the soil

The phenomena will be introduced in detail next.

Influence on the soil strength and pile capacity

Driving piles into clay will initially cause some loss in undrained shear strength because of the "remolding" (disturbance due to pile driving) of clays. After the completion of driving, the strength of the soil and pile capacity will increase because of the following:

a. Regain of undrained shear strength (structural bonds partially restored)
b. Dissipation of excess pore water pressure

As shown in Figure 7.7, the pile load capacity is regained with the elapsed time after pile driving. For some cases, it takes one hundred hours (one to several weeks) to recover the load capacity to 90% of the maximum value. For this reason, the pile load test is usually performed at one to two weeks after pile driving.

Pore pressures developed during driving

During the driving of a pile in clays, soils were subjected to shear forces along the pile surface. When the pile punched into the clay, the surrounding soils were squeezed, and thus, pore water pressure was generated. According to the excess pore water pressure measurement during pile driving shown in Figure 7.8, the maximum excess pore water pressure can be 1–2 times the effective stresses on the pile surface and 3–4 times the effective stresses at the pile base. The excess pore water pressure gradually decreases with the distance to the center of the pile, r. The distance r is normalized by the radius of the pile a, as shown in Figure 7.8. Based on field observations, the excess pore water pressure remains at the maximum value (Δu_m) within a

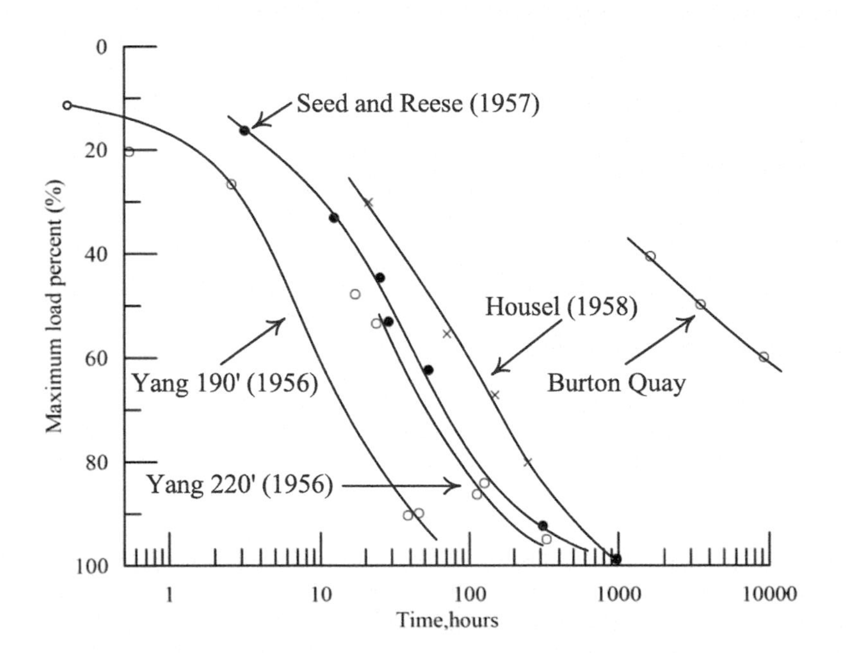

Figure 7.7 Increase in load capacity over time.

Source: Nishida (1962).

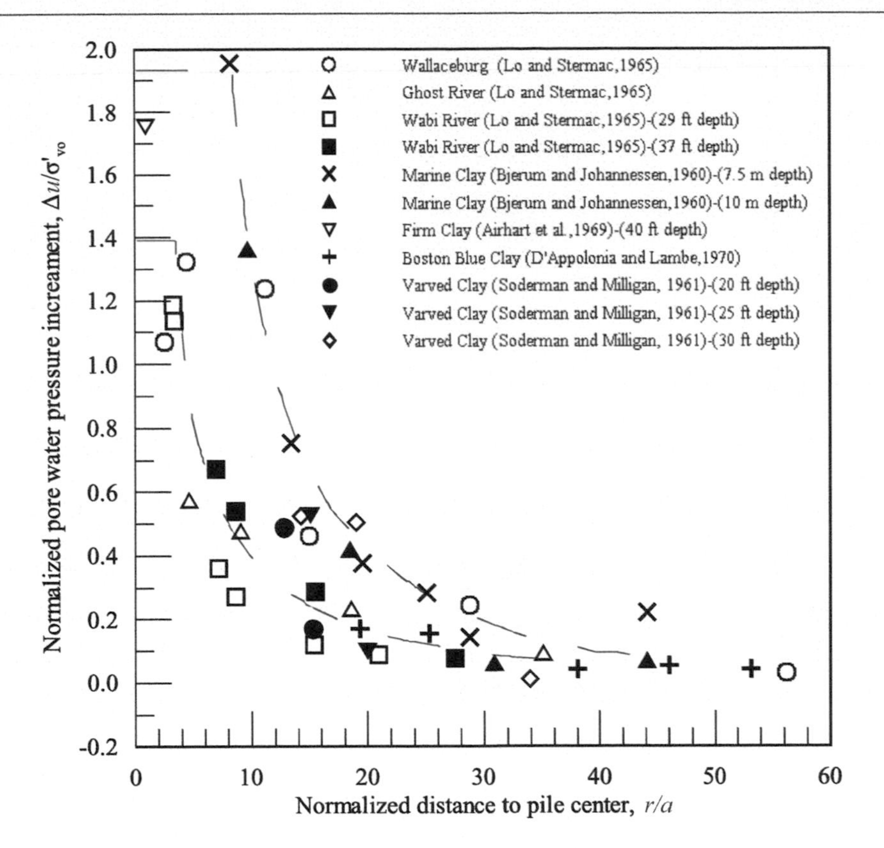

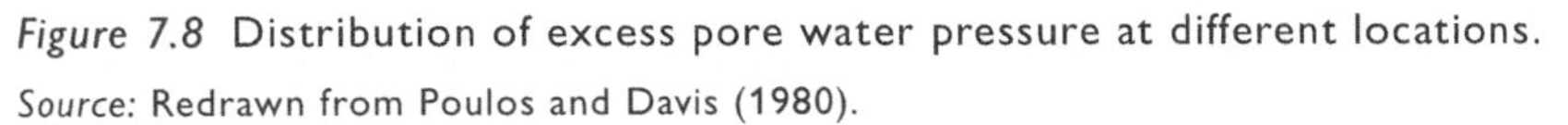

Figure 7.8 Distribution of excess pore water pressure at different locations.

Source: Redrawn from Poulos and Davis (1980).

certain distance, R, which is related to the clay type. Based on the observation shown in Figure 7.8, a simple model for the distribution of excess pore water pressure (Δu) is proposed in eq. 7.1 and shown in Figure 7.9.

Within the distance R:

$$\Delta u = \Delta u_m \tag{7.1a}$$

Beyond the distance R (Nishida, 1964; Kishida, 1967):

$$\Delta u = \Delta u_m / (r/R)^2 \tag{7.1b}$$

Where:

$R \approx 3a \sim 4a$ for insensitive clay.
$R \approx 8a$ for sensitive clay.
a = the radius of the pile.

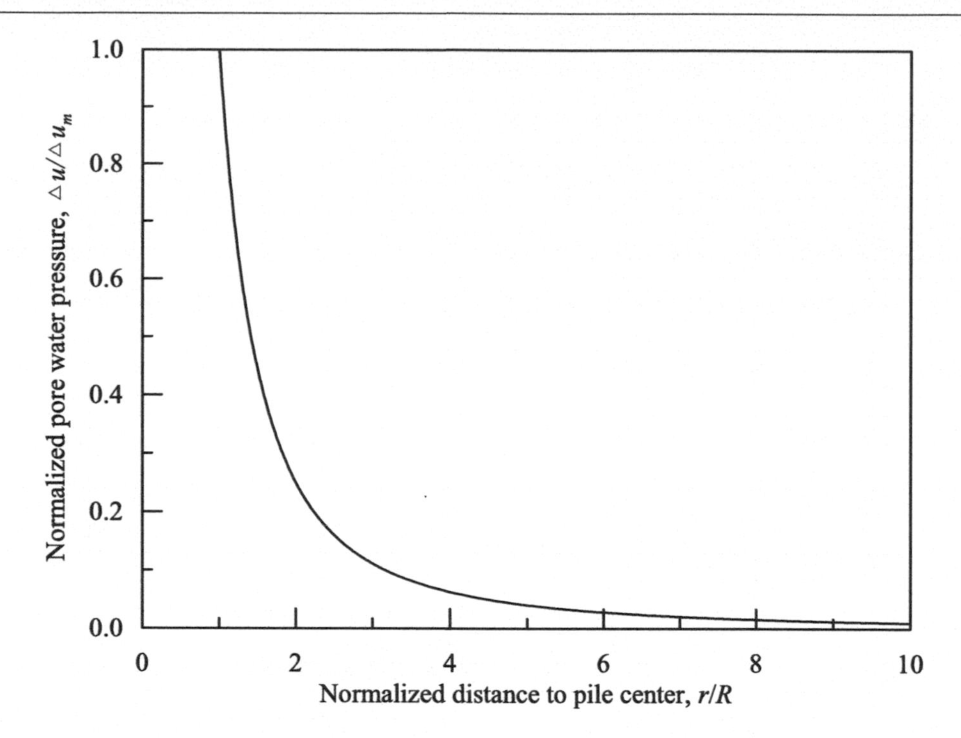

Figure 7.9 Pore pressures developed while driving beyond the distance *R*.

Notably, for pile groups, the pore pressure distributions around individual piles may be superposed, thus decreasing the pile capacity.

7.3.3 *Effect of pile driving in sands*

Single piles

Pile driving is a process of vibrating or hammer dropping that causes energy to propagate through the soil. In loose sands, the load capacity of a pile is increased due to the compaction effect. The relative density is also increased. The SPT-*N* and CPT cone resistances of the sands increase after pile driving, as shown in the field tests. The effects of driving a pile near the pile tip can be estimated as follows:

$$\phi_2' = \frac{\phi_1' + 40^\circ}{2} \quad \text{(Kishida, 1967)} \tag{7.2}$$

Where ϕ_1' and ϕ_2' are shown in Figure 7.10. Eq. 7.2 implies that when $\phi_1' \geq 40^\circ$, ϕ_2' might decrease due to pile driving. If the SPT-*N* of soils after pile driving is known, ϕ can be estimated by eq. 7.3 (Kishida, 1967) or eq. 2.13 (in Chapter 2).

$$\phi' = \sqrt{20N} + 15^\circ \tag{7.3}$$

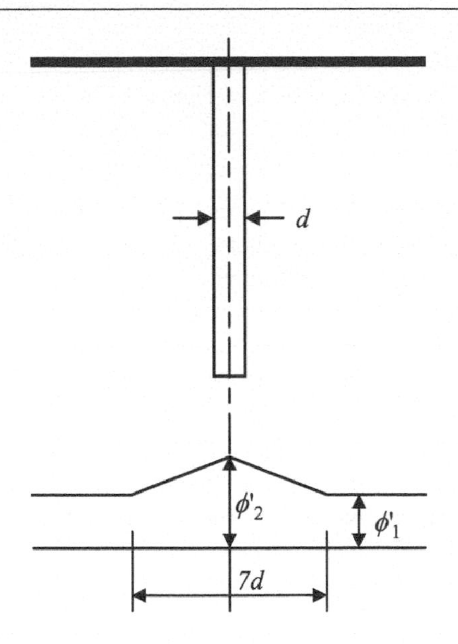

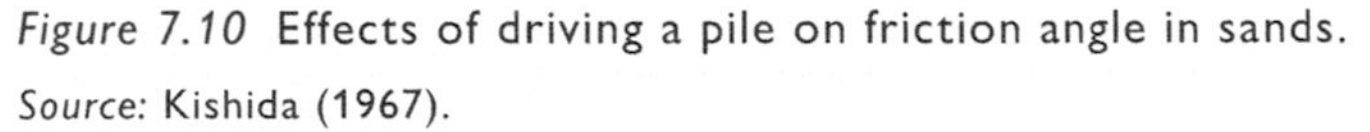
Figure 7.10 Effects of driving a pile on friction angle in sands.

Source: Kishida (1967).

Pile groups

When groups of piles are driven into loose sand, the sand around and between the piles becomes highly compacted. The densification effect due to single pile driving may overlap if the pile spacing is small enough. When the pile spacing is less than 6 times the pile diameter ($6d$), the efficiency of the group (ξ) will be greater than 1.0, which implies that the ultimate capacity of the group is greater than the sum of the single pile capacities. The group effect depends on the position of the observation points and the pile-driving sequence. These values can be estimated by applying Kishida's single pile approach (Figure 7.10 and eq. 7.2).

7.3.4 Bored and cast-in-place pile

The bored pile and cast-in-place pile need to create an opening in the ground and then insert the pile body to the design depth. Protection of the borehole wall may be needed if the ground condition is poor or the water level is high. The advantages and disadvantages are as follows.

Advantages

a. Length can be varied to suit variations in the level of the bearing stratum.
b. Material forming piles are not governed by handling or driving stress.
c. No appreciable noise or vibration.
d. No ground heaves.

Disadvantages

a. Squeezing or necking or bulges (as shown in Figure 7.11).
b. Local slumping of open bore due to groundwater seepage.

When the stability of the borehole wall is an issue during boring, the reverse circulation drilling (RCD) pile and full casing pile are better installation methods for the bored pile. The construction of the reverse circulation drilling pile is shown in Figure 7.12. During the drilling process, high-pressure air is introduced to the annulus between the inner tube and the outer rod. The air flows through the drill steel and powers the drill tool. As the air exhausts, it serves as a circulating medium by carrying the cuttings from the surface of the bit directly through the inside of the drill steel. As it exits the top of the drill stack, the air is guided into a cyclone, which slows the cuttings, separates them from the air, and collects them, while the remaining waste fluids are captured in an isolated, watertight containment bin. A rebar cage is inserted into the borehole when the design depth is reached. Concrete is poured into the borehole from the bottom, and then the pile forms.

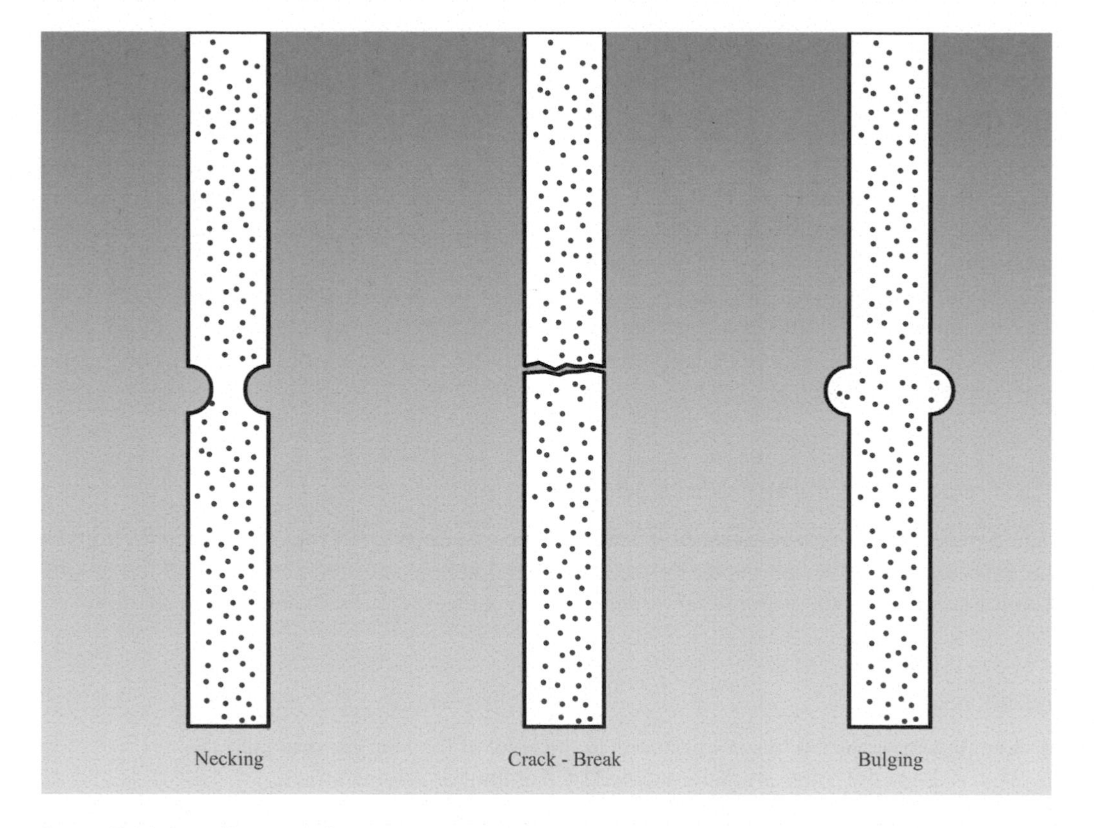

Figure 7.11 Installation defects of bored piles.

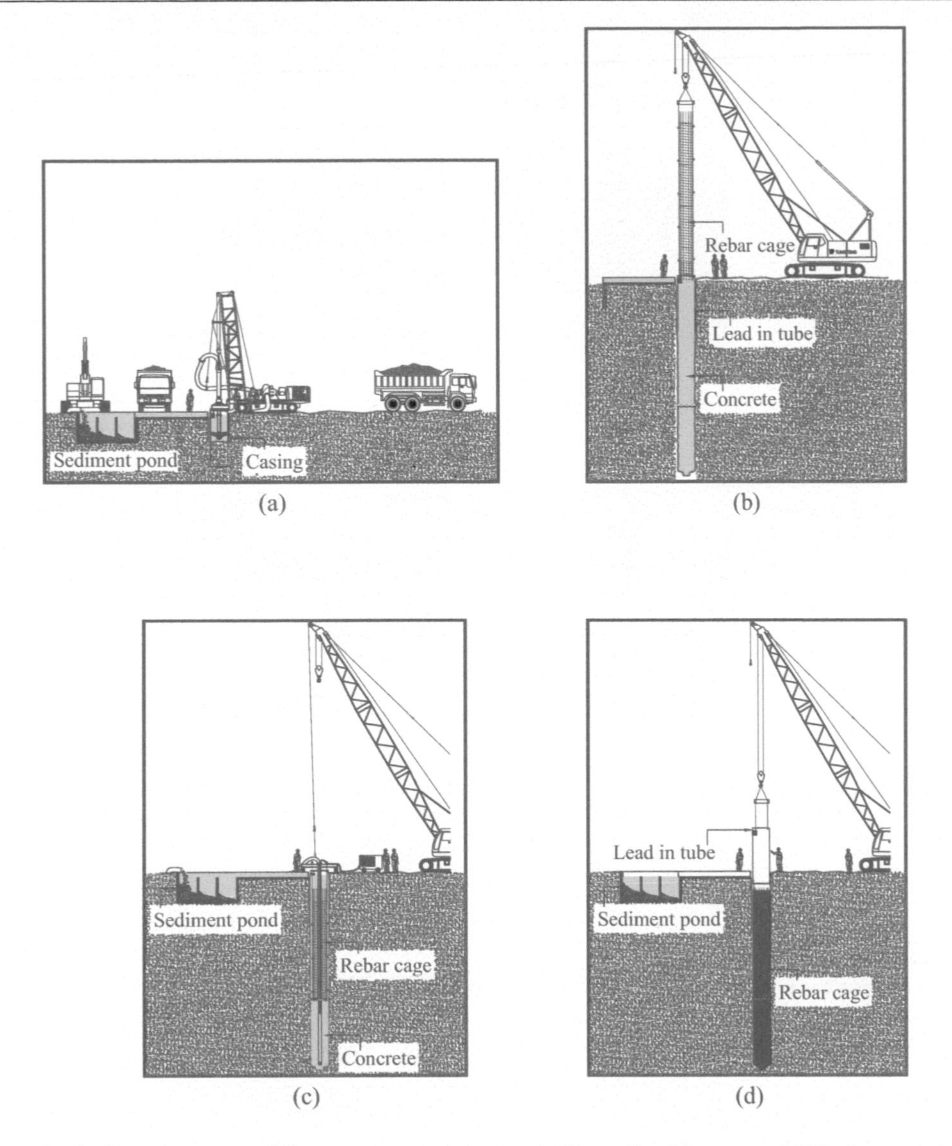

Figure 7.12 Construction of the reverse circulation drilling pile: (a) setting RCD ring and drilling, (b) installation of rebar cage, (c) bottom sludge treatment (air lift), and (d) withdrawing the casing.

Source: Adapted from Tung Feng Construction Engineering Co. Ltd.

The full casing pile is another bored pile that provides good protection of the borehole wall and prevents defects in the bored pile (as shown in Figure 7.11). Full casing means the borehole is protected by casing from the ground surface to the bottom of the borehole, as shown in Figure 7.13. The casing is welded when the total length of the pile is greater than a single casing length.

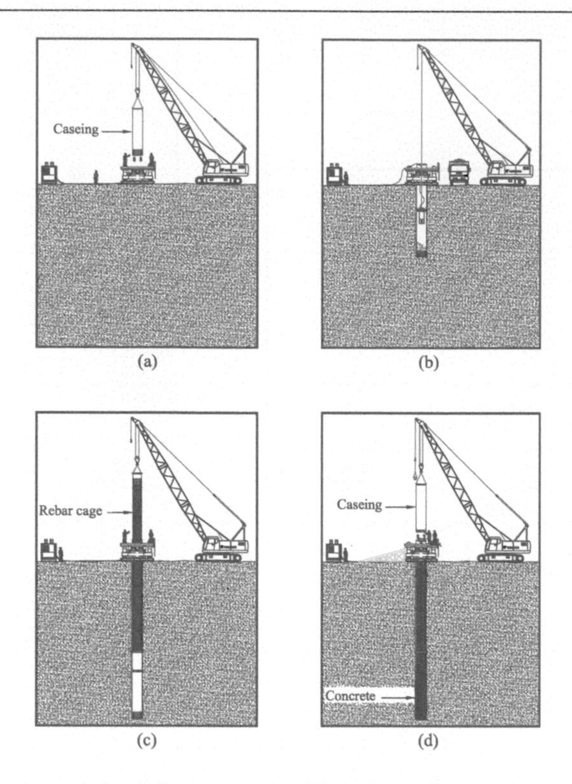

Figure 7.13 Construction of the full casing pile: (a) drilling, (b) installation of casing, (c) installation of rebar cage, and (d) pouring concrete and withdrawing casing.

Source: Adapted from Tung Feng Construction Engineering Co. Ltd.

7.3.5 Effect of installing the bored pile

Effects on the surrounding ground

When installing a bored pile into soils, the effects are different and depend on the soil types. For clayey soils, the effects are as follows.

a. Softening:
   i. Absorption of moisture from the wet concrete.
   ii. Migration of the water from the body of the clay toward the less highly stressed zone around the borehole.
   iii. Water is poured into the boring to facilitate the operation of the cutting tool.
b. Disturbance:
   Especially the clay just beneath the pile case.

For sandy soils, the borehole wall may collapse if no casing or stabilized fluid is used. Piping is another issue if there are no protection measures on the wall of the borehole.

Effects on piles

When installing a bored pile into soils, the effects on piles are:

a. Caving of the borehole, resulting in necking or misalignment of the pile.
b. Aggregate separation within the pile.
c. Buckling of the pile reinforcement.

7.4 Axial load capacity of piles

7.4.1 Load transfer mechanism

The load capacity of a pile is determined by the pile–soil interaction affected by the soil characteristics (soil type and/or existence of hard stratum) and installation method. Loads from the superstructures are transferred into the soils, and the resistance to loadings comes from both the frictional force and end bearing, as shown in Figure 7.14. The frictional force, or so-called shaft resistance, comes from the interface between the pile and surrounding soils, which is dominated by the shear strength of soils. The end bearing, or so-called base resistance, is considerable when the pile penetrates into hard strata. The net ultimate load capacity, Q_u, of a single pile is generally expressed as:

$$Q_u = Q_{su} + Q_{bu} - W \tag{7.4}$$

Where:

Q_u = net ultimate load capacity.
Q_{su} = ultimate shaft resistance (frictional resistance).
Q_{bu} = ultimate base resistance (point resistance).
W = weight of pile.

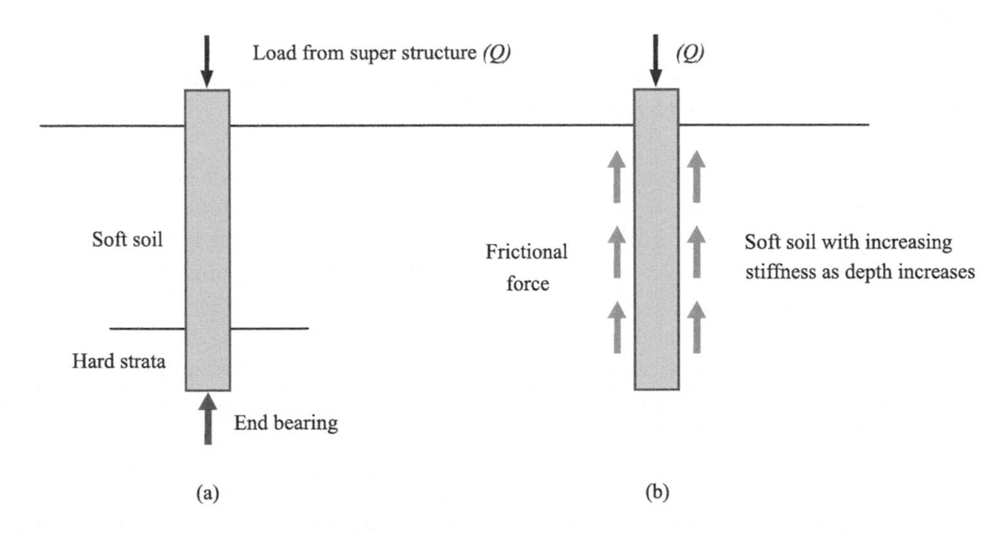

Figure 7.14 Load transfer mechanism of a pile: (a) end-bearing pile and (b) friction pile.

When a load Q is applied on the pile top and gradually increased, maximum frictional resistance will be fully mobilized when the relative displacement between the pile and the soil reaches 5–10 mm (irrespective of the pile dimension). The maximum base resistance will not be mobilized until the pile tip has moved approximate 10–25% of the pile width. In other words, the frictional resistance is mobilized faster at low settlement level than that of base resistance.

The pile is classified as a friction pile when the resistance contribution from frictional force is much greater than the end bearing, that is, $Q_{bu} \ll Q_{su}$, $Q_u \cong Q_{su}$, for instance, bored pile in thick, soft soils. On the other hand, the pile is treated as an end-bearing pile when the load resistance contribution from the end bearing is more significant than the frictional force, that is, $Q_{su} \ll Q_{bu}$, $Q_u \cong Q_{bu}$, for instance, pile tip in rocks/gravels.

7.4.2 End-bearing resistance

The ultimate base resistance, Q_{bu}, is fully activated when the soil at the bottom of the pile reaches the failure condition, that is, ultimate bearing failure; the failure surface is as shown in Figure 7.15. The pile foundations are deep foundations, and the soil fails mostly in a punching model. A triangular failure zone I is developed at the pile tip first, which is pushed downward without generating any other slip surface. In dense materials, a radial shear zone II may partially develop. Thus, the ultimate base resistance can be evaluated from bearing capacity theory similar to that in shallow foundations. All necessary shape and depth factors are included in the bearing capacity factors, that is $N_c^*\ N_q^*\ N_\gamma^*$.

$$Q_{bu} = A_b\left(cN_c^* + \sigma_{vb}N_a^* + 0.5\gamma dN_\gamma^*\right) \tag{7.5}$$

Where:

A_b = area of pile base.
c = soil cohesion.
σ_{vb} = vertical stress in soil at the pile base.
γ = unit weight of soil.

$N_c^*\ N_q^*\ N_\gamma^*$ = the bearing capacity factors considering the angle of internal friction ϕ of the soil, the relative compressibility of the soil, and the pile geometry.

d = pile diameter.

Notably, the notations used in eq. 7.5 are general terms; for example, the vertical stress σ_{vb} can represent either total stress or effective stress, depending on the ground material encountered. For saturated clay under **undrained condition (short-term),** all the properties in calculating ultimate load capacity should be values for the undrained condition. For instance, $c\,(= s_u$, undrained shear strength), $\phi\,(= 0)$, and all stresses should be **total stresses**. For drained or long-term ultimate load capacity, all parameters should correspond to the drained values, and all stresses should be **effective stresses**. The difference will be illustrated in what follows when the ultimate load capacity for clays and sands is calculated.

End-bearing resistance in clay

When a pile is driven into clayey soils, the end-bearing resistance should be calculated by undrained parameters. All the frictional angles, including the internal frictional angle ϕ and the

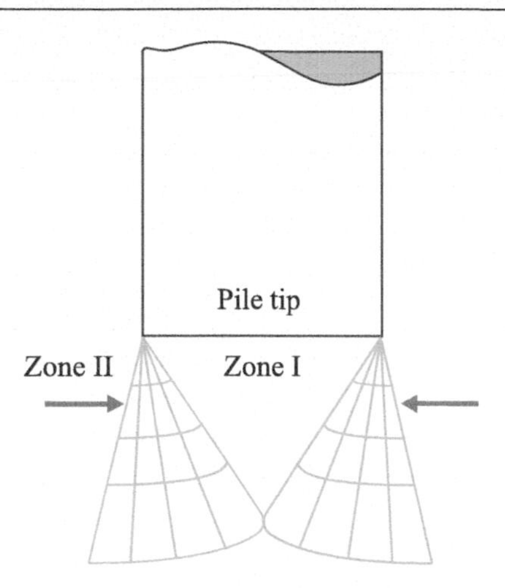

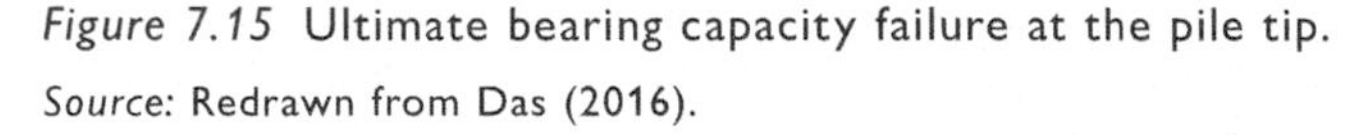

Figure 7.15 Ultimate bearing capacity failure at the pile tip.

Source: Redrawn from Das (2016).

interface frictional angle between the pile and soil δ , should be 0; the cohesion c should be the undrained shear strength of the clay s_u ; and the stresses σ_{vb} are total stresses.

If $\phi = 0$, then $N_q^* = 1$ and $N_\gamma^* = 0$.

The end-bearing capacity from eq. 7.5 is as follows:

$$Q_{bu} = A_b \left(cN_c^* + \sigma_{vb} N_a^* + 0.5\gamma d N_\gamma^* \right) = A_b (s_u N_c^* + \sigma_{vb}) \tag{7.6}$$

In many cases, $A_b \sigma_{vb} \approx W$. Thus, the net end-bearing capacity can be calculated as follows:

$$Q_{bu,net} = A_b (s_u N_c^*) \tag{7.7}$$

The bearing capacity factor, N_c^* , in eq. 7.7 has been investigated and proposed by several studies, as listed in Table 7.1. In general, $N_c^* = 9.0$ is a good value for the estimation.

End-bearing resistance in sand

When a pile is driven into sandy soils, the end-bearing resistance should be calculated with the drained parameters. All the frictional angles, including the internal frictional angle ϕ and the interface frictional angle between the pile and soil δ , should be drained values; the cohesion c should be the drained cohesion (c'); and the stresses σ'_{vb} are effective stresses. Based on field observation (Vesic, 1967a, 1967b), both the average frictional and end-bearing resistances become constant beyond a certain depth of penetration, as shown in Figure 7.16. The depth where resistances become constant is called critical depth. The normal stress on the pile surface is constant beyond the critical depth due to the arching effect. The critical depth varies between 4 to 60 times of pile width and depends on the friction angle of soils and construction method.

Table 7.1 Suggestion of bearing capacity factor N_c^*

Proposed N_c^* value	*Note*	*Reference*
6.14~9.0 ($\frac{\text{pile length}}{\text{pile diameter}} = 0 \sim \frac{\text{pile length}}{\text{pile diameter}} \geq 4$)	For circular area ($N_c^* = 9.0$ has been confirmed in tests in London clay and widely accepted in practice.)	Skempton (1951)
5.7~8.0	From model test	Sowers et al. (1961)
5.7~8.2	For expansive clay	Mohan (1963)
7.4~9.3	For insensitive clay	Ladanyi (1963)

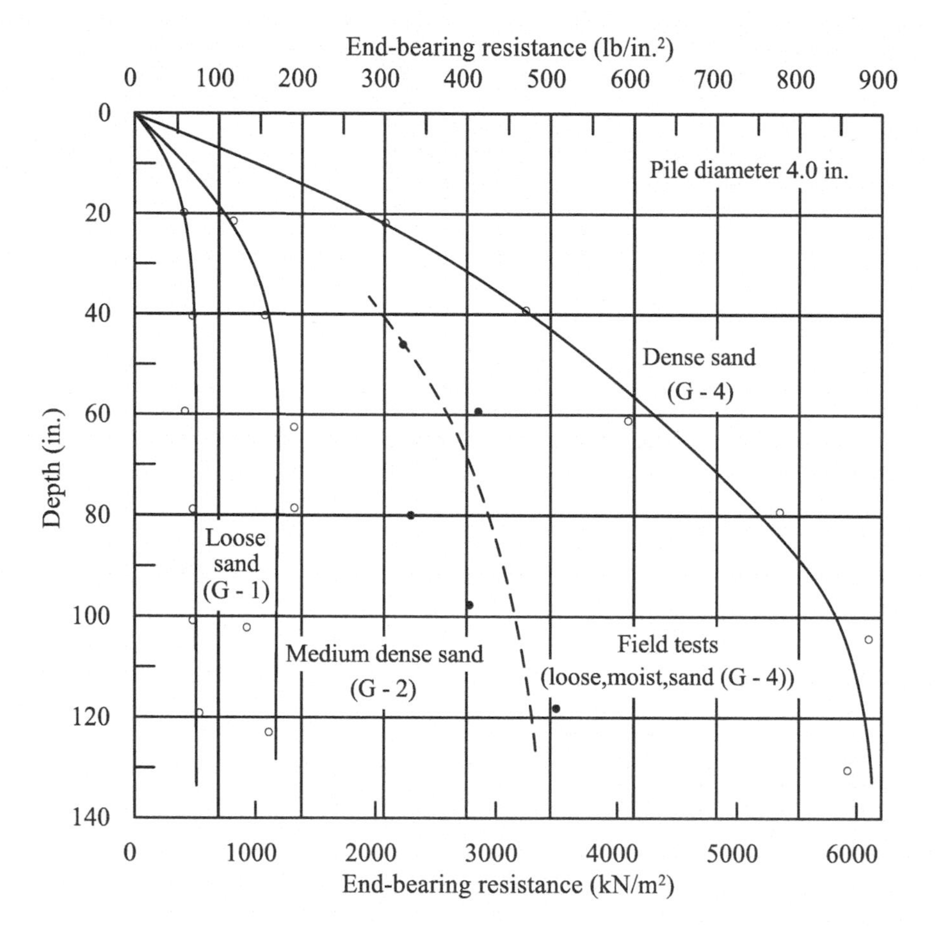

Figure 7.16 End-bearing resistance with penetration depth.

Source: Redrawn from Poulos and Davis (1980).

NAVFAC DM-7.2 (1982) suggested 20 times of pile width as the critical depth. If $c' = 0$, the net end-bearing capacity in sand can be calculated as follows:

$$Q_{bu,net} = A_b(\sigma'_{vb} N_q^* + \gamma d N_\gamma^*) \tag{7.8}$$

Where N_γ^* and N_q^* are factors in eq. 7.5 revised by Vesić. When the ratio of pile length to pile diameter exceeds 5, which is commonly seen in pile foundations, the $\gamma d N_\gamma^*$ term becomes negligible and may be ignored. The net end-bearing capacity can be calculated as follows:

$$Q_{bu,net} = A_b(\sigma'_{vb} N_q^*) \tag{7.9}$$

The coefficient N_q^* is a function of the friction angle and compressibility. The compressibility effect is defined with a rigidity index, I_r, of the soil (Vesic, 1977), as shown in eq. 7.10. N_q^* is computed with eq. 7.11.

$$I_r = \frac{\text{Shear modulus}}{\text{Shear strength}} = \frac{E}{2(1+\mu)\sigma'_{vb} \tan\phi'} \tag{7.10}$$

Where:

$E =$ modulus of elasticity of soil within the zone of influence.
$\mu =$ Poisson's ratio of soil within the zone of influence.
$\sigma'_{vb} =$ vertical effective stress at the toe elevation.

$$N_q^* = \frac{(1+2K)N_\sigma}{3}$$

$$N_\sigma = \frac{3}{3-\sin\phi'} e^{\frac{(90-\phi')\pi}{180}} \tan^2\left(45+\frac{\phi'}{2}\right) I_r^{\frac{4\sin\phi'}{3(1+\sin\phi')}} \tag{7.11}$$

Where:

$K =$ coefficient of lateral earth pressure.
$\phi' =$ effective friction angle, in degrees.

7.4.3 Frictional resistance

The ultimate frictional resistance along the pile surface, Q_{su}, is fully activated when sliding displacement between the pile and soil occurs, that is, shear failure at the interface of the pile and soil. Thus, the ultimate frictional resistance can be evaluated from the shear strength of the pile and soil interface.

$$\begin{aligned} Q_{su} &= \int_0^L U\tau_a dz \\ &= \int_0^L U(c_a + \sigma_n \tan\delta) dz \\ &= \int_0^L U(c_a + K_s \sigma_v \tan\delta) dz \end{aligned} \tag{7.12}$$

Where:

τ_a = pile–soil shear strength.
c_a = adhesion.
σ_n = normal stress between pile and soil.
δ = angle of friction between pile and soil.
σ_v = vertical stress.
K_s = coefficient of lateral pressure.
U = pile perimeter.
L = length of pile shaft.

Notably, the notations used in eq. 7.12 are general terms; for example, the vertical stress σ_v can be either total stress or effective stress, which depends on the ground material encountered. For saturated clay under **undrained condition (short-term),** all the properties in calculating ultimate load capacity should be values for the undrained condition. For instance, c_a is undrained pile–soil adhesion, ϕ (= 0), ϕ (= 0), and all stresses should be **total stresses**. For drained or long-term ultimate load capacity, all parameters should correspond to the drained values, and all stresses should be **effective stresses**. The difference will be illustrated in what follows when the ultimate load capacity for clays and sands is calculated.

Frictional resistance in clay

When a pile is driven into normal-consolidated clayey soils, the short-term frictional resistance should be calculated by undrained parameters. All the frictional angles, including the internal frictional angle ϕ and the interface frictional angle between the pile and soil ϕ_a, should be 0; the adhesion c_a should be the undrained pile–soil adhesion; and the stresses σ_v are total stresses.

If $\phi = \delta = 0$, then $\tan\delta = 0$.

The ultimate frictional resistance capacity from eq. 7.12 is:

$$Q_{su} = \int_0^L U(c_a + K_s\sigma_v \tan\delta)dz = \int_0^L U(c_a)dz \tag{7.13}$$

The pile–soil adhesion c_a is an undetermined parameter in eq. 7.13 when estimating the pile load capacity. It varies with the soil type, pile type, and installation method. It is generally determined from pile load tests and some empirical methods. Two empirical methods are introduced in the following. The undrained pile–soil adhesion c_a is determined based on the undrained shear strength of soils (α method) and the combination of mean effective vertical stress and average undrained shear strength along the pile (λ method).

α – method

In this method, the soil–pile adhesion c_a is assumed to be related to the undrained shear strength of soils c_a with a coefficient of α. That is:

$$\alpha = c_a / s_u \tag{7.14}$$

Table 7.2 Empirical correlations of α with s_u

$\frac{s_u}{P_a}$	α
≦0.1	1.00
0.2	0.92
0.3	0.82
0.4	0.74
0.6	0.62
0.8	0.54
1.0	0.48
1.2	0.42
1.4	0.40
1.6	0.38
1.8	0.36
2.0	0.35
2.4	0.34
2.8	0.34

Note: P_a = atmospheric pressure ≒100 kN/m² or 2,000 lb/ft².

Source: Interpolated values from Terzaghi et al. (1996).

Empirical correlations of α with s_u were established, as shown in Table 7.2. In general, softer clay has a higher α value. When $s_u \leq 20$ kPa, α is approximately 1.0. Once the soil–pile adhesion is determined, then the frictional resistance can be determined via eq. 7.14.

λ – method

In this method, the frictional resistance is calculated with the average pile–soil adhesion, which takes into account the mean effective vertical stress between the ground surface and pile tip and the average undrained shear strength along the pile. The average pile–soil adhesion factor is:

$$\frac{c_a}{s_{u,avg}} = \lambda\left(\frac{\sigma'_m}{s_{u,avg}} + 2\right) \text{ or } c_a = \lambda(\sigma'_m + 2s_{u,avg}) \tag{7.15}$$

Where:

σ'_m = mean effective vertical stress between the ground surface and pile tip.
$s_{u,avg}$ = average undrained shear strength along pile.
λ = coefficient of average pile–soil adhesion (see Table 7.3).

Once the average soil–pile adhesion c_a is determined, then the frictional resistance can be determined by substituting eq. 7.15 into eq. 7.13, which yields:

$$Q_{su} = \lambda\left(\sigma'_m + 2s_{u,avg}\right)A_s \tag{7.16}$$

Where:

A_S = pile surface area.

Table 7.3 Determination of the coefficient of average pile–soil adhesion

Embedment length, L (m)	λ
0	0.5
5	0.336
10	0.245
15	0.200
20	0.173
25	0.150
30	0.136
35	0.132
40	0.127
50	0.118
60	0.113
70	0.110
90	0.110

β – method

In some cases, such as piles in stiff and overconsolidated clay or long-term behavior of NC clay (refer to Section 2.2.4), the long-term pile capacity is an issue, and the frictional resistance should be calculated with the drained parameters. An effective stress approach, the β method, is thus proposed. It assumes that $c'_a = 0$ and applies the effective stress. Then, eq. 7.12 becomes:

$$Q_{su} = \int_0^L U(K_S \sigma'_v \tan\delta)dz \tag{7.17}$$

A coefficient β is thus defined as:

$$\beta = K_s \tan\delta \tag{7.18}$$

To determine the coefficient β, the following suggestions can be used.

1. For NC clay (Burland, 1973):

   $$\beta = (1 - \sin\phi')\tan\delta \tag{7.19}$$

   Where ϕ' is the effective stress friction angle for the clay.
   Thus, $\beta = 0.24 \sim 0.29$ for $\phi' = 20^\circ \sim 30^\circ$.

2. For OC clay (Burland, 1973):

   $$K_S = K_0 = (1 - \sin\phi')\sqrt{OCR} \tag{7.20}$$

   $\delta = \phi'_r$ = remolded friction angle.

3. Meyerhof (1976):

   $K_s = 1.5K_0$ for driven pile in stiff clay.

   $K_{s(\text{bored pile})} = 0.5K_{s(\text{driven pile})}$

Frictional resistance in sand

When a pile is driven into sandy soils, the frictional resistance should be calculated by drained parameters. All the frictional angles, including the internal frictional angle ϕ' and the interface frictional angle between the pile and soil δ, should be the drained values; the cohesion c should be the drained cohesion (c'); and stresses σ_v' are effective stresses. As mentioned in the calculation of the end-bearing capacity of a pile, the resistance force does not increase unlimitedly along the pile. Due to the arching effect, the normal stress on the pile surface is constant beyond a certain depth (L'), called the critical depth, as shown in Figure 7.17. The magnitude of the critical depth is normally in the range of 15 to 20 times the pile diameter. If $c' = 0$ and $c_a' = 0$, then the frictional resistance in eq. 7.12 becomes:

$$Q_{su} = \int_0^L U(K_S \sigma_v' \tan\delta) dz \tag{7.21}$$

When the effective stress σ_v' becomes constant beyond the critical depth, then the frictional resistance is approximately constant as well.

Some empirical expressions have been proposed for the coefficients K_s and $\tan\delta$ in eq. 7.21. Typically, the coefficients increase with the friction angle after pile driving. The β method is also valid for cohesionless soils, where $\beta = K_s \tan\delta$.

1. $\tan\delta$
   The coefficient of friction, $\tan\delta$, was measured with laboratory tests and correlated with the effective friction angle of the soil, ϕ'. Table 7.4 lists some typical values of δ/ϕ'.

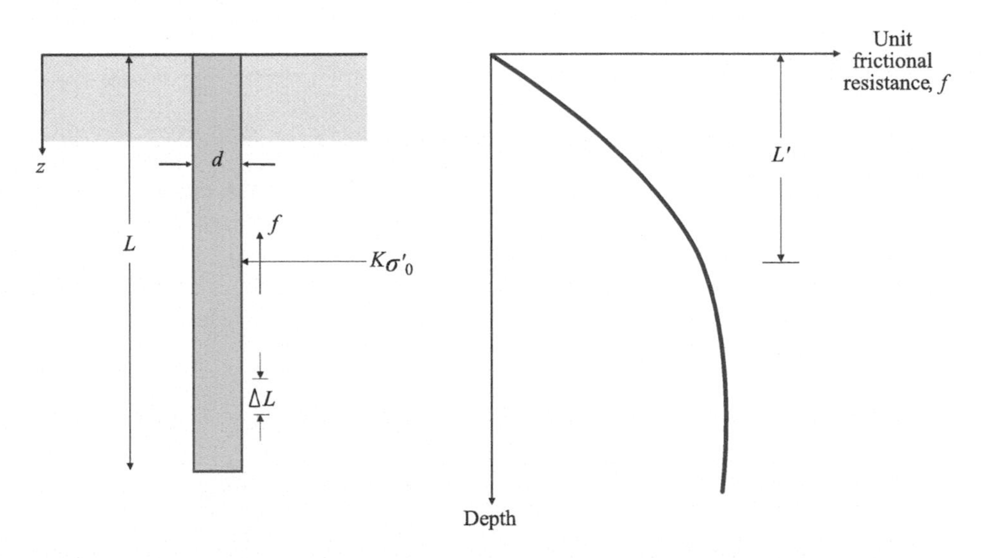

Figure 7.17 Critical depth of a pile driven in sands.

Source: Redrawn from Das (2016).

Table 7.4 Range of δ/ϕ' values for the soil–pile interface

Driven pile type	δ/ϕ'
Concrete	0.8–1.0
Rough steel (i.e., step-taper pile)	0.7–0.9
Smooth steel (i.e., pipe pile or H-pile)	0.5–0.7
Timber	0.8–0.9

Source: From Kulhawy et al. (1983) and Kulhawy (1991).

Table 7.5 Range of K_s/K_0 values for the driven pile

Method of construction	K_s/K_0
Driven pile—jetted	0.5–0.7
Driven pile—small displacement	0.7–1.2
Driven pile—large displacement	1.0–2.0

Source: From Kulhawy et al. (1983) and Kulhawy (1991).

2. K_s

   As defined earlier, K_s is the coefficient of lateral pressure, that is, the ratio between the horizontal and vertical effective stresses as shown in eq. 7.22.

$$K_S = \frac{\sigma'_h}{\sigma'_v} \tag{7.22}$$

The pile driving process will cause considerable compression to the surrounding soils and thus will result in a higher coefficient of lateral pressure, K_s, than that in the at-rest condition, K_0. The ratio of K_s/K_0 is listed in Table 7.5. The possible maximum value of K_s is K_p, the coefficient of passive earth pressure.

3. β

It is sometimes difficult to determine K_s and $\tan\delta$ separately. Thus, engineers often combine K_s and $\tan\delta$ into β, which is determined from published references or by load test results. β is approximately 0.3 for loose sands, while for dense sand, it might be as high as 1.0–1.5. Gravels can have much higher values, perhaps as large as 3. De Nicola and Randolph (1993) developed a relationship between β and D_r from static load tests on closed-end steel pipe piles in cohesionless soils.

$$\beta = 0.18 + 0.65D_r \tag{7.23}$$

Where D_r is the relative density of the soil expressed in decimal form.

7.4.4 Design with SPT data

The determination of pile load capacity presented earlier relies greatly on the reliability of soil properties obtained from lab tests on samples. In this section, the pile load capacity determined from field tests, including SPT and CPT, will be presented. However, those expressions may apply only to the regions where the data are collected. One should seek empirical expressions based on data from the local region where the pile is constructed.

SPT-N (Meyerhof, 1956; Soderberg, 1962)

A. For displacement piles in saturated sand:

$$Q_u = Q_{bu} + Q_{su} = 40N_p A_b + \frac{\bar{N}}{5} A_s \text{ (unit: force in tons, area in } m^2\text{)} \tag{7.24}$$

Where:

$\frac{\bar{N}}{5} < 10 \text{ tf/m}^2$.

N_p = standard penetration number, N, at pile base.

$\bar{N}$ = the average value of N along the pile shaft.

A_b = area of pile base.

B. For small displacement piles (e.g., steel H-pile):

$$Q_u = Q_{bu} + Q_{su} = 40N_p A_b + \frac{\bar{N}}{10} A_s \text{ (unit: force in tons, area in } m^2\text{)} \tag{7.25}$$

Where:

$\frac{\bar{N}}{5} < 5 \text{ tf/m}^2 \cdot$

A_b = net sectional area of toe.

A_s = gross surface area of shaft (area of all surfaces of flanges and web for H-piles).

7.4.5 Uplift capacity

Driven pile

The nominal uplift capacity of driven piles, $Q_{up,n}$, relies on the side friction capacity only. There is no end bearing, and the contribution from the weight of the pile is typically small. Thus, the selection of appropriate values of the side friction capacity is the most critical parameter in the design of piles subjected to uplift. According to field test results, the side friction capacity in uplift is 70–85% of the corresponding value for downward loading. The possible reasons are as follows (Tomlinson, 1969):

- The soil fabric near the pile changes due to the uplift load.
- The soil loosens due to uplift load and the shear surface.
- The Poisson effect during tensile loading decreases the pile diameter, thus slightly reducing the contact pressure.

Bored pile

Bored piles can be constructed with an enlarged (or belled) base. The purpose of an enlarged base is usually to increase the base area, and thus, the downward load is transferred through the end bearing. It also has the advantage of generating a higher upward load capacity by bearing on top of the enlarged portion. However, the belled bored piles are no longer widely used due to the safety concerns of workers and the required construction time. The alternative method is to construct a longer straight bored pile rather than an enlarged base.

Comparison between compression and tension pile load capacity

Two rectangular (barrette) piles with dimensions of $2.7\ \text{m}\times 1.0\ \text{m}\times 49.2\ \text{m}$ (width × thickness × length) were constructed at the same site in Taipei City, Taiwan. Static pile load tests for both compression and tension were performed. Figure 7.18 shows the plan of the tested pile and the soil profile. The pile load test result is shown in Figure 7.19, which clearly shows a lower frictional resistance in the tension (or uplift) pile than that in the compression pile under a similar settlement.

7.4.6 t-z curve

The *t-z* method (Kraft et al., 1981) is a widely used static analysis for developing pile load–settlement response at the pile head and vertical load distribution along the pile length. It considers three major components (side friction, end bearing, and elastic compression) and the interactions between them. Different side friction mobilization functions along the length of the pile are employed in the method for stratified soil profiles. The notation for variables t and z is inconsistent with the notation used in this book and usually causes confusion. The definitions of notations are listed here:

t: mobilized side friction resistance
z: displacement at a depth of pile
q: mobilized net end-bearing resistance

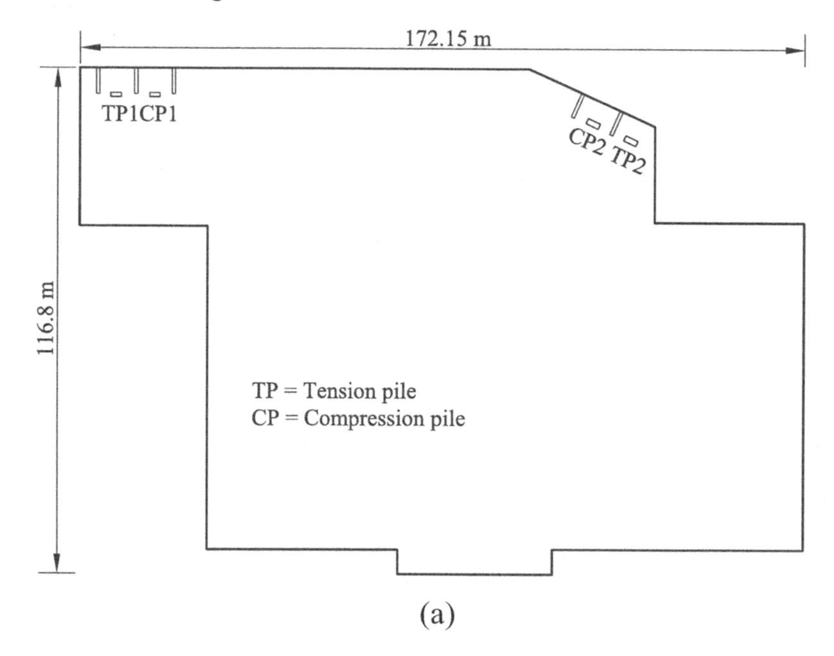

Figure 7.18 Pile load test: (a) site plan and (b) soil profile.

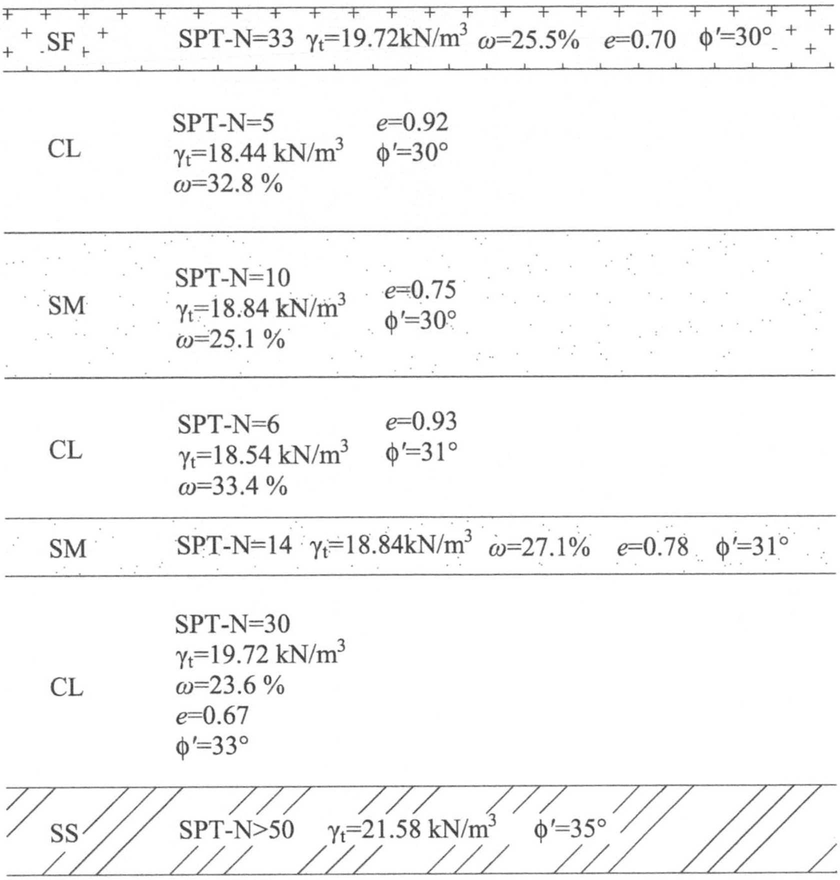

SF SPT-N=33 γ_t=19.72kN/m^3 ω=25.5% e=0.70 ϕ'=30°

CL SPT-N=5 γ_t=18.44 kN/m^3 ω=32.8 % e=0.92 ϕ'=30°

SM SPT-N=10 γ_t=18.84 kN/m^3 ω=25.1 % e=0.75 ϕ'=30°

CL SPT-N=6 γ_t=18.54 kN/m^3 ω=33.4 % e=0.93 ϕ'=31°

SM SPT-N=14 γ_t=18.84kN/m^3 ω=27.1% e=0.78 ϕ'=31°

CL SPT-N=30 γ_t=19.72 kN/m^3 ω=23.6 % e=0.67 ϕ'=33°

SS SPT-N>50 γ_t=21.58 kN/m^3 ϕ'=35°

(b)

Figure 7.18 (Continued)

The *t-z* method divides the pile into finite segments, as shown in Figure 7.20. The elastic properties for each segment are defined by the cross area, A, modulus of elasticity, E, and segment length. The side friction resistance between each segment and the soil is modeled using a nonlinear spring whose load–displacement behavior is defined by a *t-z* curve. The end bearing is also modeled using a nonlinear spring, which is defined by a *q-z* curve.

Functions for *t-z* and *q-z* curves can be found in published references, which are usually derived from static load tests or numerical analyses. If in situ load test data are available, curves may be custom-fitted and then used in a *t-z* analysis to extrapolate the load test result. An example of the *t-z* curve from the static pile load test is shown in Figure 7.21, where the site is that mentioned in Section 7.4.5. The mobilized side friction resistance can be determined by measuring the axial force taken by the pile along the pile shaft. The difference between the two sequential measurements at depth is the mobilized side friction resistance in the pile segment. To perform a *t-z* analysis, the settlement, δ, at the pile top is set as the independent variable, and an initial value is chosen. The numerical model is then solved to obtain force equilibrium and a value of the corresponding applied load, P. Repeated solutions with different δ values produce a load–settlement plot and thus calculate the ultimate pile load capacity. Commercial software

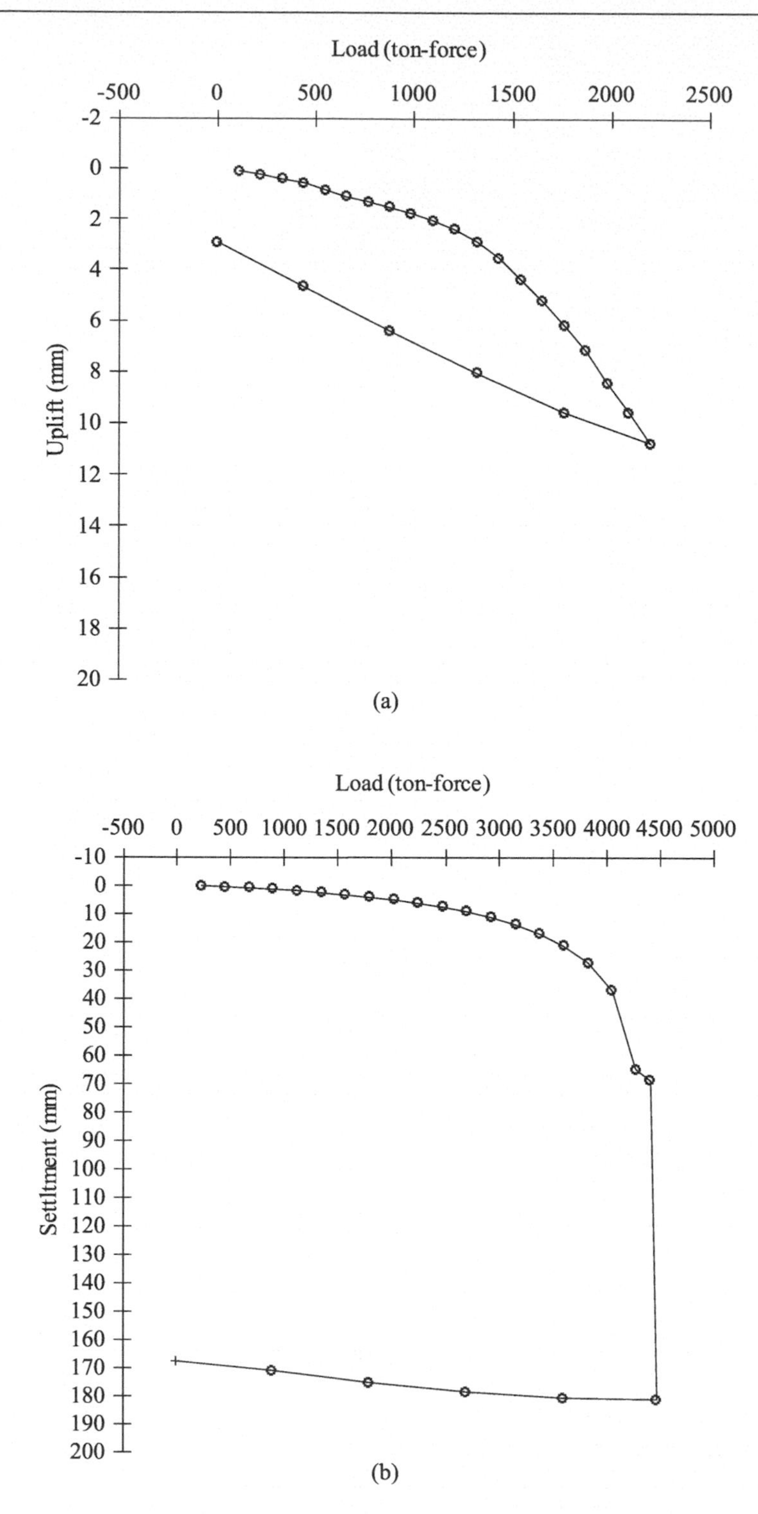

Figure 7.19 Pile load test results: (a) Tension test and (b) compression test.

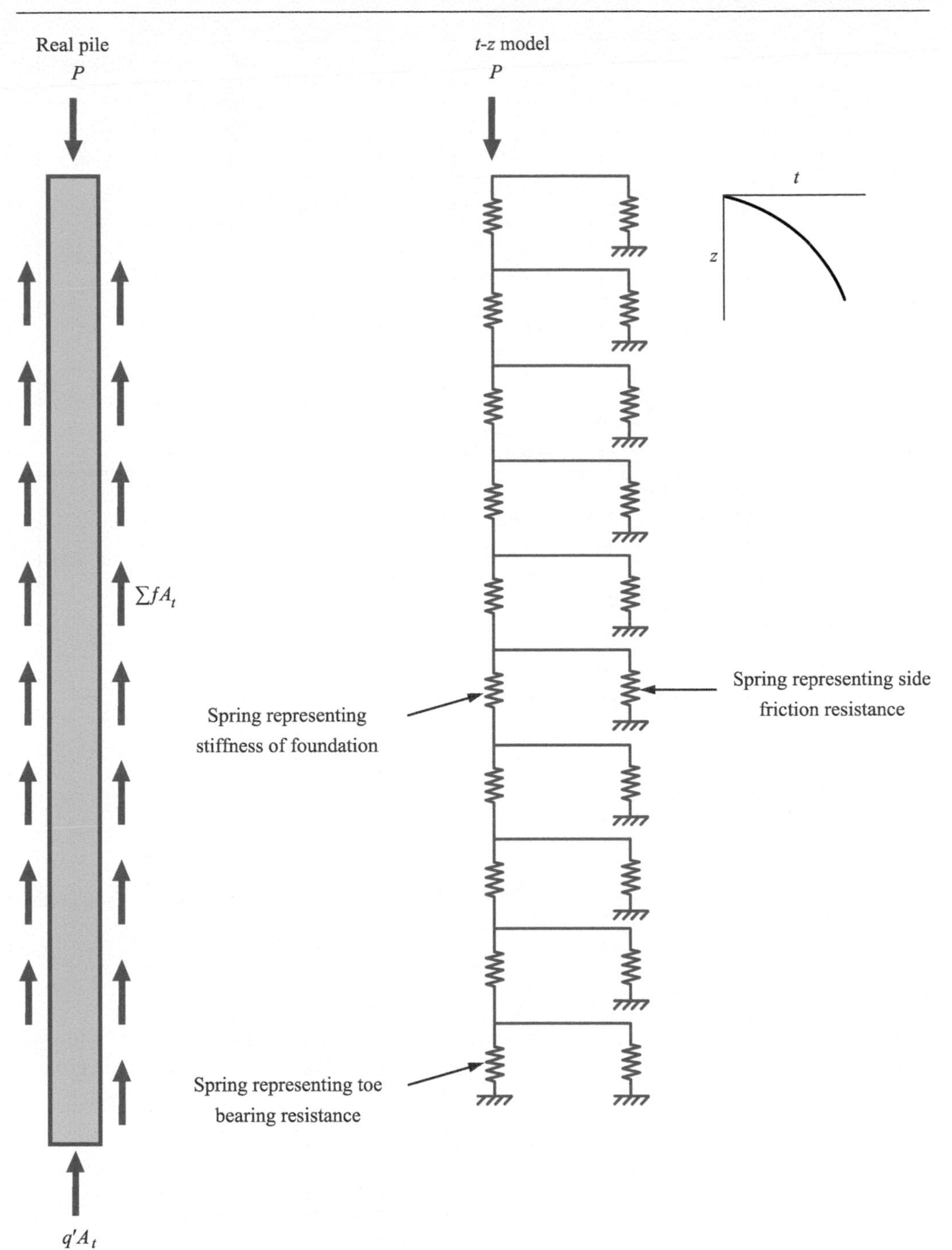

Figure 7.20 Numerical model for the *t-z* method.

Source: Redrawn from Coyle and Reese (1966).

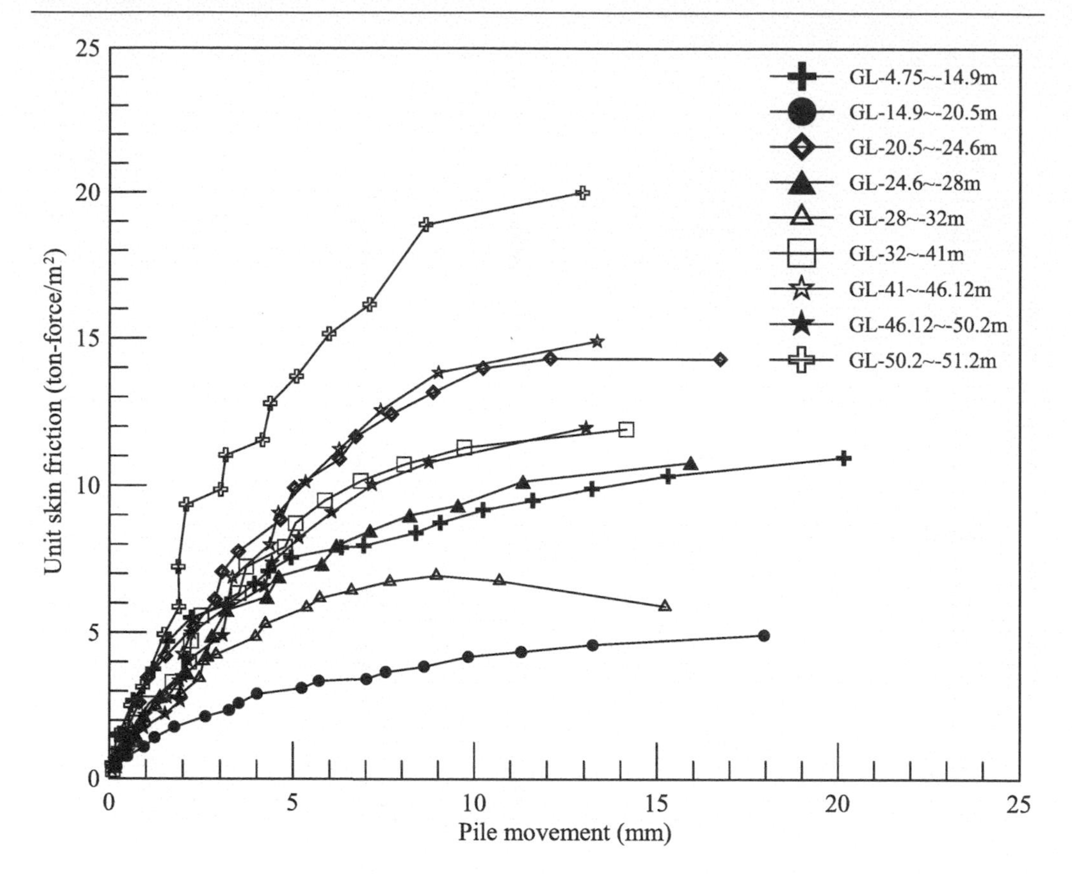

Figure 7.21 t-z curves at different depths derived from a static pile load test.

capable of performing *t-z* analysis are available. For more detailed analysis, refer to Coyle and Reese (1966).

Example 7.1: Pile load capacity in sand

- A pile 20 m in length and 0.6 m in diameter is driven into a sandy layer with the following properties: $\gamma = 18\,kN/m^3$, $\phi' = 33^\circ$, and $E_s = 35\,MPa$. The groundwater table is on the ground surface. Determine the end-bearing resistance of the pile.

Solution

The end-bearing resistance of the pile in sand can be calculated by eq. 7.9.

$$Q_{bu,net} = A_b(\sigma'_{vb} N_q^*)$$

The pile base area: $A_b = \frac{1}{4}\pi d^2 = 0.283 m^2$.

The critical depth is 20 times the pile diameter, that is, 12 m.

The effective stress at the pile base equals to that at critical depth: $\sigma'_{vb} = \gamma' z = (18 - 9.81) \times 12 =$ 98.28 kPa

$$I_r = \frac{E}{2(1+\nu)\sigma'_{vb}\tan\phi'} = \frac{35\text{ MPa}}{2\times(1+0.3)\times 98.28\text{ kPa}\times\tan 33^{\circ}} = 210.92$$

$$N_\sigma = \frac{3}{3-\sin\phi'} e^{\frac{(90-\phi')\pi}{180}} \tan^2(45+\frac{\phi'}{2}) I_r^{\frac{4\sin\phi'}{3(1+\sin\phi')}} = 138.63$$

$$N_q^* = \frac{(1+2K)N_\sigma}{3} = \frac{\left[1+2(1-\sin\phi')\right]N_\sigma}{3} = 88.29$$

$$Q_{bu,net} = A_b(\sigma'_{vb}N_q^*) = 0.283\text{ m}^2 \times 98.28\text{ kPa} \times 88.29 = 2455\text{ kN}$$

Example 7.2: Pile load capacity in clay

- A pile 20 m in length and 0.6 m in diameter is driven into clayey layers, as shown in the figure. Determine the end-bearing resistance and frictional resistance of the pile.

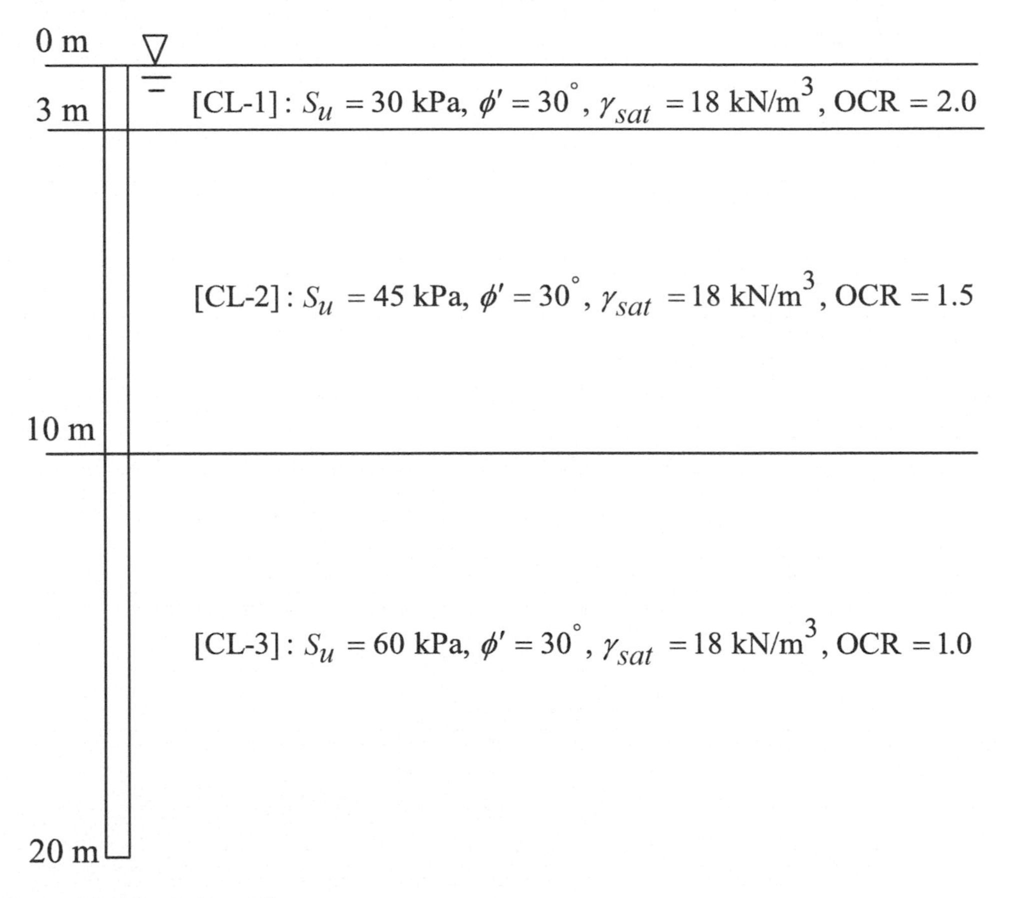

Figure EX 7.2a Soil profile.

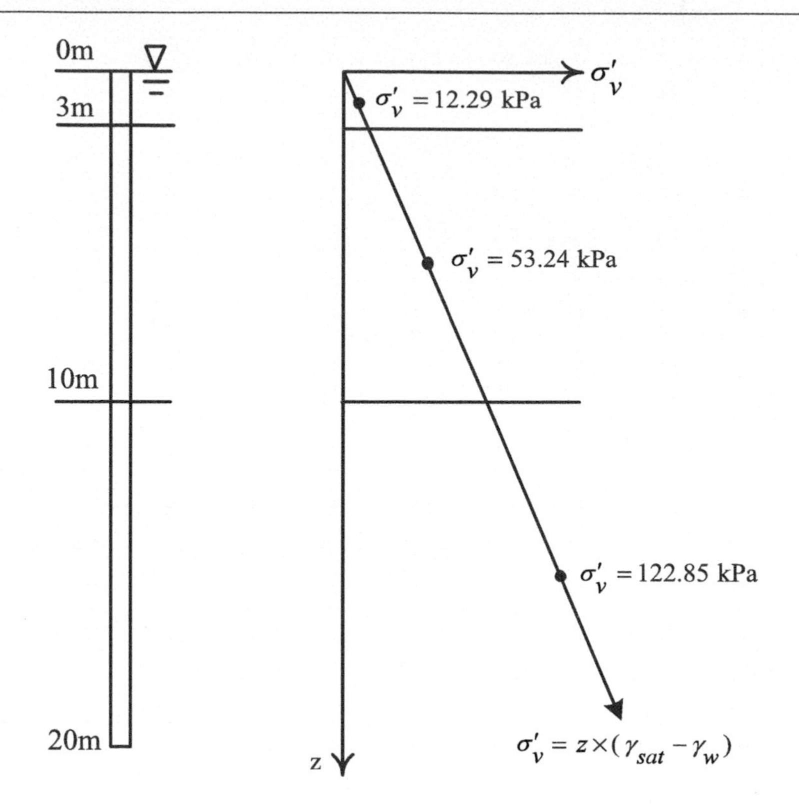

Figure EX 7.2b Vertical effective stress distribution.

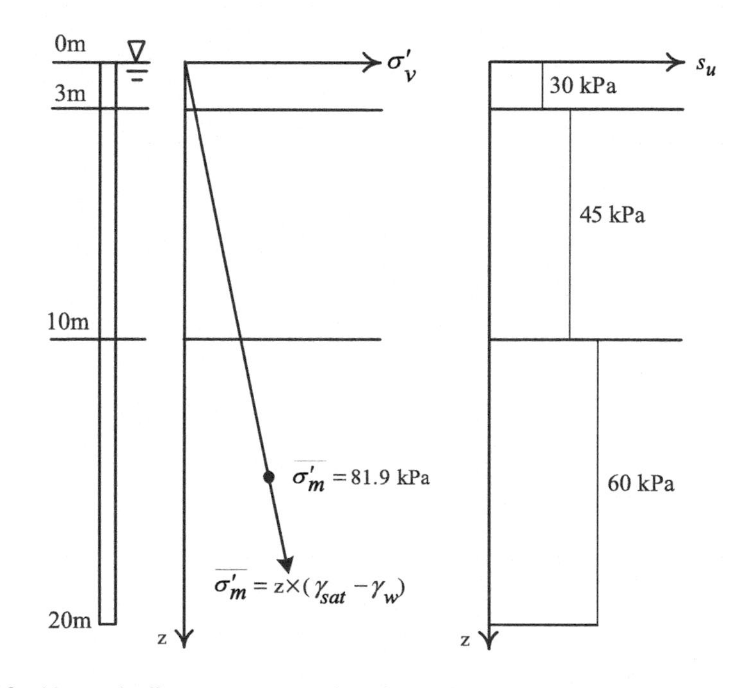

Figure EX 7.2c Vertical effective stress and undrained shear strength distribution.

Solution

1. Q_{bu}

$Q_{bu} = 9s_u A_b$

$A_b = \pi(0.3)^2 = 0.283\ \text{m}^2$

The pile tip is at GL. –20 m; thus, $s_u = 60$ kPa .

$Q_{bu} = 9s_u A_b = 9 \times 60 \times 0.283 = 152.82$ kN

2. Q_{su}

$Q_{su} = \sum f_s U \Delta L$

$U = \pi \times 0.6 = 1.885$ m

f_s can be determined by three methods, that is, the α method, β method, and λ method.

α method

Depth (m)	ΔL (m)	s_u (kPa)	α	f_s (kPa)	$f_s U \Delta L$ (kN)
0–3	3	30	0.82	24.60	139.1
3–10	7	45	0.71	31.95	421.6
10–20	10	60	0.62	37.20	701.2
					Q_{su} = 1262

β method

Depth (m)	ΔL (m)	σ'_v (kPa)	K	β	f_s (kPa)	$f_s U \Delta L$ (kN)
0–3	3	12.29	0.71	0.41	5.04	28.5
3–10	7	53.24	0.61	0.35	18.63	245.8
10–20	10	122.85	0.5	0.29	35.63	671.6
						Q_{su} = 946

λ method

$Q_{su} = \lambda\left(\sigma'_m + 2s_{u,avg}\right)A_s$; $\overline{\sigma'_m} = 81.9$ kPa

$s_{u,avg} = \frac{1}{20}\left(s_{u1} \times \Delta L_1 + s_{u2} \times \Delta L_2 + s_{u3} \times \Delta L_3\right) = 50.25$ kPa

$\lambda = 0.173$ from Table 7.3

$Q_{su} = 0.173(81.9 + 2 \times 50.25) \times 1.885 \times 20 = 1190$ kN

7.5 Lateral loaded piles

Lateral loads from superstructures applied to vertical piles will be carried by the lateral resistance of surrounding soils. When significant horizontal loads are applied to vertical piles, the pile head (or the ground surface) should not have severe movements, that is, no failure should occur at the upper part of the ground. Pile caps could be connected with horizontal concrete pads or beams to minimize movement and provide sufficient lateral resistance. Responses of single vertical piles subjected to lateral forces and evaluations of their capacities will be introduced in this section.

7.5.1 Classification of lateral piles

Two categories are used in evaluating piles subjected to lateral loads: short piles and long piles, as shown in Figure 7.22. A short pile refers to a pile without sufficient embedment length that thus behaves like a rigid body when subjected to lateral forces. The toe of a short pile tends to rotate as it is loaded laterally at the pile head. A rotation point is assumed in the pile and results in the rigid rotation of a short pile, as shown in Figure 7.22. The ultimate lateral load capacity of short piles is controlled by the soils surrounding the pile. Some soils in front of the piles yielded and thus contributed to the ultimate lateral resistance before the pile had any flexural deflections. In contrast, the long pile is the pile embedded deeply into the ground and thus places restraints on any rotations and/or horizontal movements at the pile toe. Several points in the pile with a deflection equal to zero will be found in a sufficiently long pile. The ultimate lateral capacity of long piles is thus controlled by its flexural capacity, which is governed by the pile structural design.

To classify the short pile and long pile, the deflection characteristics governed by the soil–pile interaction are employed. The pile is treated as a beam, which is loaded laterally. The responses of soil and pile interaction are solved by a nonlinear differential equation as well as equilibrium and compatibility conditions. For detailed information, refer to Chapter 8 in FHWA-HRT-04-043 (2006). The governing equation for pile deflection, y, subjected to later loads is shown in eq. 7.26. The lateral load and displacement were usually solved numerically. The eigenvalue for eq. 7.26 is β , which controls the deformation pattern of a lateral loaded pile and can be calculated by eq. 7.27. A dimensionless length factor βL (L is the pile length) is employed to classify lateral piles. When $\beta L \leq 1$, the pile is treated as a short pile or rigid pile; $\beta L > 3$ indicates a long pile. Among long piles, $3 \geq \beta L > 1$ indicates a semirigid pile, and $\beta L > 3$ indicates a nonrigid pile. The lateral load capacity is of interest to geotechnical engineers and will be introduced in the next section.

$$E_p I_p \frac{d^4 y}{dz^4} + E_s y = 0 \tag{7.26}$$

$$\beta = \sqrt[4]{\frac{E_s}{4E_p I_p}} \tag{7.27}$$

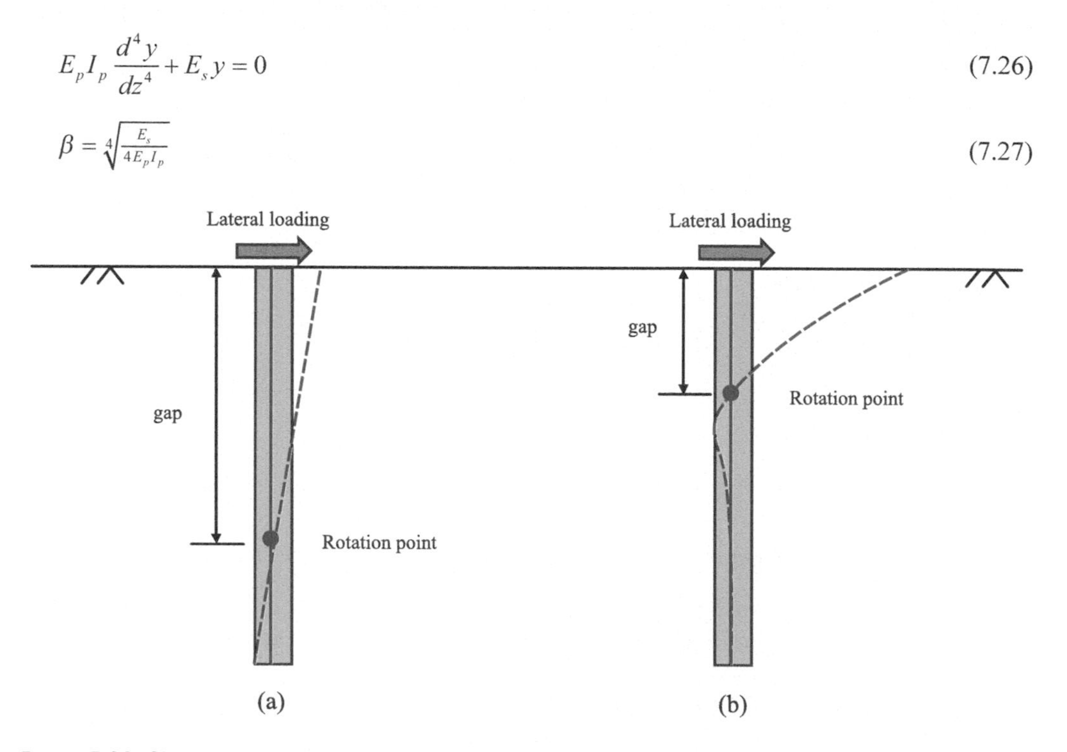

Figure 7.22 Short and long lateral piles: (a) short pile and (b) long pile.
Source: Redrawn from Zhang et al. (2016).

Where:

E_p = modulus of elasticity of the pile.
E_s = subgrade modulus of soil reactions.
I_p = moment of inertia of the pile in the direction of bending.
y = lateral deflection of the pile.
z = depth below the ground surface.

7.5.2 Lateral load capacity for rigid piles

The evaluation of the ultimate lateral load capacity for a vertical rigid pile follows the procedures proposed by Broms (1964a, 1964b, 1965), which are based on some assumptions, for example, the pile behaves like a rigid body after being pushed by a horizontal force. The prediction from Broms's method agreed reasonably with field test results, which indicates

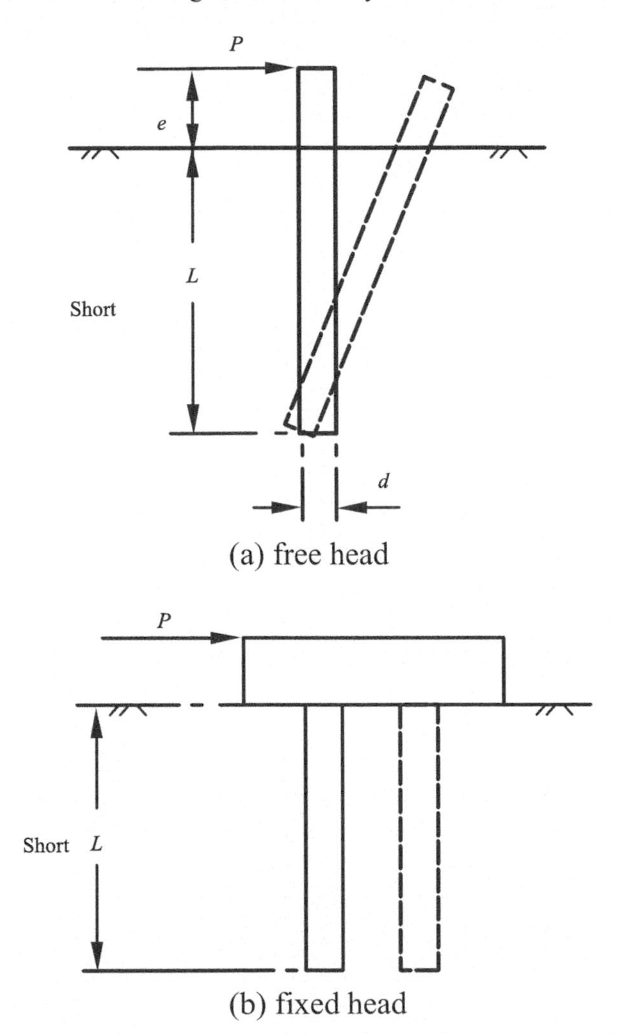

Figure 7.23 Failure mechanism for short piles.

Source: Redrawn from Broms (1964a, 1964b).

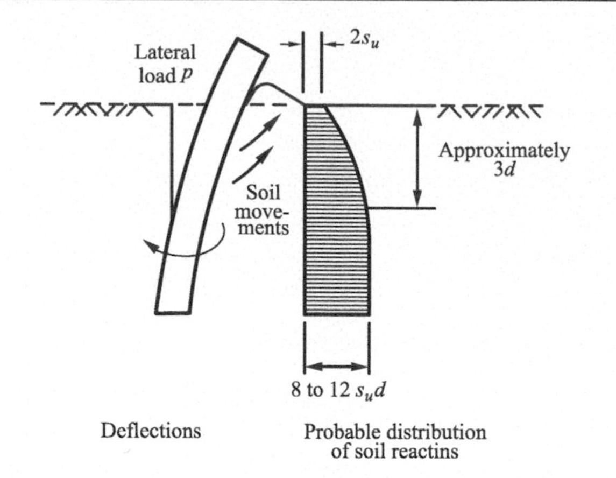

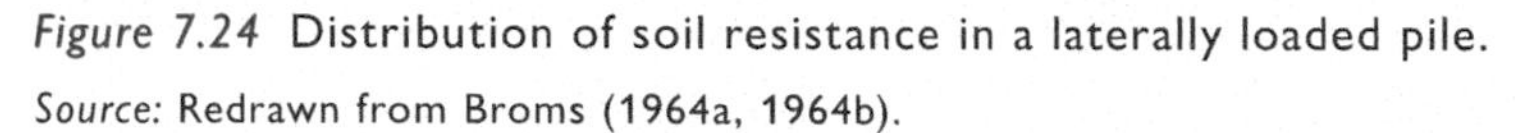

Figure 7.24 Distribution of soil resistance in a laterally loaded pile.

Source: Redrawn from Broms (1964a, 1964b).

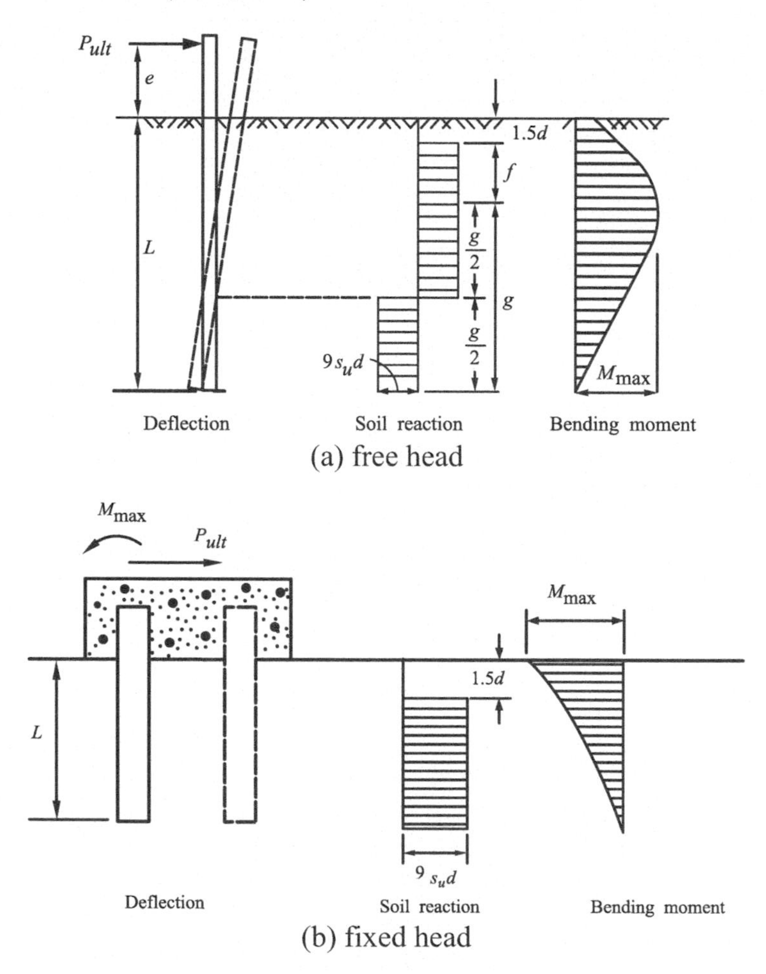

Figure 7.25 Suggested distribution of soil resistance in a laterally loaded pile in cohesive soils.

Source: Redrawn from Broms (1964a, 1964b).

that the assumptions are justified. The evaluation method is divided into piles in cohesive soils and granular soils with different pile head conditions, that is, free and fixed head. When a short rigid pile is subjected to horizontal loading, the possible failure mechanisms for both the free head and fixed head are shown in Figure 7.23. It is clear that the failure is governed by the soil strength. For the free-head rigid pile, the soils above and below rotation point A all reach the yielding condition (on different sides of the pile), as shown in Figure 7.23 (a).

Rigid piles in cohesive soils: Free head and fixed head

Based on Figure 7.24, the ultimate resistance of cohesive soils pushed by a horizontally loaded pile is in the range of 8 to 12 times the undrained shear strength (s_u). The lateral resistance reaches the value at a depth of three times the pile diameter. Therefore, Broms (1964a, 1964b, 1965) suggested a distribution of soil resistance, as shown in Figure 7.25. An ultimate soil resistance of $9s_u$ appears at a depth of 1.5 times the pile diameter. For a free-head short pile subjected to a horizontal load, Broms assumed that the maximum bending moment occurs at a depth of $1.5d+f$, as shown in Figure 7.25a, where the shear is zero. Thus, $f = P_{ult}/9s_u d$, where P_{ult} is the ultimate applied pressure at the pile head. Assuming that all soils in the pile length yield, the depth of the rotation point can be determined based on the force equilibrium. By taking moments about the pile tip and the maximum moment location, the ultimate applied pressure P_{ult} and maximum moment M_{max} can be calculated by solving eqs. 7.28 to 7.32 (Broms, 1964a, 1964b, 1965).

$$f = \frac{P_{ult}}{9s_u d} \tag{7.28}$$

$$M_{max} = P_{ult}\left(e + 1.5d + 0.5f\right) \tag{7.29}$$

$$M_{max} = 2.25 s_u d g^2 \tag{7.30}$$

$$g = L - 1.5d - f \tag{7.31}$$

$$P_{ult} = 9s_u d^2\left[\left(4\left(\tfrac{e}{d}\right)^2 + 2\left(\tfrac{L}{d}\right)^2 + 4\left(\tfrac{e}{d}\right)\left(\tfrac{L}{d}\right) + 6\left(\tfrac{e}{d}\right) + 4.5\right)^{0.5} - \left(2\left(\tfrac{e}{d}\right) + \left(\tfrac{L}{d}\right) + 1.5\right)\right] \tag{7.32}$$

For a fixed-head short pile subjected to a horizontal load, the soil resistance is shown in Figure 7.25b. Based on the force equilibrium and taking the moment at the pile tip, the ultimate applied pressure P_{ult} and maximum moment M_{max} can be calculated by eq. 7.33 and eq. 34, respectively. The dimensionless solutions are given in Figure 7.26 based on eqs. 7.32 and 7.33. For the case of piles with fixed heads, the solutions imply that the moment restraint from the pile cap is equal to that in the pile just below the cap (Broms, 1964a, 1964b, 1965).

$$P_{ult} = 9s_u d^2\left(L/d - 1.5\right) \tag{7.33}$$

$$M_{max} = P_{ult}\left(\tfrac{L}{2} + \tfrac{3d}{4}\right) = 4.5 s_u d^3\left[\left(L/d\right)^2 - 2.25\right] \tag{7.34}$$

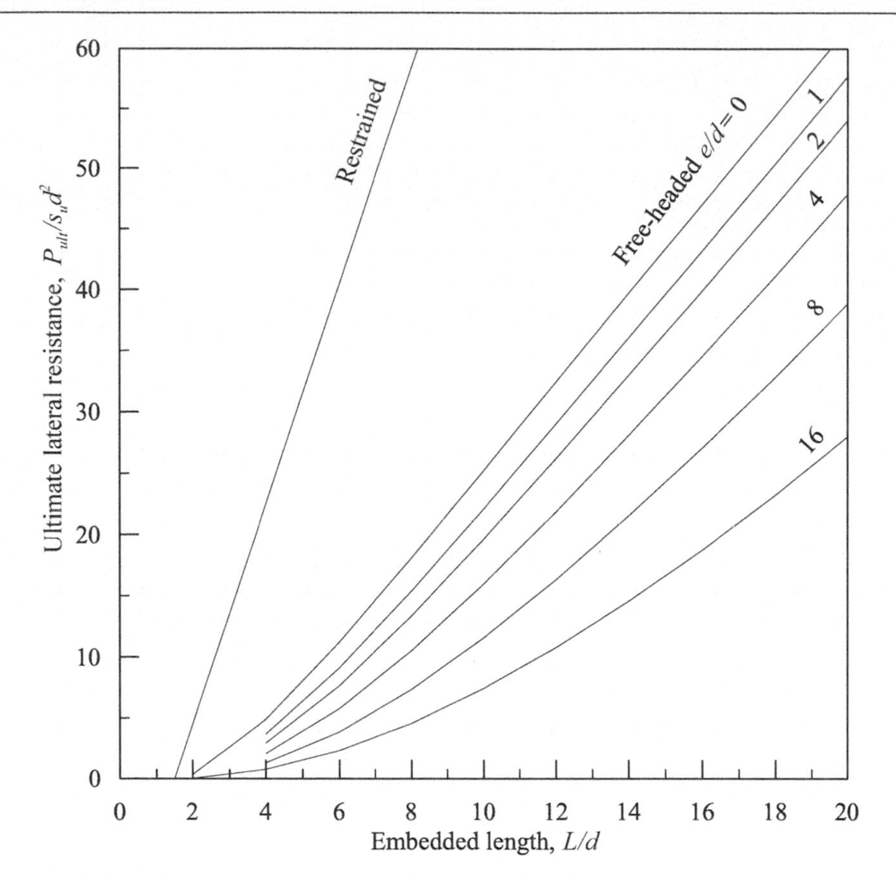

Figure 7.26 Ultimate lateral load for short piles in cohesive soils.
Source: Redrawn from Broms (1965).

Rigid piles in granular soils: Free head and fixed head

Possible failure mechanisms for short piles in granular (cohesionless) soils are shown in Figure 7.27. The following assumptions are made by Broms (1964a, 1964b):

- The active earth pressure acting on the back of the pile is neglected.
- The distribution of passive pressure along the front of the pile is equal to 3 times the Rankine passive pressure (from certain observations).
- The shape of the pile section has no influence on the distribution of the ultimate soil pressure or ultimate lateral resistance.
- The full lateral resistance is mobilized at the movement considered.

For a free-head short pile subjected to a horizontal load, Broms (1964a, 1964b) assumed that the maximum bending moment occurs at a depth of f, as shown in Figure 7.27a, where the shear will be zero. Thus, $f = \sqrt{2P_{ult}/3K_p\gamma d}$, where K_p is the coefficient of passive pressure ($K_p = (1+\sin\phi')/(1-\sin\phi')$). Based on the force equilibrium and taking the moment at the pile tip, the ultimate applied pressure P_{ult} and maximum bending moment M_{max} can be calculated by eq. 7.35 and eq. 7.36, respectively (Broms, 1964a, 1964b, 1965).

$$P_{ult} = \frac{K_p \gamma d L^2}{2(1+e/L)} \tag{7.35}$$

$$M_{max} = P_{ult}(e+L) - \frac{3K_p \gamma d L^3}{2\times 3} = P_{ult}\left(e + \tfrac{2}{3}f\right) \tag{7.36}$$

For a fixed-head short pile subjected to a horizontal load, the soil resistance is shown in Figure 7.27b. Based on the force equilibrium and taking the moment at the pile tip, the ultimate applied pressure P_{ult} and maximum bending moment M_{max} can be calculated by eq. 7.37 and eq. 7.38, respectively. Based on eqs. 7.35 and 7.37, dimensionless solutions are given in Figure 7.28 (Broms, 1964a, 1964b, 1965).

$$P_{ult} = \tfrac{3}{2} K_p \gamma d L^2 \tag{7.37}$$

$$M_{max} = K_p \gamma d L^3 \tag{7.38}$$

Example 7.4: Lateral pile load capacity

- A short pile 3 m in length and 0.6 m in diameter is subjected to a horizontal force at an elevation of 0.6 m. The soil is soft clay with $s_u = 30\text{kPa}$. Determine the ultimate lateral capacity (P_{ult}) of the pile and the maximum moment (M_{max}).

Solution

Based on the given conditions, we have $d = 0.6\text{m}$, $e = 0.6\text{m}$, $L = 3.0\text{m}$, and $s_u = 30\text{kPa}$.

$$f = \frac{P_{ult}}{9 s_u d} = \frac{P_{ult}}{162}$$

$$g = L - 1.5d - f = 2.1 - \frac{P_{ult}}{162}$$

$$M_{max} = P_{ult}(e + 1.5d + 0.5f) = P_{ult}\left(1.5 + \frac{P_{ult}}{324}\right)$$

$$M_{max} = 2.25 s_u d g^2 = 40.5\left(2.1 - \frac{P_{ult}}{162}\right)^2$$

By solving the preceding equations simultaneously, a quadratic equation can be obtained.

$$40.5 P_{ult}^2 + 66922.2 P_{ult} - 4687309.62 = 0$$

$$P_{ult} = 67.3 \text{ kN}$$

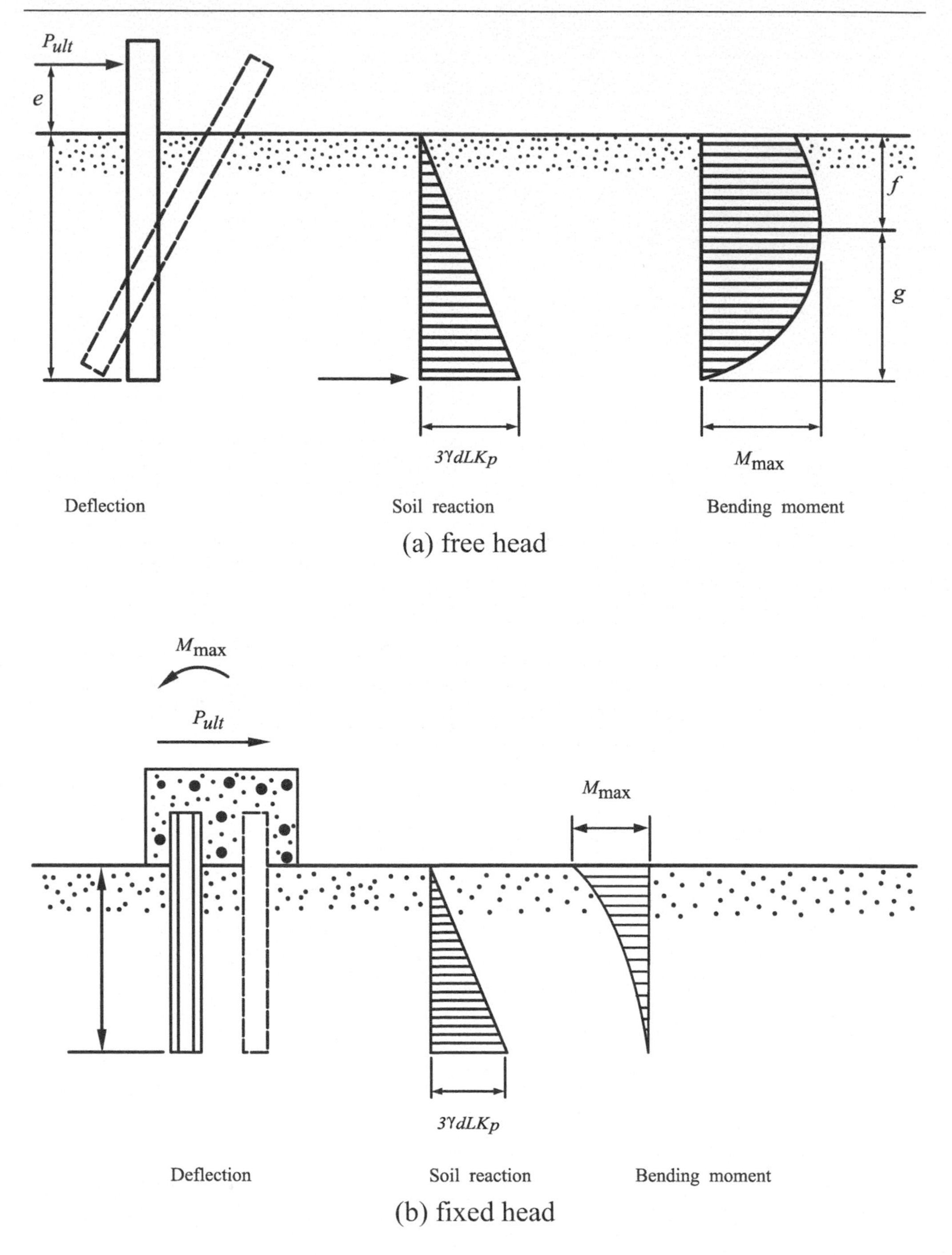

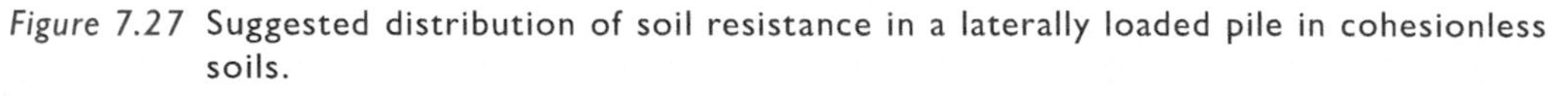

Figure 7.27 Suggested distribution of soil resistance in a laterally loaded pile in cohesionless soils.

Source: Redrawn from Broms (1964a, 1964b).

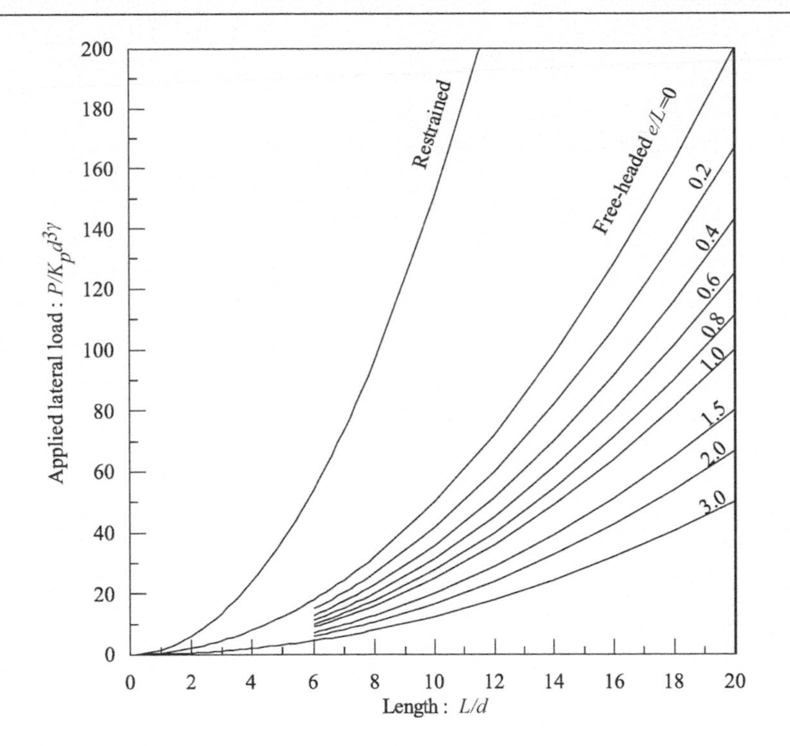

Figure 7.28 Ultimate lateral load for short piles in cohesionless soils.

Source: Redrawn from Broms (1965).

or

$$P_{ult} = 9s_u d^2\left[\left(4\left(\tfrac{e}{d}\right)^2 + 2\left(\tfrac{L}{d}\right)^2 + 4\left(\tfrac{e}{d}\right)\left(\tfrac{L}{d}\right) + 6\left(\tfrac{e}{d}\right) + 4.5\right)^{0.5} - \left(2\left(\tfrac{e}{d}\right) + \left(\tfrac{L}{d}\right) + 1.5\right)\right] = 67.3 \text{ kN}$$

$$M_{\max} = P_{ult}\left(1.5 + \frac{P_{ult}}{324}\right) = 67.3\left(1.5 + \frac{67.3}{324}\right) = 114.9 \text{ kN-m}$$

7.5.3 Lateral load capacity for nonrigid piles

For nonrigid piles, that is, long piles, rigid pile analyses do not provide accurate predictions due to simplified assumptions used in the rigid-pile method. More theoretical lateral load evaluation methods that consider the flexural rigidity of the piles, the soil's responses to the lateral force, and the soil–pile interaction have been developed. These methods require more complex analysis and the calculation of more parameters to yield a more precise prediction; thus, there are no simple closed-form solutions. The finite element method is a good numerical tool to solve the highly nonlinear soil–pile interaction problem, but it is beyond the scope of this book. Instead, the p-y method using the Winkler beam-on-elastic-foundation model is introduced here.

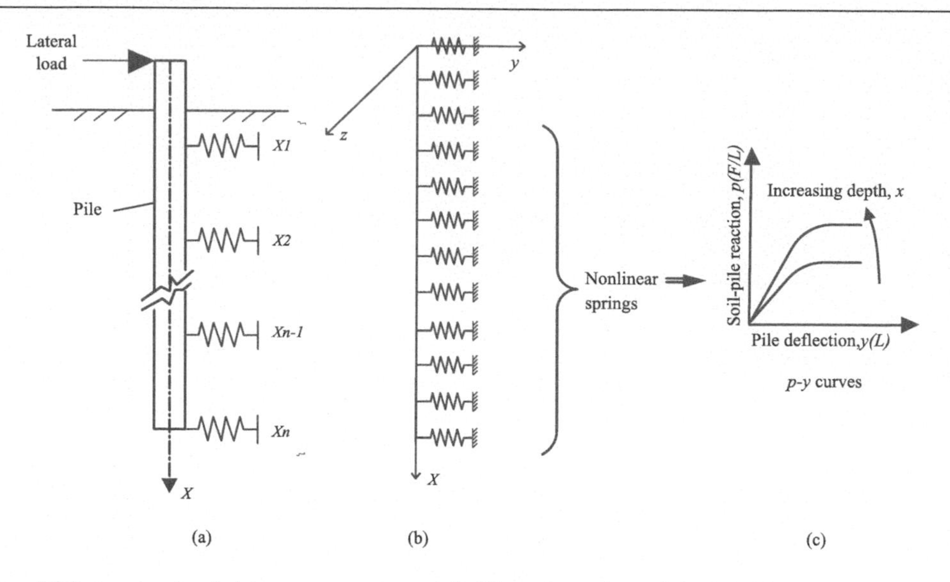

Figure 7.29 *p-y* method: (a) conceptual model, (b) analytical model, and (c) *p-y* curve.

Source: Redrawn from Reese et al. (1974).

p-y method

The *p-y* method considers the soil–pile interaction by using a series of nonlinear springs, which is similar to the method introduced for the mat foundation. The only difference is the orientation of the soil–pile interaction springs. They are in the horizontal direction, as shown in Figure 7.29a. The *p-y* method has been calibrated and verified with existing lateral pile load test data and supported by numerous commercial software programs. It is a well-accepted and preferred method for most practical design problems.

The first step of the *p-y* method is to divide the pile into finite intervals (or elements) with a node at the end of each interval. The nonlinear springs, that is, the soils, are located at each node, as shown in Figure 7.29b. The flexural rigidity of each element is defined by E_pI_p of the pile, where E_p is the modulus of elasticity of the pile and I_p is the moment of inertia of the pile in the direction of bending. The load–deformation responses of each spring are defined by a *p-y* curve, as shown in Figure 7.29c, which implies that the soil–pile reaction at any depth is proportional to the lateral displacement. Notably, the springs act independently. When applying the *p-y* model with appropriate boundary conditions, the software will apply lateral loads in increments and compute the shear, bending moment, and lateral deflection at each element.

p-y curve and its factors

The definition of the lateral load-deflection responses of soil–pile interaction, that is, *p-y* curves, are the key to the *p-y* method. In the *p-y* curve, *p* stands for the lateral soil resistance per unit length of the pile, and *y* is the lateral deflection. The *p-y* curve is usually nonlinear and is not like the linear spring in the Winker beam-on-elastic-foundation model. The shape of the *p-y* curve at a certain depth of the pile depends on the following factors:

- Soil properties
- Loading type (static, cyclic, or dynamic)

- Pile diameter and cross-sectional shape
- Interface properties between soils and piles
- Depth below the ground surface
- Pile construction method
- Group interaction effects

Although the factors of the *p-y* curves are listed here, their individual impacts are not well quantified. The *p-y* curve is usually empirically back-calculated from static load tests or using correlation based on soil properties in *p-y* software. To determine *p-y* curves in soft clays, Matlock's method (1970) is usually employed. This method is based on reference data and two lateral load tests on 324 mm diameter steel pipe piles in soils with $s_u = 38\text{kPa}$ and $s_u = 15\text{kPa}$. Matlock's *p-y* curve for static loading in soft clays is shown in eq. 7.39 and Figure 7.30. The ultimate soil resistance per unit length of pile, P_{ult}, is smaller than that calculated with eq. 7.40.

$$\frac{p}{p_{ult}} = 0.5\left(\frac{y}{y_{50}}\right)^{1/3} \tag{7.39}$$

$$P_{ult} = \left(3 + \frac{\gamma'_{avg}}{s_u} z + \frac{J}{d} z\right) s_u d \tag{7.40 a}$$

$$P_{ult} = 9 s_u d \tag{7.40 b}$$

Where:

γ'_{avg} = average effective unit weight from the ground surface to the depth is considered.
J = 0.5 for soft clay, 0.25 for medium clay.
d = pile diameter.
z = depth below the ground surface.

The deflection, y_{50}, at which $0.5P_{ult}$ occurs is:

$$y_{50} = 2.5\varepsilon_{50} d \tag{7.41}$$

ε_{50} = strain at one-half (50%) of the ultimate compressive strength of the clay.
Typical values of ε_{50} are shown in Table 7.6.

Example 7.5: p-y curve by Matlock's method

- A 400 mm diameter pile is embedded into a saturated medium clay that has an average unit weight of 15 kN/m^3 and an undrained shear strength of 30 kPa. The groundwater table is near the ground surface. Use the Matlock method to develop the static curve at a depth of 5 m.

Solution

From Table 7.6, $\varepsilon_{50} = 0.01$.

$$\gamma' = 15 - 9.8 = 5.2\ \text{kN/m}^3$$

$$y_{50} = 2.5\varepsilon_{50}d = 2.5(0.01)(400\text{mm}) = 10\text{mm}$$

$$P_{ult} = \left(3 + \frac{\gamma'_{avg}}{s_u}z + \frac{J}{d}z\right)s_u d$$

$$= \left(3 + \frac{5.2}{30}\times 5 + \frac{0.25}{0.4}\times 5\right)\times 30\times 0.4$$

$$= 84\ \text{kN/m}$$

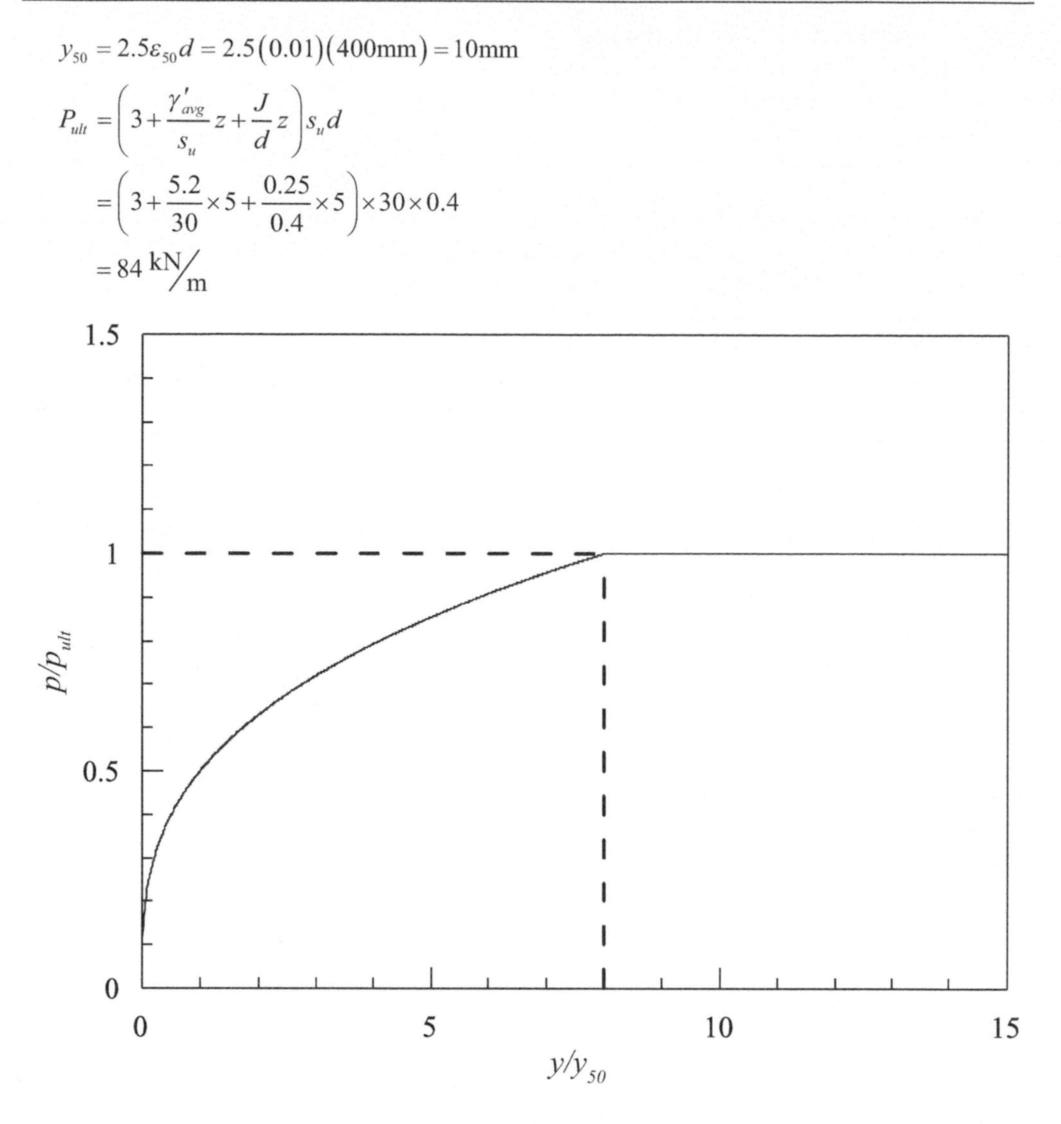

Figure 7.30 Matlock's p-y curve for static loading of a single pile in soft clays

Table 7.6 Typical ε_{50} values in clays

Consistency of clay	s_u(kPa)	ε_{50}
Very soft	<12	0.02
Soft	12–24	0.02
Medium	24–48	0.01
Stiff	48–96	0.006
Very stiff	96–192	0.005
Hard	>192	0.004

Source: Reese et al. (2006).

Alternatively:

$$P_{ult} = 9s_u d$$
$$= 108 \text{ kN/m}$$

Thus, use $P_{ult} = 84 \text{ kN/m}$.

Curve equation:

$$\frac{P}{P_{ult}} = 0.5\left(\frac{y}{y_{50}}\right)^{1/3} \leq 1;$$

$$P = 42\left(\frac{y}{10}\right)^{1/3}.$$

7.6 Pile load test

The methods of pile load capacity determination in Section 7.4 involved some assumptions, local empirical experience, and uncertainties that make predictions that do not match the real behavior of installed piles. Thus, in most large projects, a specific number of load tests must be conducted on piles. The primary reason is the unreliability of prediction methods. The vertical and lateral load-bearing capacity of a pile can be tested in the field by applying the loads to the test pile directly. Loads higher than design values, for instance, 200% of design values, may be applied. The detailed configuration and interpretation of the pile load test will be presented in this section.

7.6.1 Configuration of pile load test

A pile load test consists of the tested pile, reaction piles, reaction beams, hydraulic jacks, and monitoring systems, as shown in Figure 7.31. The function of the reaction piles and beams is to provide the required reaction force to load the tested pile; thus, the capacity of the reaction system should be greater than that of the tested pile. Loadings and displacements were measured during the test via load cells, LVDTs, strain gauges, etc.

There are two types of static load tests: stress-controlled tests and strain-controlled tests. The former applies a predetermined load (the independent variable) and measured settlements (the dependent variable). The latter increases the load to maintain a constant rate of penetration (usually vary from 0.25 to 2.5 mm/min). ASTM D1143 describes both procedures for compression piles, and ASTM D3689 covers tests on uplift (tensile) piles. There are two types of stress-controlled tests, that is, the maintained load test (ASTM D1143 Procedure B) and the quick test (ASTM D1143 Procedure A). As mentioned in ASTM D1143, test loads should be applied following one of the procedures described in the following for each test method. If feasible, the maximum applied load should reach a failure that reflects the total axial static compressive load capacity of the pile. Both methods give a load–settlement plot, and details are introduced here.

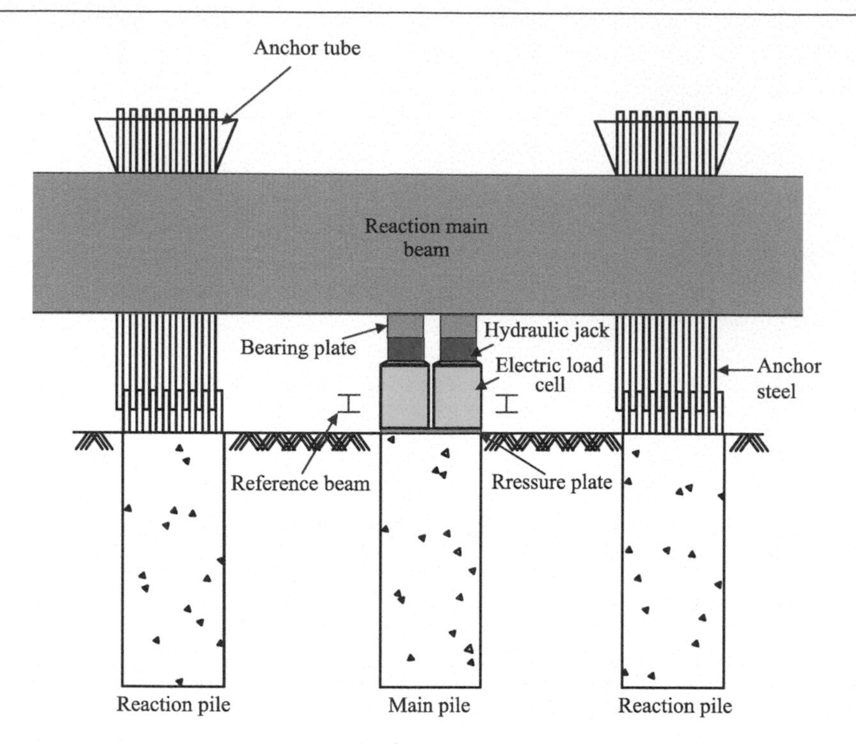

Figure 7.31 Configuration of a pile load test.

Maintain load test

In the maintain load test, the load increment is approximately 25% of the design load, and each load is maintained for two hours or until the rate of settlement is less than 0.25 mm per hour. The loading process is continued until 200% of the design load (for a proof test) or failure is reached. For the proof test, the maximum load is held for a longer period for creep effect observation. Then, the pile is progressively unloaded.

Quick test

The quick test is faster than the maintain load test. The load increment is approximately 5% of the design allowable load, and each load is held for a predetermined time interval regardless of the rate of pile displacement at the end of that interval. ASTM permits intervals of 4 to 15 minutes. The test will continue until approximately 200% of the design load or failure is reached. The quick test generally requires 4 to 6 hours to complete, while the maintain load test requires several days to complete. Thus, the quick test is commonly adopted to shorten the total test duration.

7.6.2 *Interpretation of test results*

After the pile load test, a load–settlement curve (Figure 7.32) will be obtained to determine the ultimate load capacity of the pile. Several determination methods for the ultimate load capacity are suggested and introduced in the following.

1. Davisson's method:
   It is a widely used method to determine the load capacity of a pile, such as in the AASHTO and international building code (IBC).

$$S = 4 + \frac{d}{120} + \frac{QL}{AE} \text{ (unit: mm)} \tag{7.42}$$

Where:

S = settlement at the pile head.
d = pile diameter (mm).
Q = the applied load (kN).
L = pile length (mm).
A = pile cross section area (mm^2).
E = the Young's modulus of the pile (kN/mm^2).
In eq. 7.42, the 4 mm (approximately 1.5 in) refers to the **pile head settlement contribute from shaft displacement** at the ultimate load condition, and $d/120$ (approximately $0.01d$) refers to the **pile head settlement contribute by base displacement** at the ultimate load condition. The sum of these two terms is the offset, and QL/AE is the **elastic deformation of the pile**. Thus, the ultimate load capacity is the intersection of the settlement curve and the linear line in eq. 7.42, as illustrated in Figure 7.32.

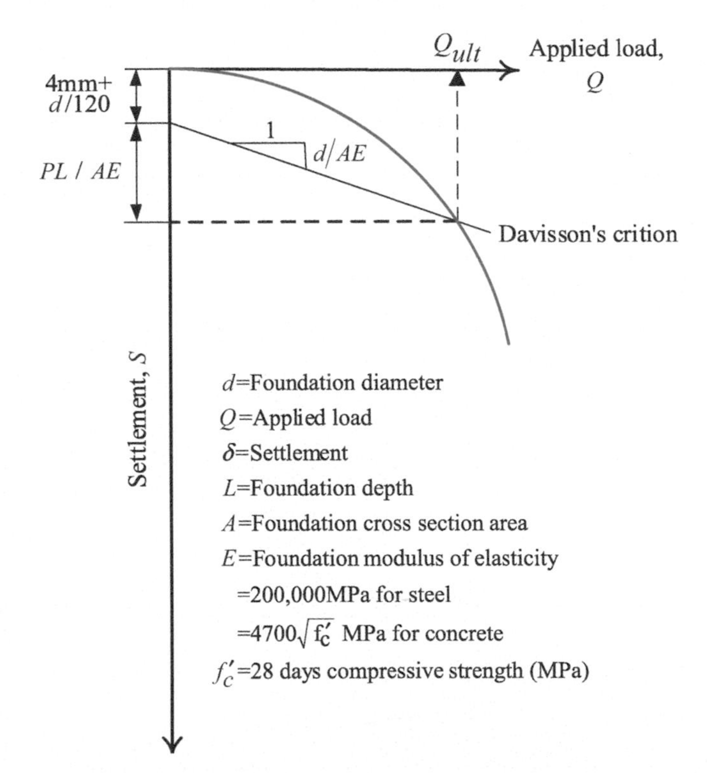

Figure 7.32 Load–settlement curve of a pile load test.

2. Fuller and Hoy method:
   The ultimate load is the load where the slope = 0.140 mm/kN on the plot of the load–settlement curve.
3. Terzaghi's method:
   The ultimate load is the load where the settlement equals 10% of the pile diameter.

Example 7.6: Interpretation of pile load test results

- A pile load test was performed on a pile 30 m in length and 0.6 m in diameter. The test result is shown in the figure that follows. Determine the ultimate load capacity of the pile.

Solution

Solutions with different methods are shown in Table EX 7.6 and Figure EX 7.6a and b.

Table EX 7.6

Method	P_u (kN)
Davisson	5827
Fuller and Hoy	5739
Terzaghi	5994

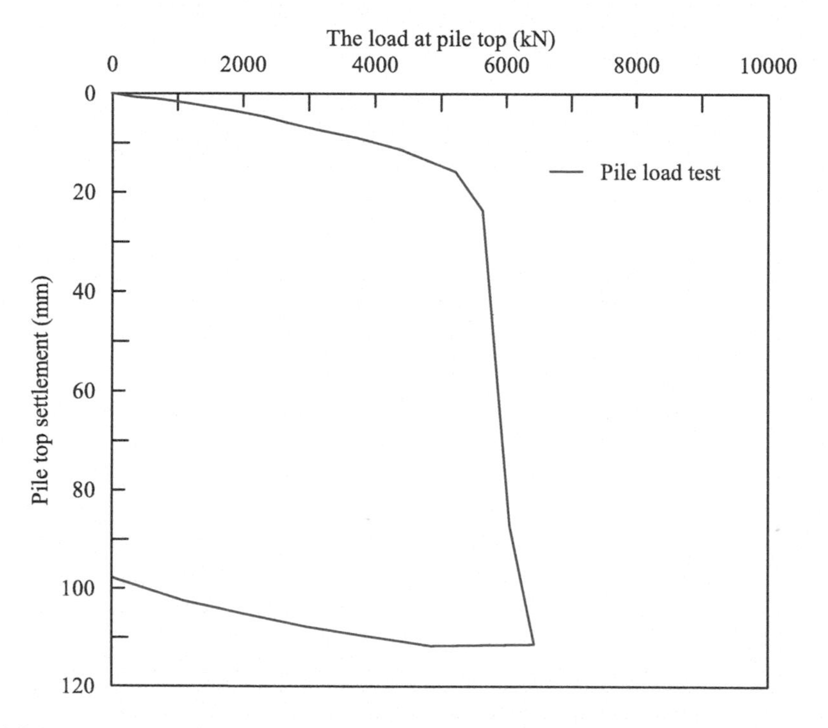

Figure EX 7.6a Load–settlement curve in the pile load test.

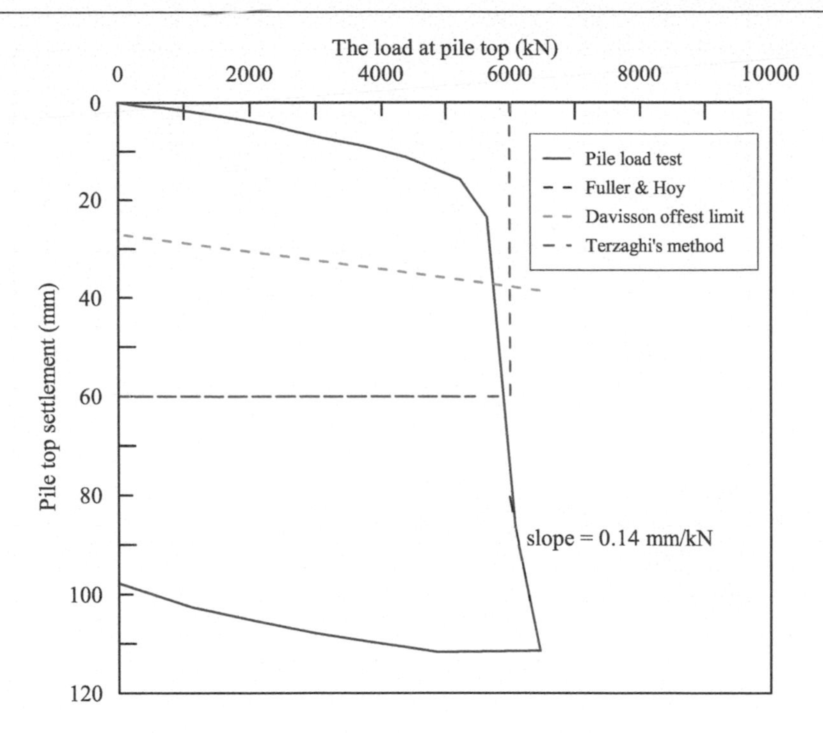

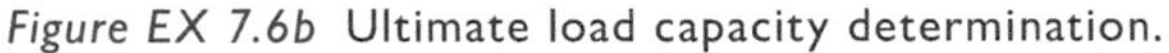
Figure EX 7.6b Ultimate load capacity determination.

7.6.3 Correlation of pile load test results with other in situ tests

The pile load test gives engineers detailed information on the pile–soil interaction behaviors, and the most important among them is the ultimate load capacity of the pile on-site. The authors collected more than 200 pile load test data points. The ultimate load capacities determined by Davisson's offset method were extracted from the data. In addition to the pile load test, other in situ test data were also collected, such as data for the standard penetration test and cone penetration test. The ultimate pile load capacity is thus correlated to the SPT N value and cone resistance q_c, which are very common in situ test results. Figure 7.33 shows the correlation between ultimate load capacity and SPT-N and q_c. The SPT-N values used in Figure 7.33a were the average between 10 pile diameters above and 4 pile diameters below the pile tip. The length-to-diameter ratio, L/d, of the pile is used as another factor in the figure. It is clear that a higher average SPT-N and a larger L/d yield a higher ultimate load capacity. However, some scatters are also observed in the data. In Figure 7.33b, the average cone tip resistances, $q_{c,avg}$, are calculated along the pile length. In the figure, it is also clear that a higher $q_{c,avg}$ and a larger L/d yield a higher ultimate load capacity.

7.7 Settlement analysis of single piles

The total settlement of a pile under a vertical working load, Q_w, is given by:

$$S_e = S_{e(1)} + S_{e(2)} + S_{e(3)} \tag{7.43}$$

Where:

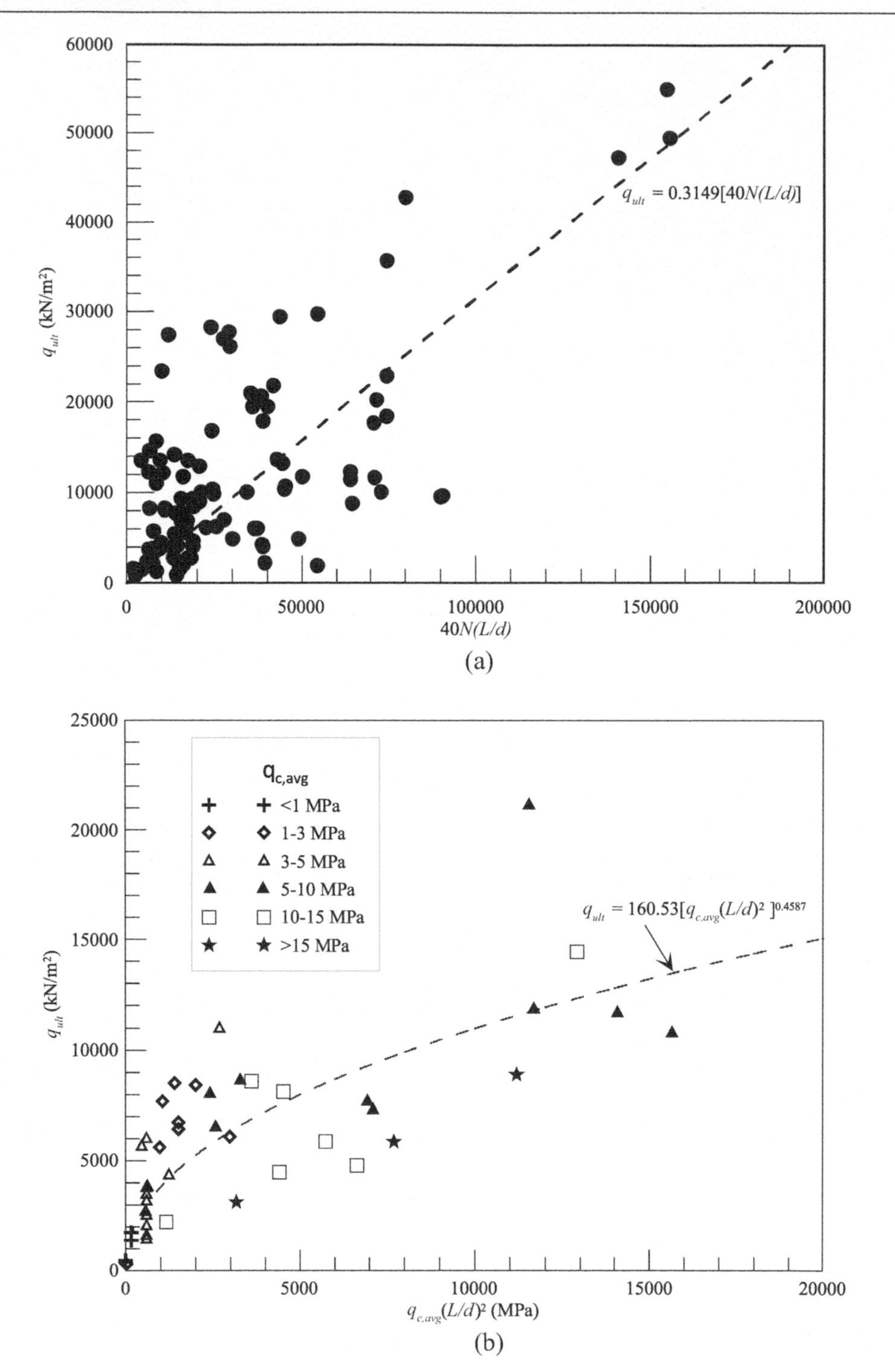

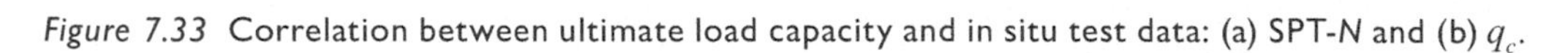

Figure 7.33 Correlation between ultimate load capacity and in situ test data: (a) SPT-*N* and (b) q_c.

$S_{e(1)}$ = elastic settlement of the pile.
$S_{e(2)}$ = settlement of the pile caused by the load at the pile tip.
$S_{e(3)}$ = settlement of the pile caused by the load transmitted along the pile shaft.

$$S_{e(1)} = \frac{(Q_{wp} + \xi Q_{ws})L}{A_p E_p} \tag{7.44}$$

Where:

Q_{wp} = load carried at the pile tip under working load conditions.
Q_{ws} = load carried by frictional resistance under working load conditions.
ξ varies between 0.5 and 0.67 and will depend on the nature of the distribution of the unit friction (skin) resistance along the pile shaft.

$$S_{e(2)} = \frac{Q_{wp} d}{A_p E_s}(1 - \mu_s^2) I_{wp} \tag{7.45}$$

Where:

d = width or diameter of the pile.
E_s = modulus of elasticity of soil at or below the pile tip.
μ_s = Poisson's ratio of soil (refer to Table 2.4).
I_{wp} = influence factor ≈ 0.85.

$$S_{e(3)} = \left(\frac{Q_{ws}}{UL}\right)\frac{d}{E_s}(1 - \mu_s^2) I_{ws} \tag{7.46}$$

Where:

U = perimeter of the pile.
L = embedment length of the pile.
I_{ws} = influence factor $= 2 + 0.35\sqrt{L/d}$ (Vesic, 1977).

7.8 Negative skin friction

When piles are driven through soft layers (such as fill layers or soft clay) into firmer layers, the piles are subjected to extra loads caused by negative skin friction in addition to the applied load on the pile head. Negative skin friction is a downward drag acting on a pile due to the downward movement, that is, settlement, of the surrounding compressible soil relative to the pile. Figure 7.34 shows the pile passing through a recently constructed cohesive soil fill. The soil below the fill is completely consolidated under its overburden pressure. The determination of negative skin friction is still being studied, and some models have been proposed. Figure 7.35 shows a neutral plane model that assumes that a natural plane exists at a certain depth where the negative and positive friction forces are identical.

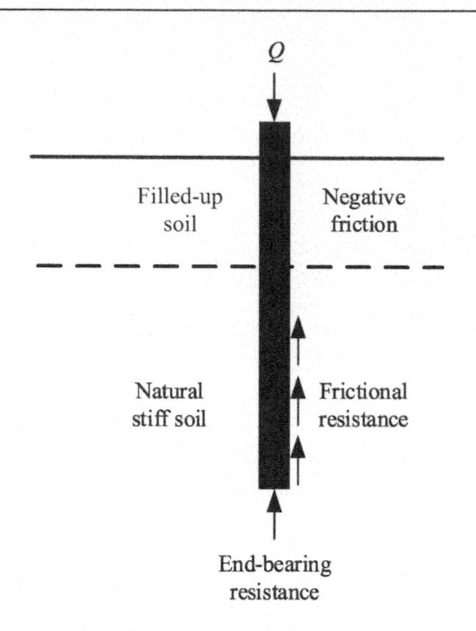

Figure 7.34 Illustration of negative skin friction.

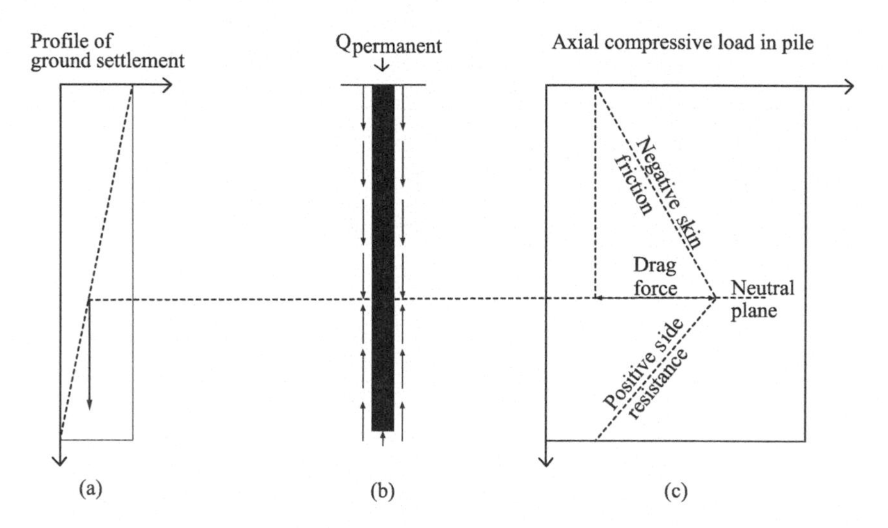

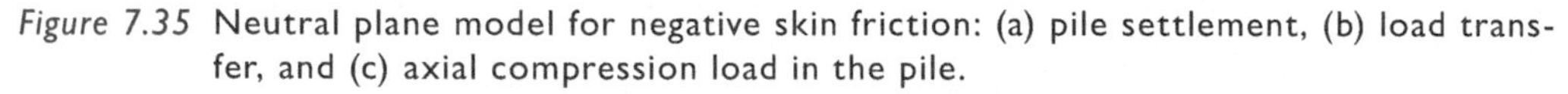

Figure 7.35 Neutral plane model for negative skin friction: (a) pile settlement, (b) load transfer, and (c) axial compression load in the pile.

Source: Redrawn from Timothy et al. (2014).

The negative skin friction force transfer to the pile depends on the pile material, the types of soils, and the amount and rate of relative movement between the soil and the pile. The soil actively settles and drags the pile downward; thus, a small relative movement on the order of 10 mm is necessary to fully trigger negative skin friction. A simple estimation of the extra load on a pile due to negative skin friction, which is related to the undrained shear strength and adhesion factor of the soil, is expressed as:

$$Q_{ns} = A_s \times \overline{s_u} \times \alpha \tag{7.47}$$

Where:

Q_{ns} = load due to negative skin friction.
A_s = circumferential area of the pile.
s_u = average undrained shear strength along the pile shaft.
α = adhesion factor.

Possible measures used to reduce the negative friction include using a smaller diameter for the pile shaft compared to the pile tip, driving the pile inside a casing and filling a viscous material between them, and coating the piles with bitumen. Bitumen coating has been proven to be an effective measure to reduce negative skin friction based on a full-scale test reported by Bjerrum et al. (1969). The negative skin friction measured at one year after pile driving on the reference pile (without bitumen coating) is approximately 3,000 kN, whereas that on the pile with bitumen coating is only 150 kN. Notably, the bitumen coating should be well protected during the pile driving process to ensure a reduction in negative skin friction.

7.9 Group pile effect

In most cases, piles are used in groups to transmit the structural load to the soil. In such a situation, a pile cap is constructed over group piles, as illustrated in Figure 7.36. Determining the load-bearing capacity of group piles is extremely complicated and has not yet been fully resolved. When piles are placed close to each other, a reasonable assumption is that the stresses transmitted by the piles to the soil will overlap and reduce the load-bearing capacity of the piles, as shown in Figure 7.37. Significant settlement or complete general shear failure of the group piles can occur even when pile load tests made on a single pile have demonstrated satisfactory performance.

A classic case history of group pile failure was reported by Terzaghi and Peck (1967). The foundation of the Charity Hospital at New Orleans consisted of approximately 20,000 timber piles that were 7.9 m long. The pile penetrated the soft clay and terminated in a 1.8 m thick dense sand that was underlain by 50 m or more soft soil. A pile load test was performed prior to construction and showed a 6.5 mm settlement under a 300 kN load. However, after the building was constructed, which is under a working load of 150 kN, the group piles settled more than 300 mm within two years. Thus, the group pile load capacity cannot be estimated by summing the load capacity of a single pile.

Ideally, the piles in a group should be spaced so that the load-bearing capacity of the group is not less than the sum of the bearing capacity of the individual piles. In practice, the minimum center-to-center pile spacing, D, is $2.5d$ and in ordinary situations is actually approximately 3 to $3.5d$. The minimum spacing, which considers different soil types and pile types, is listed in Table 7.7

Table 7.7 Minimum spacing for group piles

Pile length (m)	*Friction piles in sand*	*Friction piles in clay*	*End-bearing piles*
Less than 12 m	$3d$	$4d$	$3d$
12 to 24 m	$4d$	$5d$	$4d$
Greater than 24 m	$5d$	$6d$	$5d$

Note: d is the pile diameter or the largest side. The pile spacing is measured at the pile cutoff level, unless a racking pile is used, in which case the spacing is measured at an elevation 3 m below the pile cutoff level.

Figure 7.36 Photo of group piles.

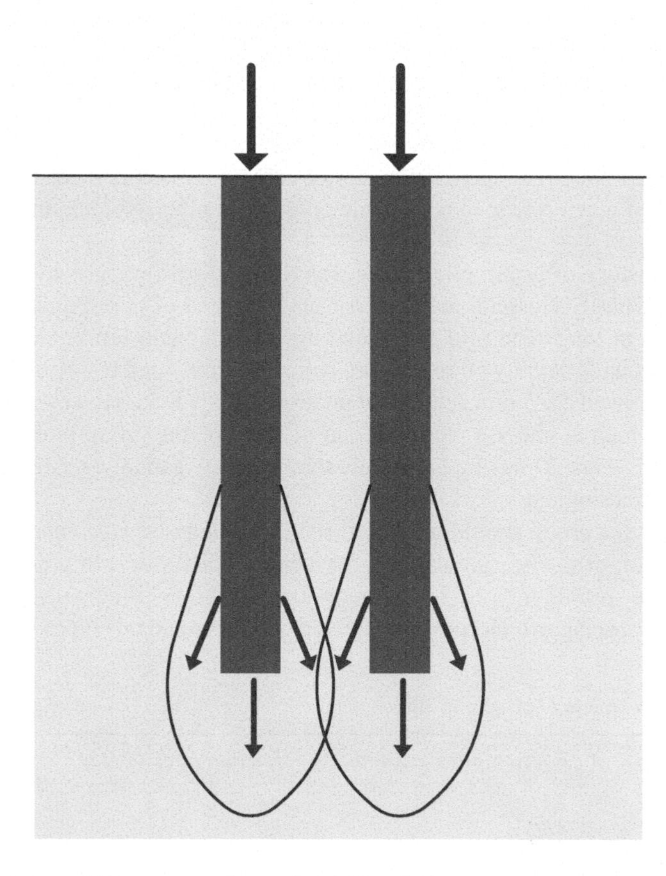

Figure 7.37 Influence of the group pile on load transfer.

Source: Redrawn from Das (2016).

Den Norske Pelekomite (1973). It is clear from the table that longer piles or piles in clays generally need more spacing to avoid reducing the group pile capacity.

An easy method to estimate the group pile capacity is the "efficiency formula." The efficiency of the load-bearing capacity of a group pile can be calculated with eq. 7.48. The efficiency factor is hardly empirical; thus, great care should be taken when using the efficiency formula.

$$\eta = \frac{Q_{g(u)}}{\sum Q_u} \tag{7.48}$$

Where:

η = the group efficiency.
$Q_{g(u)}$ = ultimate load-bearing capacity of the group pile.
Q_u = ultimate load-bearing capacity of each pile without the group effect.

In general, for a group pile that is clay, η is less than unity, approximately 0.7, as suggested by Tomlinson (1969), for a spacing of two pile diameters. For wider spacing, the piles in clay fail individually, but the efficiency ratio reaches unity at a spacing of approximately eight pile diameters. Group piles in sand tend to compact the soil around the pile; thus, the ultimate failure load of pile groups will tend to increase. Figure 7.38 shows the variation in group efficiency for a 3 × 3 group pile in sand. For loose and medium sands, the magnitude of the group efficiency can be larger than unity primarily due to the densification of sand surrounding the pile. The efficiency ratio is less than unity when pile groups are driven into dense sand.

Problems

7.1 Explain the following terms (with plots if necessary):

a. *t-z* curve
b. *p-y* curve
c. pile load test
d. replacement pile

7.2 A 600 mm diameter closed-end steel pipe pile is driven to a depth of 17 m into the following soil profile. The groundwater table is at a depth of 6.0 m. The modulus of elasticity of the well-graded sand is 35 MPa, and Poisson's ratio is 0.3. Assume that $\delta = \frac{2}{3}\phi'$. Compute the ultimate load capacity.

Depth (m)	*Soil layer*	*Unit weight* (kN/m^3)		*Friction angle (°)*
		Moisture	*Saturated*	
0–2.0	Silty sand	18.0	19.0	28
2.0–7.0	Fine to medium sand	18.5	19.5	31
7.0–16.0	Silty sand	–	19.1	29
16.0–18.0	Well-graded sand	–	19.8	35

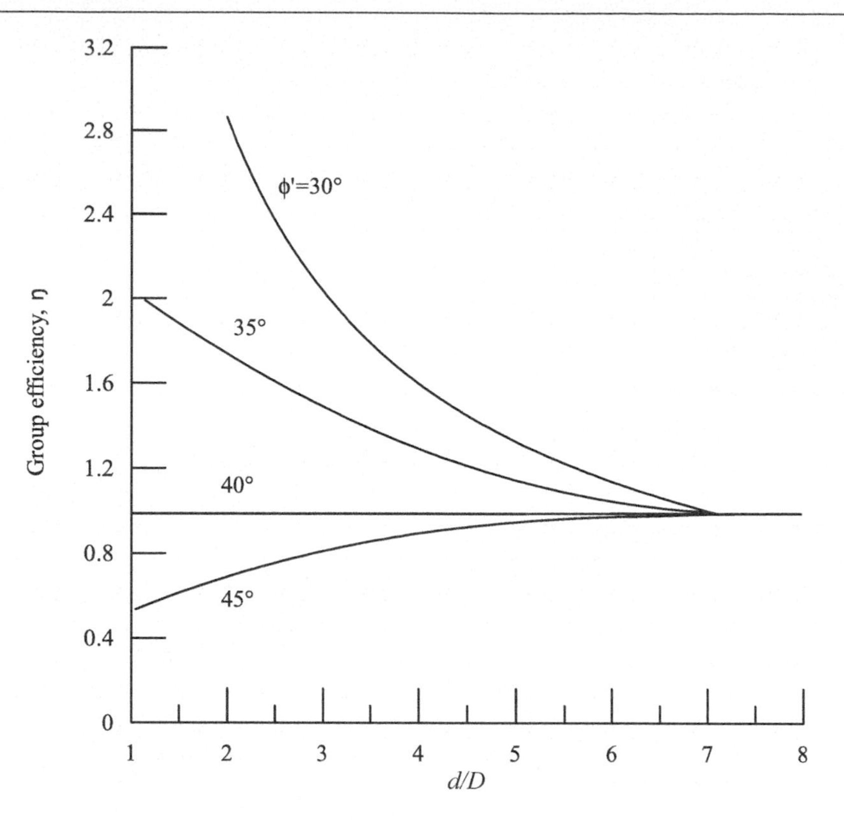

Figure 7.38 Variation in the efficiency of pile groups in sand.

Source: Based on Kishida and Meyerhof (1965).

7.3 An 800 mm diameter concrete pile is driven to a depth of 20 m into the following soil profile. The groundwater table is at the ground surface. Compute the ultimate load capacity with α, β, and λ methods.

Depth (m)	*Soil layer*	*Saturated unit weight* (kN/m^3)	*Undrained shear strength* (kPa)	*Friction angle* (°)
0–3.0	Silty clay	17.0	12	28
3.0–10.0	Clayey silt	17.5	20	29
10.0–14.0	Clay	17.8	50	30
14.0–25.0	Clay	18.1	80	31

7.4 A static pile load test is conducted on a rectangular pile with dimensions of 1.0 m × 49.0 m (diameter × length). The load–settlement curve is shown in Figure P 7.4. Please determine the ultimate load capacity by Davisson's method.

7.5 As problem. 7.4, use other methods, such as Terzaghi's method or Fuller and Hoy method, to determine the ultimate pile load capacity.

7.6 A pile with diameter of 1.2 m and length of 6.0 m is embedded in a clay layer with s_u = 30 kPa. Determine the ultimate lateral load capacity of the pile if the pile head is fixed.

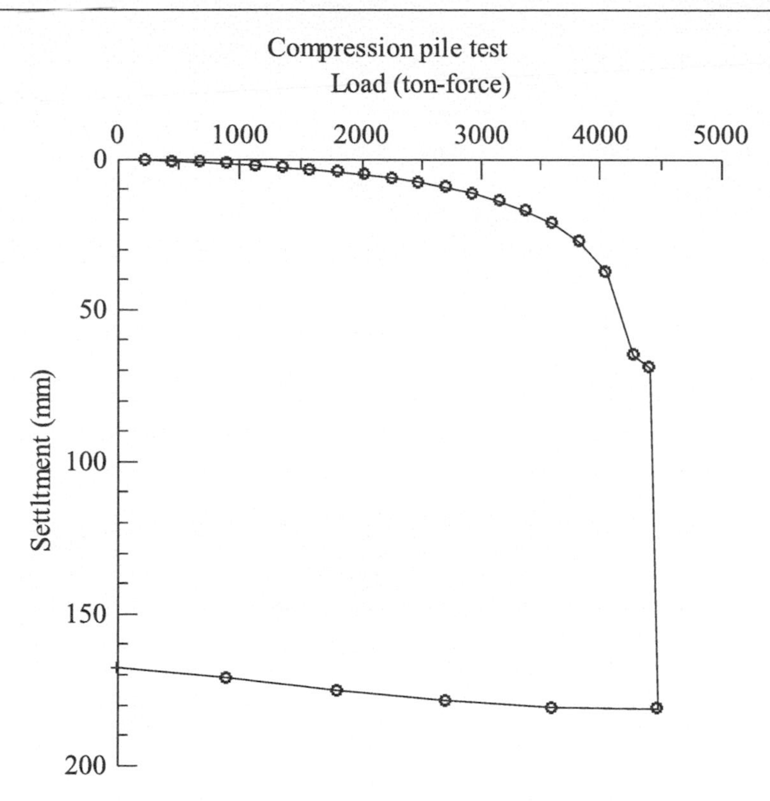

Figure P 7.4 Pile load test result.

7.7 As problem 7.6, determine the ultimate lateral load capacity of the pile if the pile head is free and the height of the lateral load is 1.0 m.

7.8 As problem 7.6 and 7.7, calculate the maximum moment for both cases.

7.9 As problem 7.6, the same pile is embedded in a sandy layer with unit weight $\gamma_{sat} = 20\ \text{kN/m}^3$ and friction angle of 32°. Determine the ultimate lateral load capacity and maximum moment of the pile if the pile head is fixed.

7.10 As problem 7.9, determine the ultimate lateral load capacity and maximum moment of the pile if the pile head is free and the height of the lateral load is 1.0 m.

7.11 A large sign is to be supported on a single post connected to a 1.2 m diameter pile drilled shaft foundation. The soil is a stiff clay with s_u = 80 kPa. The design wind load is 600 kN, and the center of the sign is 20 m above the ground. Using a factor of safety of 3.0, compute the minimum required depth of embedment, L, and the maximum moment in the drilled shaft, M_{max}.

7.12 A 600 mm diameter pile is embedded into a saturated medium clay that has an average unit weight of 18 kN/m^3 and an undrained shear strength of 50 kPa. The groundwater table is near the ground surface. Use the Matlock method to develop the static curve at a depth of 10 m.

7.13 A driven concrete pile with diameter of 1.0 m and length of 25 m is embedded in saturated sand. Given the unit weight of sand $\gamma_{sat} = 20\ \text{kN/m}^3$, drained friction angle $\phi' = 33^\circ$, and the frictional angle of sand-pile interface $\delta = 25^\circ$, estimate the ultimate frictional resistance of the pile. (Assume the effective earth pressure coefficient is 1.2.)

7.14 As problem 7.13, estimate the ultimate end-bearing capacity of the pile.

7.15 As problem 7.13. A working load on the pile is 3,500 kN. Skin resistance carries 2,100 kN of the load, and end-bearing carries the rest. Estimate the total settlement of the pile. (Assume the modulus of the elasticity of pile $E_p = 21 \times 10^6$ kPa and a uniform distribution of skin friction along the pile.)

References

Bjerrum, L., Johannessen, I.J. and Eide, O. (1969), Reduction of skin friction on steel piles to rock. *Proceeding of the 7th International Conference. in Soil Mechanics and Foundation Engineering*, Mexico, Vol. 2, pp. 27–34.

Broms, B.B. (1964a). The lateral resistance of piles in cohesive soils. *Journal of the Soil Mechanics & Foundation Divisions, ASCE*, Vol. 90, No. 2, pp. 27–63.

Broms, B.B. (1964b). The lateral resistance of piles in cohesionless soils. *Journal of the Soil Mechanics & Foundation Divisions, ASCE*, Vol. 90, No. 3, pp. 123–156.

Broms, B.B. (1965), Design of laterally loaded piles. *Journal of SMFED, ASCE*, No. SM3, pp. 79–99.

Burland, J.B. (1973), Shaft friction piles in clay—a simple fundamental approach. *Ground Engineering*, Vol. 6, No. 3.

Coyle, H.M. and Reese, L.C. (1966), Load transfer for axially loaded pile in clay. *Journal of the Soil Mechanics and Foundations Division*, ASCE, Vol. 92, pp. 1–26.

Das, B.M. (2016), *Principles of Foundation Engineering*, SI, 8th edition, Boston: Cengage Learning.

De Mello, V.F.B. (1969), Foundations of buildings in clay. State-of-the art report. In *Proceeding of the 7th International Conference. Soil Mechanics and Foundation Engineering*, Mexico City, SOA Volume, pp. 49–136.

De Nicola, A.D. and Randolph, M.F. (1993), Tensile and compressive shaft capacity of piles in sand. *Journal of the Geotechnical Engineering, ASCE*, Vol. 19, No. 12, pp. 1952–1973.

Den Norske Pelekomite (1973), Veledning ved pelefundementering. *Oslo: Norwegian Geotechnical Institute*, Veiledning No. 1.

FHWA-HRT-04-043 (2006), *A Laboratory and Field Study of Composite Piles for Bridge Substructures*, McLean: Office of Infrastructure R&D Federal Highway Administration.

Kishida, H. (1967), Ultimate bearing capacity of piles driven into loose sand. *Soils and Foundations*, Vol. 7, No. 3, pp. 20–29.

Kishida, H. and Meyerhof, G.G. (1965), Bearing capacity of pile groups under eccentric loads in sand. In *2, Proceeding of the Fifth International Conference*, Soil Mechanic and Foundation Engineering, Montreal pp. 270–274.

Kraft, L.M., Ray, R.P. and Kagawa, T. (1981), Theoretical t-z curves. *Journal of the Geotechnical Engineering Division, ASCE*, Vol. 107, No. 11, pp. 1543–1561.

Kulhawy, F.H. (1991), *Drilled Shaft Foundations, Foundation Engineering Handbook*, 2nd edition, Chap. 14. Ed. H.-Y. Fang, New York: Van Nostrand Reinhold.

Kulhawy, F.H., Trautmann, C.H., Beech, J.F., O'Rourke, T.D., McGuire, W., Wood, W.A. and Capano, C. (1983), Transmission line structure foundations for uplift-compression loading, *Rep. No. EL-2870*, Electric Power Research Institute, Palo Alto, CA.

Ladanyi, B. (1963), Evaluation of pressuremeter tests in granular soils. In *Proceeding 2nd Panamerican Conference Soil Mechanic and Foundation Engineering*, Sao Paulo, 1, pp. 3–20.

Matlock, H. (1970), Correlation for design of laterally loaded piles in soft clay. *In The Second Annual Offshore Technology Conference*, Houston, Texas: Onepetro.

Meyerhof, G.G. (1956), Penetration test and bearing capacity of cohesionless soils. *Journal of the Soil Mechanics and Foundation Division, ASCE*, Vol. 82, No. SM1, pp. 1–19.

Meyerhof, G.G. (1976), Bearing capacity and settlement of pile foundations. *Journal of Geotechnical and Geoenvironmental Engineering Division, ASCE*, Vol. 102, No. GT3, pp. 195–228.

Mohan, D., Jain, D.S. and Kumar, V. (1963), Load bearing capacity of piles. *Geotechnique*, London, Vol. 13, No. 1, pp. 76–86.

NAVFAC. (1982), *Foundations and Earth Structures Design Manual 7.2*, Department of the Navy Naval Facilities Engineering Command, Alexandria, VA.

Nishida, Y. (1964), A basic calculation on the failure zone and the initial pore pressure around a driven pile. In *Proceeding of the First Regional Asian Conference on Soil Mechanics and Foundation Engineering*, the International Society of Soil Mechanics and Foundation Engineering, Tokyo, 1, pp. 217–219.

Poulos, H.G. and Davis, E.H. (1980), *Pile Foundation Analysis and Design*, New York: Wiley, c1980.

Reese, L.C., Cox, W.R. and Koop, F.D. (1974), Analysis of laterally loaded piles in sand. *Offshore Technology Conference*, Houston, Texas, May.

Reese, L.C., Isenhower, W.M. and Wang, S.-T. (2006), *Analysis and Design of Shallow and Deep Foundations*, Hoboken, New Jersey: John Wiley and Sons.

Skempton, A.W. (1951), *The Bearing Capacity of Clays*, Vol. 1, London, Building Research Congress, pp. 180–189.

Soderberg, L.O. (1962), Consolidation theory applied to foundation pile time effects. *Geotechnique*, London, Vol. 11, No. 3, pp. 217–225.

Sowers, G.F., Martin, C.B., Wilson, L.L. and Fausold, M. (1961), The bearing capacity of friction pile groups in homogeneous clay from model studies. *Proceedings of the 5th International Conference on Soil Mechanics and Foundation Engineering*, Vol. 2, pp. 155–159.

Terzaghi, K. and Peck, R.B. (1967), *Soil Mechanics in Engineering Practice*, 2nd edition, New York: John Wiley and Sons, p. 729.

Terzaghi, K., Peck, R.B. and Mesri, G. (1996), *Soil Mechanics in Engineering Practice*, 3rd edition, New York: John Wiley and Sons, p. 549.

Timothy, C., Siegel, P.E., G.E., D.GE with Dan Brown and Associates PC (2014), Designing piles for drag force. *43rd Annual Midwest Geotechnical Conference*, October 1–October 3, 2014, Bloomington, Minnesota: United State Departments of Transportation Federal Highway Administration.

Tomlinson, M.J. (1969), *Foundation Design and Construction*, 2nd edition, New York: Wiley, p. 785.

Vesic, A.S. (1967a). A study of bearing capacity of deep foundations. *Final Report*, Project B-189, Georgia Institute of Technology, Atlanta, pp. 231–236.

Vesic, A.S. (1967b). Model testing of deep foundations in sand and scaling laws. In *Proceedings North American Conference on Deep Foundations*, Panel Discussions, Session II, Mexico City.

Vesic, A.S. (1977), Design of pile foundations. *National Cooperative Highway Research Program, Synthesis of Practice No. 42*, Transportation Research Board, Washington, DC, p. 68.

Zhang, Y., Andersen, K.H. and Tedesco, G. (2016), Ultimate bearing capacity of laterally loaded piles in clay-Some practical considerations. *Marine Structures*, Vol. 50, pp. 260–275.

Chapter 8

Slope stability

8.1 Introduction

Predicting the stability of soil and rock slopes is a classic problem for geotechnical engineers and plays an important role in designing embankments, dams, roads, tunnels, and other engineering structures. Many researchers have focused on assessing the stability of slopes (Taylor, 1948; Morgenstern, 1963; Fredlund and Krahn, 1977; Hoek and Bray, 1981; Goodman and Kieffer, 2000). However, the problem still presents a significant challenge to designers. Gravitational and seepage forces cause instability in natural and man-made slopes. The latter frequently occur in practice and may arise as a result of the construction of embankments, canals, excavations, and earth dams. Slope stability is usually analyzed using limit equilibrium procedures, where a failure surface is assumed before a factor of safety (F_s) calculation. For engineering designs, a primary concern is whether the slope is safe at a given slope angle. This chapter introduces some basic knowledge of slope stability problems for undergraduate students. The contents include the slope failure modes, stability assessments, stability charts, and stabilization.

Generally, current design practices for slope stability rely largely on the value for F_s obtained from the conventional limit equilibrium method or finite element method. The limit equilibrium method is used widely because it is relatively simple and easy to use. Conventional slope stability analyses investigate the equilibrium of a mass of soil bounded below by an assumed potential slip surface and above by the surface of the slope. Notably, simplifications are necessary in all design methods. The assumed failure mechanisms are more or less crude approximations of the actual failure mechanism. Certain assumptions are also made regarding the slope geometry and the loads acting on the slope. However, engineers sometimes misuse the limit equilibrium method or finite element method because there is a lack of knowledge about method backgrounds. In general, assumptions are a requirement because otherwise, the design methods would be overwhelmingly complex and nearly impossible to use rationally. In a robust design method, the necessary assumptions have very little influence on the end result.

8.2 Failure modes

Based on the geological structure and the stress state of soil and rock masses, certain slope failure modes appear to be more likely than others. Figure 8.1 displays some common slope failure modes. A rockfall (Figure 8.1a) refers to quantities of rock falling freely from a cliff face. This term is also used for the collapse of rock from the roof or walls of mines or quarry workings. Figure 8.1b is a slope rotational shear failure mode. For the rotational shear failure mode, the individual particles in a soil or rock mass should be very small compared to the size of the slope, and these particles are not interlocked as a result of their shape (Hoek and Bray, 1981).

DOI: 10.1201/9781003350019-8

Figure 8.1 Different slope failure modes: (a) rockfall, (b) rotational failure, (c) toppling, (d) planer slides.

Rotational shear failure generally occurs in soil slopes. The issue of whether a rock mass can be considered heavily fractured is thus mostly a matter of scale. Rotational shear failure in a large-scale slope would probably involve failure primarily along preexisting discontinuities, with some portions of the failure surface possibly going through intact rock. Additionally, rotation and translation of individual blocks in the rock mass would help create a failure surface. The resulting failure surface would follow a curved path. The failure surfaces are drawn to be relatively deep, but they could also be shallower.

Figure 8.1c shows the slope toppling failure mode. This type of failure is characterized by a successive breakdown of the rock slope. Failure could initiate by crushing of the slope toe, which in turn causes load transfer to adjacent areas that may fail. Obviously, the orientations of discontinuities and in situ stresses in relation to the rock strength are important factors governing this failure mode. The presence of discontinuities in the rock mass can result in several secondary modes of failure, as shown in Figure 8.2.

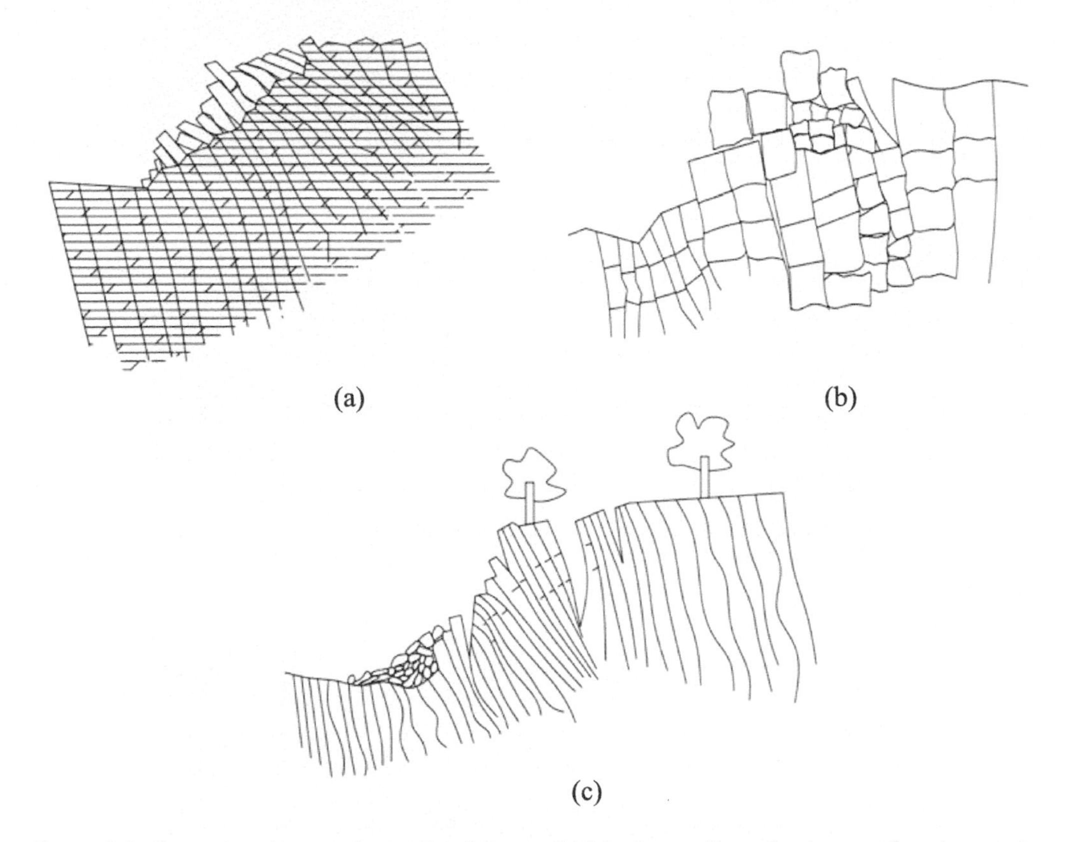

Figure 8.2 Common classes of toppling failures: (a) block toppling of columns of rock containing widely spaced orthogonal joints, (b) flexural toppling of slabs of rock dipping steeply into the face, and (c) block flexure toppling characterized by pseudocontinuous flexure of long columns through accumulated motions along numerous cross joints.

Source: Redrawn from Goodman and Bray (1976).

Figure 8.1d shows the plane shear failure mode, which could be a single discontinuity (plane failure), two discontinuities intersecting each other (wedge failure), or a combination of several discontinuities connected together (irregular failure surface). A common feature of most failure modes is the formation of a tension crack at the slope crest. For failure to occur, release surfaces must be present to define the rock block moving in the lateral direction.

A more detailed description of the governing failure mechanisms for typically occurring slope failures is given as well as those based on field observations. Notably, a real failure surface would be a combination of different failure modes. For example, there may be a combination of plane failures and rotational shear failures, with or without tension cracks at the slope crest.

8.3 Slope stability analyses

As mentioned previously, two-dimensional limit equilibrium analysis is the most popular approach to evaluate slope stability. It is known that two-dimensional solutions utilized in design will obtain a conservative evaluation for slope failure. With advances in computer hardware, evaluating slope stability using finite element method numerical analysis has become more prevalent and has gradually been widely adopted by engineers. The displacement finite element technique is widely used for predicting the load–deformation response and hence collapse of geotechnical structures. In contrast, very few numerical analyses have been performed to evaluate slope stability based on limit theorems. Limit analysis methods that produce solutions in terms of limit load can be based on either the kinematically admissible velocity field (upper bound) or the statically admissible stress field (lower bound). Due to the difficulty in computing the stress field for a slope, most studies are based on the upper bound method. In recent years, finite element upper bound (Lyamin and Sloan, 2002a; Krabbenhoft et al., 2005) and lower bound (Lyamin and Sloan, 2002b) limit analysis methods have become popular approaches in slope analyses. One of the merits of using the finite element method or finite element limit analysis method is that the techniques can generate failure mechanisms automatically.

8.3.1 Factor of safety in the limit equilibrium method

For the simplest form of limit equilibrium analysis, only the equilibrium of moment is satisfied. The sum of the forces acting to induce the sliding of parts of the slope is compared with the sum of the forces available to resist failure. The ratio between these two sums is defined as the factor of safety F_s, which is:

$$F_s = \frac{R}{D} \tag{8.1}$$

where:

R = resistance moment.
D = driving moment.

This simple definition of the safety factor can be interpreted in many ways. It can be expressed in terms of loads, forces, moments, etc. The merit of a safety factor is that the stability of a slope can be quantified by a number. According to eq. 8.1, a safety factor less than 1 indicates that failure is possible. If there are several potential failure modes or different failure surfaces that have a calculated safety factor less than 1, the slope can fail. By using the limit equilibrium method, the slip surface needs to be assumed prior to any safety factor calculation. Thus, the obtained F_s would be unreasonable if the assumed failure mechanism is far from the real one.

8.3.2 Basic limit equilibrium methods

For the general concept of the limit equilibrium method, soil masses above the potential failure surface are considered as a combination of vertical slices. Therefore, this approach is called the "procedure of slices." The actual number of slices and their width can be determined by the soil profiles, slope geometry, and assumed shape of the potential failure surface. Importantly, the assumed failure surface could significantly influence the final obtained magnitude of F_s.

Various approaches have been proposed based on the limit equilibrium concept. For the assumed potential failure surface, some methods require circular failure shapes arbitrarily. Others can work with noncircular shapes. In fact, it is very difficult to predict whether the failure surface is circular or noncircular. As discussed in Section 8.2, an actual failure surface can be a combination of different failure modes. Therefore, a reasonable assumption of a potential failure surface relies highly on the judgment made by experienced engineers without monitoring data. This issue is also a disadvantage of using the limit equilibrium method.

In general, the equilibrium of moments about the center of the circle for the entire soil mass above the assumed failure surface is considered. A circular failure surface is shown in Figure 8.3, where the overturning moment can be expressed as:

$$M_d = \sum W_i b_i \tag{8.2}$$

where W_i is the weight of the ith slice and b_i is the horizontal distance between the center of the circle and the center of the slice. Theoretically, b_i should be the distance between the center of the circle and the center of gravity of the slice. However, the difference can be neglected if sufficient numbers of slices are used.

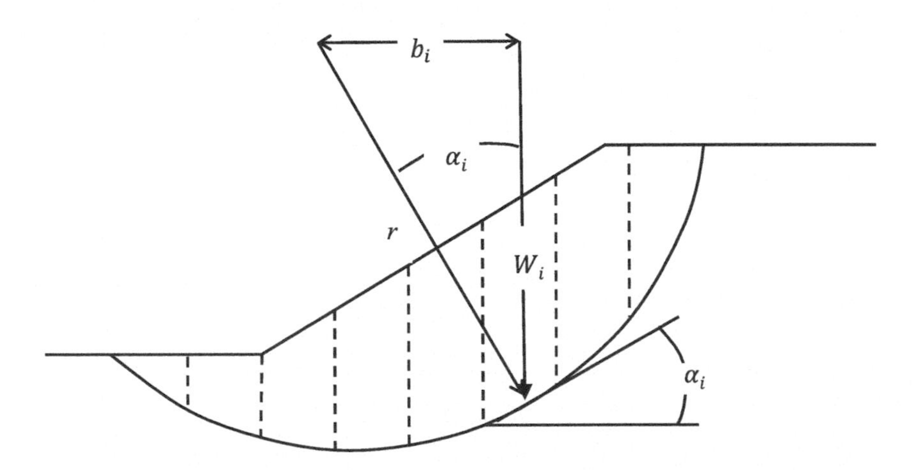

Figure 8.3 General configuration of slices for a circular failure surface.

In eq. 8.2, the moment arm b_i can be expressed by eq. 8.3, and thus, the driving moment is described by eq. 8.4.

$$b_i = r \sin \alpha_i \tag{8.3}$$

$$M_d = r \sum W_i \sin \alpha_i \tag{8.4}$$

The resisting moment is from the shear stress (τ) along the failure surface, which is on the base of each slice. Based on the Mohr–Coulomb failure criterion, the shear strength for each slice (s_i) can be expressed as:

$$s_i = c + \sigma \tan \phi \tag{8.5}$$

The resisting moment will be:

$$M_r = r \sum s_i \Delta \ell_i \tag{8.6}$$

where Δl_i is the thickness of the *i*th slice. The factor of safety of the slope can be expressed as:

$$F_s = \frac{r \sum s_i \Delta \ell_i}{r \sum W_i \sin \alpha_i} = \frac{\sum s_i \Delta \ell_i}{\sum W_i \sin \alpha_i} \tag{8.7}$$

Notably, eq. 8.7 is valid only for a circular failure surface.

8.3.3 *Infinite slope*

For cohesionless soils ($c = 0$), only the friction angle plays a role in the safety calculation. In general, the failure surface depth will not significantly influence the factor of safety. It is possible to form a small failure depth, thus meeting the requirements for an infinite slope. This condition would appear when there is a stronger soil or rock formation at shallow depths. In Figure 8.4, the sliding soil mass has weight $W = \gamma \mathrm{Lz} \cos \beta$, and thus, the shear force (V) and normal force (N) can be expressed by eqs. 8.8 and 9, respectively.

$$V = W \sin \beta = \gamma Lz \cos \beta \sin \beta \tag{8.8}$$

$$N = W \cos \beta = \gamma Lz \cos^2 \beta \tag{8.9}$$

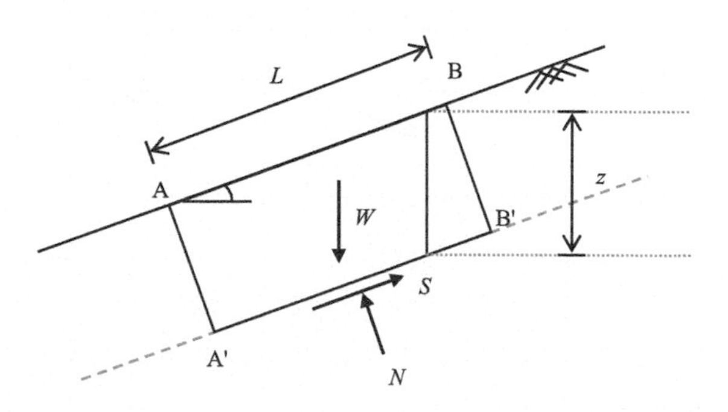

Figure 8.4 Slice with forces for an infinite slope.

From eqs. 8.8 and 8.9, the driving shear stress and normal stress can be calculated as shown in eqs. 8.10 and 8.11, respectively.

$$S_d = \gamma z \cos\beta \sin\beta \tag{8.10}$$

$$\sigma = \gamma z \cos^2\beta \tag{8.11}$$

Based on the Mohr–Coulomb failure criterion, the resisting shear strength at failure is expressed as eq. 8.12. Therefore, F_s can be calculated using eq. 8.13. For cohesionless soils ($c = 0$), F_s will be expressed simply as the ratio of $\tan\phi / \tan\beta$.

$$S = c + \gamma z \cos^2\beta \tan\phi \tag{8.12}$$

$$F_s = \frac{c + \gamma z \cos^2\beta \tan\phi}{\gamma z \cos\beta \sin\beta} \tag{8.13}$$

Figure 8.5 shows the seepage effects when the water table is at the slope surface. For this case, the sliding soil mass has weight $W = \gamma_{sat} L z \cos\beta$. The pore water pressure acts on the bottom of the typical section. By considering the effective stress theorem, eq. 8.11 based on the effective stress can be expressed by eq. 8.14. In addition, the resisting shear strength at failure, including soil cohesion, is shown in eq. 8.15. However, groundwater contributes to the saturated unit weight, which can increase the driving shear stress, as shown in eq. 8.16.

$$\sigma = \sigma - u_w = \gamma_{sat} z \cos^2\beta - \gamma_w z \cos^2\beta = \gamma' z \cos^2\beta \tag{8.14}$$

$$S_f = c + \gamma' z \cos^2\beta \tan\phi \tag{8.15}$$

$$S_d = \gamma_{sat} z \cos\beta \sin\beta \tag{8.16}$$

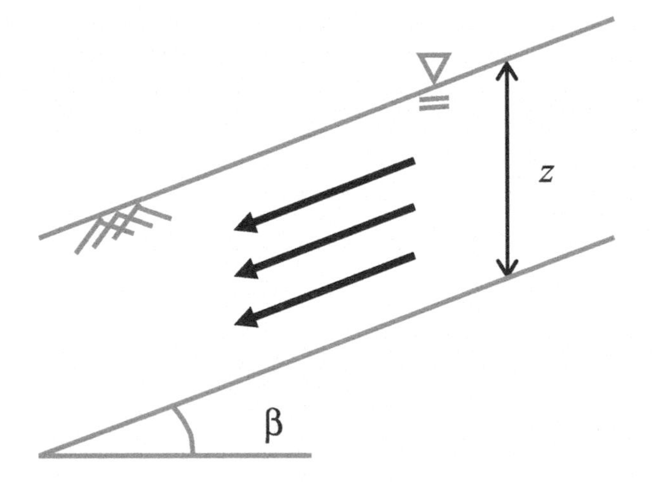

Figure 8.5 Seepage consideration for an infinite slope.

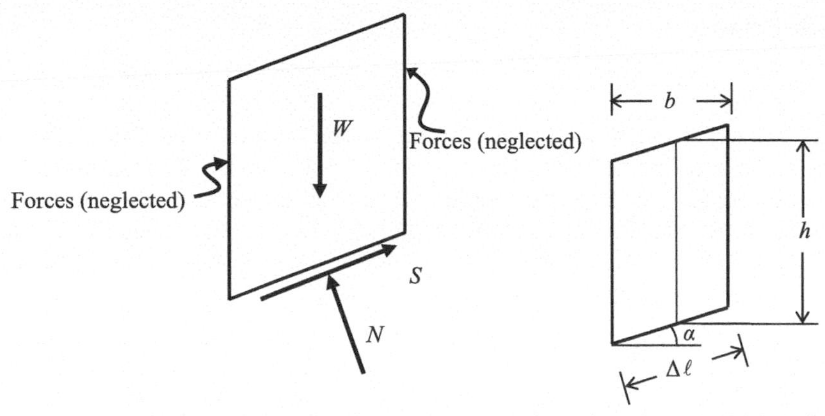

Figure 8.6 Slice with forces adopted in the ordinary method of slices.

From eqs. 8.15 and 8.16, the factor of safety can be derived as follows:

$$F_s = \frac{c + \gamma z \cos^2 \beta \tan \phi}{\gamma z \cos \beta \sin \beta} \tag{8.17}$$

Example 8.1

For the infinite slope for cohesionless slope (β = 25°) shown in Figure EX 8.1, compute F_s when $\gamma = 19\,kN/m^3$ and ϕ = 33°.

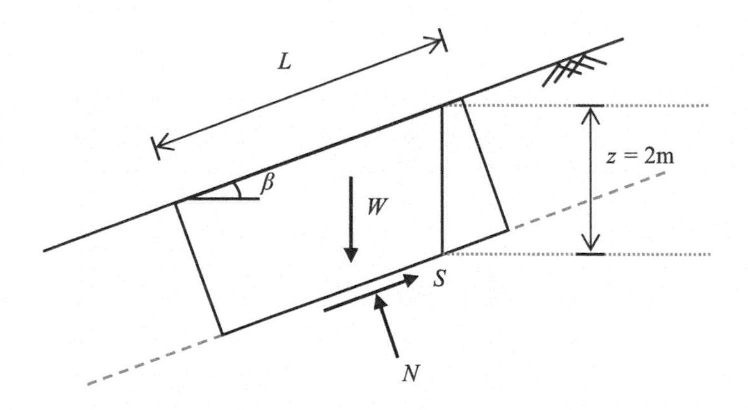

Figure EX 8.1 Infinite slope information.

Solution

F_s = tanϕ / tanβ = tan 33°/tan 25° = 1.393

8.3.4 *Finite slope*

Ordinary method of slices

The forces on the sides of slices are ignored for the ordinary method. The force conditions considered by the ordinary method of slices are shown in Figure EX 8.1. The normal force can be expressed by eq. 8.18. Bishop (1955) indicated that the normal force would exist only if the resultant force of the forces on the side of the slice acted in a direction parallel to the base of the slice. This condition is almost impossible. Moreover, interslice forces should be zero for all forces on the slices to achieve equilibrium.

$$N = W\cos\alpha \tag{8.18}$$

The normal stress on the base of a slice can be calculated by dividing the normal force by the area of the slice base ($1\times\Delta\ell_i$), as expressed by eq. 8.19.

$$\sigma = \frac{W\cos\alpha}{\Delta\ell_i} \tag{8.19}$$

From eqs. 8.18 and 8.19, the factor of safety of the slope can be expressed by eq. 8.20.

$$F_s = \frac{\sum\left(c\Delta\ell_i + W\cos\alpha\tan\phi\right)}{\sum W_i\sin\alpha_i} \tag{8.20}$$

Effective stress strength parameters should be adopted, as shown in eq. 8.25, if the soil mass is in the drained condition.

$$F_s = \frac{\sum\left(c'\Delta\ell_i + W\cos\alpha\tan\phi'\right)}{\sum W_i\sin\alpha_i} \tag{8.21}$$

The weight of the slice can be expressed in eq. 8.22, and thus, the factor of safety is shown in eq. 8.23. Note that eq. 8.23 will lead to a significantly low effective stress ($\sigma'=\gamma h\cos^2\alpha - u$), even negative, when the pore water pressure is larger or the failure surface is steep.

$$W = \gamma hb = \gamma h\Delta\ell\cos\alpha \tag{8.22}$$

$$F_s = \frac{\sum\left[c'\Delta\ell_i + \left(W_i\cos\alpha - u\Delta\ell_i\cos^2\alpha\right)\tan\phi'\right]}{\sum W_i\sin\alpha_i} \tag{8.23}$$

Based on the ordinary method of slices, a better expression for the factor of safety can be obtained by considering the effective slice weight, W', which is expressed as:

$$W' = W - ub \tag{8.24}$$

From eqs. 8.18, 8.21, and 8.24, the factor of safety derived from moment equilibrium is shown in eq. 8.25. This equation does not calculate negative effective stresses if the pore water pressure is less than the total vertical pressure.

$$F_s = \frac{\sum\left[c'\Delta\ell + (\sigma - u)\tan\phi'\right]}{\sum W\sin\alpha} \tag{8.25}$$

Example 8.2

For the plane strain slope shown in Figure EX 8.2, compute F_s using the ordinary method of slices. Note that for a real-world problem, you would have to choose a failure circle and the number of slices and then measure all the parameters you are given for slice information. For the plane strain slope shown in the following, the factor of safety is computed using the ordinary method of slices.

Slice information for example 8.2:

Slice	*Ave. h (m)*	*Ave. h_w (m)*	*a (degree)*
1	0.20	0.20	-24
2	0.60	0.60	-14
3	1.35	1.35	-11
4	2.40	2.40	-3
5	3.40	3.20	0
6	4.35	3.60	3.5
7	5.25	3.80	11.5
8	5.60	3.80	14
9	5.25	3.70	24
10	4.75	3.40	29
11	4.20	3.10	32.5
12	3.50	2.60	38.5
13	2.50	1.70	46
14	1.25	0.60	57

*h_w is the water height in a slice.

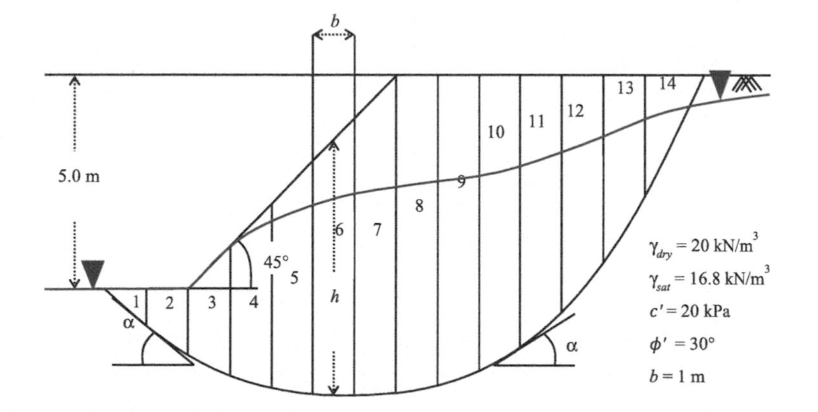

Figure EX 8.2 Slope information.

Solution

The only trick here is to correctly calculate the total weight of each slice, given that part of every slice is dry and a part is saturated. Namely:

$$W_i = b\left(h_{wi}\gamma_{sat} + \left(h_i - h_{wi}\right)\gamma_{dry}\right)$$

Additionally, note that $\Delta\ell_i = \dfrac{b}{\cos\alpha_i}$ and $u_i = h_{wi}\gamma_w$.

Solution information for example 8.1:

Slice	h	h_w	α	b	Sin α	Cos α	c'xΔl	$u = \gamma_w h_w$	u × Δl	W sin α	c' + c' (Wcos a −ul)tan ϕ'
1	0.2	0.2	−24	1	−0.41	0.91	16	2	2	−1.63	16.97
2	0.6	0.6	−14	1	−0.24	0.97	15	6	6	−2.90	17.49
3	1.35	1.35	−11	1	−0.19	0.98	15	13	13	−5.15	20.02
4	2.4	2.4	−3	1	−0.5	1.00	15	24	24	−2.51	23.89
5	3.4	3.2	0	1	0.00	1.00	15	31	31	0.00	28.09
6	4.35	3.6	5.5	1	0.10	1.00	15	35	35	8.11	32.81
7	5.25	3.8	11.5	1	0.20	0.98	15	37	38	20.01	37.26
8	5.6	3.8	14	1	0.24	0.97	15	37	38	25.70	39.00
9	5.25	3.7	24	1	0.41	0.91	16	36	40	40.69	35.22
10	4.75	3.4	29	1	0.48	0.87	17	33	38	43.96	32.14
11	4.2	3.1	32.5	1	0.54	0.84	18	30	36	43.24	29.37
12	3.5	2.6	38.5	1	0.62	0.78	19	26	33	41.78	26.42
13	2.5	1.7	46	1	0.72	0.69	22	17	24	34.13	24.85
14	1.25	0.6	57	1	0.84	0.54	28	6	11	19.22	28.15
								Total		264.65	391.66

$F_s = 391.66/264.65 = 1.480$

Simplified Bishop procedure

For the simplified Bishop method, all forces applied to a slice are displayed in Figure 8.7. The normal interaction forces between adjacent slices are assumed to be collinear. In addition, no shear stresses between slices are assumed. The force equilibrium equation in the vertical direction can be written as:

$$N\cos\alpha + S\sin\alpha - W = 0 \tag{8.26}$$

From eq. 8.5, the shear force expressed by the shear strength and factor of safety can be written as eq. 8.27.

$$S = \frac{s_i\Delta\ell}{F_s} \tag{8.27}$$

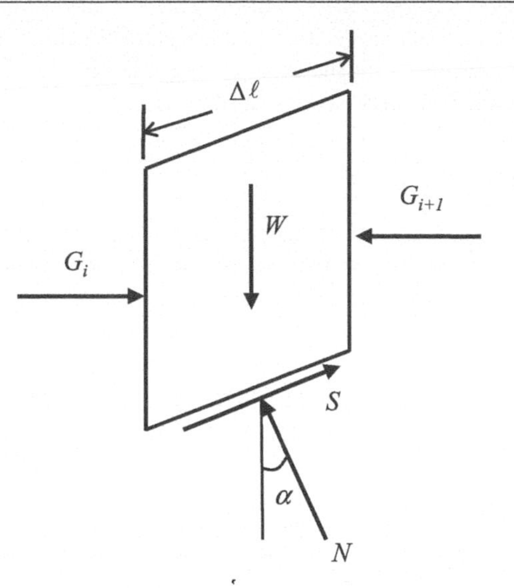

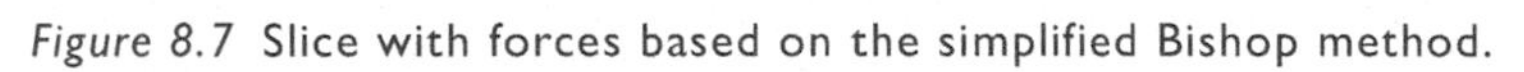

Figure 8.7 Slice with forces based on the simplified Bishop method.

By considering effective stresses, we can write:

$$S = \frac{1}{F_s}\left[c'\Delta\ell + \left(N - u\Delta\ell\right)\tan\phi'\right] \tag{8.28}$$

From eqs. 8.26 and 8.28, the normal force, N, can be presented by eq. 8.29 based on the Mohr–Coulomb failure criterion.

$$N = \frac{W - \left(1/F_s\right)\left(c'\Delta\ell\text{-}u\Delta\ell\tan\phi'\right)\sin\alpha}{\cos\alpha + \dfrac{\left(\sin\alpha\tan\phi'\right)}{F_s}} \tag{8.29}$$

The effective normal stress on the base of the slice is:

$$\sigma' = \frac{N}{\Delta\ell} - u \tag{8.30}$$

By combining eqs. 8.29 and 8.33, the factor of safety can be written as:

$$F_s = \frac{\sum\left[\dfrac{c'\Delta\ell\cos\alpha + \left(W - u\Delta\ell\cos\alpha\right)\tan\phi'}{\cos\alpha + \dfrac{\left(\sin\alpha\tan\phi'\right)}{F_s}}\right]}{\sum W\sin\alpha} \tag{8.31}$$

In eq. 8.31, F_s appears on both sides. To use the simplified Bishop procedure, the F_s on the right-hand side should be assumed for the trial. The iterative process can converge on F_s. Commonly, a computer program is used to address problems.

Inclusion of additional known forces

In general, there are additional known driving and resisting forces. For example, in Figure 8.8, the pseudostatic method is used to simulate seismic loading. Moreover, additional forces could result from slope reinforcement. Their effects must be included in the equilibrium equation to compute the factor of safety. The simplified Bishop procedure is employed for illustration to include the additional forces. For the overall moment equilibrium about the center of a circle, the two components are the soil weight and shear strength. Therefore, equilibrium can be expressed in eq. 8.32. Counterclockwise and clockwise moments are considered positive and negative, respectively.

$$\gamma \sum \frac{s_i \Delta \ell_i}{F_s} - \gamma \sum W_i \sin \alpha_i = 0 \tag{8.32}$$

As shown in Figure 8.8, if there are seismic forces (kW_i) and forces from soil reinforcement (T_i), the equilibrium equation can be expressed as:

$$\gamma \sum \frac{s_i \Delta \ell_i}{F_s} - \gamma \sum W_i \sin \alpha_i - \sum k W_i d_i + \sum T_i h_j = 0 \tag{8.33}$$

where k is the seismic coefficient, d_i is the vertical distance between the center of gravity of the slice and the center of the circle, and h_j is the moment arm of the reinforcement force

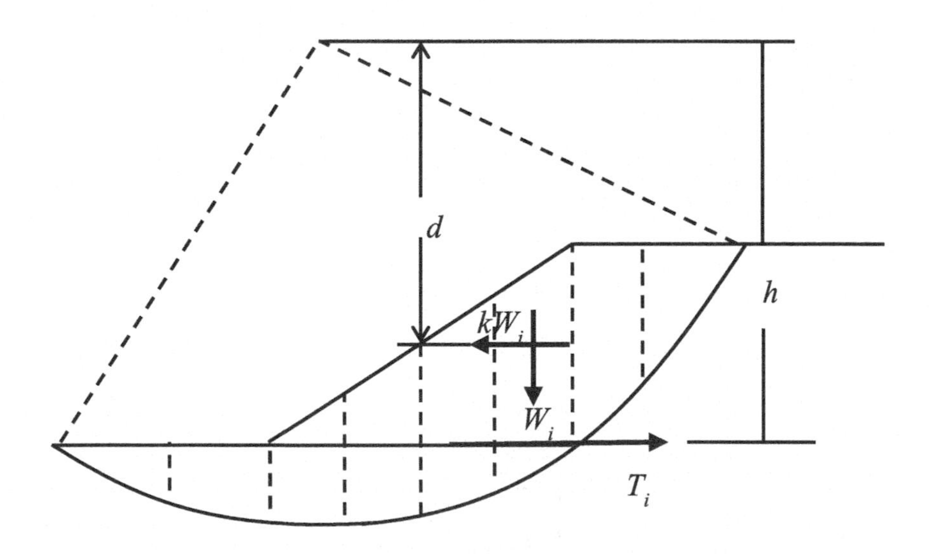

Figure 8.8 Slope with additional known seismic and reinforcement forces.

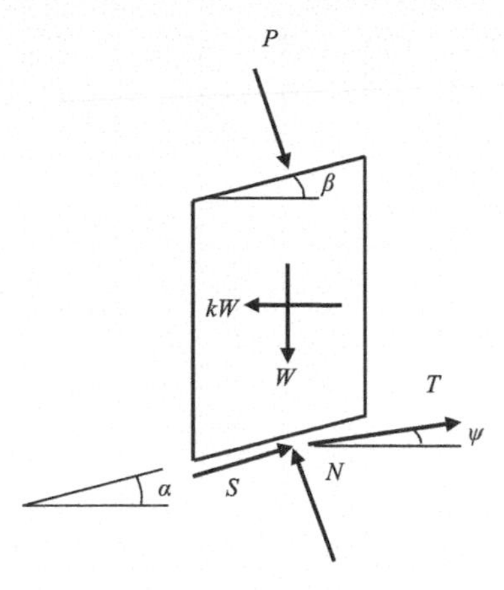

Figure 8.9 Individual slice with additional known forces.

about the center of the circle. Notably, the reinforcement force is not horizontal, as shown in Figure 8.9. The magnitude of k is generally determined by design codes.

In eq. 8.33, the forces represented by the last two summations involve only known quantities, and thus, it is convenient to replace them by a single moment, M_n. An additional moment because of a force, P, is from the known forces. Eq. 8.33 can be represented by eq. 8.34.

$$\gamma\sum\frac{s_i\Delta l_i}{F_s}-\gamma\sum W_i\sin\alpha_i+M_n=0 \tag{8.34}$$

The factor of safety that satisfies moment equilibrium is shown in eq. 8.35.

$$F_s=\frac{\sum s_i\Delta\ell_i}{\sum W_i\sin\alpha_i-\dfrac{Mn}{r}} \tag{8.35}$$

Based on the effective stresses, the factor of safety can be expressed as eq. 8.36 when the summation of each slice is considered.

$$F_s=\frac{\sum\left[c'+(\sigma-u)\tan\phi'\right]\Delta\ell}{\sum W\sin\alpha-\dfrac{Mn}{r}} \tag{8.36}$$

In Figure 8.9, the known forces are included in each slice. They are a seismic force (kW), a water force (P), and a reinforcement force (T). The force P is applied perpendicular to the top of the slice, and the reinforcement force is inclined at an angle, ψ, from the horizontal. To determine the normal force in eq. 8.36, the summations of vertical forces in Figure 8.10 can be presented in eq. 8.37.

$$N\cos\alpha+S\sin\alpha-W-P\cos\beta+T\sin\psi=0 \tag{8.37}$$

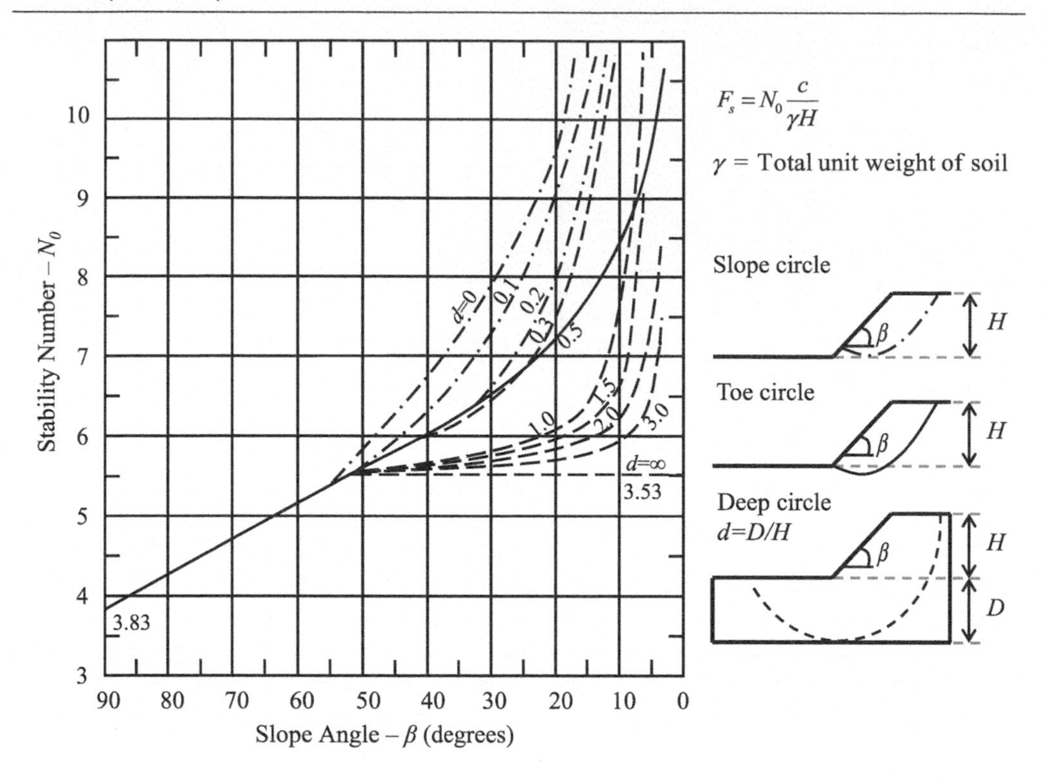

Figure 8.10 Taylor's chart.

Source: Redrawn form Taylor (1937).

where β is the inclination of the top of the slice. Notably, the seismic force in Figure 8.9 is horizontal, and therefore, it makes no contribution to the vertical force equilibrium. In fact, the vertical force can be taken into account if needed. To simplify the contribution of the known forces in eq. 8.37, F_v represent the vertical components of all known forces, as shown in equation 8.38. The only exception is the slice weight.

$$F_v = -P\cos\beta + T\sin\psi \tag{8.38}$$

Therefore, the summation of vertical forces in eq. 8.37 can be expressed as:

$$N\cos\alpha + S\sin\alpha - W + F_v = 0 \tag{8.39}$$

Based on the Mohr–Coulomb failure criterion and eq. 8.28, the normal force can be solved and presented as:

$$N = \frac{W - F_v \left(\frac{1}{F_s}\right)\left(c'\Delta\ell \text{-} u\Delta\ell\tan\phi'\right)\sin\alpha}{\cos\alpha + \frac{\left(\sin\alpha\tan\phi'\right)}{F_s}} \tag{8.40}$$

The normal force in eq. 8.29 combines with eq. 8.36 for the factor of safety, and then the factor of safety can be expressed as:

$$F_s = \frac{\sum \left[\dfrac{c'\Delta\ell\cos\alpha + (W - F_v - u\Delta\ell\cos\alpha)\tan\phi'}{\cos\alpha + \dfrac{(\sin\alpha\tan\phi')}{F_s}} \right]}{\sum W\sin\alpha - \dfrac{Mn}{r}} \tag{8.41}$$

A summary of the very basic limit equilibrium methods for analyzing slopes is shown in Table 8.1, where the required inputs and their limitations are included. For strength parameter selections (drained or undrained condition), hydraulic conductivity can be considered. Practically, rainfall-induced slope failures generally take a few days to allow water infiltration into a certain depth (failure surface) of soil. For the greatest magnitude of clay hydraulic conductivity, rainfall will most likely not reach this depth easily. Therefore, drained parameters are suggested for slope stability analyses under rainfall.

Table 8.1 Available limit equilibrium method for homogeneous slope stability evaluations

Analysis method	*Application(s)*	*Limitations*
Cohesionless soils (sands, $c = 0$)		
Infinite slope	Moist (γ_m), submerged (γ'), seepage (γ_{sat}), homogeeous	plane failure, same F_s for moist and submerged sands
Cohesive soils (clays, silts, sand mixed with clays/silts, $c > 0$)		
Infinite slope	Moist (γ_m), submerged (γ'), seepage (γ_{sat}), homogeneous	plane failure, limited to very long slopes
Wedge (Culmann)	Moist (γ_m), submerged (γ'), homogenous	plane failure, acceptable for near-vertical slopes, iterate for F_s if $\phi \neq 0$
Taylor's charts	Moist (γ_m), submerged (γ'), homogenous, rapid drawdown	circular failure, iterate for F_s if, $\phi \neq 0$ nonrigorous solution, short-term total stress analysis with $c_u - \phi_u$ (or $c - \phi$)
Ordinary method of slices (Fellenius)	Moist (γ_m), submerged (γ'), seepage (γ_{sat}), nonhomogenous profile, external loads	circular failure, nonrigorous solution, long- or short-term *analysis**
Simplified method of slices (Bishop)	Moist (γ_m), submerged (γ'), Seepage (γ_{sat}), nonhomogenous profile, external loads	circular failure, iterate for F_s if $\phi \neq 0$, semirigorous solution, long- or short-term *analysis**
Bishop-Morgenstern and Barnes tables	Moist (γ_m), submerged (γ'), seepage (γ_{sat}), homogenous, simplified method of slices	Circular failure, long-term effective stress analysis with $c' - \phi'$
Cousins charts	Moist (γ_m), submerged (γ'), seepage (γ_{sat}), homogenous, simplified method of slices	Circular failure, long-term effective stress analysis with $c' - \phi'$
Morgenstern rapid drawdown	Seepage (γ_{sat}), homogenous	Short-term effective stress drawdown analysis with $c' - \phi'$

*For slice methods: Long-term effective stress analysis with $c' - \phi'$ and $u \geq 0$ (smallest F_s for cuts in saturated clay); short-term total stress analysis using $c_u - \phi_u$ or $c - \phi$ and $u = 0$ (smallest F_s for embankments on saturated clay).

8.4 Slope stability charts

Stability charts for soil slopes were first produced by Taylor (1937), and they continue to be used extensively as design tools and draw the attention of many investigators (Morgenstern, 1963; Zanbak, 1983; Michalowski, 2002; Qian et al., 2015). The presented chart solutions cover a range of slope and soil conditions. They provide an easy and simple way of obtaining the factor of safety. They are still useful for many conditions. Because stability charts are still produced based on the methods in Section 8.3, only some of the most basic chart solutions will be introduced in this section.

Figure 8.10 shows the chart solutions proposed by Taylor (1937) for homogeneous cohesive soil, such as clay. In this case, the soil has only cohesive strength (c), with $\phi = 0$, which is the undrained condition (short-term). In Figure 8.10, d is the depth factor, and N_0 is the stability number. Notably, N_0 is dimensionless. For a two-dimensional purely cohesive slope, the slope failure mode is generally a deep circle when the slope is gentle. In other words, the slope failure surface will be controlled by the bottom rigid layer, such as the rock formation. Therefore, the stability number can be influenced by the depth factor. For steep slopes, toe failure is the common failure mode.

For drained conditions, $c' = 0$ should be adopted for pure sand, which rarely appears alone in slopes. In practice, soil compaction will be generally applied when constructing fill slopes. Because of overconsolidation from compaction, the soil could have both cohesion and friction angles, known as cohesive-frictional ($c' - \phi'$) soil. In addition, the soils in natural slopes are rarely pure components of clay or sand. They will generally have slight overconsolidation, unsaturated effects, or rock particles in slopes. Therefore, $c' - \phi'$ soil is also more appropriate for describing soil strength. For the preceding cases, Taylor's chart considering uniform undrained shear strength is no longer applicable. A number of charts have been published based on the effective stress theorem, but they usually apply to very specific conditions. One of the more useful charts was presented by Hoek and Bray (1981), who considered different groundwater conditions. Two examples are shown in Figure 8.11, where the effects of tension cracks are included. Figure 8.12

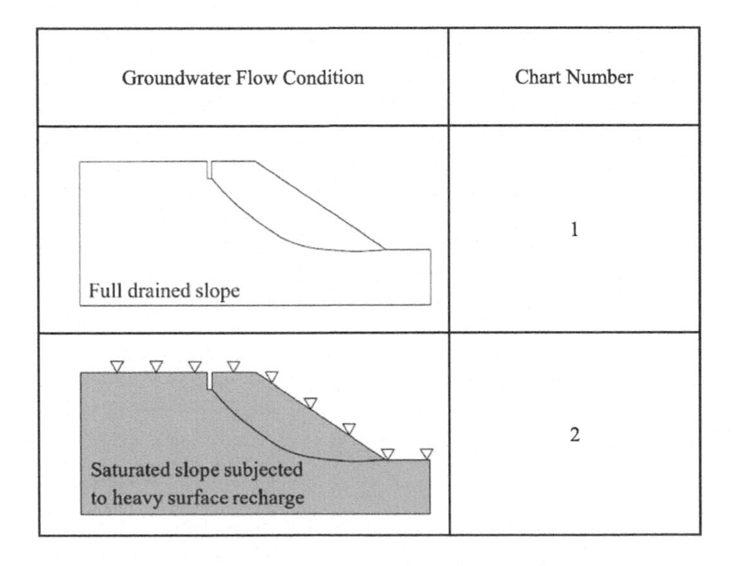

Figure 8.11 Water table and tension crack assumptions for Hoek-Bray charts.

Source: Redrawn form Hoek and Bray (1981).

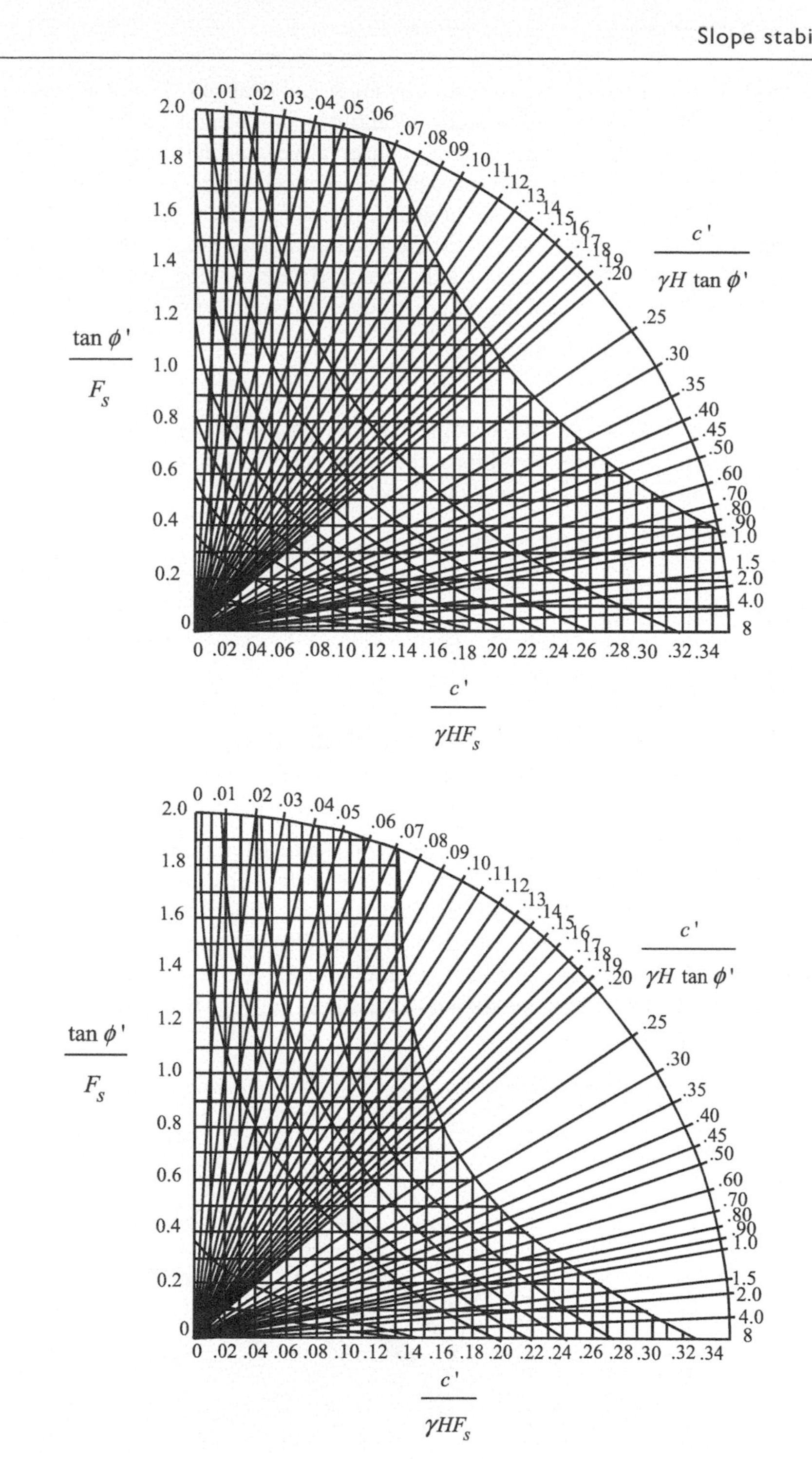

Figure 8.12 Hoek-Bray charts.

Source: Redrawn form Hoek and Bray (1981).

shows the stability charts relevant to Figure 8.11 by Hoek and Bray (1981). For a given slope, the slope F_s can be assessed when the soil strength (c' and ϕ'), unit weight, and slope height (H) are known. Using the saturated unit weight here is a conservative consideration.

In fact, the safety factor is highly relevant to the failure mechanism. As mentioned previously, the deep circles are the main failure mode of gentle slopes for purely cohesive soils. However, the general failure mode for $c'-\phi'$ soil is toe circles. For a given slope angle, the depth of the failure surface decreases with increasing friction angle. Therefore, the bottom rigid layer does not influence the factor of safety and the failure mode if it is deep enough.

Example 8.2

A 10 m high slope with a 20° slope angle has the parameters $c' = 2\text{kPa}$, $\phi' = 25^\circ$, and $\gamma_{sat} = 16\text{kN/m}^3$. Under long-term conditions, the water table is at the ground surface for distances greater than 40 m behind the toe of the slope.

Solution

Using Figure 8.13c, calculate $\dfrac{c'}{\gamma H \tan\phi'} = \dfrac{2}{(16)(10)(\tan 25^\circ)} = 0.027$.

For a slope angle of $20°$, read off the chart.

Either $c'/\gamma H F_s = 0.009$ or $\tan\phi'/F_s = 0.318$ is obtained.

Hence, $F_s \approx 1.41$.

In addition, three-dimensional boundary effects will affect failure surface development, as shown in Figure 8.13. Generally, they make the failure surface shallower. The effects are more obvious for purely cohesive soils (Lim et al., 2015) because the failure surface is deeper for purely cohesive soils than for cohesive-frictional soils. This difference also means that three-dimensional effects are relatively significant in purely cohesive soil slopes. Gens et al. (1988) indicated that the difference in the slope stability assessment between two- and three-dimensional

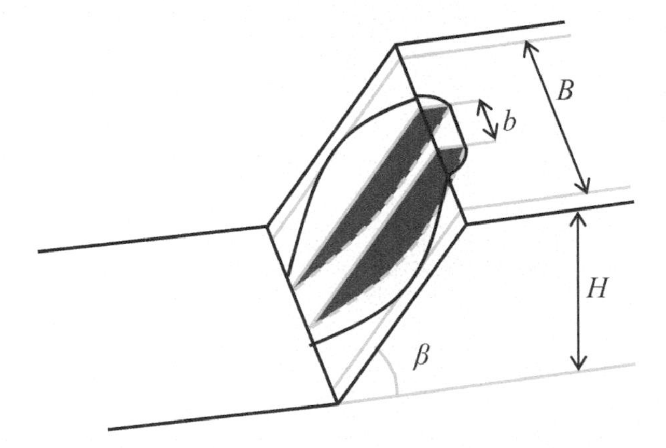

Figure 8.13 Three-dimensional slope failure mechanism modified with a plane insert.

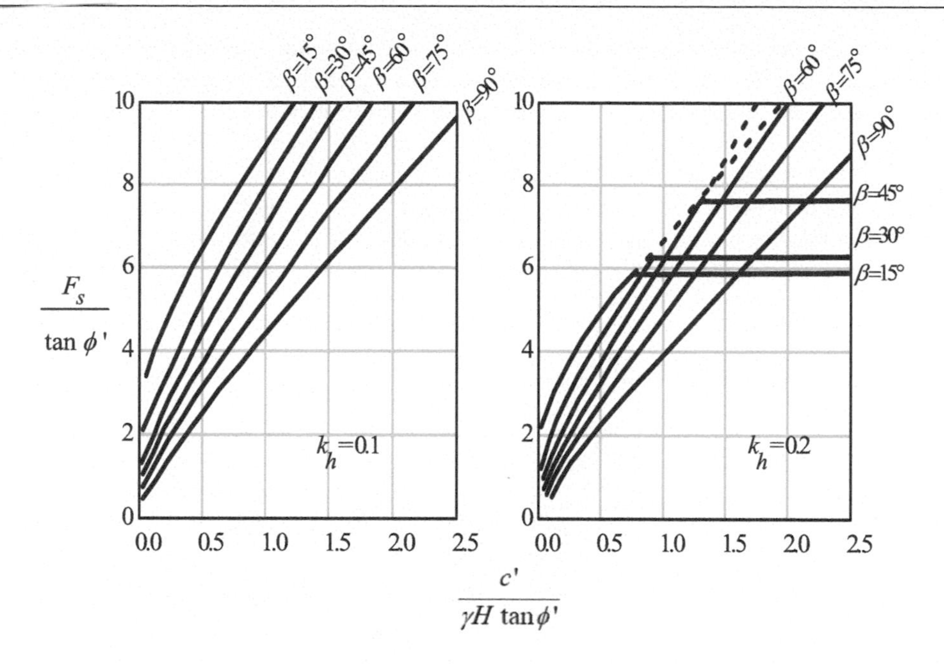

Figure 8.14 Stability charts considering seismic effects.

Source: Redrawn from Michalowski (2002).

analyses can range from 3% to 30% and an average of 13.9%. As indicated by Li et al. (2009), the differences in the factor of safety between two-dimensional and three-dimensional analyses can be up to several hundred percent. This phenomenon is observed much less in cohesive-frictional soil slopes, particularly for $B/H \geq 5$. Therefore, engineers should be aware of this issue when assessing three-dimensional slope stability.

Michalowski (2002) adopted the pseudostatic method to account for seismic force and then produced stability charts for cohesive-frictional soil slopes without a groundwater table. The chart solutions are displayed in Figure 8.14 for various horizontal seismic coefficients (k_h). In fact, the pseudostatic method is a simplified approach that would not be appropriate for slopes with a high groundwater table because an earthquake can generate high excess pore water pressure. Notably, the range of the predicted failure surface increases with increasing k_h when the Mohr–Coulomb failure criterion is applied to the slope material (Li et al., 2020). Many commercial limit equilibrium software programs should set up searching boundaries. To obtain reasonable critical failure mechanisms, it is important to adjust the search boundary properly.

8.5 Stabilization methods

For slope stability problems, an appropriate stabilization method is determined by considering potential failure mechanisms, as shown in Figure 5.20. Specifically, the slope reinforcement should be installed over the range of potential failure surfaces. The failure surface development makes it more difficult to pass through the slope reinforcement. Therefore, the factor of safety increases after slope reinforcement. In Chapter 5, many reinforced slope stabilization methods, such as anchor and geosynthetics in Figures 5.25b and 5.28, respectively, have been well introduced. In this section, only two commonly used methods will be briefly included.

Figure 8.15 is a configuration of the drilled shaft. The retaining wall can have a certain space (S_p) between two piles. Notably, the space between two piles is also helpful for drainage. The soil arching effects shown in Figure 8.16 can be utilized to increase the factor of safety. After inserting the piles, the horizontal soil stress is significantly different. The horizontal stresses will be concentrated at both side piles after installation. Of course, the distance between the two piles increases, and then the soil arching effects decrease. The drilled shaft behaves like a cantilever wall. In general, the pile toe will be installed into the bottom rigid layer to prevent sliding. Therefore, the drilled shaft alone would not be suitable for slopes with a very thick soil layer.

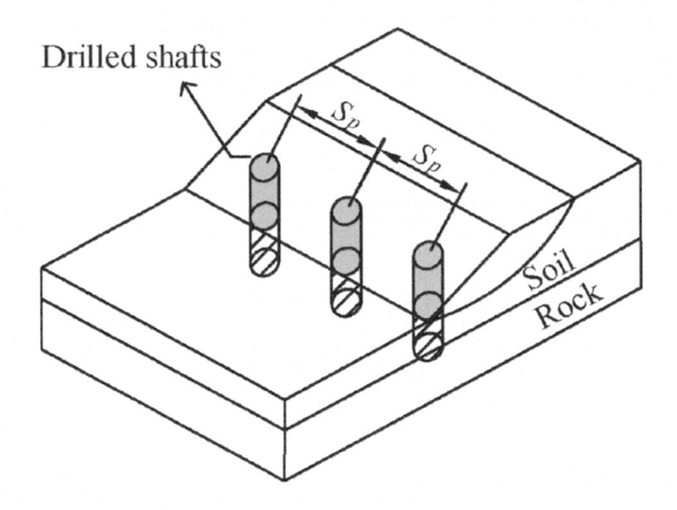

Figure 8.15 Pile retaining wall.

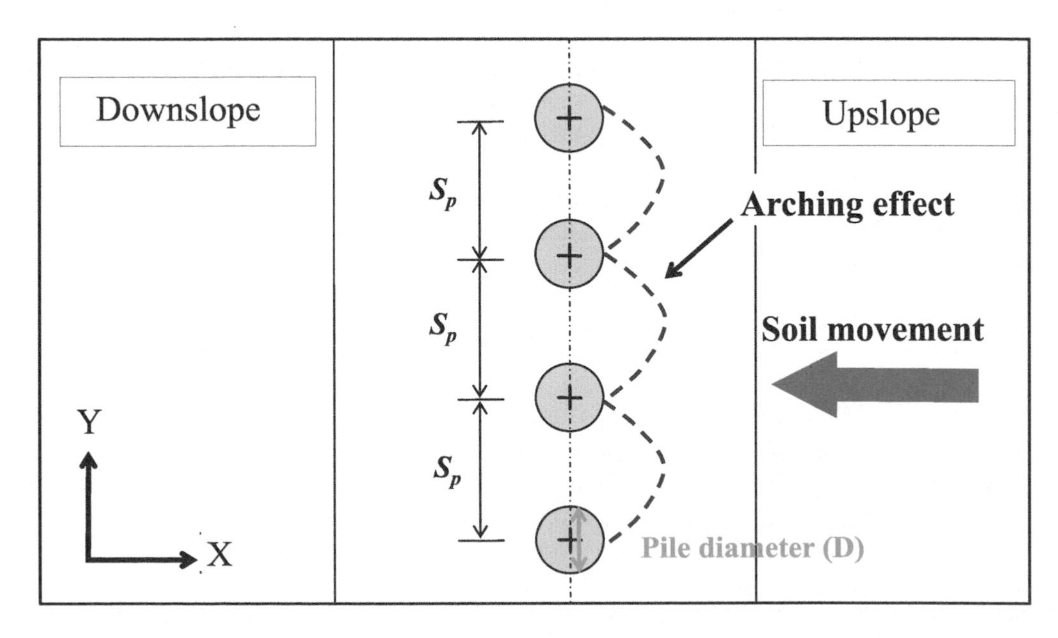

Figure 8.16 Soil arching effect (top view).

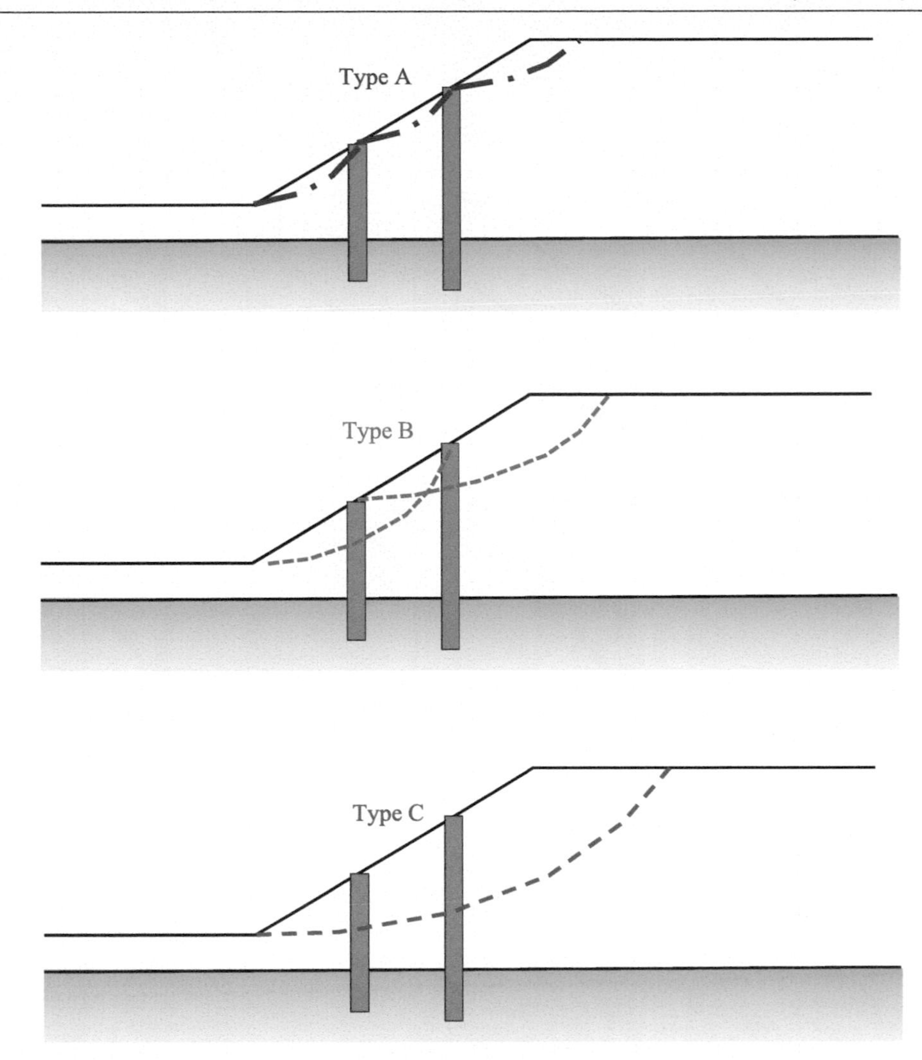

Figure 8.17 Potential failure mechanisms with multiple drilled shaft.

In some cases, a two-row pile could be adopted for a slope, as shown in Figure 8.17, where three types of failure surfaces can be obtained based on Li et al. (2020). Type A failure surfaces are restricted by piles. Type B and Type C failure surfaces pass through a single-row and a two-row pile, respectively. For Type A, Figure 8.18 shows that the slope can be divided into three parts between shafts. Thus, each part can be assessed individually, implying that F_s increases because of the reduction in the slope height. In addition, Li et al. (2022) proposed slope stability charts (Figure 8.18), which are suitable for Type A and Type B failure surfaces. However, their solutions are not applicable to slopes with Type C failure surfaces. Theoretically, the most effective location of drilled shafts is near the middle of the slope, because the failure surface development could be suppressed.

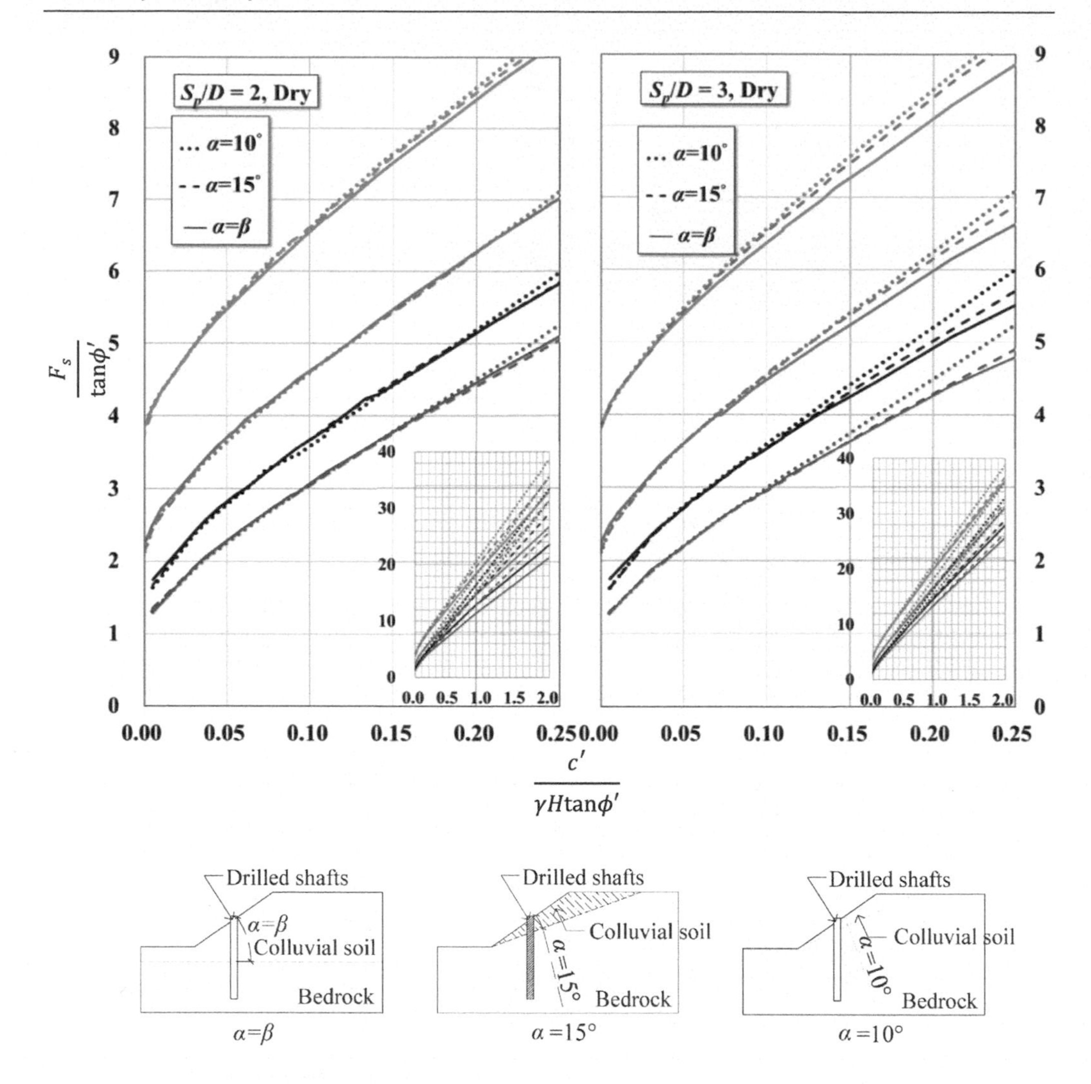

Figure 8.18 Stability chart for the results obtained under various S_p/D ratios (dry slopes).

Figure 8.19 shows the use of a drainage well to collect and lower the groundwater table. For some potential large-scale landslides, commonly used reinforced methods would not be very effective because the reinforcements have difficulty reaching ranges beyond the potential sliding slope masses. Thus, the general slope reinforcement scale is much smaller than the potential sliding mass, so drainage wells could be more useful than general slope reinforcements. In particular, the groundwater table can rise quickly during rainfall, and thus, the slope is critical. The horizontal drainage pipe is useful to slow down the rise in water pressure. This feature contributes directly to the factor of safety because of the increasing soil effective strength. We investigated a range of soil slope parameters, which showed that H is the main controlling factor for F_s when $H \geq 20$ m and there is a fairly high water level. This issue is commonly faced by geotechnical engineers in Taiwan, where there are a large number of colluvium slopes. This

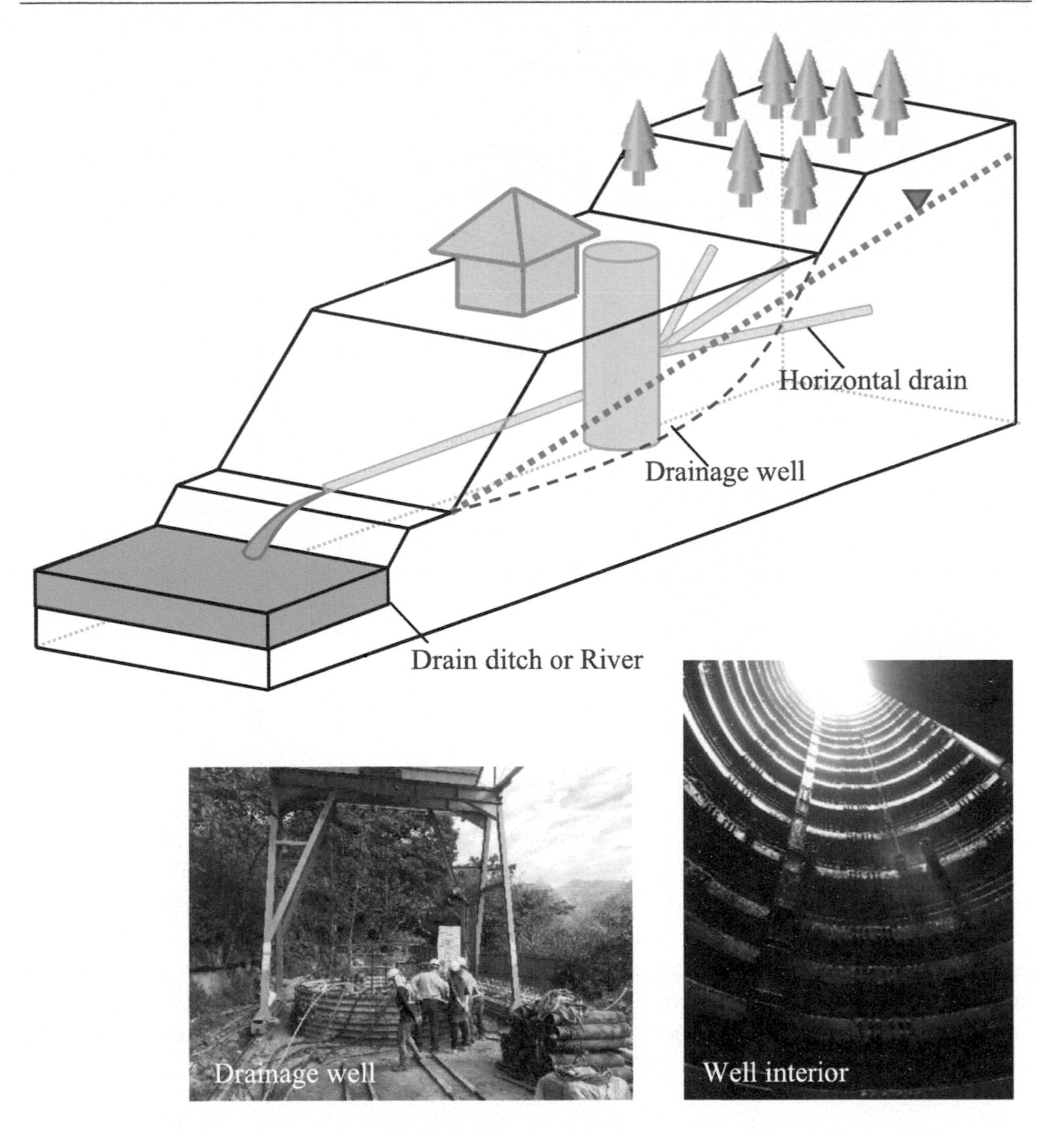

Figure 8.19 Configuration of drainage well.

feature occurs because changes in natural soil strength have only limited effects on F_s. For some large-scale landslides, an F_s of 1.5, the general static design requirement, is difficult to satisfy. Therefore, maintaining a relatively low water level by the drainage well is an advisable approach.

In addition to the stabilization methods, long-term monitoring on the slope is required. For example, an inclinometer can be used to measure soil movements and identify the moving direction, as shown in Figure 8.20. Figure 8.20 also shows various monitoring techniques which have been frequently applied recently. In general, they are used to measure groundwater table, surface or underground movements.

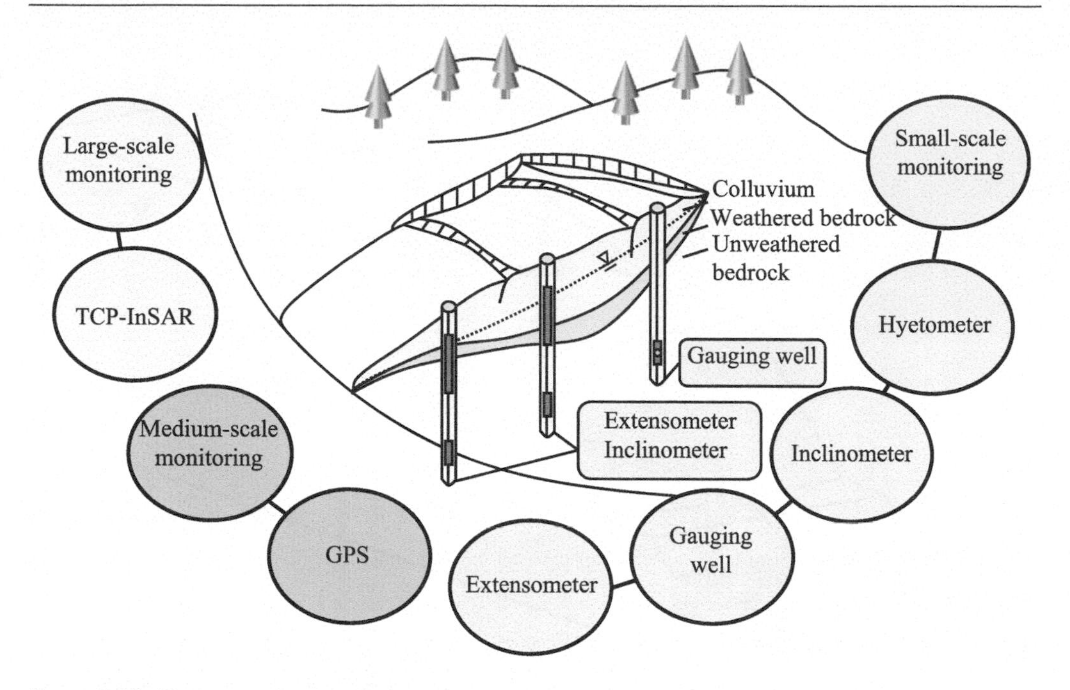

Figure 8.20 Slope monitoring techniques.

8.6 Summary and general comments

For the past few decades, many researchers have dedicated their time and efforts to soil mechanics, and therefore, we can obtain better understandings of soil strength and behavior. They also enhance knowledge of slope stability problems. Thus far, our lessons are still learned from experience. Perhaps some of these experiences are bad. Investigations of past slope failures help us understand potential factors that can cause slope instability. However, it is believed that we have not captured the entire phenomenon. In particular, global warming brings more extreme weather conditions, which will create major challenges for civil engineers in the near future. Reducing the losses of lives and property requires advances in this area.

Problem

8.1 For the slope shown in Figure P8.1, determine the factor of safety.

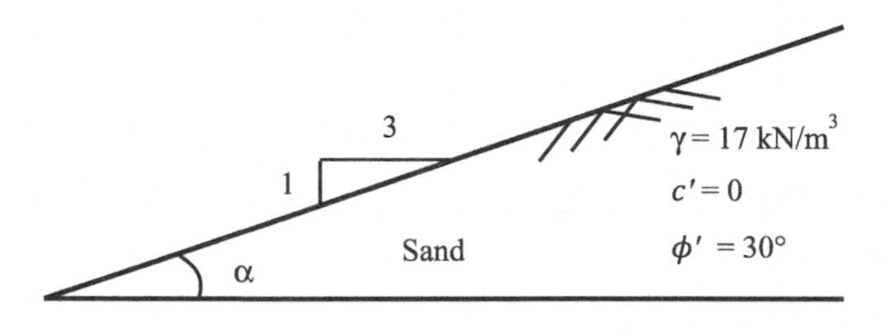

Figure P8.1

8.2 Figure P8.2 shows a 10 m high slope in saturated clay. For the soil, assume $\gamma = 19$ kN/m^3, $\phi' = 20^\circ$, and $c' = 20$ kN/m^2. Determine F_s using the ordinary method of slices.

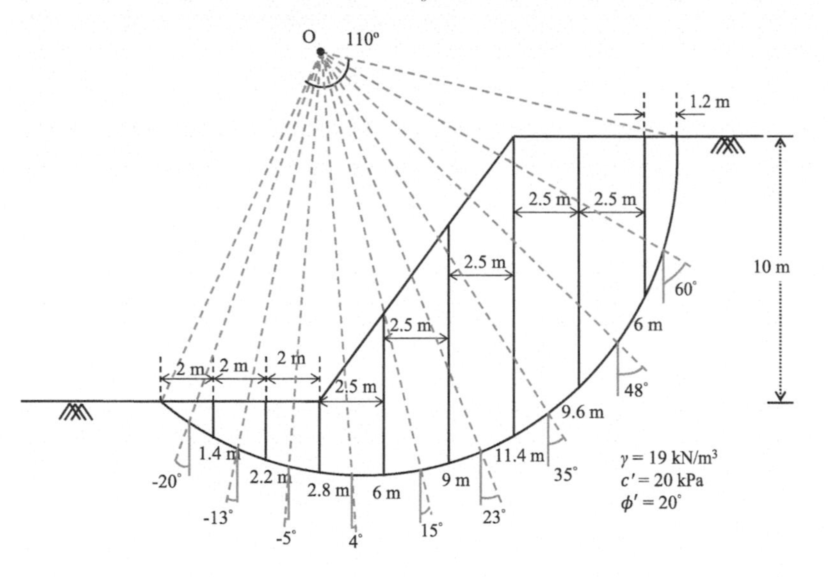

Figure P8.2

8.3 Refer to problem 8.2. For the same slope shown in Figure P8.2, compute the factor of safety using the simplified Bishop method of slices.

8.4 Figure P8.3 shows a slope with dimensions similar to those in Figure P8.2. However, there is steady-state seepage. The phreatic line is shown in Figure P8.3. The other parameters remain the same. Determine F_s using the ordinary method of slices.

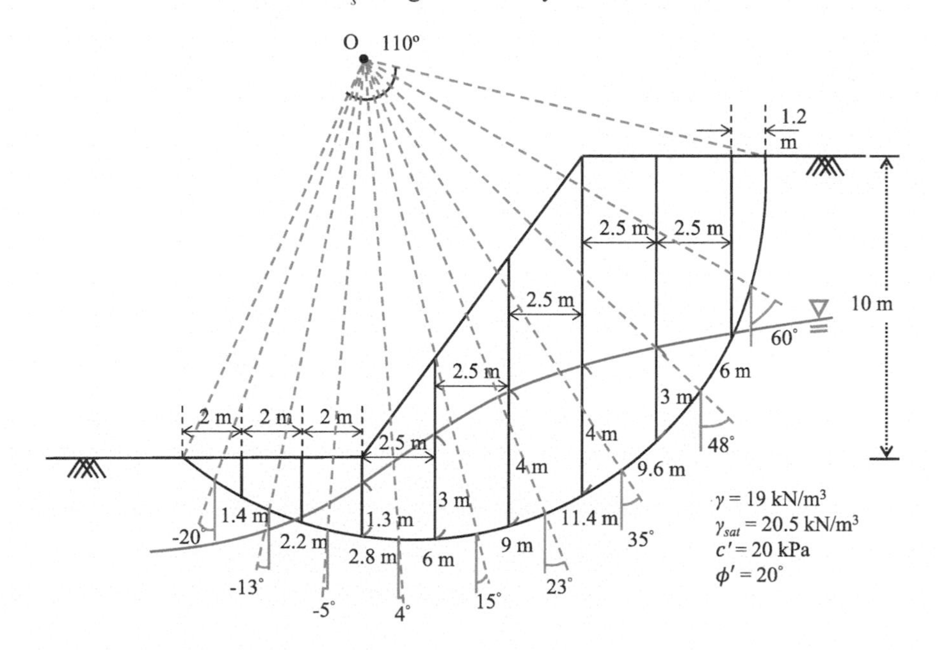

Figure P8.3

8.5 A 15 m high slope with a face angle of 40° is to be excavated in soil with density $\gamma = 16\ \text{kN/m}^3$, a cohesive strength of 40 kPa, and a friction angle of 30°. Find the factor of safety of the slope, assuming that it is a saturated slope subjected to heavy surface recharge.

8.6 A slope with $\beta = 60°$ is to be constructed with a soil that has $\phi' = 20°$ and $c' = 25\,\text{kPa}$. The unit weight of the compacted soil will be 19 kN/m³.

   a. Find the critical height of the slope using Michalowski's solution.
   b. If the height of the slope is 15 m, determine the factor of safety using Michalowski's solution.

References

Bishop, A.W. (1955), The use of the slip circle in the stability analysis of slopes. *Geotechnique*, Vol. 5, No. 1, pp. 7–17.

Fredlund, D.G. and Krahn, J. (1977), Comparison of slope stability methods of analysis. *Canadian Geotechnical Journal*, Vol. 14, No. 3, pp. 429–439.

Gens, A., Hutchinson, J.N. and Cavounidis, S. (1988), Three-dimensional analysis of slides in cohesive soils. *Geotechnique*, Vol. 38, No. 1, pp. 1–23.

Goodman, R.E. and Bray, J.W. (1976), Toppling of rock slopes. *Proceedings of the Specialty Conference on Rock Engineering for Foundations and Slopes, ASCE*, Vol. 2, pp. 201–234.

Goodman, R.E. and Kieffer, D.S. (2000), Behavior of rock in slope. *Journal of Geotechnical and Geoenvironmental Engineering*, Vol. 126, No. 8, pp. 675–684.

Hoek, E. and Bray, J.W. (1981), *Rock Slope Engineering*, London: Institute of Mining and Metallurgy.

Krabbenhoft, K., Lyamin, A.V., Hjiaj, M. and Sloan, S.W. (2005), A new discontinuous upper bound limit analysis formulation. *International Journal for Numerical Methods in Engineering*, Vol. 63, No. 7, pp. 1069–1088.

Li, A.J., Jange, A., Lin, H.D. and Huang, F.K. (2020), Investigation of slurry supported trench stability under seismic condition in purely cohesive soils. *Journal of GeoEngineering*, Vol. 15, No. 4, pp. 195–203.

Li, A.J., Lin, H.D. and Wang, W.C. (2022), Investigations of slope reinforcement with drilled shafts in colluvium soils. *Geomechanics and Engineering*, Vol. 31, No. 1, pp. 71–86.

Li, A.J., Merifield, R.S. and Lyamin, A.V. (2009), Limit analysis solutions for three dimensional undrained slopes. *Computers and Geotechnics*, Vol. 36, No. 8, pp. 1330–1351.

Lim, K., Li, A.J. and Lyamin, A.V. (2015), Three-dimensional slope stability assessment of two-layered undrained clay. *Computers and Geotechnics*, Vol. 70, pp. 1–17.

Lyamin, A.V. and Sloan, S.W. (2002a), Lower bound limit analysis using non-linear programing. *International Journal for Numerical Methods in Engineering*, Vol. 55, No. 5, pp. 573–611.

Lyamin, A.V. and Sloan, S.W. (2002b), Upper bound limit analysis using linear finite elements and non-linear programing. *International Journal for Numerical Methods in Engineering*, Vol. 26, No. 2, pp. 181–216.

Michalowski, R.L. (2002), Stability charts for uniform slopes. *Journal of Geotechnical and Geoenvironmental Engineering*, Vol. 128, No. 4, pp. 351–355.

Morgenstern, N. (1963), Stability charts for each slopes during rapid drawdown. *Geotechnique*, Vol. 13, pp. 123–131.

Qian, Z.G., Li, A.J., Merifield, R.S. and Lyamin, A.V. (2015), Slope stability charts for two-layered purely cohesive soils based on finite-element limit analysis methods. *International Journal of Geomechanics*, Vol. 15, No. 3.

Taylor, D.W. (1937), Stability of earth slopes. *Journal of the Boston Society of Civil Engineering*, Vol. 24, pp. 197–246.

Taylor, D.W. (1948), *Fundamentals of Soil Mechanics*, New York: John Wily and Sons, pp. 406–479.

Zanbak, C. (1983), Design charts for rock slopes susceptible to toppling. *Journal of Geotechnical Engineering, ASCE*, Vol. 109, No. 8, pp. 1039–1062.

Solutions

Chapter 1

1.1

Table S1.1

Test	*Purpose*
Soil physical properties tests	Obtain grain distribution, moisture content, unit weight, liquid limit, and plastic limit of soil and classify soil type based on these properties.
Triaxial unconsolidated undrained test (UU)	Obtain undrained shear strength (s_u) of clayey soil.
Triaxial consolidated undrained test (CU)	Obtain total strength parameter (c, ϕ) and effective strength parameter (c', ϕ').
Oedometer test	Obtain compression index and swell index (C_c, C_s).
Direct shear test	Obtain (c', ϕ') of sandy soil.
Permeability test	Obtain the value of hydraulic conductivity (k) of soil.
Dynamic triaxial test	Obtain stress–strain relationship, stress path, and liquefaction resistance strength.

1.3

Ans:

Clayey soil.

1.5

Ans:

Table S1.5

Depth (m)	*N1*	*N2*	*N3*	*N-value*
1.5	2	3	2	5
3.0	2	2	4	6
4.5	1	3	3	6
6.0	2	3	4	7
7.5	3	4	4	8

1.7

Ans:

$$\text{Ar} = (\text{split spoon}) = \frac{d_0^2 - d_i^2}{d_i^2} = \frac{50.8^2 - 34.9^2}{34.9^2} = 111.9\%$$

$$\text{Ar (Thin-walled tube)} = \frac{d_0^2 - d_i^2}{d_i^2} = \frac{76.2^2 - (76.2 - 3.1)^2}{(76.2 - 3.1)^2} = 8.7\%$$

Chapter 3

3.1

For this sandy layer:

$c = c' = 0$
$\phi = \phi' = 34°, N_c = 52.64, N_q = 36.5, N_\gamma = 38.04$ (see Table 3.1)

a. 3m wide strip footing

Eq. 3.5 is used.

$$q_u = (1)(0)(52.64) + (1)(18.5)(36.5) + (0.5)(18.5)(3)(38.04) = 1{,}730.86 \text{ kN/m}^2$$

b. 3 m × 3 m square footing

Eq. 3.8 is used.

$$q_u = (1.3)(0)(52.64) + (1)(18.5)(36.5) + (0.4)(18.5)(3)(38.04) = 1{,}519.74 \text{ kN/m}^2$$

c. Circular footing of 3 m in diameter

Eq. 3.9 is used.

$$q_u = (1.3)(0)(52.64) + (1)(18.5)(36.5) + (0.3)(18.5)(3)(38.04) = 1{,}308.62 \text{ kN/m}^2$$

3.3

Eq. 3.10 is used.

For this sandy layer:

$c = c' = 0$
$\phi = \phi' = 34°, N_c = 42.16, N_q = 29.44, N_r = 31.15$

a. 3m wide strip footing

$B = 3$ m, $L = \infty$, $D_f = 1$ m
$F_{cs} = 1$, $F_{qs} = F_{rs} = 1$
$F_{cd} = 1.125$, $F_{qd} = F_{rd} = 1.063$
$F_{ci} = 1$, $F_{qi} = F_{ri} = 1$
$q = 18.5$ kN/m^2

$$q_u = cN_cF_{cs}F_{cd}F_{ci} + qN_qF_{qs}F_{qd}F_{qi} + \frac{1}{2}\gamma BN_rF_{rs}F_{rd}F_{qi}$$
$$= (0)(42.16)(1)(1.125)(1) + (18.5)(29.44)(1)(1.063)(1) + (0.5)(18.5)(3)(31.15)(1)(1.063)(1)$$
$$= 1{,}497.8 \text{ kN/m}^2$$

b/c. 3 m × 3 m square footing; circular footing of 3 m in diameter

$B = 3$ m, $L = 3$ m, $D_f = 1$ m
$F_{cs} = 1.707$, $F_{qs} = F_{rs} = 1.354$
$F_{cd} = 1.125$, $F_{qd} = F_{rd} = 1.062$
$F_{ci} = 1$, $F_{qi} = F_{ri} = 1$

$$q_u = cN_cF_{cs}F_{cd}F_{ci} + qN_qF_{qs}F_{qd}F_{qi} + \frac{1}{2}\gamma BN_rF_{rs}F_{rd}F_{qi}$$
$$= (0)(42.16)(1.707)(1.125)(1) + (18.5)(29.44)(1.354)(1.063)(1) + (0.5)(18.5)(3)(31.15)(1.354)(1.063)(1)$$
$$= 2{,}028 \text{ kN/m}^2$$

3.5

a. $c = 0$, $\phi = 34^\circ$, $\gamma = 18.5 \text{ kN/m}^3$

$$N_q = \tan^2(45 + \phi/2)e^{\pi \tan\phi} = \tan^2(45 + 34/2)e^{\pi \tan 34} = 29.44$$
$$N_c = \cot\phi(N_q - 1) = \cot 34(29.44 - 1) = 42.16$$
$$N_\gamma = \tan 1.4\phi(N_q - 1) = \tan 1.4 \times 34(29.44 - 1) = 31.15$$
$$F_{cs} = 1 + 0.2(B/L)\tan^2(45 + \phi/2) = 1 + 0.2(3/4)\tan^2(45 + 34/2) = 1.531$$
$$F_{qs} = F_{\gamma s} = 1 + 0.1(B/L)\tan^2(45 + \phi/2) = 1 + 0.1(3/4)\tan^2(45 + 34/2) = 1.265$$
$$F_{cd} = 1 + 0.2(D_f/B)\tan(45 + \phi/2) = 1 + 0.2(1/3)\tan(45 + 34/2) = 1.125$$
$$F_{qd} = F_{\gamma d} = 1 + 0.1(D_f/B)\tan(45 + \phi/2) = 1 + 0.1(1/3)\tan(45 + 34/2) = 1.063$$
$$F_{ci} = F_{qi} = 1 - (\beta/90^\circ)^2 = 1$$
$$F_{\gamma i} = 1 - (\beta/\phi)^2 = 1$$
$$q_u = cN_cF_{cs}F_{cd}F_{ci} + qN_qF_{qs}F_{qd}F_{qi} + 0.5\gamma BN_\gamma F_{\gamma s}F_{\gamma d}F_{\gamma i}$$
$$= 0 + 18.5 \times 1 \times 29.44 \times 1.265 \times 1.063 \times 1 + 0.5 \times 18.5 \times 3 \times 31.15 \times 1.265 \times 1.063 \times 1$$
$$= 1894.75 \text{ kN/m}^2$$
$$q_{u(net)} = 1894.75 - 18.5 \times 1 = 1876.25 \text{ kN/m}^2$$

b. $q_{b(net)} = 6000/(3\times4) - 18\times1 = 482$

$FS = 1876.25/482 = 3.893$

3.7

$B = 3$ m, $L = 4$ m, $Q = 6{,}000$ kN, $M_L = 1{,}200$ kN-m, $e_L = 1{,}200/6{,}000 = 0.2 < 4/6$
$q_{b,max/min} = [(6{,}000)/(3 \times 4)](1 \pm (6)(0.2)/3) = 700\ \text{kN/m}^2/300\ \text{kN/m}^2$

Using eq. 3.22:

$B' = \min(3 - 2 \times 0.2, 4) = 2.6$ m, $L' = \max(3 - 2 \times 0.2, 4) = 4$ m
$c = c' = 0$
$\phi = \phi' = 34°$, $N_q = 29.44$, $N_c = 42.16$, $N_\gamma = 31.15$
$F_{cs} = 1 + (0.2)(2.6/4)\tan^2(45 + 34/2) = 1.46$
$F_{qs} = F_{\gamma s} = 1 + (0.1)(2.6/4)\tan^2(45 + 34/2) = 1.23$
$F_{cd} = 1 + (0.2)(1/3)\tan(45 + 34/2) = 1.125$
$F_{qd} = F_{\gamma d} = 1 + (0.1)(1/3)\tan(45 + 34/2) = 1.063$
$F_{ci} = F_{qi} = F_{\gamma i} = 1$
$q_u = (0)(42.16)(1.46)(1.125)(1) + (1)(18.5)(29.44)(1.23)(1.063)(1) + (1/2)(18.5)(2.6)(31.15)(1.23)(1.063)(1) = 1{,}691.63\ \text{kN/m}^2$

3.9

Eq. 3.24 is used.

$B = 3$ m, $L = 4$ m, $H = 5–1 = 4$ m, $D_f = 1$ m
$m = L/B = 4/3$, $n = H/(B/2) = 8/3$
$I_f = 0.862$ (refer to Table 3.4)

$$I_s = F_1 + \frac{1-2\mu_s}{1-\mu_s}F_2 = 0.398$$

$$S_{i(rigid)} = 0.93S_{i(center)} = (0.93)(250)(4)(3/2)\left(\frac{1-(0.3)^2}{12000}\right)(0.398)(0.862) = 0.0363\ \text{m}$$

$= 36.3$ mm

3.11

$$\sigma_0' = (18-9.81)\times(1+4/2) = 24.57\ \text{kN/m}^2 < p_c' = 80\ \text{kN/m}^2$$

$$\Delta\sigma_{ave}' = \frac{200\times3\times4}{(3+2)\times(4+2)} = 80$$

$$\sigma_0' + \Delta\sigma_{ave}' = 104.57 > p_c' = 80$$

$$S_c = \frac{C_s H_c}{1+e_0}\log\frac{p_c'}{\sigma_0'} + \frac{C_c H_c}{1+e_0}\log\frac{\sigma_0' + \Delta\sigma_{ave}'}{p_c}$$

$$= \frac{0.05\times 4}{1+0.8}\log\frac{80}{24.57} + \frac{0.25\times 4}{1+0.8}\log\frac{104.57}{80} = 121.6 \text{ mm}$$

3.13

1. Settlement controlled.

Use eq. 3.24.

$$S_i = q_{b(net)}(\alpha B')\frac{1-\mu_s^2}{E_s}I_s I_f$$

$\mu_s = 0.3$

For the center of the footing, $\alpha = 4$:

$m = L/B = 4/3 = 1.33$, $n = H/(B/2) = 4/(3/2) = 2.67$ => $F_1 = 0.36$, $F_2 = 0.0669$
$I_s = 0.36 + [(1 - 2(0.3))/(1 - 0.3)](0.0669) = 0.398$
$B/L = 3/4 = 0.75$, $D_f/B = 1/3 = 0.33$ => $I_f = 0.862$
$S_{i\ (rigid)} = (0.93)q_{all(net)}(4)(3/2)[(1 - 0.3^2)/12{,}000](0.398)(0.862) = 0.04 \text{ m} = 40 \text{ mm}$
$q_{all(net)} = 275.53 \text{ kN/m}^2$

2. Bearing capacity controlled.

$c = c' = 0 \text{ kN/m}^2$, $\phi = \phi' = 34°$, $\gamma = \gamma' = 18.5 \text{ kN/m}^2$, $N_q = 29.44$, $N_c = 42.16$, $N_\gamma = 31.15$
$F_{cs} = 1 + (0.2)(3/4)\tan^2(45 + 34/2) = 1.531$
$F_{qs} = F_{\gamma s} = 1 + (0.1)(3/3)\tan^2(45 + 34/2) = 1.265$
$F_{cd} = 1 + (0.2)(1/3)\tan(45 + 24/2) = 1.125$
$F_{qd} = F_{\gamma d} = 1 + (0.1)(1/3)\tan(45 + 24/2) = 1.063$
$F_{ci} = F_{qi} = 1$
$F_{\gamma i} = 1$
$q_u = (0)(42.16)(1.531)(1.125)(1) + (1)(18.5)(29.44)(1.265)(1.063)(1) + (1/2)(18.5)(3)(31.15)(1.265)(1.063)(1) = 1894.75 \text{ kN/m}^2$
$q_{u\ (net)} = 1894.75 - 18.5(1) = 1876.25 \text{ kN/m}^2$
$q_{all(net)} = 1876.25/2 = 938.125 \text{ kN/m}^2$
=> Considering conditions 1 and 2, choose the smaller one.
$q_{all(net)} = 275.53 \text{ kN/m}^2$ (The allowable settlement governs!)

Chapter 4

4.1

For $\phi' = 32°$, $K_0 = 1 - \sin\phi' = 0.47$
At $z = 0 \text{ m}$, $\sigma_h' = 0$

At $z = 2.5$ m, $\sigma_h' = 16 \times 2.5 \times 0.47 = 18.8$ kN/m^2

At $z = 9.5$ m, $\sigma_h' = (16 \times 2.5 + 7 \times 20 - 9.81 \times 7) \times 0.47 = 52.3$ kN/m^2

$u_w = 9.81 \times 7 = 68.7$ kN/m^2

Total lateral force = 512.8 kN/m^2, $\bar{z} = 2.885$ m

4.3

a. $K_a = 1$, $z_c = 40/17 = 2.353$ m

b. Assume that tension cracks are formed and no water is in the tension cracks.

At $z_c = 2.353$ m, $\sigma_a = 0$,

At z = 4 m$^-$, $\sigma_a = 4 \times 17 - 2 \times 20 = 28$ kN/m^2

At z = 4 m$^+$, $\sigma_a = 4 \times 17 - 2 \times 30 = 8$ kN/m^2

At z = 9 m, $\sigma_a = 4 \times 17 + 5 \times 18 - 2 \times 30 = 158 - 60 = 98$ kN/m^2

c. $P_{a1} = 28 \times 1.647 / 2 = 23.06$ kN/m

$P_{a2} = 8 \times 5 = 40$ kN/m

$P_{a3} = (98 - 8) \times 5 / 2 = 225$ kN/m

$P = P_{a1} + P_{a2} + P_{a3} = 288.06$ kN/m

$\bar{z} = 2.09$ m

4.5

$\phi' = 34^\circ$, $K_a = 0.283$

At z = 0 m, $\sigma_a' = 0$ kPa

At z = 2 m, $\sigma_a' = 9.06$ kPa, $u_w = 0$ kPa

At z = 6 m, $\sigma_a' = 20.59$ kPa $u_w = 9.81 \times 4 = 39.24$ kPa

$P_{a1} = 9.06 \times 2 / 2 = 9.06$ kN/m

$P_{a2} = 9.06 \times 4 = 36.22$ kN/m

$P_{a3} = (20.59 - 9.06) \times 4 / 2 = 23.07$ kN/m

$P_W = 9.81 \times 4 \times 4 / 2 = 78.48$ kN/m

$P = P_{a1} + P_{a2} + P_{a3} + P_W = 146.83$ kN/m

$\bar{z} = 1.70$ m

4.7

For clay, $\phi = 0$, $K_a = 1$

a. $z_c = 40/16 = 2.5$ m

b. Assume that tension cracks are formed and water is in the tension cracks.

At z = 0 m, $\sigma_a = 0$ kPa , $u_w = 0$

At z = 2.5 m, $\sigma_a = 2.5\times16-2\times20 = 0$kPa , $u_w = 9.81\times2.5 = 24.53$kPa

At z = 4 m$^-$, $\sigma_a = 1.5\times16 = 24$ kPa

At z = 4 m$^+$, $\sigma_a = 4\times16-2\times25 = 64-50 = 14$ kPa

At z = 8 m, $\sigma_a = 4\times16+4\times18-2\times25 = 136-50 = 86$ kPa

c.

$$P_{a1} = 24\times1.5/2 = 18 \text{ kN/m}$$

$$P_{a2} = 14\times4 = 56 \text{ kN/m}$$

$$P_{a3} = (86-14)\times4/2 = 144 \text{ kN/m}$$

$$P = P_{a1} + P_{a2} + P_{a3} = 218 \text{ kN/m}$$

$$\bar{z} = \frac{P_{a1}\times(1.5/3+4)+P_{a2}\times4/2+P_{a3}\times4/3}{P} = 1.77 \text{ m}$$

4.9

$\phi_1' = 30^\circ$ $\phi_2' = 32^\circ$, $K_{a1} = 0.333, K_{a2} = 0.307$

a. At z = 0 m, $\sigma_a' = 0$

At z = 2 m$^-$,$\sigma_a' = \gamma_1'\times z\times K_{a1} = 6.13$ kPa

At z = 2 m$^+$,$\sigma_a' = \gamma_1'\times z\times K_{a2} = 5.64$ kPa

At z = 5 m ,$\sigma_a' = 15.95$ kPa , $u_w = 9.81\times5 = 49.05$ kPa

b.

$$P_{a1} = 6.13\times2/2 = 6.13 \text{ kN/m}$$

$$P_{a2} = 5.64\times3 = 16.92 \text{ kN/m}$$

$$P_{a3} = (15.95-5.64)\times3/2 = 15.47 \text{ kN/m}$$

$$P_w = 49.05\times5/2 = 122.63 \text{ kN/m}$$

$$P = P_{a1} + P_{a2} + P_{a3} + P_w = 161.15 \text{ kN/m}$$

$$\bar{z} = 1.66 \text{ m}$$

4.11

For sand, $\phi' = 32^\circ$, $K_{a,\text{sand}} = 0.307$

For clay, $\phi = 0$, $K_{a,\text{clay}} = 1$

At z = 0 m, $\sigma_a' = 0$kPa , $u_w = 0$ kPa

At z = 3 m$^-$, $\sigma_a' = \gamma_1'\times z\times K_{a,\text{sand}} = 9.39$ kPa , $u_w = 9.81\times3 = 29.43$ kPa

At z = 3 m$^+$, $\sigma_a = \gamma_1\times z\times K_{a,\text{clay}} - 2s_u = 20$ kPa

At z = 7 m , $\sigma_a = 60+4\times16-40 = 84$ kPa

$P_{a1} = 9.39 \times 3/2 = 14.09$ kN/m
$P_{a2} = 20 \times 4 = 80$ kN/m
$P_{a3} = (84-20) \times 4/2 = 120$ kN/m
$P_w = 29.43 \times 3/2 = 44.15$ kN/m
$P = P_{a1} + P_{a2} + P_{a3} + P_w = 266.22$ kN/m
$\bar{z} = 2.75$ m

4.13

$c' = 0$ kPa , $\phi'=34^o$, $\gamma=16$ kN/m^2 , $\gamma_{sat} = 20$ kN/m^2

When $\delta = \phi'/2 = 17^o$, $\theta=0^o$, $\beta = 0^o$, $K_a = 0.26$, $K_{ah} = 0.25$

z = 0 m, $\sigma'_h = 0$ kPa
z = 2 m, $\sigma'_h = 2 \times 16 \times 0.25 = 8$ kPa
z = 6 m, $\sigma'_h = \left[2 \times 16 + 4 \times (20 - 9.81)\right] \times 0.25 = 18.19$ kPa , $u_w = 4 \times 9.81 = 39.24$ kPa
$P_{ah} = 0.5 \times 2 \times 8 + 8 \times 4 + 0.5 \times 4 \times (18.19 - 8) + 0.5 \times 4 \times 39.24 = 138.9$ kN/m
$\bar{z} = 1.68$ m

When $\delta = 2\phi'/3 = 22.67^o$, $\theta=0^o$, $\beta = 0^o$, $K_a = 0.25$, $K_{ah} = 0.23$

z = 0 m, $\sigma'_h = 0$ kPa
z = 2 m, $\sigma'_h = 2 \times 16 \times 0.23 = 7.36$ kPa
z = 6 m, $\sigma'_h = \left[2 \times 16 + 4 \times (20 - 9.81)\right] \times 0.23 = 16.73$ kPa
$u_w = 4 \times 9.81 = 39.24$ kPa
$P_{ah} = 0.5 \times 2 \times 7.36 + 7.36 \times 4 + 0.5 \times 4 \times (16.73 - 7.36) + 0.5 \times 4 \times 39.24 = 134$ kN/m
$\bar{z} = 1.66$ m

4.15

$\phi_u = 0^o$, $K_{p1} = K_{p2} = 1$
z = 0 m , $\sigma_p = 0 + 2 \times 20 = 40$ kPa
z = 4 m$^-$, $\sigma_p = (4 \times 17 \times 1) + 2 \times 20 \times \sqrt{1} = 108$ kPa
z = 4 m$^+$, $\sigma_p = (4 \times 17 \times 1) + 2 \times 30 \times \sqrt{1} = 128$ kPa
z = 9 m , $\sigma_p = [4 \times 17 + 5 \times 18] \times 1 + 2 \times 30 \times \sqrt{1} = 218$ kPa
$P_p = 40 \times 4 + (108 - 40) \times 4/2 + 128 \times 5 + (218 - 128) \times 5/2 = 1161$kN/m

$$\bar{z} = \frac{\left[160 \times 7 + 136 \times (5+4/3) + (640 \times 2.5) + (225 \times 5/3)\right]}{1161} = 3.41 \text{ m}$$

4.17

$c' = 0$ kPa , $\phi'=34^o$, $\gamma=16$ kN/m^2 , $\gamma_{sat} = 20$ kN/m^2
$K_p = 3.537$

$z = 0\ \text{m}$, $\sigma'_p = 0$ kPa
$z = 2\ \text{m}$, $\sigma'_p = (2 \times 16 \times 3.537) = 113.2$ kPa
$z = 6\ \text{m}$, $\sigma'_p = [2 \times 16 + 4 \times (20 - 9.81)] \times 3.537 = 257.4$ kPa
$u_w = 4 \times 9.81 = 39.24$ kPa
$P_p = 0.5 \times 2 \times 113.2 + 113.2 \times 4 + 0.5 \times 4 \times (257.4 - 113.2) + 0.5 \times 4 \times 39.24 = 932.7$ kN/m
$\bar{z} = 2.1$ m

4.19

$c' = 0$ kPa, $\phi' = 34^o$, $\gamma = 16$ kN/m^2, $\gamma_{sat} = 20$ kN/m^2

When $\delta = \phi'/2 = 17^o$, $\theta = 0^o$, $\beta = 0^o$ $K_p = 6.767$, $K_{ph} = 6.471$

$z = 0\ \text{m}$, $\sigma'_h = 0$
$z = 2\ \text{m}$, $\sigma'_h = 2 \times 16 \times 6.471 = 207.1$ kPa
$z = 6\ \text{m}$, $\sigma'_h = [2 \times 16 + 4 \times (20 - 9.81)] \times 6.471 = 470.8$ kPa
$P_{ph} = 2 \times 207.1/2 + 207.1 \times 4 + 4(470.8 - 207.1)/2 + 4(4 \times 9.81)/2$
$= 1641.4$ kN/m

$$\bar{z} = \frac{\left[(207.1 \times 4.67) + (828.4 \times 2) + (527.4 \times 4/3) + (78.48 \times 4/3)\right]}{1641.4}$$

$= 2.09$ m

When $\delta = 2\phi'/3 = 22.67^o$, $\theta = 0^o$, $\beta = 0^o$

$K_p = 8.946$, $K_{ph} = 8.255$

$z = 0\ \text{m}$, $\sigma'_h = 0$ kPa

$z = 2\ \text{m}$, $\sigma'_h = 2 \times 16 \times 8.255 = 264.2$ kPa

$z = 6\ \text{m}$, $\sigma'_h = [2 \times 16 + 4 \times (20 - 9.81)] \times 8.255 = 598.5$ kPa

$P_{ph} = 2 \times 264.2/2 + 264.2 \times 4 + 4(598.5 - 264.2)/2 + 4(4 \times 9.81)/2$

$= 2068.1$ kN/m

$$\bar{z} = \frac{\left[(264.2 \times 4.67) + (1056.8 \times 2) + (668.6 \times 4/3) + (78.48 \times 4/3)\right]}{2068.1}$$

$= 2.1$ m

4.21

$K_0 = 1 - \sin\phi' = 0.47$
$z = 0$ to 9.5 m
$\sigma_{h0} = 30 \times 0.47 = 14.1$kPa
$z = 0\ \text{m}$, $\sigma'_{h0} = 0$ kPa

$z = 2.5$ m, $\sigma'_{h0} = 2.5 \times 16 \times 0.47 = 18.8$ kPa

$z = 9.5$ m, $\sigma'_{h0} = [2.5 \times 16 + 7 \times (20 - 9.81)] \times 0.47 = 52.3$ kPa

$u_w = 7 \times 9.81 = 68.67$ kPa

$P_0 = 0.5 \times 2.5 \times 18.8 + 7(18.8 + 52.3)/2 + 7 \times 68.67/2 + 14.1 \times 9.5 = 646.7$ kN/m

$$\bar{z} = \frac{\left[(23.5 \times 7.83) + (131.6 \times 3.5) + (117.34 \times 7/3) + (240.35 \times 7/3) + (133.95 \times 9.5/2\right]}{646.7}$$

$= 3.27$ m

4.23

$s_{u1} = 20$ kN/m^2, $\gamma_1 = 17$ kN/m^2; $s_{u2} = 30$ kN/m^2, $\gamma_2 = 18$ kN/m^2

$\phi_u = 0^o$, $K_p = 1$

$z = 0$ m, $\sigma_p = 2 \times 20 \times \sqrt{1} + 40 = 40 + 40 = 80$ kPa

$z = 4$ m$^-$, $\sigma_p = (4 \times 17 \times 1) + 1 \times 40 + 2 \times 20 \times \sqrt{1} = 68 + 40 + 40 = 148$ kPa

$z = 4$ m$^+$, $\sigma_p = (4 \times 17 \times 1) + 1 \times 40 + 2 \times 30 \times \sqrt{1} = 68 + 60 + 40 = 168$ kPa

$z = 9$ m, $\sigma_p = [4 \times 17 + 5 \times 18] \times 1 + 60 + 40 = 158 + 60 + 40 = 258$ kPa

$P_p = 0.5 \times 4 \times (80 + 148) + 0.5 \times 5 \times (258 + 168) = 1521$ kN/m

$\bar{z} = 3.7$ m

Chapter 5

5.1

$$\beta = \tan^{-1}\left(\frac{5}{(12-8)/2}\right) = 68.2^\circ$$

$\rightarrow \beta \leq \phi'$

In this condition, the construction of earth retaining structures is necessary.

5.3

ANS: a: (1); b: (1)

The figure depicts a reinforced wall with wraparound facing. The wall facing is vegetated for natural appearance. Although the reinforced wall with vegetation on the facing looks like a natural slope, if you look closer, you still can observe the wraparound configuration and reinforcement layers from the wall facing. The reinforced wall is classified as an internally stabilized fill wall. The stability of internally stabilized walls is provided by internal resistance through the reinforcement tensile force mobilized by soil–reinforcement interaction.

5.5

Calculate Rankine's active earth pressure coefficient

$$K_a = \frac{1}{3}$$

Horizontal force components

$$P_a = \frac{1}{2} \times 18 \times \frac{1}{3} \times 5^2 = 75.0 \text{ kN/m}$$

$$P_{ah} = P_a = 75.0 \text{ kN/m}$$

Vertical force components

$$W_1 = 0.5 \times (5 - 0.3) \times 24 = 56.4 \text{ kN/m}$$

$$W_2 = 2.5 \times (5 - 0.3) \times 18 = 211.5 \text{ kN/m}$$

$$W_3 = 0.3 \times 3.5 \times 24 = 25.2 \text{ kN/m}$$

$$P_{av} = 0 \text{ kN/m}$$

$$\sum W_i + P_{av} = V = 293.1 \text{ kN/m}$$

$$\text{Moment}_i = W_i \times \text{Moment Arm}$$

$$M_1 = 56.4 \times 0.75 = 42.3 \text{ kN} \cdot \text{m/m}$$

$$M_2 = 211.5 \times 2.25 = 475.9 \text{ kN} \cdot \text{m/m}$$

$$M_3 = 25.2 \times 1.75 = 44.1 \text{ kN} \cdot \text{m/m}$$

$$\sum M_R = 562.3 \text{ kN} \cdot \text{m/m}$$

$$M_D = 75 \times \frac{5}{3} = 125.0 \text{ kN} \cdot \text{m/m}$$

Calculate the factor of safety against overturning.

$$F_o = \frac{\sum M_R}{\sum M_D} = \frac{562.3}{125.0} = 4.50 > 2 \quad \text{ok}$$

Calculate the factor of safety against sliding.

$$F_s = \frac{293.1 \times \tan(20°)}{75.0} = 1.42 < 1.5 \quad \text{NG!}$$

5.7

Calculate Coulomb's active earth pressure coefficient

$$\alpha = \beta = 10°, \theta = 0°, \delta = \frac{2}{3}\phi' = 20°$$

$$K_a = 0.43$$

Horizontal force components

$$P_a = \frac{1}{2} \times 18 \times 0.2973 \times 5.1^2 = 100.66 \text{ kN/m}$$

$$P_{ah} = P_a \cos(10) = 99.13 \text{ kN/m}$$

Vertical force components

$$W_1 = 0.5 \times (5 - 0.3) \times 24 = 56.4 \text{ kN/m}$$

$$W_2 = 0.5 \times 2.5 \times 0.1 \times 18 = 2.25 \text{ kN/m}$$

$$W_3 = 2.5 \times (5 - 0.3) \times 18 = 211.5 \text{ kN/m}$$

$$W_4 = 0.3 \times 3.5 \times 24 = 25.2 \text{ kN/m}$$

$$P_{av} = 100.66 \times \sin(10°) = 17.48 \text{ kN/m}$$

$$\sum W_i + P_{av} = V = 312.83 \text{ kN/m}$$

$$\text{Moment}_i = W_i \times \text{Moment Arm}$$

$$M_1 = 56.4 \times 0.75 = 42.3 \text{ kN} \cdot \text{m/m}$$

$$M_2 = 2.25 \times 2.67 = 6.0 \text{ kN} \cdot \text{m/m}$$

$$M_3 = 211.5 \times 2.25 = 475.9 \text{ kN} \cdot \text{m/m}$$

$$M_4 = 25.2 \times 1.75 = 44.1 \text{ kN} \cdot \text{m/m}$$

$$M_{av} = 17.48 \times 3.5 = 61.18 \text{ kN} \cdot \text{m/m}$$

$$\sum M_R = 629.48 \text{ kN} \cdot \text{m/m}$$

$$M_D = 99.13 \times \frac{5.1}{3} = 168.52 \text{ kN} \cdot \text{m/m}$$

Calculate the factor of safety against overturning.

$$F_o = \frac{\sum M_R}{\sum M_D} = \frac{629.48}{168.52} = 3.74 > 2 \quad \text{ok}$$

Calculate the factor of safety against sliding.

$$F_s = \frac{312.83 \times \tan(20°)}{99.13} = 1.15 < 1.5 \quad \text{NG!}$$

5.9

Answer to part A

Calculate maximum reinforcement load at the top and bottom layers.
Top layer:

$$\frac{K_r}{K_a} = 1 \text{ for geosynthetics}$$

$$\sigma_h = \frac{K_r}{K_a} \times K_a \times (\gamma z + q) - 2c\sqrt{K_a} = -4.3 \text{ kPa}$$

$$T_{max} = 0 \text{ kN/m}$$

Due to the soil a tension (negative lateral earth pressure) occurs at the top of the wall; therefore, no reinforcement tensile force is mobilized ($T_{max} = 0$ kN/m).

Bottom layer:

$$\sigma_h = 36 \text{ kPa}$$

$$T_{\max} = \sigma_h \times S_v = 36 \times 0.5 = 18 \text{ kN/m}$$

Calculate long-term reinforcement tensile strength.

$$T_{al} = \frac{T_{ult}}{RF_{CR} \times RF_D \times RF_{ID}} = \frac{150}{2 \times 1.5 \times 1.5} = 33.33 \text{ kN/m}$$

Calculate F_{br} for all reinforcement layers.

Table S5.9a

No. of layer	z (m)	σ'_v (kPa)	σ_h (kPa)	T_{max} (kN/m)	F_{br}
16 (top layer)	0.25	4.20	−4.31	0	–
15	0.75	12.60	−1.62	0	–
14	1.25	21.00	1.06	0.53	62.57
13	1.75	29.40	3.75	1.88	17.74
12	2.25	37.80	6.44	3.22	10.34
11	2.75	46.20	9.13	4.57	7.29
10	3.25	54.60	11.82	5.91	5.63
9	3.75	63.00	14.51	7.25	4.59
8	4.25	71.40	17.20	8.60	3.87
7	4.75	79.80	19.89	9.94	3.35
6	5.25	88.20	22.57	11.29	2.95
5	5.75	96.60	25.26	12.63	2.64
4	6.25	105.00	27.95	13.98	2.38
3	6.75	113.40	30.64	15.32	2.17
2	7.25	121.80	33.33	16.67	2.00
1 (bottom layer)	7.75	130.20	36.02	18.01	1.85

The bottom layer is the most critical for reinforcement breakage because the lateral earth pressure increases linearly with depth.

Bottom layer:

$$F_{br} = \frac{T_{al}}{T_{\max}} = \frac{33.33}{18.01} = 1.85 > 1.5 \text{ ok}$$

Answer to part B

Calculate pullout resistance at the top and bottom layers.

Top layer:

$$P_r = C \times F^* \times \alpha \times \sigma'_v \times L_e = 2 \times 0.4 \times 0.8 \times 4.2 \times 4.36 = 0.31 \text{ kN/m}$$

where

$$C = 2$$

$$F^* = \frac{2}{3} \tan 31^\circ = 0.40$$

$$\alpha = 0.8$$

$$L_e = 4.5 - \frac{(8-0.25)}{\tan(45+\frac{31}{2})} = 0.12 \text{ m}$$

Bottom layer:

$$P_r = 2\times 0.4\times 0.8\times 130.2\times 4.36 = 363.19 \text{ kN/m}$$

Calculate F_{po} for all reinforcement layers.

Table S5.9b

No. of layer	*z (m)*	*L_e (m)*	*σ'$_v$ (kPa)*	*P_r (kN/m)*	*T_{max} (kN/m)*	*F_{po}*
16 (top layer)	0.25	0.12	4.20	0.31	0	–
15	0.75	0.40	12.60	3.21	0	–
14	1.25	0.68	21.00	9.15	0.53	17.20
13	1.75	0.96	29.40	18.14	1.88	9.66
12	2.25	1.25	37.80	30.16	3.22	9.36
11	2.75	1.53	46.20	45.23	4.57	9.91
10	3.25	1.81	54.60	63.34	5.91	10.72
9	3.75	2.10	63.00	84.49	7.25	11.65
8	4.25	2.38	71.40	108.68	8.60	12.64
7	4.75	2.66	79.80	135.91	9.94	13.67
6	5.25	2.94	88.20	166.19	11.29	14.72
5	5.75	3.23	96.60	199.51	12.63	15.79
4	6.25	3.51	105.00	235.87	13.98	16.88
3	6.75	3.79	113.40	275.27	15.32	17.97
2	7.25	4.08	121.80	317.71	16.67	19.06
1 (bottom layer)	7.75	4.36	130.20	363.19	18.01	20.17

The top layer is not the most critical for reinforcement pullout because of no reinforcement tensile force mobilized at the top of the wall.

As shown in the table, the 12th layer is the most critical for reinforcement pullout due to the combined effect of the short reinforcement embedment length and low overburden pressure.

The 12th layer:

$$F_{po} = \frac{P_r}{T_{\max}} = \frac{30.16}{3.22} = 9.36 > 1.5 \text{ ok}$$

5.11

Calculate long-term reinforcement tensile strength.

$$T_{al} = \frac{120}{2.5\times 1.5\times 1.2} = 26.7 \text{ kN/m}$$

$$F_{br} = \frac{T_{al}}{T_{\max}}$$

$$T_{\max} = \frac{T_{al}}{F_{br}} = \frac{26.7}{1.5} = 17.8 \text{ kN/m}$$

$$\sigma_h = \frac{K_r}{K_a} \times K_a \times (\gamma z + q) - 2c\sqrt{K_a} = 5.27z - 5.65$$

Determine the vertical spacing of the reinforcement.

$$T_{\max} = \sigma_h \times S_v$$

$$S_v = \frac{T_{\max}}{\sigma_h} = \frac{17.8}{5.27z - 5.65}$$

The bottom layer is the most critical for reinforcement breakage.

at $z = 6$ m: $S_v = 0.68$ m
select $S_v = 0.6$ m for design

Determine the length of the reinforcement.

$$F_{po} = \frac{P_r}{T_{\max}} = \frac{C \times F^* \times \alpha \times \sigma_v' \times L_e}{\sigma_h \times S_v} = \frac{2 \times \frac{2}{3} \times 0.8 \times \sigma_v' \times L_e}{\sigma_h \times 0.7} = \frac{0.916 \times \sigma_v' \times L_e}{\sigma_h}$$

$$L_e = \frac{\frac{1.5}{0.916} \times \sigma_h}{\sigma_v'} = L - \frac{(H - z)}{\tan(45 + \frac{\phi'}{2})}$$

$$L = \frac{1.638 \times \sigma_h}{\sigma_v'} + \frac{(H - z)}{\tan(45 + \frac{\phi'}{2})}$$

$$L = \frac{1.638 \times (0.32 \times 16.5 \times z - 2 \times 5 \times \sqrt{0.32})}{16.5 \times z} + \frac{(6 - z)}{1.767} = \frac{-16.5z^2 + 114.283z - 16.373}{29.156z}$$

The top layer is not the most critical for reinforcement pullout because of no reinforcement tensile force mobilized at the top of the wall. In the calculation, several reinforcement layers are evaluated to identify the most critical layer.

at $z = 0.3$ m: $L = 1.87$ m
at $z = 0.9$ m: $L = 2.79$ m
at $z = 1.5$ m: $L = 2.69$ m
at $z = 2.1$ m; $L = 2.46$ m
select $L = 2.8$ m for design

5.13

Calculate Rankine's active and passive earth pressure coefficients.

$$K_{ah1} = \tan^2(45 - \frac{30}{2}) = 0.333$$

$$K_{ah2} = \tan^2(45 - \frac{20}{2}) = 0.490$$

$$K_{ph2} = \tan^2(45 + \frac{20}{2}) = 2.040$$

Calculate lateral earth forces.

$$P_{au1}=\frac{1}{2}\times18\times3^2\times0.333=26.973\ \text{kN/m}$$

$$P_{au2}=\left(18\times3\times0.49-2\times10\times\sqrt{0.49}\right)\times\left(H_0+3\right)$$
$$=12.46H_0+37.38$$

$$P_{au3}=\frac{1}{2}\left((18\times3+20(H_0+3))\times0.49-2\times10\times\sqrt{0.49}-\left(18\times3\times0.49-2\times10\times\sqrt{0.49}\right)\right)\times\left(H_0+3\right)$$
$$=4.9H_0^{\ 2}+29.4H_0+44.1$$

$$P_{pu1}=2\times10\times\sqrt{2.04}H_0$$
$$=28.57H_0$$

$$P_{pu2}=\frac{1}{2}\left(20\times H_0\times2.04\right)H_0$$
$$=20.4H_0^2$$

Determine the wall embedment length to satisfy the required F_S.
Take moment on point O.

$$F_S=\frac{M_r}{M_d}\geq1.5$$

$$=\frac{P_{p1}\left(\frac{1}{2}H_0\right)+P_{p2}\left(\frac{1}{3}H_0\right)}{P_{a1}(4+H_0)+P_{a2}\left(\frac{1}{2}H_0+3\right)+P_{a3}\left(\frac{1}{3}H_0+3\right)}$$

$$H_0\geq9\ m$$

$$H_p=1.2H_0=10.8\ \text{m}$$

Compare the forces above and below point O to ensure the wall embedment length is sufficient.

P_{pu1}=257.13 kN/m

P_{pu2}=1652.40 kN/m

$P_{au1}=26.973$ kN/m

$P_{au2}=149.52$ kN/m

P_{au3}=705.6 kN/m

$\sigma_{al1}=74.2$ kPa

$\sigma_{al2}=91.84$ kPa

σ_{pl1}= 628.32 kPa

σ_{pl2}= 701.76 kPa

$$P_{pl} = 1197.08 \text{ kN/m}$$

$$P_{al} = 149.44 \text{ kN/m}$$

$$P_{pl} - P_{al} = 1047.65 \text{ kN/m}$$

$$R = P_{pu} - P_{au} = 1027.4 \text{ kN/m}$$

$$P_{pl} - P_{al} > R \text{ ok}$$

5.15

Answer to part A

Assume the wall length is less than the depth of groundwater level.

Calculate Rankine's active and passive earth pressure coefficients.

$$K_{ah} = K_a = \tan^2\left(45 - \frac{35}{2}\right) = 0.271$$

$$K_{ph} = K_p = \tan^2\left(45 + \frac{35}{2}\right) = 3.69$$

Calculate lateral earth forces.

$$\sigma_a = K_a \gamma (H_e + H_p) = 32.25 + 4.607 H_p$$

$$\sigma_p = K_a \gamma H_p = 3.69 \times 17 \times H_p = 62.73 H_p$$

$$P_a = \frac{1}{2} \times (32.25 + 4.607 H_p) \times \left(H_p + H_e\right)$$

$$= 2.304 H_p{}^2 + 32.253 H_p + 112.875 \text{ kN/m}$$

$$P_p = \frac{1}{2} \times 62.73 H_p \times H_p$$

$$= 31.365 H_p^2 \text{ kN/m}$$

Determine the wall embedment length to satisfy the required F_S.

$$F_S = \frac{M_r}{M_d} = \frac{P_p \times L_p}{P_a \times L_a} \geq 1.5$$

$$= \frac{P_p \times \left(\frac{2}{3} H_p + H_e - H_s\right)}{P_a \times \left[\frac{2}{3}\left(H_p + H_e\right) - H_s\right]} \geq 1.5$$

$$H_p \geq 2.7 \text{ m}$$

The wall length is 9.7 m, which is still above the groundwater level (H_w = 10 m). The assumption is valid.

Answer to part B

Summary of resultant forces of the active and passive earth pressures.

$$\sigma_a = 44.69 \text{ kPa}$$
$$\sigma_p = 169.371 \text{ kPa}$$
$$P_a = 216.74 \text{ kN/m}$$
$$P_p = 228.65 \text{ kN/m}$$

Determine the anchor load.

Take moment on point O, and solve equation to obtain T.

$$T = \frac{\left(216.74 \times \frac{7+2.7}{3}\right) - \left(\frac{228.65}{1.5} \times 0.9\right)}{\cos 15 \times (7+2.7-1)}$$
$$= 67.06 \text{ kN/m}$$

Solve T again based on the horizontal force equilibrium.

$$T = \frac{216.74 - \frac{228.65}{1.5}}{\cos 15} = 66.57 \text{ kN/m}$$

For the anchor load with anchor horizontal spacing of 3 m:

$$F_T = 67.06 \times 3 = 201.19 \text{ kN}$$

Answer to part C

The length of the tie rod should extend beyond the potential failure surface to achieve adequate anchorage.

$$L \cos\alpha \geq \left(H_e + H_p - H_s - L \sin\alpha\right) \tan(45 - \frac{\phi'}{2})$$

For simplicity, ignore the $L\sin\alpha$ in the calculation.

$$L \geq 4.68 \text{ m}$$

Chapter 6

6.1

a. Load factor method

Coulomb Theory: $K_a = 0.3, K_p = 6.13$

$$K_{a,h} = K_a \cos\delta = 0.3 \times \cos(20.1) = 0.2817$$
$$K_{p,h} = K_p \cos\delta = 6.13 \times \cos(20.1) = 5.7566$$

At GL-5.5 m:

$$\sigma'_{a,h} = \sigma'_v \times K_{a,h} = 88 \times 0.2817 = 24.79 \text{ kPa}$$

At GL-9.0 m:

$$\sigma'_{a,h} = \sigma'_v \times K_{a,h} = 144 \times 0.2817 = 40.56 \text{ kPa}$$

At GL-17.0 m:

$$\sigma'_{a,h} = \sigma'_{v,a} \times K_{a,h} = 225.52 \times 0.2817 = 63.53 \text{ kPa}$$

$$\sigma'_{p,h} = \sigma'_{v,p} \times K_{p,h} = 81.52 \times 5.76 = 469.28 \text{ kPa}$$

$$F_b = \frac{P_{p,h}L_p}{P_{a,h}L_a} = \frac{16581.23 + 2772.96}{151.84 + 64.394 + 2433.6 + 811.6 + 2772.96} = 3.1$$

b. *Strength reduction method*

By trial and error, try $F_b = 2.05$.

$$\phi_r = \tan^{-1}\left(\frac{\tan\phi}{F_b}\right) = \tan^{-1}\left(\frac{\tan(30)}{2.05}\right) = 15.73°$$

Coulomb Theory: $K_a = 0.518, K_p = 2.226$

$$K_{a,h} = K_a \cos\delta = 0.518 \times \cos(10.54) = 0.509$$

$$K_{p,h} = K_p \cos\delta = 2.226 \times \cos(10.54) = 2.188$$

At GL-5.5 m:

$$\sigma'_{a,h} = \sigma'_v \times K_{a,h} = 88 \cdot 0.509 = 44.79 \text{ kPa}$$

At GL-9.0 m.

$$\sigma'_{a,h} = \sigma'_v \times K_{a,h} = 144 \times 0.509 = 73.296 \text{ kPa}$$

At GL-17.0 m.

$$\sigma'_{a,h} = \sigma'_{v,a} \times K_{a,h} = 225.52 \times 0.509 = 114.79 \text{ kPa}$$

$$\sigma'_{p,h} = \sigma'_{v,p} \times K_{p,h} = 81.52 \times 2.188 = 178.366 \text{ kPa}$$

$$\frac{P_{p,h}L_p}{P_{a,h}L_a} \approx 1.0$$

Hence, $F_b = 2.05$.

6.3

a. Load factor method

Coulomb Theory: $K_a = 0.3, K_p = 6.13$

$$K_{a,h} = K_a \cos\delta = 0.3 \times \cos(20.1) = 0.28$$
$$K_{p,h} = K_p \cos\delta = 6.13 \times \cos(20.1) = 5.76$$

At GL-5.5 m:

$$\sigma'_{a,h} = \sigma'_v \times K_{a,h} = 88 \times 0.2817 = 24.79 \text{ kPa}$$

At GL-9.0 m:

$$\sigma'_{a,h} = \sigma'_v \times K_{a,h} = 144 \times 0.2817 = 40.56 \text{ kPa}$$

At GL-(9 + H_p):

$$\sigma'_{a,h} = \sigma'_{v,a} \times K_{a,h} = \left(144 + 10.19H_p\right) \times 0.2817 = 40.56 + 2.87H_p$$
$$\sigma'_{p,h} = \sigma'_{v,p} \times K_{p,h} = 10.19H_p \times 5.76 = 58.69H_p$$
$$F_b = \frac{P_{p,h}L_p}{P_{a,h}L_a} = 1.20$$

By trial and error, it can be obtained that $H_p = 2.454$ m.

b. Strength factor method

$$F_b = 1.2,\ \phi_r = \tan^{-1}\left(\frac{\tan\phi}{F_b}\right) = \tan^{-1}\left(\frac{\tan(30)}{1.2}\right) = 25.693°$$

Coulomb Theory: $K_a = 0.35, K_p = 4.3$

$$K_{a,h} = K_a \cos\delta = 0.35 \times \cos(17.2) = 0.334$$
$$K_{p,h} = K_p \cos\delta = 4.3 \times \cos(17.2) = 4.107$$

At GL-5.5 m:

$$\sigma'_{a,h} = \sigma'_v \times K_{a,h} = 88 \times 0.334 = 29.39 \text{ kPa}$$

At GL-9.0 m:

$$\sigma'_{a,h} = \sigma'_v \times K_{a,h} = 144 \times 0.334 = 48.096 \text{ kPa}$$

At the wall bottom:

$$\sigma'_{a,h} = \sigma'_{v,a} \times K_{a,h} = \left(144 + 10.19H_p\right) \times 0.334 = 48.096 + 3.4H_p$$

$$\sigma'_{p,h} = \sigma'_{v,p} \times K_{p,h} = 10.19H_p \times 4.107 = 41.85H_p$$

$$\frac{P_{p,h}L_p}{P_{a,h}L_a} \approx 1.0$$

By trial-and-error method, $H_p = 3.06$ m.

6.5

According to Rankine's earth pressure theory, $\phi = 0^\circ$, $K_a = K_p = 1.0$.

a. Load factor method

Active earth pressure

At GL-5 m:

$$\sigma_v = (2\times16+3\times17) = 83 \text{ kPa}, \ \sigma_a = 83\times1-2\times20 = 43 \text{ kPa} ;$$

At GL = –9 m:

In clay 2 layer: $\sigma_v = (2\times16+7\times17) = 151$ kPa, $\sigma_a = 151\times1-2\times20 = 111$ kPa ;

In clay 3 layer: $\sigma_v = (2\times16+7\times17) = 151$ kPa, $\sigma_a = 151\times1-2\times30 = 91$ kPa ;

At GL-20 m:

$$\sigma_v = (2\times16+7\times17+11\times18) = 349 \text{ kPa}, \ \sigma_a = 349\times1-2\times30 = 289 \text{ kPa} ;$$

Passive earth pressure

At GL-9 m:

$$\sigma_v = 0 \text{ kPa}, \ \sigma_a = 0+2\times30 = 60 \text{ kPa}$$

At GL-20 m:

$$\sigma_v = (11\times18) = 198 \text{ kPa}, \ \sigma_a = 198\times1+2\times30 = 258 \text{ kPa}$$

$$\text{P}_{a1} = 43\times4 = 172 \text{ kN}, \ \text{L}_{a1} = \frac{4}{2} = 2 \text{ m}$$

$$\text{P}_{a2} = (111-43)\times4\times\frac{1}{2} = 136 \text{ kN}, \ \text{L}_{a2} = 4\times\frac{2}{3} = 2.67 \text{ m}$$

$$\text{P}_{a3} = 91\times11 = 1001 \text{ kN}, \ \text{L}_{a3} = 4+\frac{11}{2} = 9.5 \text{ m}$$

$$P_{a4} = (289-91)\times 11\times\frac{1}{2} = 1089 \text{ kN},\ L_{a4} = 4+11\times\frac{2}{3} = 11.33 \text{ m}$$

$$P_{p1} = 60\times 11 = 660 \text{ kN},\ L_{p1} = 4+\frac{11}{2} = 9.5 \text{ m}$$

$$P_{p2} = (258-60)\times 11\times\frac{1}{2} = 1089 \text{ kN},\ L_{p2} = 4+11\times\frac{2}{3} = 11.33 \text{ m}$$

$$M_r = \sum P_p L_p = 660\times 9.5+1089\times 11.33 = 18612 \text{ kN-m/m}$$

$$M_d = \sum P_a L_a = 172\times 2+136\times 2.67+1001\times 9.5+1089\times 11.33 = 22558 \text{ kN-m/m}$$

$$F_b = \frac{M_r}{M_d} = 0.83$$

b. Strength factor method

By trial and error, try $F_b = 0.76$.

Active earth pressure

At GL-5 m:

$$\sigma_v = (2\times 16+3\times 17) = 83 \text{ kPa},\ \sigma_a = 83\times 1-2\times\frac{20}{0.76} = 30.37 \text{ kPa}$$

At GL-9 m:

In clay 2 layer: $\sigma_v = (2\times 16+7\times 17) = 151 \text{ kPa},\ \sigma_a = 151\times 1-2\times\frac{20}{0.76} = 98.37 \text{ kPa}$

In clay 3 layer: $\sigma_v = (2\times 16+7\times 17) = 151 \text{ kPa},\ \sigma_a = 151\times 1-2\times\frac{30}{0.76} = 72.05 \text{ kPa}$

At GL-20 m:

$$\sigma_v = (2\times 16+7\times 17+11\times 18) = 349 \text{ kPa},\ \sigma_a = 349\times 1-2\times\frac{30}{0.76} = 270.05 \text{ kPa}$$

Passive earth pressure

At GL-9 m:

$$\sigma_v = 0 \text{ kPa},\ \sigma_a = 0+2\times\frac{30}{0.76} = 78.95 \text{ kPa}$$

At GL-20 m:

$$\sigma_v = (11\times 18) = 198 \text{ kPa},\ \sigma_a = 198\times 1+2\times\frac{30}{0.76} = 276.95 \text{ kPa}$$

$$P_{a1} = 30.37 \times 4 = 121.48 \text{ kN}, \ L_{a1} = \frac{4}{2} = 2 \text{ m}$$

$$P_{a2} = (98.37\text{-}30.37) \times 4 \times \frac{1}{2} = 136 \text{ kN}, \ L_{a2} = 4 \times \frac{2}{3} = 2.67 \text{ m}$$

$$P_{a3} = 72.05 \times 11 = 792.55 \text{ kN}, \ L_{a3} = 4 + \frac{11}{2} = 9.5 \text{ m}$$

$$P_{a4} = (270.05\text{-}72.05) \times 11 \times \frac{1}{2} = 1089 \text{ kN}, \ L_{a4} = 4 + 11 \times \frac{2}{3} = 11.33 \text{ m}$$

$$P_{p1} = 78.95 \times 11 = 868.45 \text{ kN}, \ L_{p1} = 4 + \frac{11}{2} = 9.5 \text{ m}$$

$$P_{p2} = (276.95\text{-}78.95) \times 11 \times \frac{1}{2} = 1089 \text{ kN}, \ L_{p2} = 4 + 11 \times \frac{2}{3} = 11.33 \text{ m}$$

$$M_r = \sum P_p L_p = 868.45 \times 9.5 + 1089 \times 11.33 = 20592 \text{ kN-m/m}$$

$$M_d = \sum P_a L_a = 121.48 \times 2 + 136 \times 2.67 + 792.55 \times 9.5 + 1089 \times 11.33 = 20477 \text{ kN-m/m}$$

$\sum P_p L_p / \sum P_a L_a$ =20592/20477 ≈ 1.0. Hence, $F_b = 0.76$.

6.7

a. Load factor method

Rankine's earth pressure theory, $\phi = 0^\circ$, $K_a = K_p = 1.0$.

Active earth pressure

At GL-6 m:

$$\sigma_v = (2 \times 18 + 4 \times 19) = 112 \text{ kPa}, \ \sigma_a = 112 \times 1 - 2 \times 25 = 62 \text{ kPa}$$

At GL-10 m:

In clay 2 layer: $\sigma_v = (2 \times 18 + 8 \times 19) = 188$ kPa, $\sigma_a = 188 \times 1 - 2 \times 25 = 138$ kPa
In clay 3 layer: $\sigma_v = (2 \times 18 + 8 \times 19) = 188$ kPa, $\sigma_a = 188 \times 1 - 2 \times 30 = 128$ kPa

At GL-21 m:

$$\sigma_v = (2 \times 18 + 8 \times 19 + 18 \times 11) = 386 \text{ kPa}, \ \sigma_a = 386 \times 1 - 2 \times 30 = 326 \text{ kPa}$$

Passive earth pressure

At GL-9 m:

$$\sigma_v = 0 \text{ kPa}, \ \sigma_a = 0 + 2 \times 30 = 60 \text{ kPa}$$

At GL-20 m:

$$\sigma_v = (11 \times 18) = 198 \text{ kPa}, \ \sigma_a = 198 \times 1 + 2 \times 30 = 258 \text{ kPa}$$

$$P_{a1} = 62 \times 4 = 248 \text{ kN}, \ L_{a1} = \frac{4}{2} = 2 \text{ m}$$

$$P_{a2} = (138\text{-}62) \times 4 \times \frac{1}{2} = 152 \text{ kN}, \ L_{a2} = 4 \times \frac{2}{3} = 2.67 \text{ m}$$

$$P_{a3} = 128 \times 11 = 1408 \text{ kN}, \ L_{a3} = 4 + \frac{11}{2} = 9.5 \text{ m}$$

$$P_{a4} = (326\text{-}128) \times 11 \times \frac{1}{2} = 1089 \text{ kN}, \ L_{a4} = 4 + 11 \times \frac{2}{3} = 11.33 \text{ m}$$

$$P_{p1} = 60 \times 11 = 660 \text{ kN}, \ L_{p1} = 4 + \frac{11}{2} = 9.5 \text{ m}$$

$$P_{p2} = (258 - 60) \times 11 \times \frac{1}{2} = 1089 \text{ kN}, \ L_{p2} = 4 + 11 \times \frac{2}{3} = 11.33 \text{ m}$$

$$M_r = \sum P_p L_p = 660 \times 9.5 + 1089 \times 11.33 = 18612 \text{ kN-m/m}$$

$$M_d = \sum P_a L_a = 248 \times 2 + 152 \times 2.67 + 1408 \times 9.5 + 1089 \times 11.33 = 26619 \text{ kN-m/m}$$

$$F_b = \frac{M_r}{M_d} = 0.7$$

b. Strength factor method

According to Rankine's earth pressure theory, $\phi = 0^{\circ}$, $K_a = K_p = 1.0$.
By trial and error, try $F_b = 0.6$.

$s_{u1}/F_b = 12/0.6 = 20$ kPa, $s_{u2}/F_b = 25/0.6 = 41.7$ kPa, $s_{u3}/F_b = 30/0.6 = 50$ kPa

Active earth pressure

At GL-6 m:

$$\sigma_v = (2 \times 18 + 4 \times 19) = 112 \text{ kPa}, \ \sigma_a = 112 \times 1 - 2 \times 41.7 = 28.6 \text{ kPa}$$

At GL-10 m:

In clay 2 layer: $\sigma_v = (2 \times 18 + 8 \times 19) = 188$ kPa, $\sigma_a = 188 \times 1 - 2 \times 41.7 = 104.6$ kPa
In clay 3 layer: $\sigma_v = (2 \times 18 + 8 \times 19) = 188$ kPa, $\sigma_a = 188 \times 1 - 2 \times 50 = 88$ kPa

At GL-21 m:

$$\sigma_v = (2 \times 18 + 8 \times 19 + 18 \times 11) = 386 \text{ kPa}, \ \sigma_a = 386 \times 1 - 2 \times 50 = 286 \text{ kPa}$$

Passive earth pressure

At GL-9 m:

$$\sigma_v = 0 \text{ kPa}, \ \sigma_a = 0 + 2 \times 50 = 100 \text{ kPa}$$

At GL-20 m:

$$\sigma_v = (11\times18) = 198 \text{ kPa}, \ \sigma_a = 198\times1 + 2\times50 = 298 \text{ kPa}$$

$$\text{P}_{a1} = 28.6\times4 = 114.4 \text{ kN}, \ \text{L}_{a1} = \frac{4}{2} = 2 \text{ m}$$

$$\text{P}_{a2} = (104.6\text{-}28.6)\times4\times\frac{1}{2} = 152 \text{ kN}, \ \text{L}_{a2} = 4\times\frac{2}{3} = 2.67 \text{ m}$$

$$\text{P}_{a3} = 88\times11 = 968 \text{ kN}, \ \text{L}_{a3} = 4 + \frac{11}{2} = 9.5 \text{ m}$$

$$\text{P}_{a4} = (286\text{-}88)\times11\times\frac{1}{2} = 1089 \text{ kN}, \ \text{L}_{a4} = 4 + 11\times\frac{2}{3} = 11.33 \text{ m}$$

$$\text{P}_{p1} = 100\times11 = 1100 \text{ kN}, \ \text{L}_{p1} = 4 + \frac{11}{2} = 9.5 \text{ m}$$

$$\text{P}_{p2} = (298-100)\times11\times\frac{1}{2} = 1089 \text{ kN}, \ \text{L}_{p2} = 4 + 11\times\frac{2}{3} = 11.33 \text{ m}$$

$$\text{M}_r = \sum P_p L_p = 1100\times9.5 + 1089\times11.33 = 22788 \text{ kN-m/m}$$

$$\text{M}_d = \sum P_a L_a = 114.4\times2 + 152\times2.67 + 968\times9.5 + 1089\times11.33 = 22169 \text{ kN-m/m}$$

$\sum P_p L_p / \sum P_a L_a \approx 1.0$. Therefore, $F_b = 0.6$.

6.9

a. Load factor method

Active earth pressure

At GL-5 m:

$$\sigma_a = 83 - 2\times10 = 63\text{kN/m}^2$$

At GL-13 m (layer 2):

$$\sigma_a = 219 - 2\times20 = 179\text{kN/m}^2$$

At GL-13 m (layer 3):

$$\sigma_a = 219 - 2\times30 = 159\text{kN/m}^2$$

At GL-20 m:

$$\sigma_a = 345 - 2\times30 = 285\text{kN/m}^2$$

Passive earth pressure

At GL-9 m:

$$\sigma_p = 0 + 2\times30 = 60\text{kN/m}^2$$

At GL-20 m:

$$\sigma_p = 198 + 2 \times 30 = 258\text{kN/m}^2$$

$$M_d = 504 \times 4 + 464 \times \frac{16}{3} + 1113 \times 11.5 + 441 \times \left(7 \times \frac{2}{3} + 8\right) = 22{,}876\text{kN-m/m}$$

$$M_r = 660 \times \left(\frac{11}{2} + 4\right) + 1089 \times \left(11 \times \frac{2}{3} + 4\right) = 18{,}612\text{kN-m/m}$$

$$F_b = \frac{M_r}{M_d} = 0.81$$

b. Strength factor method

According to Rankine's earth pressure theory, $\phi = 0^\circ$, $K_a = K_p = 1.0$.
By trial and error, try $F_b = 0.75$.

$s_{u1}/F_b = 10/0.75 = 13.33$ kPa, $s_{u2}/F_b = 20/0.75 = 26.67$ kPa, $s_{u3}/F_b = 30/0.75 = 40$ kPa

Active earth pressure

At GL-5 m:

$$\sigma_a = 83 - 2 \times 26.67 = 29.66\text{kN/m}^2$$

At GL-13 m (layer 2):

$$\sigma_a = 219 - 2 \times 26.7 = 165.66\text{kN/m}^2$$

At GL-13 m (layer 3):

$$\sigma_a = 219 - 2 \times 40 = 139\text{kN/m}^2$$

At GL-20 m:

$$\sigma_a = 345 - 2 \times 40 = 265\text{kN/m}^2$$

Passive earth pressure

At GL-9 m:

$$\sigma_p = 0 + 2 \times 40 = 80\text{kN/m}^2$$

At GL-20 m:

$$\sigma_p = 198 + 2\times 40 = 278\text{kN/m}^2$$

$$M_r = \sum P_p L_p = 880\times 9.5 + 1089\times 11.33 = 20698 \text{ kN-m/m}$$

$$M_d = \sum P_a L_a = 237.28\times 4 + 544\times 5.3 + 973\times 11.5 + 441\times 12.7 = 20622 \text{ kN-m/m}$$

$$\sum P_p L_p / \sum P_a L_a \approx 1.0$$

Therefore, $F_b = 0.7$.

c. *Slip circle method*

Assuming the failure surface passes the wall toe, the radius of the failure surface is equal to 15 m.

$$\alpha_1 = \tan^{-1}\frac{15}{4} = 1.31 \text{ rad} = 75.1^\circ,\ \alpha_2 = \cos^{-1}\frac{8}{15} = 1 \text{ rad} = 57.8^\circ,\ \alpha_3 = \sin^{-1}\frac{8}{15} = 0.56 \text{ rad} = 32.2^\circ$$

The weight of the soil column is $W = 2\times 15\times 16 + 7\times 15\times 17 = 2265$ kN/m

$$F_b = \frac{M_r}{M_d} = \frac{\int_0^{\pi/2.4} 15\times 30\times(15\times d\theta) + \int_0^{\pi/3.1} 15\times 30\times(15\times d\theta) + \int_0^{\pi/5.58} 15\times 20\times(15\times d\theta)}{2265\times 15/2} = 1.06$$

d. *Terzaghi's method*

D = 16 m. According to Terzaghi's method, the radius of the failure surface is $B/\sqrt{2} = 20/\sqrt{2} = 14.14 < D$.

$$F_b = \frac{Q_u}{W - s_{u1}H_e} = \frac{5.7 s_{u2} B/\sqrt{2}}{(\gamma H_e + q_s)B/\sqrt{2} - s_{u1}H_e} = \frac{5.7\times 25\times 20/\sqrt{2}}{(16\times 2 + 17\times 7)20/\sqrt{2} - (10\times 2 + 20\times 7)}$$
$$= 1.02$$

e. *Bjerrum and Eide's method*

The radius of the failure surface is:

$$\frac{B}{\sqrt{2}} = \frac{20}{\sqrt{2}} = 14.14 \text{ m}$$

$L/B = \infty$, $H_e/B = 9/20$, *then get the* $N_c = 5.9$ *from the figure.*

$$F_b = \frac{N_c \times S_u}{\gamma\times H_e + q_s} = \frac{5.9\times s_{u3}}{\gamma\times H_e} = \frac{5.9\times 30}{2\times 16 + 7\times 17} = 1.17$$

6.11

Draw a flow net.

$$B = 60 \text{ m},\ H_p = 8 \text{ m},\ \Delta H_w = 9 \text{ m},\ N_d = 18, N_f = 6$$

Terzaghi's method:

$$i_{avg} = \frac{\frac{(0.22+0.36)\Delta H_w}{2}}{H_p} = \frac{\frac{(0.22+0.36)9}{2}}{8} = 0.33$$

The uplift force:

$$U = \frac{1}{2}H_P^2 i_{avg}\gamma_w = \frac{1}{2}\times 8^2 \times 0.33 \times 9.81 = 103.6$$

The weight of soil column:

$$W' = \frac{1}{2}H_P^2(\gamma_{sat} - \gamma_w) = \frac{1}{2}\times 8^2 \times (20 - 9.81) = 326.1$$

$$F_s = \frac{W'}{U} = \frac{326.1}{103.6} = 3.15$$

Harzar's method:

$$L = 1.84 \text{ m},\ \Delta h = 9/18 = 0.5,\ i_{\text{exit(max)}} = 0.5/1.84 = 0.272$$

$$i_{cr} = \frac{\gamma'}{\gamma_w} = \frac{10}{9.81} = 1.22$$

$$F_s = \frac{i_{cr}}{i_{\text{max(exit)}}} = \frac{1.22}{0.272} = 4.49$$

6.13

$$u_w = 9.81 \times 25 = 245.25 \text{ kPa}$$

$$\sigma_v = (11+5)\times 18 = 288 \text{ kPa}$$

$$F_{up} = \frac{\sigma_v}{u_w} = \frac{288}{245.25} = 1.17$$

6.15

a. Draw the apparent earth pressure

$$s_{u,avg} = \frac{10\times 2 + 20\times 7}{2+7} = 17.8kPa,\ \gamma_{avg} = \frac{16\times 2 + 17\times 7}{2+7} = 16.8kPa$$

$$\frac{\gamma H_e}{s_u} = \frac{16.8 \times 9}{17.8} = 8.5 > 4 \text{ (soft to medium clay)}$$

$$N_b = \frac{\gamma H_e}{s_b} = \frac{16.8 \times 9}{30} = 5.04 \le 5.14 \ (m = 1.0)$$

$$K_a = \left(1 - m\frac{4s_u}{\gamma H_e}\right) = \left(1 - \frac{4 \times 17.8}{16.8 \times 9}\right) = 0.53$$

$$p_{a,1} = K_a \gamma H_e = 0.53 \times 16.8 \times 9 = 80.14kPa$$

$$p_{a,2} = K_a \gamma H_e = 0.30 \times 16.8 \times 9 = 45.36kPa$$

Choose $p_a = 80.14kPa$

b. The horizontal strut spacing, L = 5 m

With the half method:

$$Q_1 = (2 + 1.5 + 1.25) \times 0.5 \times p_a \times L$$

$$Q_1 = (2 + 1.5 + 1.25) \times 0.5 \times 80.136 \times 5 = 951.615kPa$$

$$Q_2 = (1.5 + 1.5) \times p_a \times L$$

$$Q_2 = (1.5 + 1.5) \times 80.136 \times 5 = 1202.04kPa$$

b.

Active earth pressure distribution is assumed: $K_a = K_p = 1$.

Depth of tension crack: $z_c = \frac{2s_u}{\gamma_{sat}} = \frac{2 \times 10}{16} = 1.25\,m$

Active side:

At $GL - 1.25m, \ \sigma_a = 0kPa$

At $GL - 2m(\text{upper}), \ \sigma_a = \sigma_v K_a - 2s_u\sqrt{K_a} = (2 \times 16) - 2(10) = 12kPa$

At $GL - 2m(\text{lower}), \ \sigma_a = \sigma_v K_a - 2s_u\sqrt{K_a} = (2 \times 16) - 2(20) = -8kPa \approx 0kPa$

At $GL - 9m(\text{upper}), \ \sigma_a = \sigma_v K_a - 2s_u\sqrt{K_a} = 32 + (7 \times 17) - 2(20) = 111kPa$

At $GL - 9m(\text{lower}), \ \sigma_a = \sigma_v K_a - 2s_u\sqrt{K_a} = 32 + (7 \times 17) - 2(30) = 91kPa$

At $GL - 20m, \ \sigma_a = \sigma_v K_a - 2s_u\sqrt{K_a} = 151 + (11 \times 18) - 2(30) = 289kPa$

Passive side:

$$\text{At } GL-9m,\ \sigma_p = \sigma_v K_a + 2s_u\sqrt{K_a} = 0 + 2(30) = 60kPa$$
$$\text{At } GL-20m,\ \sigma_p = \sigma_v K_a + 2s_u\sqrt{K_a} = (11\times18) + 2(30) = 258kPa$$

Location of the assumed support:

$$l = \frac{P_a l_a}{P_p} - s = 8.9\,\text{m}$$

The maximum bending moment will occur at the location of the shear force equal to 0.
The maximum bending moment is equal to 508.85 kN-m

The section modulus, $S = \frac{M_{x,\max}}{\sigma_{all}} = \frac{508.85}{170\times1000} = 2.99\times10^{-3} m^3$.

The size and type of the sheet pile can then be selected according to the product manual.

c. The strut load at the second level of the strut Q_2 = 3117.79kN. The horizontal strut spacing L = 5 m. The uniform pressure acting on the wale:

$$p = 3117.79 / L = 3117.79 / 5 = 623.558\,\text{kN/ m}$$

If the simply supported model is adopted:

$$M_{\max} = \frac{1}{8} pL^2 = \frac{1}{8}\times623.558\times5^2 = 1948.62kN\text{-}m$$

The section modulus: $S = \frac{M_{\max}}{\sigma_{all}} = \frac{1948.62}{170\times1000} = 1.15\times10^{-2} m^3$

If the fixed end beam model is employed:

$$M_{\max} = \frac{1}{12} pL^2 = \frac{1}{12}\times623.558\times5^2 = 1299.08kN\text{-}m$$

The section modulus: $S = \frac{M_{\max}}{\sigma_{all}} = \frac{1299.08}{170\times1000} = 7.64\times10^{-3} m^3$

6.17

Calculate the system stiffness as:

$$S_w = \frac{EI}{\gamma_w h_{avg}^4} = \frac{65000}{9.81\times3^4} = 80$$

The factor of safety against basal heave using Terzaghi's method F_b = 1.22.
According to Figure 6.22, we can find $\delta_{hm} / H_e = 1.0\%$.
The maximum wall deflection δ_{hm} = 0.01 × 9,000 = 90 mm.
The maximum surface settlement $d_{vm} \approx 0.75\delta_{hm}$ = 68 mm.

Let d represent the distance away from the wall.
According to Clough and O'Rourke's method:

$d = 0$, $\delta_v = 68$ mm
$d = 0.75H_e = 6.75$ m, $\delta_v = 68$ mm
$d = 2H_e = 18$ m, $\delta_v = 0$

Chapter 7

7.1

a. t-z curve

t: mobilized side friction resistance
z: displacement at a depth of pile

The t-z method is a widely used static analysis for developing pile load–settlement response at the pile head and vertical load distribution along the pile length.

b. p-y method

p: stands for the lateral soil resistance per unit length of the pile
y: the lateral deflection

It's a method for determination of later load capacity of piles. It considers the soil–pile interaction by using a series of nonlinear springs, which is similar to the method introduced for the mat foundation.

c. In situ pile load test

In order to obtain or estimate the load–deformation relationship of a single pile in actual use, to obtain a judgment for judging the supporting force of the foundation pile or the integrity of the pile body.

d. Replacement pile

Remove soils by boring first, and then fill the borehole (either with casing or without casing) with concrete.

7.3

Known: d = 0.8 m

$$A_b = \frac{1}{4}\pi d^2 = 0.5 \text{ m}^2$$

$$Q_{bu} = 9s_u A_b = 9 \times 80 \times 0.5 = 361.73 \text{ kN}$$

α method:

$$Q_{su} = \alpha \times s_u \times A_s$$

$$Q_{su(total)} = 88.67 + 323.55 + 341.63 + 651.11 = 1404.96 \text{ kN}$$

$$Q_u = Q_{bu} + Q_{su(total)} = 361.73 + 1404.96 = 1766.69 \text{ kN......Ans}$$

β method:

$Q_{su(total)} = \sum \sigma' K_s \tan\delta A_s = 700.72$ kN

$Q_u = Q_{bu} + Q_{su(total)} = 361.73 + 700.72 = 1062.45$ kN......Ans

λ method:

$d = 20$ m , $\lambda = 0.173$

$$\sigma'_{v,avg} = \frac{10.79\times3+48.49\times7+91.38\times4+132.23\times6}{20} = 76.54 \text{ kPa}$$

$$C_{u,avg} = \frac{12\times3+20\times7+50\times4+80\times6}{20} = 42.8 \text{ kPa}$$

$$Q_{su} = \lambda(\sigma'_{v,avg} + 2\times C_{u,avg})A_s = 0.173(76.54+2\times42.8)\times0.8\pi\times20 = 1409.96 \text{ kN}$$

$Q_u = Q_{bu} + Q_{su} = 361.73 + 1409.96 = 1771.69$ kN......Ans

7.5

Terzaghi's method: P_{ult} at $s_{10\%}$ diameter

$d = 1\ m$

$d_{10\%} = 0.1\ m = 100\ mm$

$\therefore\ P_{ult} \approx 4400$ tf

7.7

use table (7.32)

$$P_{ult} = 9\times s_u\times d^2\left[(4\times(\frac{e}{d})^2 + 2\times(\frac{L}{d})^2 + 4\times(\frac{e}{d})(\frac{L}{d}) + 6\times(\frac{e}{d}) + 4.5)^{0.5} - (2\times(\frac{e}{d}) + (\frac{L}{d}) + 1.5)\right]$$

$$= 9\times30\times1.2^2\left[(4\times(\frac{1}{1.2})^2 + 2\times(\frac{6}{1.2})^2 + 4\times(\frac{1}{1.2})(\frac{6}{1.2}) + 6\times(\frac{1}{1.2}) + 4.5)^{0.5} - (2\times(\frac{1}{1.2}) + (\frac{6}{1.2}) + 1.5)\right]$$

$= 279.3$ kN

7.9

$$K_P = \tan^2\left(45 + \frac{\phi'}{2}\right) = \tan^2\left(45 + \frac{32°}{2}\right) = 3.25$$

$\gamma' = \gamma_{sat} - \gamma_w = 20 - 9.8 = 10.2$ kN/m^3

$$P_{ult} = \frac{3}{2}\times K_p\times\gamma'\times d\times L^2$$

$$= \frac{3}{2}\times3.75\ \times10.2\times1.2\times6^2$$

$= 2478.6$ kN

$$M_{\max} = K_P\times\gamma'\times d\times L^3$$

$= 3.25\times10.2\times1.2\times6^3$

$= 8592.5$ kN$\cdot$m

7.11

Assumption: Rigid free-head pile

$$f=\frac{P_u}{9s_u d}=\frac{600}{9\times 80\times 1.2}=0.694\text{ m}$$

Take moment at pile tip,

$$P_u(L+e)-\frac{1}{2}(9s_u d)(L-1.5d)^2+9s_u dz^2=0$$

$$L=1.5\times 1.2+0.694+2z$$

$$z=\frac{1}{2}(L-2.494)$$

Substitute z to the equation,

$$-252L^2+1257.36L+11719.92=0$$

$$\therefore L=9.756\ m$$

Take moment at max. depth $(1.5d+f)$

$$M_{design}=P_u(e+1.5d+f)-\frac{1}{2}(9s_u d)f^2=13.288\text{ MN}\cdot\text{m}$$

$$FS=\frac{M_{max}}{M_{design}}=3$$

$$\therefore M_{max}=3\times 13.288=39.865\text{ MN}\cdot\text{m}$$

7.13

consider the sand arch effect, use L′= 15D =15 m

$$Z=15\ (m)\ ,\ f_s=Ks\times\tan(\delta)\times\sigma'_{v,15D}$$
$$=1.2\times\tan(25°)\times(20-9.81)\times 15$$
$$=85.53\text{ kN/m}$$

$$Q_s=\Sigma f_s A_s=\frac{85.53}{2}\times\pi\times 1\times 15+85.53\times\pi\times 1\times 10=4702.26\text{ kN}$$

7.15

$A_p=\frac{\pi}{4}\times 12=0.785\text{ m}^2$ $\quad\xi=0.67\quad\mu_s=0.3$ (from table 2.4) $\quad I_{wp}=0.85$

$U=\pi\qquad L=25\text{ m}\qquad I_{ws}=2+0.35\sqrt{\frac{L}{d}}=2+0.35\sqrt{\frac{25}{1}}=3.75$

total settlement $S_{e(total)} = S_{e(1)} + S_{e(2)} + S_{e(3)}$

$$S_{e(1)} = \text{Elastic settlement} = \frac{(Q_{wp} + \xi Q_{ws}) \times L}{A_p \times E_p} = \frac{(1400 + 0.67 \times 2100) \times 25}{0.7854 \times 21 \times 10^6}$$

$$= 4.255 \times 10^{-3} \text{ m} = 4.255 \text{ mm}$$

$$S_{e(2)} = \text{By the load at the tip} = \frac{(Q_{wp}) \times d}{A_p \times Es} \times (1 - \mu_s^2) \times I_{wp} = \frac{(1400) \times 1}{0.7854 \times 35000} \times (1 - 0.3^2) \times 0.85$$

$$= 3.9394 \times 10^{-2} \text{ m} = 39.394 \text{ mm}$$

$$S_{e(3)} = \text{By the load transmitted along the pile shaft} = \left(\frac{Q_{ws}}{U \times L}\right) \times \frac{d}{Es} \times (1 - \mu_s^2) \times I_{ws}$$

$$= \left(\frac{2100}{\pi \times 25}\right) \times \frac{1}{35000} \times (1 - 0.3^2) \times 3.75$$

$$= 2.607 \times 10^{-3} \text{ m} = 2.607 \text{ mm}$$

$$S_{e(total)} = 4.255 + 39.394 + 2.607 = 46.256 \text{ mm}$$

Chapter 8

8.1

$$F_s = \frac{\tan \phi'}{\tan \alpha} = \frac{\tan 30^\circ}{\frac{1}{3}} = 1.732$$

8.3

Table S8.3a

1	2	3	4	5	6	7	8	9	10
Slice no.	*b (m)*	*Weight (kN/m)*	*a*	*sin a*	*cos a*	*c'b*	*Wsina*	*Wtan ϕ*	*(8) + (9)*
1	1.2	68.4	60	0.87	0.50	24	59.24	24.90	84.13
2	2.5	370.5	48	0.74	0.67	50	275.34	134.85	410.19
3	2.5	498.75	35	0.57	0.82	50	286.07	181.53	467.60
4	2.5	484.5	23	0.39	0.92	50	189.31	176.34	365.65
5	2.5	356.25	15	0.26	0.97	50	92.20	129.66	221.87
6	2.5	209	4	0.07	1.00	50	14.58	76.07	90.65
7	2	95	−5	−0.09	1.00	40	−8.28	34.58	26.30
8	2	68.4	−13	−0.22	0.97	40	−15.39	24.90	9.51
9	2	26.6	−20	−0.34	0.94	40	−9.10	9.68	0.58
–	–	–	–	–	–	Total	883.97	–	–

Table S8.3b

1	1st try $F_s = 1.5$		2nd try $F_s = 1.3$		3rd try $F_s = 1.28$		4th try $F_s = 1.27$		5th try $F_s = 1.26$	
Slice no.	m_a	$(10)/m_a$	m_a	$(10)/m_a$	m_a	$(10)/m_a$	m_a	$(10)/m_a$	m_a	$(10)/m_a$
1	1.32	63.74	1.45	58.18	1.46	57.59	1.47	57.29	1.48	56.99
2	1.41	291.49	1.52	269.72	1.53	267.39	1.54	266.20	1.55	265.01
3	1.44	323.78	1.54	303.57	1.55	301.37	1.56	300.25	1.56	299.12
4	1.42	256.84	1.50	243.60	1.51	242.14	1.51	241.39	1.52	240.64
5	1.38	160.64	1.44	153.54	1.45	152.75	1.46	152.35	1.46	151.94
6	1.29	70.45	1.33	68.10	1.34	67.83	1.34	67.69	1.34	67.56
7	1.18	22.27	1.21	21.75	1.21	21.69	1.21	21.66	1.22	21.63
8	1.07	8.91	1.08	8.79	1.08	8.78	1.08	8.77	1.08	8.77
9	0.95	0.61	0.96	0.61	0.96	0.61	0.96	0.61	0.96	0.61
	Total	1,198.74	Total	1,127.86	Total	1,120.14	Total	1,116.22	Total	1,112.28

1. First trial and error.

$$F_s = \frac{(\Sigma \frac{1}{m_\alpha})c'b + W\tan\phi'}{(\Sigma W \sin\alpha)} = \frac{1198.74}{883.97} = 1.36$$

2. Second trial and error.

$$F_s \frac{(\Sigma \frac{1}{m_\alpha})(c'b + W\tan\phi')}{(\Sigma W \sin\alpha)} = \frac{1127.86}{883.97} = 1.28$$

3. Third trial and error.

$$F_s = \frac{(\Sigma \frac{1}{m_a})(c'b + W\tan\phi')}{(\Sigma W \sin\alpha)} = \frac{1120.14}{883.97} = 1.27$$

4. Fourth trial and error.

$$F_s = \frac{(\Sigma \frac{1}{m_a})(c'b + W\tan\phi')}{(\Sigma W \sin\alpha)} = \frac{1116.22}{883.97} = 1.26$$

5. Fifth trial and error.

$$F_s = \frac{(\Sigma \frac{1}{m_a})(c'b + W\tan\phi')}{(\Sigma W \sin\alpha)} = \frac{1112.28}{883.97} = 1.26 \text{ (OK)}$$

Factory of safety = 1.26

8.5

$$\frac{c}{\gamma H \tan\phi'} = \frac{40}{16 \times 15 \times \tan 30} = 0.29$$

From Figure 8.13, for the 40°, the corresponding value of $\frac{\tan\phi'}{F_s} = 0.95$.

Factory of safety = 0.61

Index